Practical Computed Tomography (CT) Guide for the Small Animal Orthopaedic and Neurosurgeon

More clinics and large practices are investing in computed tomography (CT) equipment, meaning veterinary surgeons are often confronted with unfamiliar technical challenges like positioning and handling of the equipment. Even after obtaining CT images, the interpretation can often be confusing, particularly when surgeons are not used to reading axial and transverse images.

This book is split into two parts: first, it focuses on the principles of CT, and then, it explores the specific procedures by body region. This book is packed with over 200 high-quality images, practical protocols, and easy-to-locate information. It provides much-needed support in choosing the right equipment, providing the exact protocol, and reaching a final diagnosis.

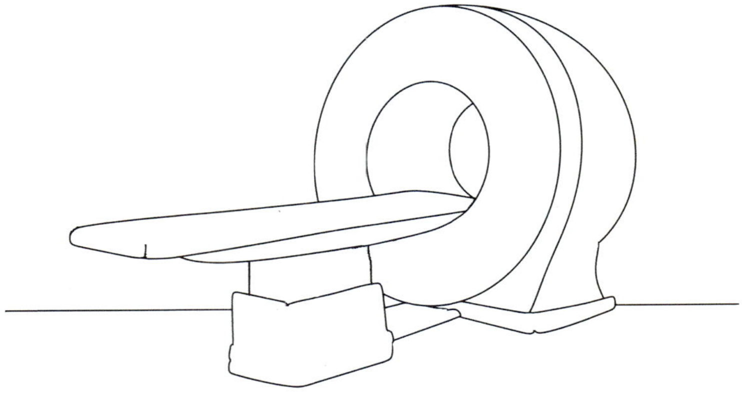

Practical Computed Tomography (CT) Guide for the Small Animal Orthopaedic and Neurosurgeon

Edited by
INGRID GIELEN AND
HENRI VAN BREE

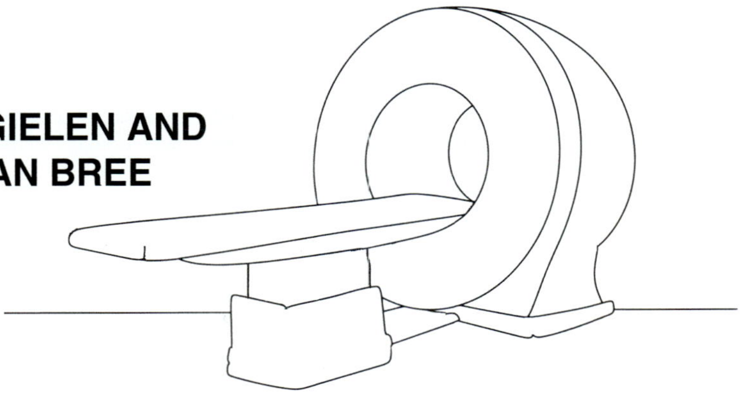

CRC Press
Taylor & Francis Group
Boca Raton London New York

CRC Press is an imprint of the
Taylor & Francis Group, an **informa** business

First edition published 2026
by CRC Press
2385 NW Executive Center Drive, Suite 320, Boca Raton FL 33431

and by CRC Press
4 Park Square, Milton Park, Abingdon, Oxon, OX14 4RN

CRC Press is an imprint of Taylor & Francis Group, LLC

ISBN: 9781482225884 (hbk)
ISBN: 9781041145202 (pbk)
ISBN: 9781315118130 (ebk)

DOI: 10.1201/9781315118130

Typeset in Janson
by codeMantra

MIX
Paper | Supporting responsible forestry
FSC® C013056
www.fsc.org

Printed and bound in Great Britain by
TJ Books, Padstow, Cornwall

Dedication

'For my dad, Dr. Jan Gielen, who taught me what strength and perseverance mean. You will always remain in my heart.'
Ingrid

'To my mentor and friend, Prof. Dr. Francis Verschooten, Your guidance and friendship have shaped who I am today. This book is as much your success as it is mine.'
Henri

Epigraph

"Success is not achieved by simply starting, but by persistently finishing what you've started."

–By Peter Hollins

CONTENTS

FOREWORD I

Radiology has traditionally been the principal non-invasive diagnostic modality used by veterinary surgeons for the investigation of skeletal injuries and disorders. In recent years the convenience and quality of radiographic images greatly improved with the transition from hard films to digital images. However, the inherent limitations of two-dimensional radiographic images have expedited the demand for computed tomographic imaging for imaging of more complex skeletal problems. Computed tomography is certainly not new, but the increased affordability and access have made its usage routine nowadays. But, without the assistance of diagnostic imaging specialists, many surgeons are not comfortable in interpreting the array of transverse, axial and multiplanar reconstructed computed tomographic images that are generated. So, the launch of this new book entitled *Practical Computed Tomography Guide for the Small Animal Orthopaedic and Neurosurgeon* edited by Ingrid Gielen and Henri van Bree is both timely and valuable.

This new book has been very carefully assembled with comprehensive input by a team of specialists. It begins with the basic physics to financial decision-making necessary for getting computed tomography into your practice. This is followed by a comprehensive chapter explaining the range of applications of computed tomography that are beyond the capabilities of radiology. After this, the unique injuries and disorders of each of the major joints of the appendicular skeleton are illustrated with images. Then follows imaging of long bones, muscle, spine, brachial and lumbo-sacral plexus, and brain. The layout of the book makes it possible for readers to quickly identify the information they need but it is sufficiently in-depth for surgeons to acquire a greater level of expertise in the interpretation of computed tomography images in practice with more careful study. The authors are warmly congratulated on this wonderful new contribution of knowledge and expertise to this field.

Kenneth A Johnson
MVSc, PhD, FACVS, Diplomate ACVS & ECVS
Emeritus Professor of Orthopaedics, University of Sydney
Specialist Small Animal Surgeon, SASH Alexandria, Sydney

FOREWORD II

Praktische CT für Kleintierorthopäden & Neurochirurgen

Die Spezialisierung in der Kleintiermedizin hat in den letzten Jahrzehnten immer weiter zugenommen. Dadurch war es möglich gerade durch den Einsatz neuer Bildgebender Verfahren (u.a. CT und MRT) bei Hund und Katze die Diagnostik von Erkrankungen in der Orthopädie und Neurologie so zu verbessern, dass die Quote erfolgreicher operativer Eingriffe in diesen Fachgebieten immer weiter zugenommen hat.

Gerade bei schwierigen chirurgischen Fragestellungen im Bereich des Skelettsystems ist eine exakte Bildgebende Diagnostik von entscheidender Bedeutung, um eine präzise Planung (inklusive Instrumenten- und Implantatwahl) und damit die bestmögliche Operationsart auswählen zu können.

Die CT ist für komplexe orthopädische (u.a. Tarsal- und Carpalgelenksverletzungen, Trauma des Beckenrings oder Deformationen der Vorder- und oder der Hintergliedmaßen) und neurologische Pathologien (u.a. Diskopathie bei chondrodytrophen Rassen, Deformationen der Wirbelsäule, Wirbelsäulentraumata, knöcherne Schädelverletzungen) der Golden Standard in der Diagnostik bei Mensch und Tier.

Die Autoren, die diesen praktischen Leitfaden geschrieben haben, sind ausgewiesene Spezialisten auf dem Gebiet der Computertomographie im Bereich Orthopädie und Neurologie beim Kleintier.

Aus meiner Sicht ist dieses Buch schon jetzt ein absolutes Standardwerk für Orthopäden, Neurologen und Bildgeber in der Kleintiermedizin und sollte in keiner Klinik oder Praxis fehlen.

Prof. Dr. Dr. h.c. mult. Martin Kramer
Leiter der Klinik für Kleintiere, Chirurgie & Bildgebung im Fachbereich Veterinärmedizin der Justus-Liebig-Universität Gießen
Präsident der Deutschen Veterinärmedizinischen Gesellschaft (DVG)
Dip. ECVDI, FTA-Röntgenologie und anderer Bildgebender Verfahren

English Version is available on the following page

English Version

Practical CT for Small Animal Orthopaedic & Neurosurgeons

Specialisation in small animal medicine has continued to increase in recent decades. This has made it possible to improve the diagnosis of orthopaedic and neurological diseases in dogs and cats, particularly through the use of new imaging techniques (including CT and MRI), with the result that the rate of successful surgical procedures in these fields has continued to rise.

Particularly in difficult surgical cases involving the skeletal system, accurate imaging diagnostics are crucial for precise planning (including the selection of instruments and implants) and thus for choosing the best possible type of surgery.

CT is the gold standard in diagnostics for humans and animals for complex orthopaedic (including tarsal and carpal joint injuries, trauma to the pelvic ring or deformities of the front and/or hind limbs) and neurological pathologies (including discopathy in chondrodystrophic breeds, deformities of the spine, spinal trauma, bony skull injuries) in humans and animals.

The authors who wrote this practical guide are proven specialists in the field of computed tomography in orthopaedics and neurology in small animals.

In my opinion, this book is already an absolute standard work for orthopaedists, neurologists and imaging specialists in small animal medicine and should be available in every clinic or practice.

Prof. Dr. Dr. h.c. mult. Martin Kramer
Head of the Clinic for Small Animals, Surgery & Imaging in the Department of Veterinary Medicine at Justus Liebig University Giessen
President of the German Veterinary Medical Association (DVG)
Dip. ECVDI, FTA radiology and other imaging techniques

ACKNOWLEDGEMENTS

We are extremely grateful to several people without whose support and help, the completion of this book would not have been possible.

We would like to express our profound gratitude to Prof. Dr. Francis Verschooten, who supported us from the beginning with this project and enabled us to fully develop the CT unit within our department.

In addition, we wish to extend our sincere thanks to Dr. Annemie Van Caelenberg, who helped us with collecting the images and correcting the texts. Special thanks goes to Marnix Verdonck, our technician, who adapted the drawings for veterinary use and helped in the past with all technical issues.

Last but not least, we wish to thank all the colleagues who, by providing us with all the interesting cases, enabled us to gain the expertise necessary to finish this project.

ABOUT THE EDITORS

Prof. Dr. Ingrid Gielen graduated from Ghent University in Belgium in July 1995 and joined the staff as Assistant at the Department of Medical Imaging and Small Animal Orthopaedics, Faculty of Veterinary Medicine, Ghent University, Belgium.

She received a master's degree in Laboratory Animal Science from Ghent University in 2000. Her PhD thesis was completed with the title "Computed Tomography in the Diagnosis and Treatment of Canine Tarsocrural Osteochondrosis" in 2003. She was President of the European Association of Veterinary Diagnostic Imaging (EAVDI), 2004–2006.

Since 2004, Dr. Gielen has been Division Head at the Faculty of Veterinary Medicine, Ghent University, Belgium and, since 2015, Visiting Professor at the Department of Radiology and Radiation Hygiene, Faculty of Veterinary Medicine, University of Belgrade, Serbia. She is a board member of the International Elbow Working Group and certified scrutinizer canine hip & elbow dysplasia (FCI/IEWG).

She was invited as a speaker at many international conferences on medical imaging and orthopaedics and neurology and is the author of more than 185 peer-reviewed publications in national and international journals on medical imaging and orthopaedics. She has a particular interest in imaging techniques in joint diseases and neurologic diseases.

Prof. Dr. Henri van Bree graduated from Ghent University in Belgium in 1974. Since 1991, he has been a Full Professor in medical imaging and orthopaedic surgery at the Department of Medical Imaging and Small Animal Orthopaedics, Faculty of Veterinary Medicine, University of Ghent, Belgium. He completed his PhD at the University of Utrecht, Holland, Department of Radiology on "comparative imaging in the canine shoulder". From 2001 to 2015, he was the Head of the Department of Medical Imaging and Small Animal Orthopaedics.

Prof. van Bree is a Diplomate of both the European College of Veterinary Diagnostic Imaging (ECVDI) and the European College of Veterinary Surgeons (ECVS). He received the Richard-Völker-Medaille from the DGK-DVG – Kleintiere in 2014 and has authored about 250 publications on medical imaging and orthopaedics. He has been an invited speaker at about 150 international conferences on medical imaging and small animal arthroscopy. He is the CEO of VetMedImage, a teleradiology service. His current research interest is comparative imaging in small animal joint disease.

CONTRIBUTORS

Amalia Agut
Veterinary Teaching Hospital
University of Murcia
Murcia, Spain

Jean François Bardet
Selarl Bardet Neuilly sur Seine
Neuilly sur Seine, France

Henri van Bree
Faculty of Veterinary Medicine
Ghent University
Merelbeke, Belgium
And
VetMedImage
Belgium

Annemie van Caelenberg
Faculty of Veterinary Medicine
Ghent University, Merelbeke, Belgium

Nuria Corzo-Menéndez
VetOracle teleradiology
Milton Keynes, England, UK

Steven De Decker
Royal Veterinary College
University of London
London, UK

Cristian Falzone
Pedrani Veterinary Clinic
Zugliano, Italy

Ingrid Gielen
Faculty of Veterinary Medicine
Ghent University
Merelbeke, Belgium
And
Faculty of Veterinary Medicine
University of Belgrade
Belgrade, Serbia

Wolfgang Henninger
Kleintierzentrum Breitensee
Wien, Austria

Stef Janssens
BioScan
Oud-Turnhout, Belgium

Stephan Kaiser
Fachtierarzt für Radiologie und Chirurgie
Fachtierarztpraxis am Erzberg
Braunschweig, Germany

Ingmar Kiefer
Fachtierarzt für Kleintiere und Radiologie
Clinic for Small Animals
Leipzig University
Leipzig, Germany

Michael Kowaleski
Tufts University Cummings School of Veterinary Medicine
North Grafton, Massachusetts

Rafael Latorre
Veterinary Teaching Hospital
University of Murcia
Murcia, Spain

Eberhard Ludewig
Fachtierarzt für Kleintiere und Radiologie
University of Veterinary Medicine
Vienna, Austria

Pierre Moissonnier
Alfort National Veterinary School
Maisons-Alfort, France

Christian Niesterok
Fachtierarzt für Kleintiere
Steinbergkirche Animal Health Center,
Steinbergkirche, Germany

Gian Luca Rovesti
Ad Maiora srl
Emilia-Romagna, Italy

Peter Smith
Hamilton Specialist Referrals
High Wycombe, UK

Perry Sprawls
Emory University,
Atlanta, GA
And
Sprawls Educational Foundation, http://www.sprawls.org

Cees Verlooij
Canon Medical Systems
Zoetermeer, The Netherlands

Tim Waelbers
boa vets for vets
Maaseik, Belgium

PART I
Principles of Computed Tomography

1

CHAPTER 1

INTRODUCTION AND OVERVIEW

Perry Sprawls

Computed tomography (CT) is a valuable imaging modality because of its unique features and image characteristics that are distinctly different from those of conventional radiography. It is an X-ray imaging method but uses a different and more complex method for producing an image. There are three distinct phases in the production and display of CT images, as illustrated in Figure 1.1.

Here, each of the phases is introduced which will be considered in much more detail later. There are adjustable protocol factors associated with each phase that are used to control image characteristics and radiation dose. Knowledge of the physics principles described in this chapter will enable the clinical staff to develop imaging protocols that optimize each imaging procedure to meet the clinical objectives for each examination. Each image has specific quality characteristics. These characteristics are controlled through a somewhat complex relationship with the protocol factors.

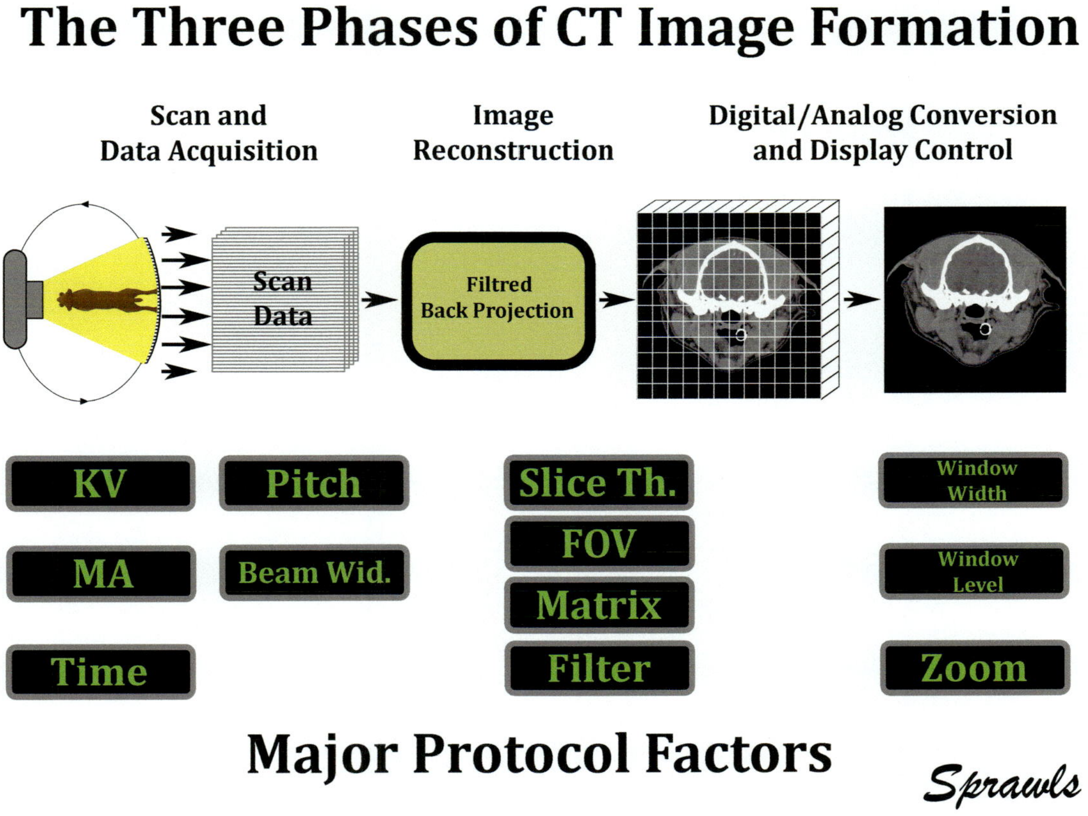

Figure 1.1 The three phases of the computed tomography (CT) process.

DOI: 10.1201/9781315118130-2

CHAPTER 2

CT IMAGE CHARACTERISTICS

Perry Sprawls

Knowledge of the specific characteristics of CT images is the foundation for effective imaging to meet the wide range of clinical requirements. The value, and sometimes the challenge, of CT imaging is that the various image characteristics can be changed and adjusted for each procedure. It is the combination of individual image characteristics that affects the visibility of specific anatomical structures and the different tissues, including signs of pathology.

2.1 TOMOGRAPHIC, DIGITAL AND A MATRIX OF VOXELS

CT images differ from conventional radiographs in several respects. They are tomographic or images of individual slices within the body rather than a more composite image of an entire anatomical region formed by projecting an X-ray beam through the body and casting shadows of the internal anatomy. CT imaging has several advantages. It permits close up of the images, within a few millimetres, visualization of each anatomical site and removes interfering overlying structures.

In the CT imaging process, each anatomical slice is divided into a matrix of voxels (volume elements), as illustrated in Figure 2.1.

CT imaging is a sampling process, with each voxel being a small sample of tissue. The size of the voxel is a critical factor determining the quality of the image and the visibility of anatomical and pathologic features. Voxel size is controlled by a combination of protocol factors adjusted to control the imaging procedure.

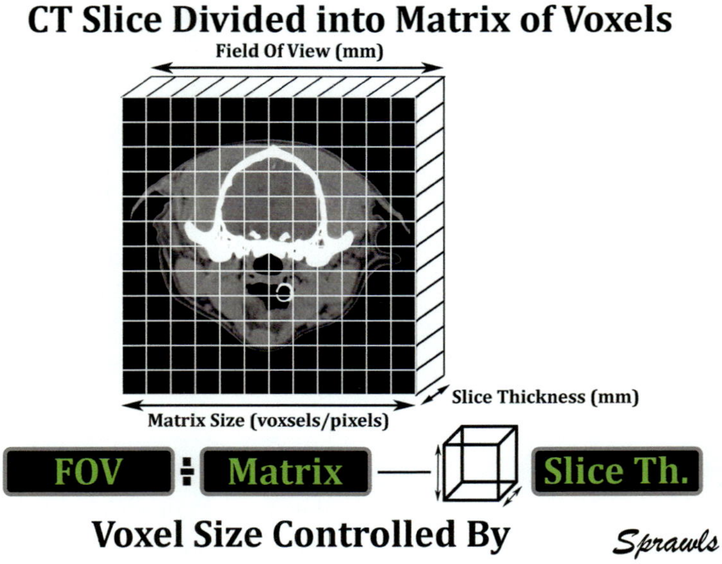

Figure 2.1 A slice of an anatomical region that is displayed in an image as it is divided into voxels.

DOI: 10.1201/9781315118130-3

2.1.1 Matrix size

Matrix size is the number of voxels that the slice is divided into in each direction. A matrix size of 512 is typical but can be varied to optimize the image for specific clinical requirements.

2.1.2 Field of view (FOV)

Field of view is the dimension of the anatomical area displayed in the image.

2.1.3 Slice thickness

Slice thickness is generally the largest dimension of a voxel, ranging from 2 to 5 mm, in most procedures and is also associated with voxel size and can be adjusted over a considerable range.

2.2 CONTRAST AND VISUALIZATION OF THE SOFT TISSUE, COMPARED TO RADIOGRAPHY

Other than producing tomographic images, CT is superior to radiography owing to its very high contrast sensitivity, especially for visualizing the soft tissues. A quick review of contrast sensitivity is displayed in Figure 2.2.

For enabling visibility of different anatomical structures and tissues, they must be physically different from each other, thus creating physical contrast. CT imaging converts and transfers the physical contrast within the body into visible contrast displayed in an image. Another critical characteristic is the ability of the imaging process to visualize the physical contrast within the body, characteristic of contrast sensitivity. The most significant source of physical contrast that leads to visibility in CT imaging is the difference in physical density (specific gravity). Another factor is the difference in the chemical composition that contributes to contrast of bone and iodine contrast media.

In Figure 2.2, different objects and materials are arranged along a scale in order of their contrast relative to the surrounding soft tissue. It is the contrast sensitivity of an imaging procedure that extends the range of visibility with respect to the physical contrast of the tissues. CT has both a very-high and adjustable contrast sensitivity.

2.2.1 CT numbers

The characteristics such as high contrast sensitivity and ability to visualize the differences or contrast among the soft tissues are inherent in the way CT imaging creates images. During the scanning and image forming process, the physical density or major difference in chemical atomic number, Z (presence of calcium, iodine, etc.) is measured and translated into a CT number for each tissue voxel, as illustrated in Figure 2.3.

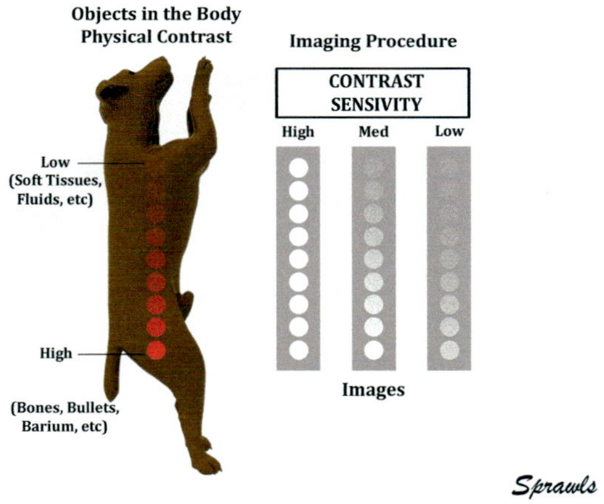

Figure 2.2 The concept of contrast sensitivity and its effect on visualizing low-contrast tissues.

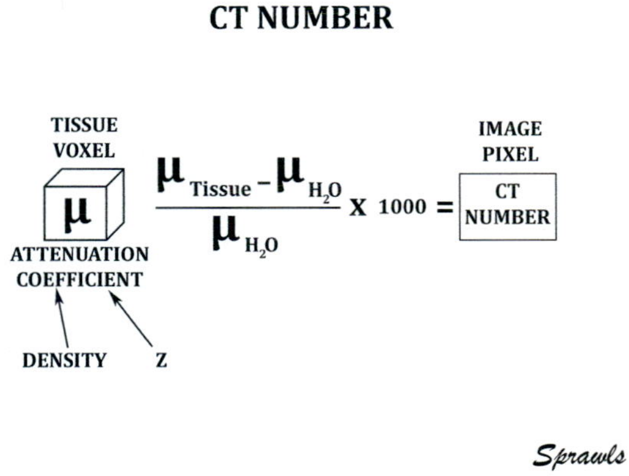

Figure 2.3 The relationship of CT number values for each image pixel to the X-ray attenuation coefficient values of the tissue in each voxel.

During the scanning process, X-ray beam is attenuated or absorbed at a rate determined by the density and Z of the tissue through which it is passing. Mathematically, the rate of attenuation is expressed by the value of the attenuation coefficient, u, measured by the passage of the X-ray beam. Then, during the image reconstruction phase, the value for each voxel is used to calculate a CT number value using the given formula.

CT number values are calculated based on the attenuation of X-rays in a tissue relative to water that has an assigned value of 0. CT systems are calibrated using water as a reference. Any tissue or material that has a higher X-ray attenuation rate, typically because of higher density, than water will have positive CT number values, and those that are less dense than water will have negative values. The CT number value for each pixel is not displayed directly in the image but determines the brightness of the pixel.

2.2.2 The CT number scale and windowing

The high contrast sensitivity and visualization of soft tissues are derived from the ability to select a limited range of CT numbers (tissue densities) and expand or display them over a large range of brightness (black to white) in the image.

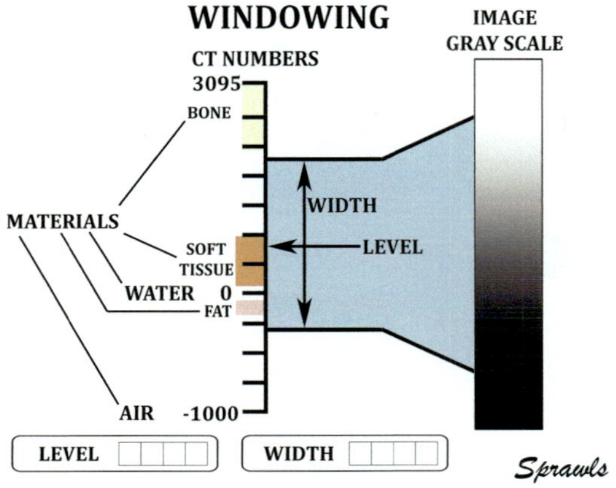

Figure 2.4 The concept of windowing in which a small range of CT number values is selected for display over the full brightness or grey-scale range.

This is the process of windowing, illustrated in Figure 2.4.

The CT numbers range from a value of –1,000 for air to approximately +3,000 for dense bone, with water having a CT number of 0. They are expressed in Hounsfield units (HU), named after Sir Godfrey Hounsfield, the engineer who developed the first clinical CT system. Since soft tissues have density values that are not considerably different from water (1gm/cc), their CT numbers are clustered in a small range around the zero value of water.

The window is the range of CT numbers that will be displayed over the full image brightness range, as illustrated. The window *width* and *level* are two adjustable factors that control the image contrast and the contrast sensitivity for visualizing specific tissues or other materials in the body. Setting the window width to a small or narrow value increases the contrast in the displayed image.

If the window is set to cover the full CT scale from –1,000 to +3,000, there would be almost no visible tissue contrast in the image, just a grey area. It is the ability to window and select a small range of CT numbers (tissue densities) for display that yields the high contrast sensitivity.

2.3 DETAIL, BLURRING, EFFECT ON VISIBILITY, COMPARED TO RADIOGRAPHY, AND CONTROLLING FACTORS

Size limits the visibility of structures and objects within a body. Visibility of the smaller objects is limited by blurring observed during the imaging process. There is some blurring that is a natural occurrence in all imaging processes, including our human vision. Blurring limits our ability to visualize details in the form of small things such as print. Figure 2.5 illustrates the effect of blurring on the visibility of details.

Specific sources of blur, sometimes called un-sharpness or resolution, are observed in all imaging procedures, and they can usually be controlled to some extent. In radiography, the

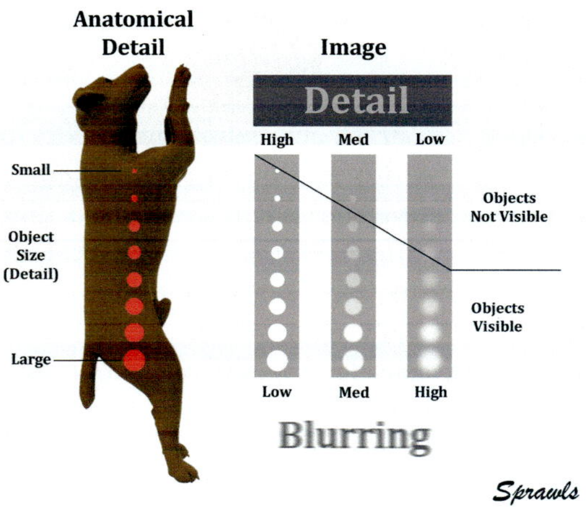

Figure 2.5 The effect of blurring on the visibility of detail.

size of the X-ray tube focal spot is a source of blurring and the selection of the smaller focal spot, when possible, contributes to improved visibility of details.

CT is a more complex process than radiography, and there are more sources of blurring, as illustrated in Figure 2.6.

Importantly, blurring is cumulative. The detail in the image is limited by the sum of the blur occurrences from all of the sources. In general, the source of the largest blur, especially if it is significantly larger than the others, becomes the controlling factor. By reducing the blur from one of the sources in case of overwhelming by the others does not have much importance. Also, changing factors such as voxel size to reduce blurring can have other detrimental effects on image quality, such as increasing visual noise.

2.4 NOISE, CHARACTERISTICS, EFFECT ON VISIBILITY, SOURCE AND CONTROLLING FACTORS

Noise is an undesirable characteristic present to some extent in all clinical imaging methods, which can reduce visibility, as illustrated in Figure 2.7.

Noise reduces the visibility of objects or structures in the body that have low contrast. This is especially significant in visualization of various soft tissues. We recall that in CT

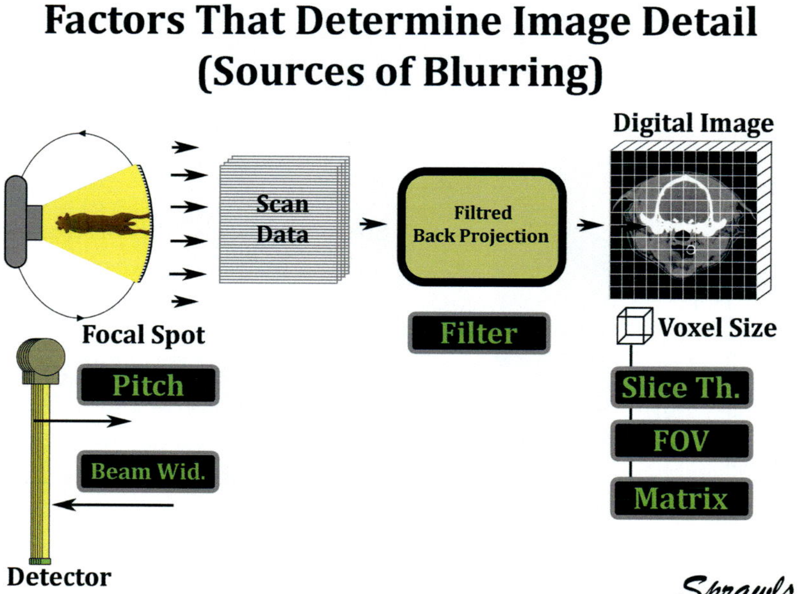

Figure 2.6 The adjustable protocol factors associated with each phase of the CT process that determines image detail.

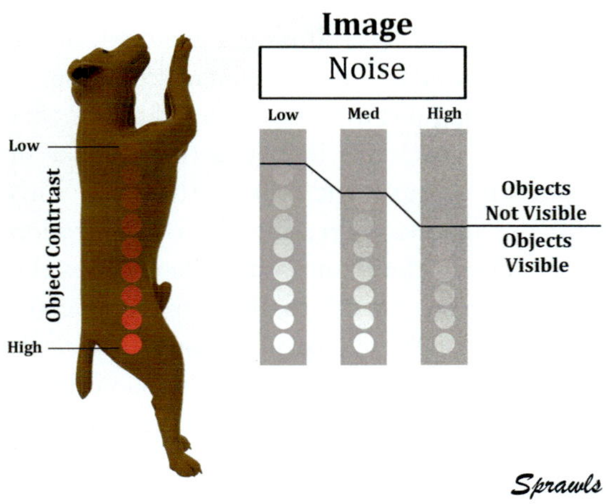

Figure 2.7 The effect of visual noise on the visibility of low-contrast objects within the body.

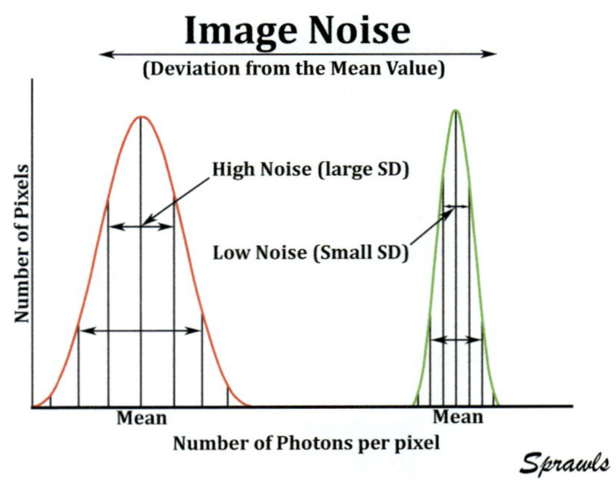

Figure 2.8 The general relationship of noise, as expressed by the SD, to the number of photons per tissue voxel or corresponding image pixel.

imaging, the source of contrast is the physical difference in tissue densities. This difference, or contrast, is often quite small between normal and pathologic tissues.

In all X-ray imaging methods (radiography, fluoroscopy and CT), the major source of image noise is the structure of the X-ray beam. An X-ray beam is not a continuous flow of energy but consists of many individual small units, namely, photons. The photons, or quanta of energy, are randomly distributed spatially and temporally within the body. It is this random statistical nature of the X-ray photons as they interact with the tissue in each voxel that is the source of image noise. With knowledge of the statistics, it is possible to control the noise.

The number of photons that are collected within each tissue voxel is assumed to be randomly distributed according to a Poisson statistical distribution. The variation from voxel to voxel, even when receiving the same X-ray exposure or dose, can be expressed by the value of the standard deviation (SD). The SD is used as a measure of the noise. As described later, all CT systems provide the operator with the capability to calculate and display SD values for areas or regions-of-interest (ROI) within an image to determine the amount of noise.

A characteristic of a Poisson distribution which the X-ray photons follow is that the value of the SD, and therefore the noise, is related to the average number or mean value of photons per voxel. This is an inverse relationship with the SD, as a fraction or percentage, which decreases with an increase in the number of photons, as illustrated in Figure 2.8.

This now brings us to one of the most significant, and perhaps critical, issues in CT and other types of X-ray imaging. Since each photon is a quantum of energy, the number of photons absorbed in a voxel corresponds to the radiation dose to that tissue. This relationship is illustrated in Figure 2.9.

In setting up a CT imaging protocol, there are two goals: to produce an image with adequate

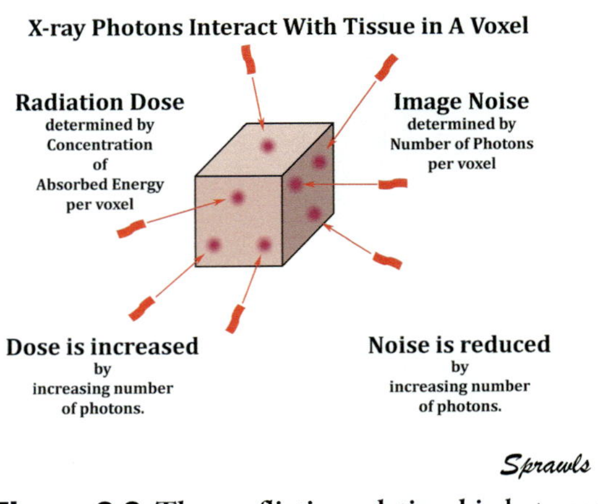

Figure 2.9 The conflicting relationship between radiation dose to the tissue and image noise.

quality (reduced noise) and to have the radiation dose as low as possible. As is observed, there is a conflict between these two goals. It also gets more complex when the size of the voxel must be considered. Voxel size is one of the major contributing factors to image blurring, which limits the visibility of detail. As shown in Figure 2.10, voxel size has a controlling effect on two of the major image characteristics, detail and noise.

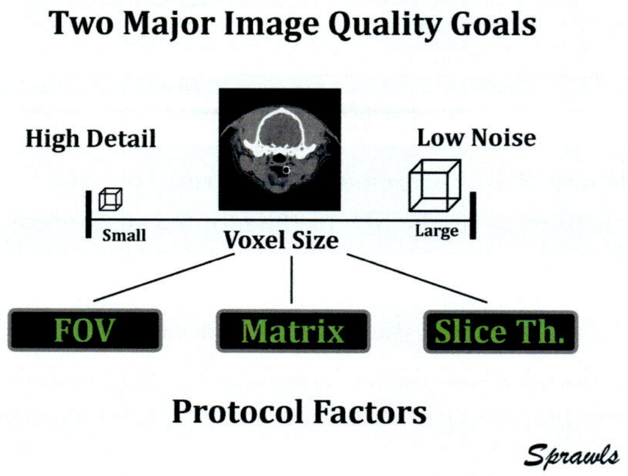

Figure 2.10 The effect of voxel size on both image detail and image noise.

This relationship is based on the fact that increasing voxel size increases the number of photons it captures and reduces the noise. This does not involve increasing the concentration of photons in the tissue, which determines the radiation dose. Therefore, there are three competing and opposing goals that must be considered when setting up a protocol: good visibility of detail (reduced blurring), low noise and a substantially low radiation dose.

2.5 ARTEFACTS AND GENERAL CHARACTERISTICS

Artefacts in CT images are most often in the form of streaks or "ghosts" that are not actual objects in the body. Common causes are metal objects or some bone structures.

CHAPTER 3

THE CT IMAGING PROCESS AND EQUIPMENT OVERVIEW

Perry Sprawls

The computed tomography (CT) imaging process consists of three distinct phases. There are adjustable protocol factors associated with each phase. The combination of these factors determines the characteristics of images for specific procedures. Some of the most significant factors are identified below, before dealing with each phase in more detail in the subsequent chapters. Unlike radiography, this is not yet processed in the form of an image.

After the scanning phase, the second phase is the mathematical process of image reconstruction. This transforms the scan data into digital images. This process actually creates the anatomical slices and the tissue voxels described previously.

The third and final phase is the conversion of the digital images into visible (called analogue) images for viewing.

DOI: 10.1201/9781315118130-4

CHAPTER 4

THE SCANNING PHASE

Perry Sprawls

The first phase is when the X-ray beam is scanned around and along the patient's body. The X-ray beam is produced at the focal spot within the X-ray tube (not shown) and then after passing through the body is measured by an array of detectors, as illustrated in Figure 4.1.

As the beam scans around, measurements are made at different angles, typically at approximately one-degree increments, of the radiation passing through the body along each line or ray. This set of data from each angle makes up one view. Here, the term view does not relate to any type of image view but is just a set of data from a specific X-ray beam direction. One view is made up of many rays. A ray is the pathway from the X-ray tube focal spot to an individual X-ray detector element after passing through the body. As the radiation passes along each ray, it is attenuated in proportion to the density of the tissues encountered along the way. The radiation measured by each detector represents the 'sum' of the density values along the ray. This is recorded and stored as one data point. One data set required to produce one image by the reconstruction process is made up of many points, generally the product of the number of views (typically 360) and the number of rays within a view.

4.1 DESCRIPTION

The two different scanning motions of the X-ray beam in relation to the body are illustrated in Figure 4.2.

The scanning around the body produces the data used to create images of individual slices. If images of more than one slice are required, the X-ray tube and detectors are moved relative to the length of the body. Actually, it is the body that is moved through the beam. This motion of the body relative to the beam can occur in two ways.

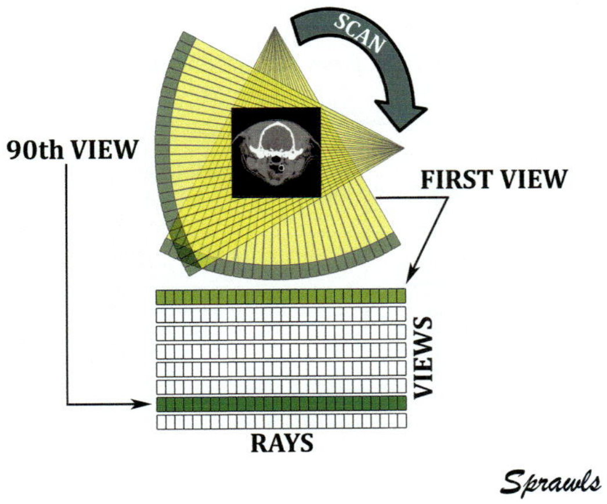

Figure 4.1 The scanning around the body to produce different views.

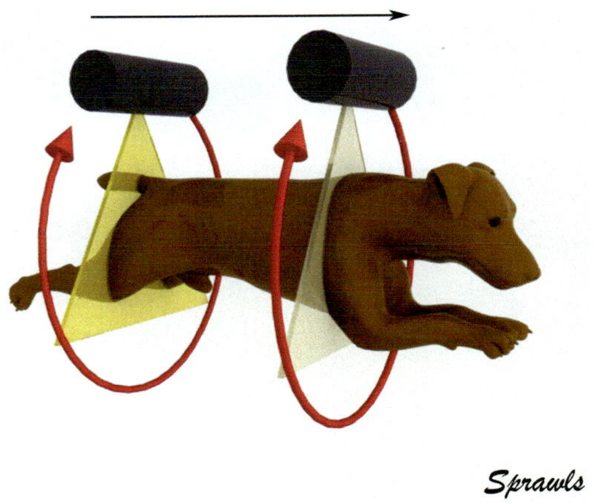

Figure 4.2 The two scan motions, around and along the length of the body.

DOI: 10.1201/9781315118130-5

4.1.1 Step-by-step scanning

This was the first scanning method developed and continues to be used for many procedures and especially when an image of just one slice is required. This is illustrated in Figure 4.3.

In this method, the body is kept stationary and not moving when the X-ray tube and beam is scanned around the body. If images of additional slices are required, the body is moved to that position and is scanned again. The scan in each slice position produces a data set from which the image will be reconstructed. There are several limiting characteristics of this method that are overcome by using the helical or spiral method described below. Since there is a data set created for each body position, the thickness and position of the imaged slice are determined at the time of scanning. If multiple slice images are required, the method progresses at a relatively less speed. That is because the movement of the body is stopped in each slice position, it is scanned, and then moved to the next position, and that slice is then scanned. The rotation of the X-ray tube around the body is stopped between each scan when the body is being moved. It is a step-by-step scanning method compared to the spiral/helical, which is a continuous scanning method.

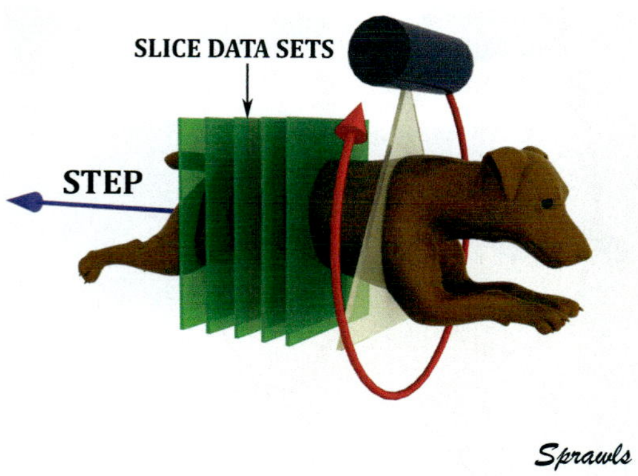

Figure 4.3 The step-by-step, or slice-by-slice scanning method.

4.1.2 Spiral or helical scanning

Spiral or helical scanning is a more advanced technology that increases the data with respect to the number, thickness, and position of slices that can be imaged in one scanning procedure. The concept is illustrated in Figure 4.4.

With spiral scanning, the X-ray tube and beam is rotated continuously as the body is moved, as shown above. The track of the beam on the body would be a spiral or helical pattern, thus the name. The advantage of spiral scanning is that it produces a continuous data set extending over a volume region of the body and not limited to just one slice position. This offers more options and flexibility during image reconstruction. Images of slices from anywhere within the scanned region and data set can be reconstructed, as illustrated in Figure 4.5.

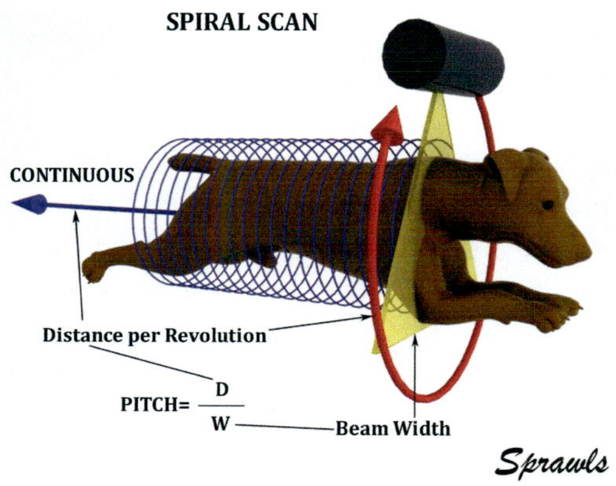

Figure 4.4 The concept of spiral scanning.

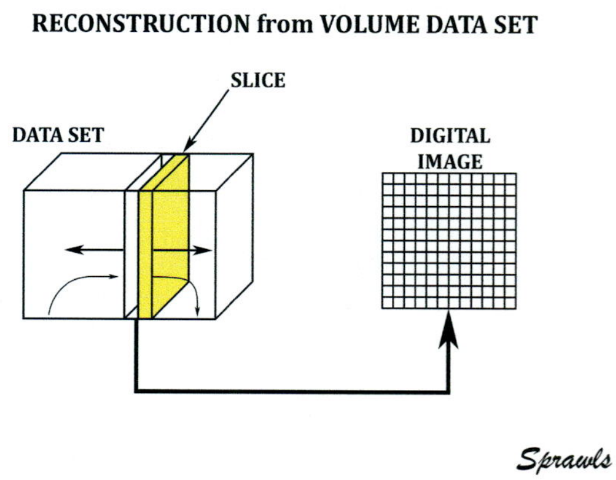

Figure 4.5 The possible relationships of an imaged slice to the data set obtained with a spiral scan.

The slice thickness, position, and orientation can be selected and adjusted at the time of reconstruction. Another significant advantage is that after a spiral scan and a volume data set is collected, there is no limit to the number of image reconstructions that can be made. The reconstruction process can be repeated to produce images with different positions, orientations, and thicknesses as needed. This is done without any additional exposure to the body. There is also the opportunity to produce three-dimensional (3D) images if the appropriate reconstruction software is available.

A major advantage of spiral compared to step-by-step scanning is the significant increase in speed for covering an extended anatomical region. This is because the X-ray tube rotation and body movement are continuous compared to the step-by-step method.

The speed with which the body is moved relative to the beam rotation is controlled by the adjustable protocol factor, the pitch.

The pitch value is defined in terms of the distance the body moves during each rotation of the X-ray beam around the body. Specifically, the value of the pitch factor is the distance the body moves during one rotation divided by the width of the X-ray beam, as illustrated in Figure 4.4. A major issue is selecting the most appropriate pitch factor value for each scanning procedure.

Increasing the pitch value causes the body to be moved faster relative to the beam rotations and has the advantage of decreasing the scan time to cover a specific anatomical section and also decreases the radiation dose if no other factors are changed. However, increasing pitch has the disadvantage of decreasing image quality in two possible ways.

It can increase the width of the X-ray beam, which then increases blurring in the slice thickness direction. This is because a longer segment of the body is scanned and data collected during each rotation. That can limit details in the slice thickness direction and to some extent limit the advantage of thinner slices. Also, if no other factors are changed, increasing the pitch will increase image noise because a lower concentration of radiation, or dose, is absorbed in the tissue.

The selection of the optimum pitch value for a specific procedure is usually based on clinical experience combining the requirements for visualizing the various anatomical structures with knowledge of the principles of helical scanning described here.

4.2 PROTOCOL FACTORS AND VARIABLES ASSOCIATED WITH SCANNING

As described earlier, there are three distinct phases of the CT imaging process and certain protocol factors associated with each phase. The protocol factors that control the scanning phase are shown in Figure 4.6.

The major significance of these factors is that they control two important things. One is the quality of the data acquired during the scanning, and the other is the radiation dose delivered to the body. As will be discussed later, there are other factors associated with other phases that have an effect on both image quality and dose. However, they cannot be used to change the limitations on image quality or the dose that has already been delivered to the body.

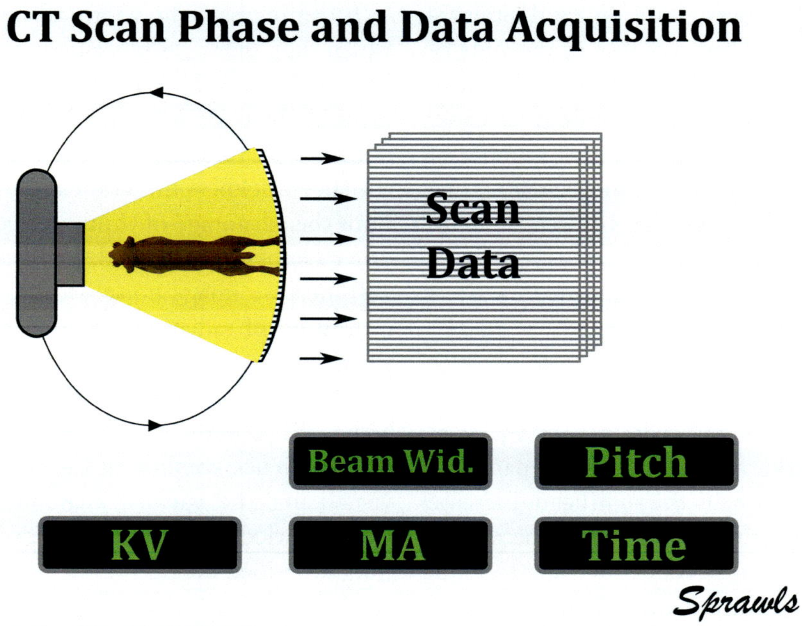

Figure 4.6 The protocol factors that control the scan phase and data collection.

CHAPTER 5

IMAGE RECONSTRUCTION

Perry Sprawls

The second phase of the CT imaging process is image reconstruction. It is a completely mathematical process that uses the data collected during the first phase to produce or reconstruct a digital image.

5.1 CONCEPT AND DESCRIPTION

The standard method used for image reconstruction is known as filtered back projection. The concept of filtered back projection is illustrated in Figure 5.1.

First, the general concept of image reconstruction by back projection will be described, followed by a discussion of the more significant issues of the adjustable protocol factors that must be considered for each procedure. As discussed earlier, when the X-ray tube and beam rotate around the body, each tube position creates one view. Typically, several hundred views are produced in one rotation. The data from one view, as illustrated in Figure 5.1, are not sufficient to reconstruct or produce a good image, but one view is illustrated to develop the concept.

The illustration shows a round simulated body section that contains two objects. The data collected from one view are not a complete image but more like a shadow or profile of the objects. If such data are then back-projected to form an image, as shown on the right side,

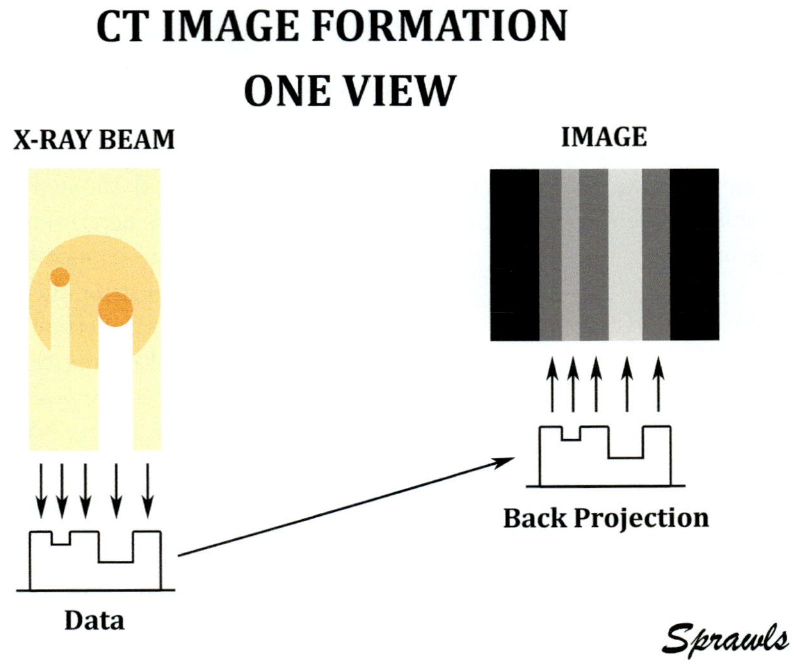

Figure 5.1 A very simple illustration of the CT imaging process.

DOI: 10.1201/9781315118130-6

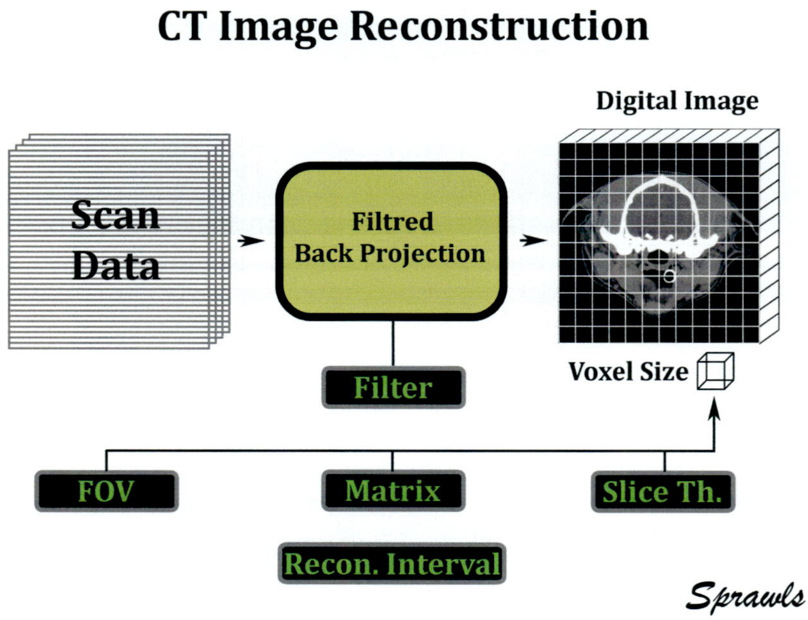

Figure 5.2 The major protocol factors associated with the reconstruction process.

the resulting image would show only streaks or shadows. However, when the data from many views are back-projected and combined, a complete high-quality image can be formed.

There are actually two mathematical functions involved in CT image reconstruction: back-projection process and digital filtering procedure: filtered back-projection for the reconstruction process. In clinical imaging, adjustments to the back-projection process are generally unnecessary, but the filter is one of the significant adjustable protocol factors that has an effect on image quality.

5.2 PROTOCOL FACTORS ASSOCIATED WITH RECONSTRUCTION

When a patient is scanned in the first phase, the data are collected and stored in the computer memory. Then, during the reconstruction phase, data are formed into images by back-projection.

Several adjustable protocol factors are associated with the reconstruction phase, as illustrated in Figure 5.2.

The adjustable protocol factors that are associated with the reconstruction phase depend on the type of scanning used during the acquisition phase, the step-by-step (slice-by-slice), or the helical/spiral volume method.

5.3 SLICE THICKNESS

The selection of slice thickness is very different between the two scanning methods. As discussed previously, with the step-by-step method, the slice thickness is determined and set at the time of the scan and cannot be changed after that, except by repeating the scan. However, with the helical/spiral method, the actual slice thickness is selected during the reconstruction phase, along with the slice position and orientation, as illustrated in Figure 4.5. When a scan is performed with this method and a volume data set is acquired, the image slices are formed by the reconstruction process. A significant advantage is that the reconstruction process can be repeated to form images with different slice thicknesses if desired and without rescanning or additional radiation exposure.

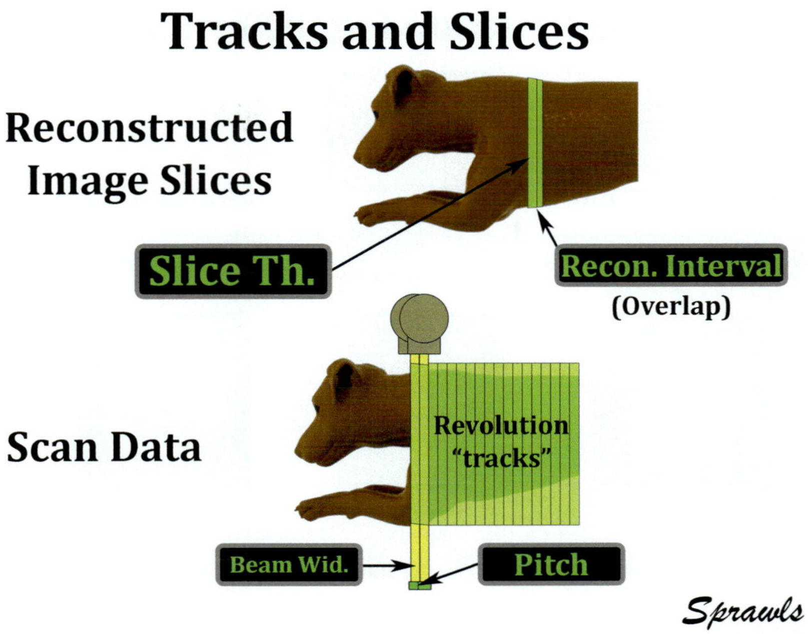

Figure 5.3 The difference between X-ray beam width and slice thickness.

Slice thickness is a major factor in determining image quality with respect to both image detail and visual noise. As illustrated first in Figure 2.1, slice thickness is generally the most adjustable factor in determining the tissue voxel size. Voxel size is a major factor in optimizing image quality with respect to detail and noise and balancing with radiation exposure and dose.

The reason for using thinner slices is to produce smaller voxels and improve the visibility of detail. Even though with the helical/spiral method the slice thickness can be adjusted during the reconstruction phase and after the scanning and acquisition is complete, there is a limit to the image detail that can be achieved by opting for thinner reconstructed slices. Here, it is important to distinguish between image slice thickness and the width of the X-ray beam during the acquisition phase. In the step-by-step method, image slice thickness and X-ray beam width are essentially the same since it is a slice-by-slice process. The difference between beam width and slice thickness with helical/spiral scanning is illustrated in Figure 5.3.

With helical/spiral scanning, the data set extends continuously over a volume segment of the body but is actually divided into segments or "tracks" formed by each revolution of the X-ray tube and beam around the body. Because the body is moving, these segments are not sharply defined but are somewhat blended together. The effective width of each track is determined by the combination of two adjustable protocol factors, beam width and the value of the pitch. The significance is that the width of the track determines the anatomical detail visible in the image. In effect, the width of the track blurs the image in the axial direction. Even though the actual image slice thickness is set during the reconstruction, adjusting for a thin slice cannot provide increased image detail because of the blurring in the data.

The selection of reconstructed image slice thickness must consider the beam width that is used in the scan acquisition. Producing very thin slices with relatively large beam width cannot provide increased detail but will actually deteriorate the image quality by increasing the noise.

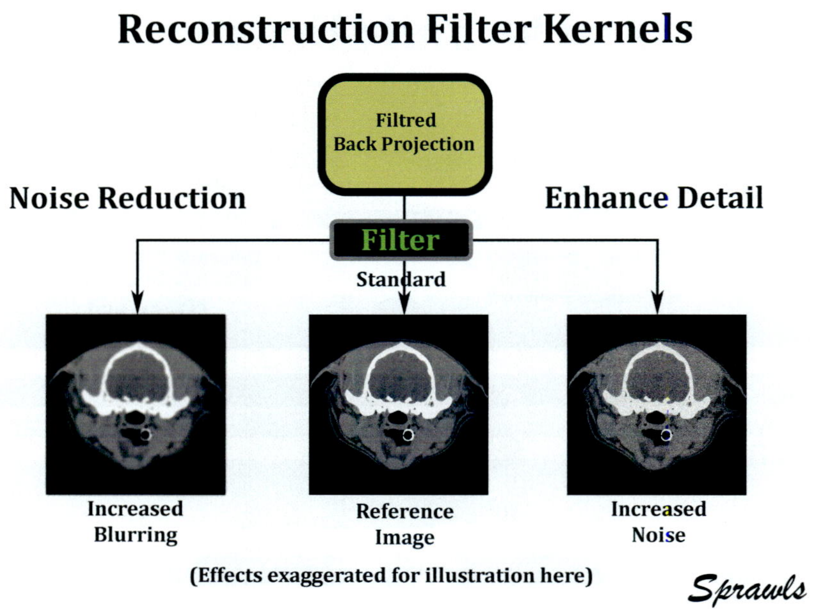

Figure 5.4 Two image characteristics, detail and noise, that can be changed by the selection of specific filter algorithms or kernels.

5.4 MATRIX SIZE AND FIELD OF VIEW

The ratio of the field of view (FOV) to the matrix size determines the size of the tissue voxel in the plane of the image, as illustrated in Figure 2.1. Both factors are associated with the reconstruction process. While an appropriate FOV is selected to provide the necessary anatomical coverage, the effect on the image quality must also be considered. Reducing the FOV with other factors remaining unchanged reduces voxel size.

5.5 RECONSTRUCTION FILTERS (ALGORITHMS AND KERNELS)

Filtered back-projection is the conventional method used for reconstruction of CT images. The term filtered refers to the digital image processing that is combined with the mathematical back-projection. The filters are in the form of a computer program, sometimes referred to as algorithms, that is used to change some of the image characteristics. The selection of the appropriate filter or algorithm is an important step in optimizing an imaging protocol. The typical CT system will have a variety of filters that are often associated with the imaging procedures for different anatomical regions and organ systems. While these are often already preset in the protocols, it is helpful to understand some of the general filter functions and characteristics. Two of the major filter functions are illustrated in Figure 5.4.

The filter can be used to improve the image quality by reducing noise and enhancing detail, but not at the same time. Blurring is used to reduce noise in an image and has the undesirable effect of also reducing the visibility of detail, as discussed previously.

Low noise and high detail are both desirable image characteristics to enhance the visibility. The challenge in CT imaging is that they have conflicting characteristics. Making changes to improve one generally deteriorates the other. This was illustrated in Figure 2.10 as it relates to the selection of voxel size. Small voxels are required to increase image detail, but using small voxels increases the noise. That same compromise between noise and detail must be made with the selection of a filter.

CHAPTER 6

IMAGE DISPLAY AND VIEWING

Perry Sprawls

The third and final phase of the CT imaging process is the conversion of digital images produced in the reconstruction phase into visible images or analogue images. There are several adjustable factors associated with this that can have a significant effect on the visibility within various anatomical regions. The general process and associated factors are illustrated in Figure 6.1.

6.1 DIGITAL-TO-ANALOGUE (GREY-SCALE) CONVERSION

Digital to analogue is the computer term describing the conversion between the two ways that images are represented or recorded. This is an electronic process significant for adjustable factors that control the appearance of the viewed image.

The reconstruction process produces an image in which the pixel has a CT number that relates to the X-ray attenuation (generally tissue density) of the corresponding tissue voxel, as illustrated in Figure 2.3. The CT number values range from –1,000 for air to approximately +2,500 for dense bone. If this full range of CT numbers is converted to brightness values and displayed, it would yield a very low visible contrast for most anatomy, especially soft tissues. This is because the difference in CT numbers among soft tissues is very small, especially in relation to the full CT number scale. Therefore, in order to have sufficient visible contrast, it is necessary to select a small range from within the CT number scale and display it over the full brightness or grey-scale range. That is the function of the window. The ability to select a small range of CT numbers for display is a major factor contributing to the high contrast sensitivity provided by CT, especially for visualizing differences in soft tissues.

6.2 WINDOWING

The window represents the range of CT numbers selected for display and is controlled by two factors, as illustrated in Figure 6.1, the window level and window width.

The window width functions as a contrast control for the displayed image. Reducing window width increases image contrast and contrast sensitivity for visualizing differences in soft tissues. For comparison, if the window width is set to cover the full CT number range, the resulting image would be dull grey with very little contrast.

Notably, reducing window width is an increase in visible noise. Noise is actually a variation in CT numbers produced by the statistical characteristic of X-ray–photon interactions. Changing the window width does not change the actual noise value, just the contrast and visibility of the noise, as displayed in an image.

The window level control is used to position the window or range of CT numbers to be displayed in the image at various locations within the large CT number range. For example, placing the window near the zero CT number point (the value for water) provides good visualization of soft tissue contrast, whereas higher window level provides better visualization of bone structures. Appropriate window settings for different clinical procedures are developed from a general knowledge of the concepts of windowing and experience. Importantly, the use of the

DOI: 10.1201/9781315118130-7

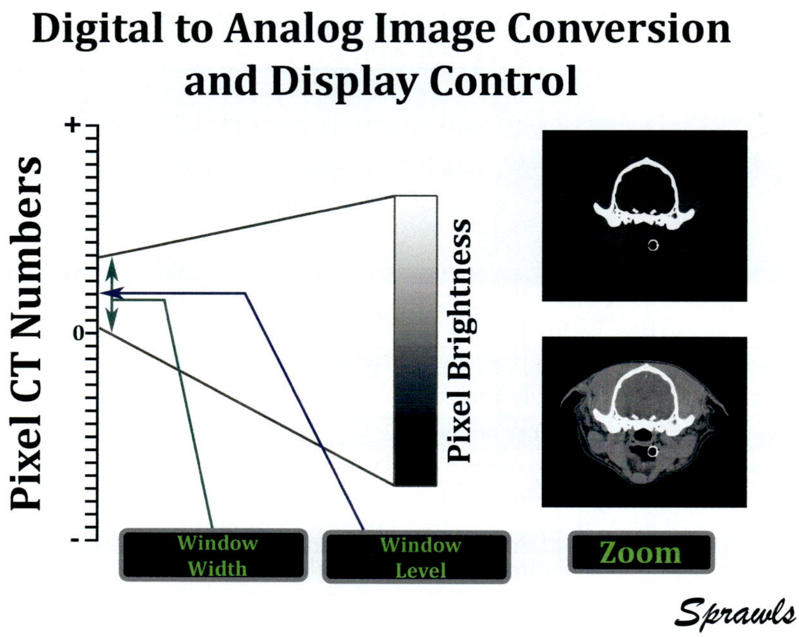

Figure 6.1 The conversion of image pixel CT numbers into visible brightness or shade of grey, as controlled by the window.

window divides the CT number scale into three segments. Visualization of tissue contrast is possible only within the window segment. There is no visibility of contrast or contrast sensitivity for tissues with CT numbers in the segments, either below or above the window.

The most effective way to learn about the window is to actually sit at the console and experiment with it.

6.3 ZOOM/MAGNIFICATION

The concept of zooming is familiar to us in photography, where most modern cameras have a zoom function. Zooming in reduces the field of view and produces a magnified image. Magnification is related to the field of view. During the image display and viewing phase, there is typically a zoom or magnification function available. The important thing is to distinguish between this function and the setting of the field of view in the reconstruction phase. Both have the feature of adjusting magnification, but they do not have the same effect on the visibility of detail.

During the reconstruction phase, the size of tissue voxels and corresponding image pixels is set, and as described previously, this is one of the limiting factors on image detail. At this point, reducing the field of view reduces voxel/pixel size and potentially improves the visibility of detail. In principle, the image detail is locked in with the reconstructed image and cannot be enhanced later.

Using the zoom/magnification function when viewing an image might provide some advantage in visualization of some features, but it cannot increase the image detail. When the zoom function is used, it does not change the reconstructed image pixel size, which limits the visibility of image detail. A smaller region of the reconstructed image is now magnified to fill the viewing area, but it does not reduce the original image pixel size and the related image detail. An interesting exercise to demonstrate this point is to adjust the zoom/magnification to maximum value. Typically, the original image pixels will become magnified and visible, and the image begins to look blurry.

CHAPTER 7

RADIATION DOSE

Perry Sprawls and Wolfgang Henninger (Veterinary issues)

Since CT is an X-ray procedure, it deposits absorbed energy in the patient's body and also potentially exposes clinical staff to radiation that is scattered from the patient. There is value in reducing the amount of radiation used in each procedure, but there are significant limits on the amount of dose that can be reduced. A major factor in CT, and other forms of X-ray imaging, is that image quality and visibility are very dependent on the level of radiation used to produce the image. In general, increasing image quality and visibility requires the use of more radiation. This somewhat complex relationship must be considered in the optimization of imaging protocols.

7.1 HOW IT IS EXPRESSED: THE SPECIAL QUANTITIES USED IN CT

A factor adding to the complexity is the variety of quantities and methods used to express the radiation dose. The general relationship is illustrated in Figure 7.1.

Dose is the radiation quantity used to express the concentration of energy absorbed in tissue at a specific point or location within a body. The unit, Gray (Gy), is defined as the absorption of 1 Joule of energy per kilogram of tissue. That is a rather large unit and used for radiation therapy, and the unit mGy is more appropriate for diagnostic procedures.

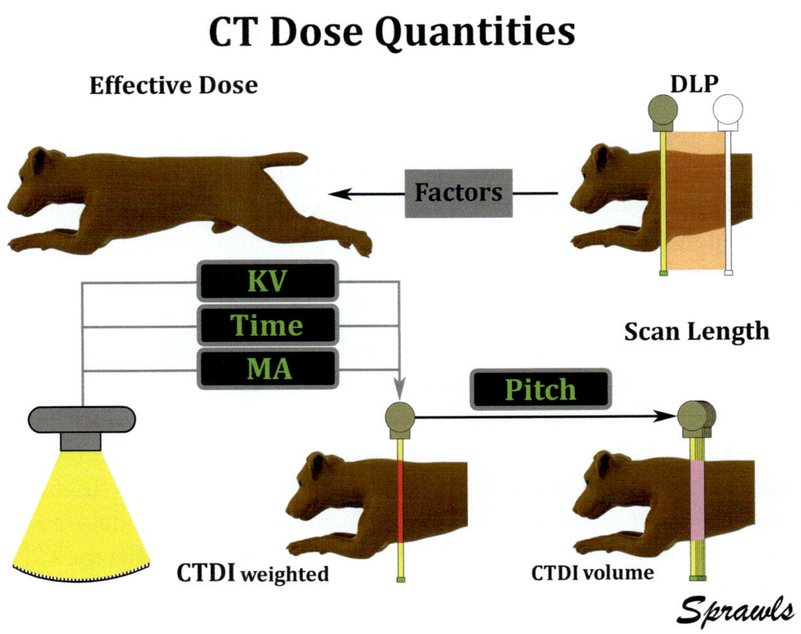

Figure 7.1 The various quantities used to express the radiation dose delivered during the CT procedure.

DOI: 10.1201/9781315118130-8

There is no practical method for measuring the actual dose absorbed in tissue during a CT procedure. This would require the insertion of a dosimeter directly into the body at the point of interest. Therefore, the dose to internal tissues or organs is determined indirectly through external measurements and calculations.

The computed tomography dose index (CTDI) is the special unit for expressing the dose in CT. Since it is not a directly measured quantity, it is classified as a dose estimate. While it is not a completely accurate dose value for a specific procedure, it is useful for general dose information and comparing dose within different imaging methods and equipment.

The established procedure is that CTDI values for a specific CT machine are determined by making measurements in a phantom, a cylinder of material with the same x-ray absorption characteristics as soft tissue. This is illustrated in the lower left of Figure 7.1.

Dosimeters, electronic devices that measure dose, are inserted in several locations within the phantom. The dose values measured in the different locations are mathematically combined to provide an "average" within the slice, which is designated as the CTDI *weighted* value. This process is calibrating the specific machine to determine the produced dose.

This measurement is made at a specific kVp and mAs value that might be different from that used for a patient. By taking into account the differences in the factors used for the phantom and a patient, the CTDI value for a patient can be determined. Many CT machines actually make this calculation and display the value as the technical factors are selected before a scan. It must be recognized that the obtained CTDI value is an estimate, not a precise dose value. With spiral/helical scanning with a pitch different from a value of 1, the actual dose will be different from that calculated as just described. If all other factors remain unchanged, the dose is decreased by increasing the pitch. This is because increasing the pitch has the effect of moving the body faster through the x-ray beam and reducing the concentration of the radiation at each location.

While the dose expressed by the CTDI describes the concentration of radiation energy absorbed in an individual slice of tissue, it does not provide information about the total radiation energy imparted to the body. This requires consideration of the area of the body covered by the scanning X-ray beam. In CT, an indication of the relative total energy for procedures is expressed by the dose area product (DLP). The value is obtained by multiplying the CTDI (representing the dose) by the length of the body area scanned, as illustrated. Some CT machines calculate and display values of the DLP when the factors are selected to set up a procedure. The significance of DLP compared to the CTDI is that it indicates total energy imparted to the body.

If there is an interest in the potential biological effect by the radiation from a specific procedure, it is necessary to consider additional factors. These are the anatomical locations of the scanned region and the organs involved along with the relative radiation sensitivity of the various organs and tissues. If these have been combined into a set of conversion factors, a research project, they can then be used to calculate the so-called effective dose. The effective dose is a radiation quantity representing the equivalent whole-body exposure.

Within general clinical practice, the CTDI and DLP are the radiation quantities used to monitor exposure in specific procedures. The values do not provide highly accurate dose information for individual cases but are useful for comparing different CT machines and scanning procedures with respect to radiation exposure.

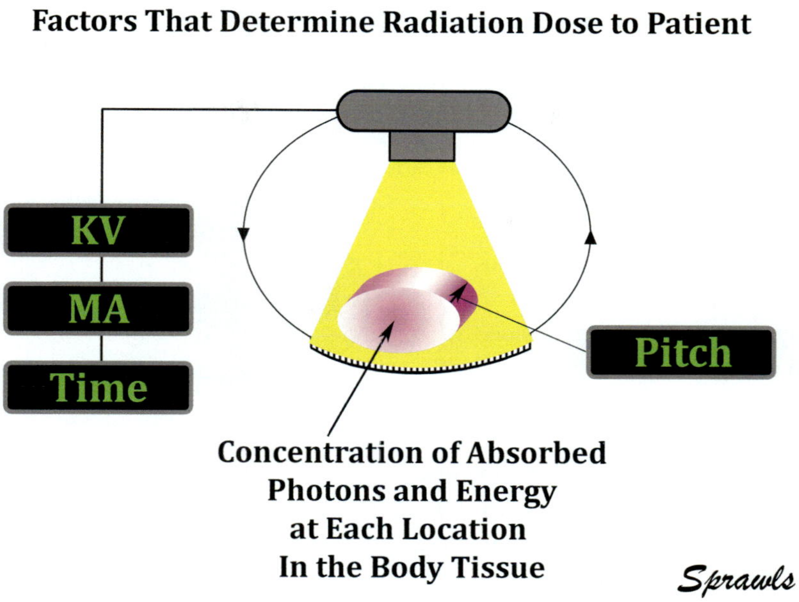

Figure 7.2 Factors that determine the dose in a specific scanning procedure.

7.2 FACTORS THAT AFFECT THE DOSE

While it is appropriate practice to keep the radiation dose as low as "reasonably achievable", the ALARA principle, the major issue in CT imaging is balancing the dose with the requirements for image quality by using *optimized imaging protocols*.

We will begin by reviewing the factors that actually determine the dose and then see how these factors also affect image quality. For a specific patient and imaging procedure, the dose (concentration of absorbed radiation energy) is determined by the combination of protocol factors shown in Figure 7.2.

7.3 RADIATION DOSE: VETERINARY ISSUES

7.3.1 Patient and personnel safety

Responsible veterinary radiologists need to ensure the safety of their patients, clients, and colleagues when they conduct CT examinations. Although patient safety is important during the scanning procedure, it is less important than the legislative regulations that minimize ionizing radiation compared with human medicine. This does not mean that anyone should be careless, but it does focus on more pressing topics, such as the entire procedure, which is usually much more time-consuming in veterinary medicine than in human medicine.

Safety of medical personnel in CT imaging is a critical component of surveillance. It is strongly recommended that at least one technician is responsible for the proper functioning of the CT scanner; furthermore, the technical support by the manufacturer maintaining hardware and software of the machine shall be regularly used. This is often supported by various forms of contracts by the manufacturer but can also be booked individually.

The five commandments for the responsible technician are as follows:

1. Check that the system's warm-up and self-check are correctly performed every time the scanner is used;
2. Check that the scanner's air- conditioning/water cooling system is properly functioning during working hours;

3. Check the door and/or window contacts regularly (otherwise the scanner won't work);
4. Keep every person out of the examination room while the tube is emitting radiation;
5. Have at least one set of lead apron and thyroid shield ready for anyone who has to enter the examination room unexpectedly (in case the radiation does not stop automatically when the door opens) or anyone else has to stay inside the examination room with an unstable patient during scanning.

However, despite the general rule of staying out of the room where the scanner is housed during radiation, a special situation may occur that some medical personnel or an anaesthesist has to stay with the animal; this includes a very unstable patient who needs special assistance during anaesthesia or when some sort of intervention is performed and assistance is needed during a CT-guided biopsy or CT-guided treatment. A person who is fully dressed in a lead apron and thyroid shield and probably eye protection glasses has two safe positions: close to either of the gantry's short sides. (Figure 7.3). The scanner scatters its radiation cloud in the shape of an asymmetric leaf; therefore, someone facing the gantry is less exposed because the equipment within the gantry shields this position.

The technicians' duties also cover the workflow in the scanning procedure of the patient. The correct patient data have to be taken from the patient's data bank or have to be keyed manually into the CT system. This will help find the correct images later in a picture archiving system (PACS) for comparison or follow-up purposes. This leads to the question: How to deal with the patient's and owner's data? Every information fed into the scanner mask will appear in the image header, although there are possibilities to decide which one is on display. A decision needs to be made about openness or privacy with owners' and animals' names and the amount of information displayed in a single image. If it is decided, for example, only to display the breed and name of the patient or

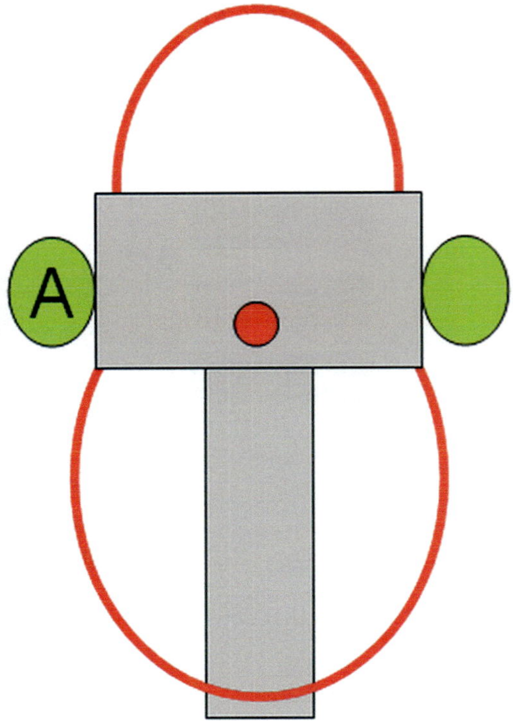

Figure 7.3 **Relatively "safe" position of a shielded person beside a CT scanner during a necessary stay within the examination room while the tube is emitting radiation (A—anaesthesist/technician, labelled in red—position of the tube and area of scattered radiation).**

a patient's hospital number, this information should be connected to the general patient data archive to ensure individual identification.

Safe anaesthesia, which is crucial for the patient, is essential for a state-of-the-art CT exam. A profound trained veterinary anaesthetist is recommended. An anaesthetist examines the animal beforehand to determine the anaesthetic risk and care is taken that the intravenous (IV) access is performed safely with knowledge how to administer narcotics and IV fluids correctly; and how to handle emergency situations, including having an emergency kit ready at the site, since many patients are older and often have multiple diseases. An anaesthetist is also responsible for providing the correct amount of warmed iodine contrast medium to be administered intravenously, either manually or by an injection pump. Furthermore, experience is required to monitor all the equipment and use and to read the curves; including the inspection for a full oxygen bottle and a full vapour with iso- or sebofIurane, and so on.

7.3.2 CT scan parameters and their impact on the radiation dose

The CT scan parameters for each determined region can be preset for each CT scanner and individually modified as required. The impact of the radiation dose always must be balanced according to the ALARA principle.

After nearly 30 years of CT scanning in veterinary medicine (and about 50 years in human medicine), there are many sophisticated CT scanners available on the market for veterinary use. Single-slice scanners are rarely in use nowadays, these days the main focus is multi-slice scanning from 4 up to 64 slices (or even more) per tube rotation. While the "stop&go" incremental/sequential procedure is acceptable, the technical progress made helical/spiral scanning the favourite technique. By means of this technique, a digitized volume of the scanned body region is acquired and can be expressed in transverse slices with a defined thickness. The accuracy of the image information depends on several factors like the beam collimation representing the original slice thickness (ranging from 0.5 up to 2 mm or more) and the table feed represented by a number called "pitch". Pitch is a factor of the beam collimation broken by the velocity of the table in mm/rotation of the tube. This results in values less than 1 (overlapped scanning) or equal to 1 (no increment, slice on slice) or higher than 1 up to 3 (extended helical scanning). By choosing the various pitch modalities, the accuracy of the image information depends grossly; smaller pitch values represent higher image accuracy because more than 1 full (360°) tube rotation can be used for image reconstruction. For comparison in pitch 2 or 3, only half or less of the information gained during tube rotation can be used for image reconstruction. This is quite important when tiny body parts are scanned, e.g., medial coronoid process of the elbows, middle and inner ear, and small objects in general.

Image quality is directly linked to the radiation dose, and this means the amount of mA represents the crucial parameter in CT scanning. The reconstructed digitized image has a defined matrix and a defined depth of grey shades. This leads to a certain amount of noise within the image. Each voxel represents the density of the scanned tissue, and the accuracy of calculation for a representing pixel mainly depends on the amount of mA used for the scanning procedure. Higher mA will increase the image quality and reduce image noise, but constantly, high mA will heat the tube more quickly and prolong interscan delays to cool down the tube. Therefore, the manufacturers invented the mA modulation technique to use higher mA in higher attenuating body parts and less in those with less absorbing tissue (e.g. lungs). In veterinary medicine, the varying body size of the patients also saves X-ray tubes from getting hot too quickly.

The patient's radiation dose will increase with fixed or maximum mA settings and also with a rather narrow collimation (pitch less than 1). The patient's radiation dose can be reduced by using the mA modulation technique and increased pitch (e.g., 1.5 or 2). Using an mA saving technique will also heat the tube more slowly and shorten interscan delays.

kV is used for adapting tissue volume and density. For clinical use, kV values are mainly fixed between 110 and 130 kVp. Indications to increase kVp might be very large and/or obese patients or very dense tissue, e.g., teeth of rabbits or small mammals.

The patient's radiation dose could be substantially increased by increasing the mA settings compared to kV.

Various CT systems do calculate a CT dose index (CTDI value) so that the absorbed radiation dose in mGy/kg body weight can be estimated; however, these values are used for orientation in human medicine and have limited practical value in animals.

CHAPTER 8

IMAGE QUALITY AND DOSE OPTIMIZATION

Perry Sprawls

The optimization of a protocol with respect to image quality and dose is a seemingly complex relationship because of the interaction of several factors.

8.1 RELATIONSHIP OF IMAGE NOISE TO DOSE

Noise as an image quality characteristic is most closely related to and dependent on dose. As discussed previously and illustrated in Figure 2.9, both dose and noise are determined by the concentration or quantity of photons absorbed in each tissue voxel. The noise can be decreased by increasing the number of photons, which increases the dose. The factors involved in optimizing this relationship are shown in Figure 8.1.

With all methods of X-ray imaging, including CT, visual noise is the image quality factor that limits the reduction of dose. However, some factors can change the noise-to-dose relationship and reduce noise for a specific dose value, though this comes with some compromise. The KV, MA, time, and pitch, as illustrated in Figure 8.1, affect both the dose and noise because they control the concentration of photons deposited in tissue. The digital image processing filter can be set to reduce noise without changing the dose, but has an accompanying limitation of adding some blurring to the image and reduces detail. This must be considered

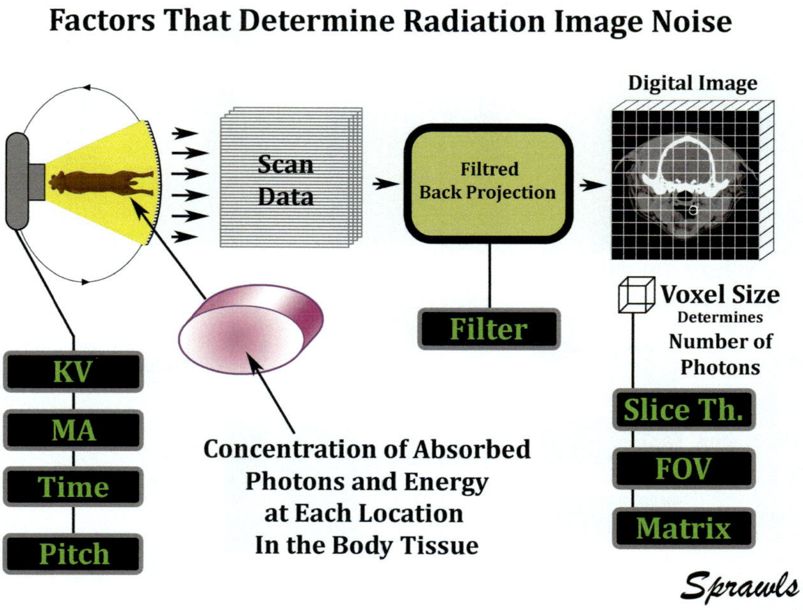

Figure 8.1 Protocol factors that have an effect on both dose and noise.

DOI: 10.1201/9781315118130-9

when selecting the appropriate filter algorithms for specific procedures.

The most significant factor that affects the noise-to-dose relationship is voxel size because the noise is determined by the *number* of photons absorbed in each voxel and that is also determined by voxel size. For a specific dose (concentration of photons) to the tissue, the noise can be reduced by increasing the voxel size.

8.2 THE DETAIL AND NOISE BALANCE

Now is focussed on the most significant relationship in optimizing imaging protocols: the relationship between noise and detail along with the dependence on dose, as illustrated in Figure 8.2.

Detail has an indirect effect on dose and involves noise. The following illustration shows how this occurs. The left image that was produced with a relatively low but adequate dose. If the detail is enhanced by changing any of the three factors, slice thickness, matrix size, of field of view, an increase in noise is seen. Reducing the noise back to the original level requires an increase in the dose, typically by increasing the mA value.

8.3 OPTIMIZED PROTOCOLS

An optimized protocol for a specific procedure provides adequate visibility of the anatomical structures and signs of pathology or injury with the lowest reasonable radiation dose. This becomes somewhat complex both because of the conflicting image quality characteristics, especially between detail and noise, and the dependence of image quality on the radiation dose.

Knowledge of the imaging principles and quality characteristics, as developed here, is useful for a better understanding of the protocols used and how they can be changed if necessary. This is best applied in the clinical setting by critically viewing images for quality and comparing the protocol factors used to produce them.

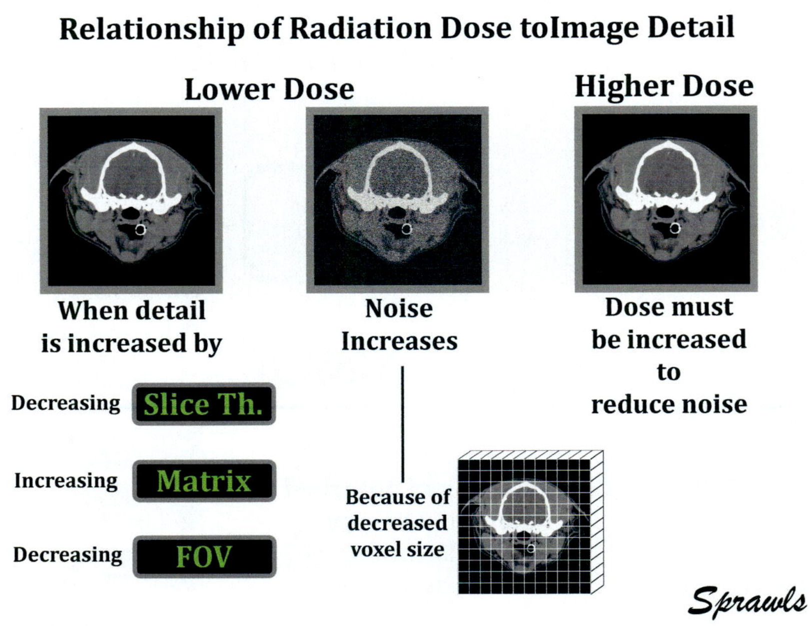

Figure 8.2 The effect of image detail on dose.

CHAPTER 9

PATIENT PREPARATION

Henri van Bree, Ingrid Gielen, and Tim Waelbers
(Anesthesia/sedation protocols)

9.1 PATIENT PREPARATION

Before any CT examination is performed, it is essential that the original order requisition of the referring veterinarian is available in order to plan the appropriate examination. Because the patient has to remain completely still during the CT examination, usually a heavy sedation or a short general anaesthesia is performed. In preparation for the anaesthesia, the patient must refrain from consuming food from the evening before. A general clinical workup – including a range of blood tests – depending on the age of the animal and the information supplied by the referring veterinarian should be carried out before the CT examination. The staff should be informed about the eventual occurrence of kidney disease as this may increase the risk of an unusual adverse effect. In patients suspected of having neoplastic disease, a chest radiograph is also required prior to anaesthesia. Checking haemostasis is necessary before CT-guided biopsies are performed. Anaesthetic machines must provide controlled respiration as in some circumstances it is important to control breathing. Anaesthetic and monitoring equipment should not pass through the gantry. Metal objects may affect CT imaging and should be removed prior to the exam.

The areas for performing eventual arthro-CT or taking biopsies should be clipped and aseptically prepared prior to the CT procedure.

Intravenous (IV) fluids are always administered prior to the CT examination until the patient is discharged. Especially in older animals having existing health issues or in patients having a contrast study, IV fluids assist in flushing any contrast media through the kidneys after the scan.

Intravenous contrast should be administered on the opposite side of the pathologic process as pooling of contrast may hinder the interpretation of lesions. Although very rare in veterinary patients, the staff should be notified in case of a history of allergic reactions to contrast media.

The attention should be drawn to proper patient positioning, which will be addressed in the following chapters. Patients may be placed into the scanner in ventral or dorsal position, headfirst or feet first. The CT scanning system needs information on how the patient is positioned so that left and right markers are annotated correctly.

Often, straps are used to fix the patient on the CT table. Body parts should be immobilized using wedge sponges, Velcro bands and foam cushions and blocks (Figure 9.1).

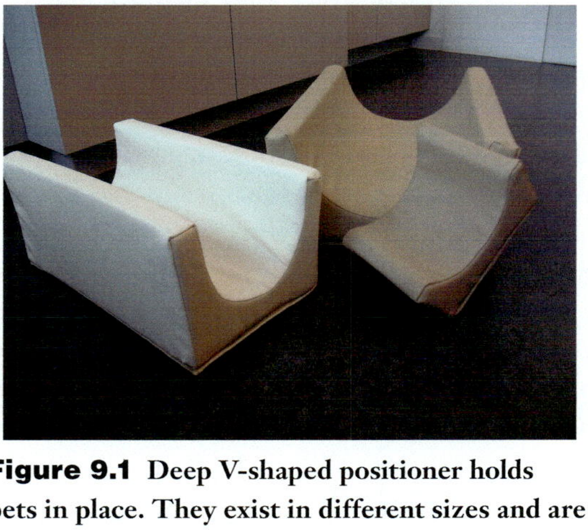

Figure 9.1 Deep V-shaped positioner holds pets in place. They exist in different sizes and are commercially available.

DOI: 10.1201/9781315118130-10

Before bringing the patient to the examination room, it is advised to perform calibrations and tube warm-up procedures of the CT device.

9.2 ANAESTHESIA/ SEDATION PROTOCOLS

The use of CT is limited in veterinary medicine because the patient needs to be immobilized during the procedure and thus requires sedation or general anaesthesia. Since there is no such thing as a "small sedation" or "anaesthesia light", anaesthesia-related risks must be decreased significantly by adhering to normal anaesthesia procedures, including reviewing the patient's clinical history and medical records, preparation of the anaesthetic record, pre-anaesthetic physical examination, venous access and prevention of hypothermia. Furthermore, the anaesthetic protocol should be tailored to the patient's specific needs. Anaesthesia should always be induced in a room used solely for that purpose, preferably one adjacent to the CT scanner, with standard anaesthetic, resuscitation and transport equipment so that the patient can be anaesthetized easily and quickly. The monitoring equipment in the CT unit must conform to the same standards as the operating theatre equipment. Because access to the anaesthetized patient can sometimes be restricted during the scanning process, every staff member must be meticulously organized and trained to minimize any anaesthesia-related risks.

This chapter provides an overview of sedation or anaesthesia protocols. The anaesthetist must always choose a protocol that is tailored to each patient's specific needs; however, specific use of each protocol is beyond the scope of this chapter.

Sedation is generally considered safer than general anaesthesia, but it depends on the patient and may not always be the case. Since patients need to be immobilized during the scanning process, a deep level of sedation is needed to ensure the patient does not move because the risk of radiation implies the patient cannot be restrained manually. When a patient is sedated, a fully equipped anaesthesia workstation with airway management and resuscitation equipment should be available.

Several protocols can be used to sedate small animals; however, this chapter will cover only the IV administration route, which achieves the most reliable level of sedation. Venous access is mandatory for patient safety and to administer contrast medium when required. Acepromazine is most commonly used in small animals, but when combined with an opioid, it increases the sedation level because its effect can be unpredictable. The kappa opioid receptor agonist butorphanol, which profoundly increases the level of sedation compared with that of acepromazine alone, is a popular choice. Butorphanol has an antagonistic effect when combined with mu opioid agonists like methadone, morphine and fentanyl, so when deep analgesia is required (for example, for a presurgical CT scan or with herniated disc disease), we recommend combining acepromazine with a mu opioid agonist like methadone or morphine. The alpha-2-agonists medetomidine and dexmedetomidine are also frequently used for sedation in companion animals; they are extremely potent sedatives that can cause severe cardiovascular side effects and must be used with caution in young, healthy animals only. Like acepromazine, alpha-2-agonists can be combined with opioids to increase the level of sedation. When this combination is used, severe bradycardia may occur. When muscle relaxation is insufficient, the benzodiazepine midazolam can also be added to the protocol.

Midazolam must be administered following sedation with the alpha-2-agonist due to its possible side effects that include excitement, behavioural changes and nervousness. The effects of alpha-2-agonists can be completely reversed with the alpha-2-anatgonist atipamezole, which should be administered intramuscularly.

Sedation is easier to perform than general anaesthesia but has limited use because the sedative effect wears off during a long imaging protocol, resulting in movement of the patient, resulting in an inadequate and blurred image. Also, imaging of the abdominal or thoracic region may be hindered by respiratory movements. Finally, the inability to secure a patent airway (for example, in brachycephalic breeds, the head and neck may need to be flexed for correct positioning) is an important concern in sedation. General anaesthesia is therefore frequently used for immobilization during CT scanning. The products listed in Table 9.1 are used for premedication, generally in the lower dose ranges. Anaesthesia can be induced by administration of a volatile agent through a mask, although increased patient stress and pollution of the working environment limit the application of this method in companion animals. Anaesthesia can also be induced with several injectable agents that should all be administered "to effect" (i.e., until endotracheal intubation is possible) to minimize cardiovascular and respiratory side effects. Premedication with an alpha-2-agonist is very effective to decrease the dosage of the injectable agent needed to obtain endotracheal intubation. Other drug combinations can also be used—propofol, alfaxolone and midazolam combined with ketamine are used routinely (Table 9.2). All the advantages, disadvantages and side effects are outside the scope of this chapter.

Anaesthesia can be maintained with injectable agents like propofol and alfaxolone administered as a continuous rate infusion (Table 9.3), although volatile agents like isoflurane and sevoflurane are more frequently used for maintenance in companion animals. The airway should always be secured by endotracheal intubation, and oxygen should be supplied while the depth of anaesthesia has to be checked regularly. If the level is too deep, the infusion rate or vaporizer setting must be adjusted to limit any side effects.

Even in patients under general anaesthesia, respiratory movements can cause motion artefacts. If ventilation support is immediately available, apnoea can be induced, and the radiologist can scan for 60 seconds; then the patient must be mechanically ventilated until spontaneous ventilation is restored. Several cycles can be used until the scanning process is completed.

Apnoea can be achieved with several drugs (Table 9.3) or by inducing hypocapnia (i.e., decreasing the level of carbon dioxide in the blood) by temporarily hyperventilating the patient mechanically. When apnoea is induced, an experienced anaesthetist is needed to support the ventilation and monitor the patient closely until spontaneous respiration is restored.

The need for sedation or general anaesthesia during the scanning process in companion animals clearly limits its use compared with human medicine. As with every anaesthetic procedure,

Table 9.1 **Drugs used for sedation during CT scan in dogs and cats**

	SPECIES	ROUTE OF ADMINISTRATION	DOSE IN MG/KG
Acepromazine	Dog	IV	0.01–0.05
	Cat	IV	0.01–0.1
Butorphanol	Dog and cat	IV	0.1–0.4
Methadone	Dog and cat	IV	0.1–0.5
Morphine	Dog and cat	IV slow	0.1–0.5
Medetomidine	Dog and cat	IV	0.002–0.02
Dexmedetomidine	Dog and cat	IV	0.001–0.01
Midazolam	Dog and cat	IV	0.2–0.4

Table 9.2 **Injectable agents used in dogs and cats for induction and maintenance of anaesthesia**

	SPECIES	ROUTE OF ADMINISTRATION	DOSE IN MG/KG
Propofol (induction)	Dog	IV	4–6 to effect
	Cat	IV	6–8 to effect
Propofol (maintenance)	Dog	IV	0.05–0.4 mg/kg/min
	Cat	IV	0.05-0.2 mg/kg/min (≤30 minutes)
Alfaxolone (induction)	Dog	IV	1–3 to effect
	Cat	IV	2–5 to effect
Alfaxolone (maintenance)	Dog	IV	4–7 mg/kg/hour
	Cat	IV	5–8 mg/kg/hour
Midazolam (M)+ ketamine (K)	Dog and cat	IV	M: 0.2–0.4 K: 5–10 to effect

Table 9.3 **Drugs used to induce apnoea in companion animals under general anaesthesia (a dedicated anaesthetist and ventilatory support should be available)**

	SPECIES	ROUTE OF ADMINISTRATION	DOSE IN MG/KG
Fentanyl	Dog and cat	IV	0.001–0.002
Midazolam	Dog and cat	IV	0.2–0.3
propofol	Dog and cat	IV	1–2
ketamine	Dog and cat	IV	0.5–1

it is important to minimize the risks by choosing a protocol that is tailored to the patient's specific needs and monitoring the patient closely to full recovery.

FURTHER READING

Arlachov Y, Ganatra RH. 2012. Sedation/anaesthesia in paediatric radiology. *British Journal of Radiology* 85:e1018–e1031.

Hasebroock KM, Serkova NJ. 2009. Toxicity of MRI and CT contrast agents. *Expert Opinion on Drug Metabolism & Toxicology* 5(4):403–416.

Romans LS. 2011. Patient preparation. In: Romans LS (ed) *Computed Tomography for Technologists: A Comprehensive Text*, Wolters Kluwer Health/ Lippincott Williams and Wilkins, Philadelphia, PA, Chapter 11, p. 110–119.

Seymour C, Duke-Novakovski T. 2007. *BSAVA Manual of Canine and Feline Anaesthesia and Analgesia*. Second edition. British Small Animal Veterinary Association Gloucester UK. Veterinary Anesthesia & Analgesia Support Group (VASG). www.vasg.org.

CHAPTER 10

IMAGE INTERPRETATION AND DIAGNOSIS

Wolfgang Henninger and Jean François Bardet
(Advantages of the 3D-scanner)

10.1 HOW TO START!

As with all imaging analyses, interpretation needs an organised and comprehensive approach.

Nearly all modern Digital Imaging and Communications in Medicine (DICOM) viewers allow to display multiple windows of the region at the same time on big screen layouts; the recommended screen diameter is about 28–30 inches (70–75 cm).

When a CT study is loaded into the viewing system, the screen layout has to be adapted to see different windows of the region at the same time. Therefore, it is recommended to place the various transverse images always in the same order. First on the top left of the screen, the bone/lung window is placed, followed by the contrast soft tissue window, and next to it the plain soft tissue window. With two rows and three or four columns of your screen layout, the bottom row for multiplanar reformatted views such as sagittal or horizontal orientation can be used. it is also recommend to use always the same window settings for bone (e.g., WL+400 and WW4000) or lung (e.g., WL-450 and WW2500) window, soft tissue contrast window (e.g., WL+50 and WW350) and plain soft tissue window (e.g., WL+50 and WW350) (Figure 10.1 – general screen layout). The image orientation is always the same as the international regulations within the colleges proposed; the patient's right side is orientated on the image's left, the patient's dorsal side on top and the ventral side on the bottom. Sagittal images are orientated as a "lateral radiograph" with the head to the left side. Dorsal/horizontal projections are orientated as VD views in "supine" position, and for the appendicular skeleton, the same procedure is done as for the corresponding radiograph (the lateral aspect of the joint is on the left side of the image).

Each series of images, whether transverse or reformatted, shall be scrolled back and forth in every window as each single image is connected to its predecessor and follower. To save time, it is

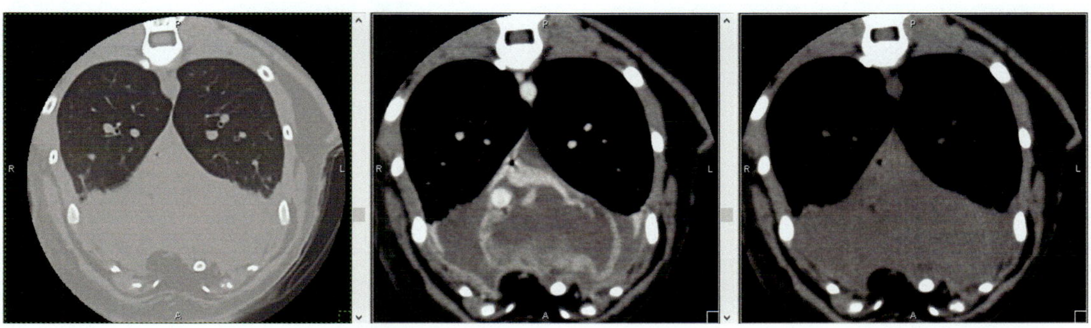

Figure 10.1 **Screen layout for evaluation of transverse CT image series (left: high contrast bone or lung window, middle: contrast-enhanced soft tissue window, right: plain soft tissue window): Pleuritis (mixed breed Dog fs, 12 years., courtesy Dr. Gilli & Deutsch, Bruck-Oberaich, Austria)**

DOI: 10.1201/9781315118130-11

also recommended to look at those images first where the pathology is suspected. Nevertheless, all images of each series have to be visualised to understand all varying aspects of the pathology and also not to miss any incidental findings. In complex regions, dividing the entire scan region into various compartments can be systematically helpful. Therefore, it is recommended to compare right and left joint images always in the same orientation on the screen: two rows at the screen layout and three columns for the different windows, where each joint is orientated at the same projection, with the top row for the right side and the lower row for the left side; it is remarkable to mirror the bottom row so that the image corresponding to the top row in orientation and level is visualised. It is absolutely necessary to compare the appearance of each section at the same level with the other side to detect any abnormal size, shape, contour, structure and density (Figure 10.2 – joint screen layout).

The importance of proper window levelling in search of a lesion has to be strengthened. Shift of the soft tissue window onto a fat window (e.g., WL-50 and WW250) is recommended when fat stranding is suspected. In addition, a lung window (WL-450 and WW2500) is extremely helpful in detecting gas within soft tissues or abdomen after perforation, in general seen on the oesophagus or digestive tract due to bacterial gas accumulation (Figure 10.3). As mentioned before, multiplanar reformation of CT images in both sagittal and dorsal planes is also extremely helpful to get detailed image information of the altered tissue at any part of the body (Figure 10.4).

10.2 INTERPRETATION OF SPINAL CT IMAGES

Systematic studies are warranted even in case of rapid tissue changes. The entire images of each series should be investigated to get an overview

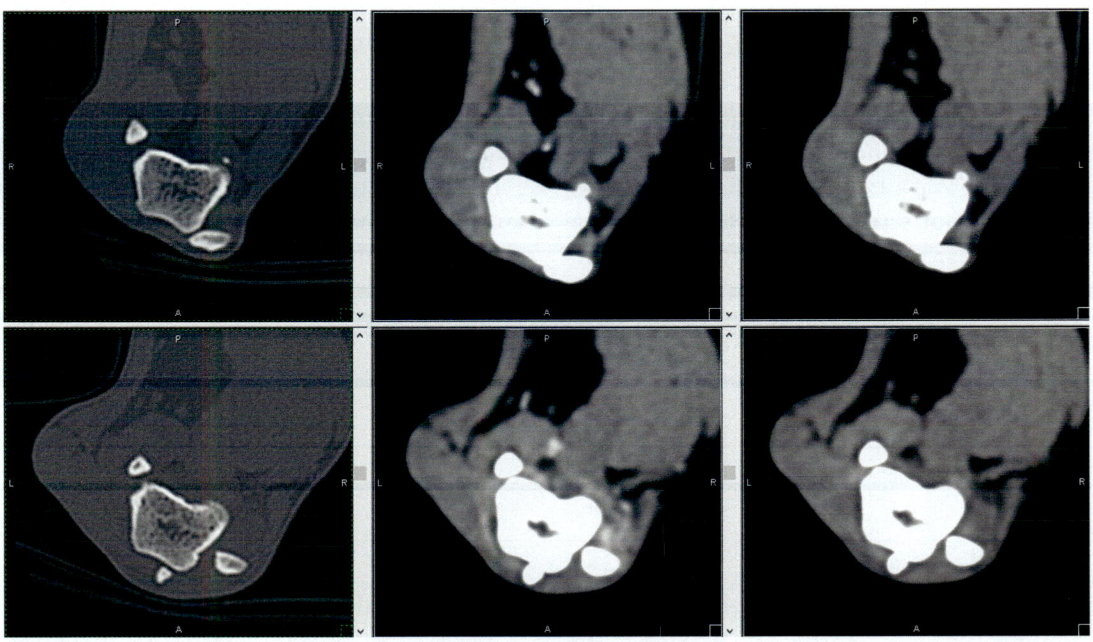

Figure 10.2 Screen layout for evaluation of transverse joint CT image series – upper row right joint, lower row left joint mirrored (left: high contrast bone window, middle: contrast-enhanced soft tissue window, right: plain soft tissue window): bilateral medial patellar luxation with patellar fracture left (DSH mc, 4 years., courtesy Dr. Gilli & Deutsch, Bruck-Oberaich, Austria.)

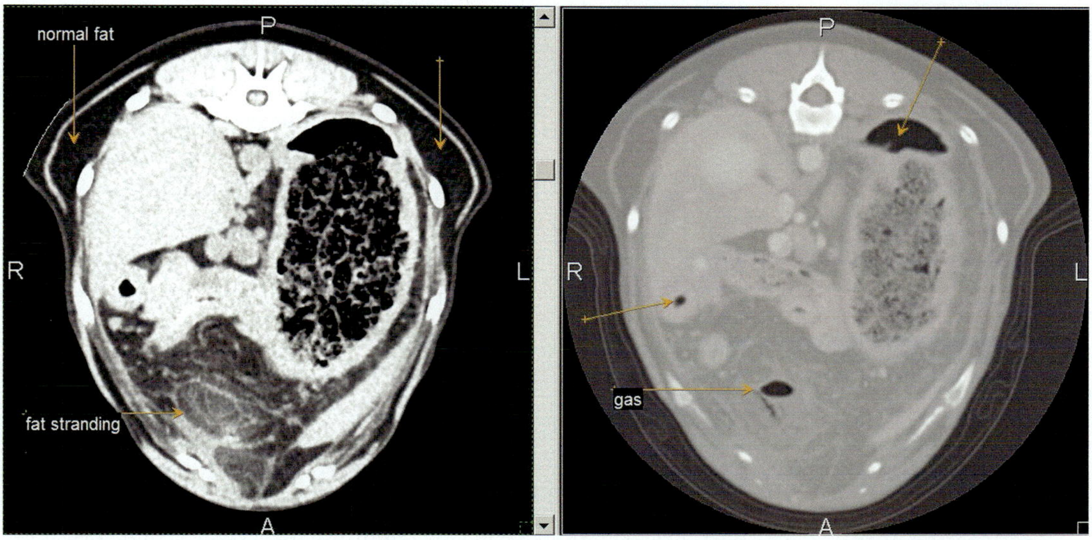

Figure 10.3 Transverse CT images of the cranial abdomen/epigastrium (left fat window; right lung window): fat stranding and suspected small bowel perforation due to a wooden foreign body. (Mixed breed dog mc, 10 years.)

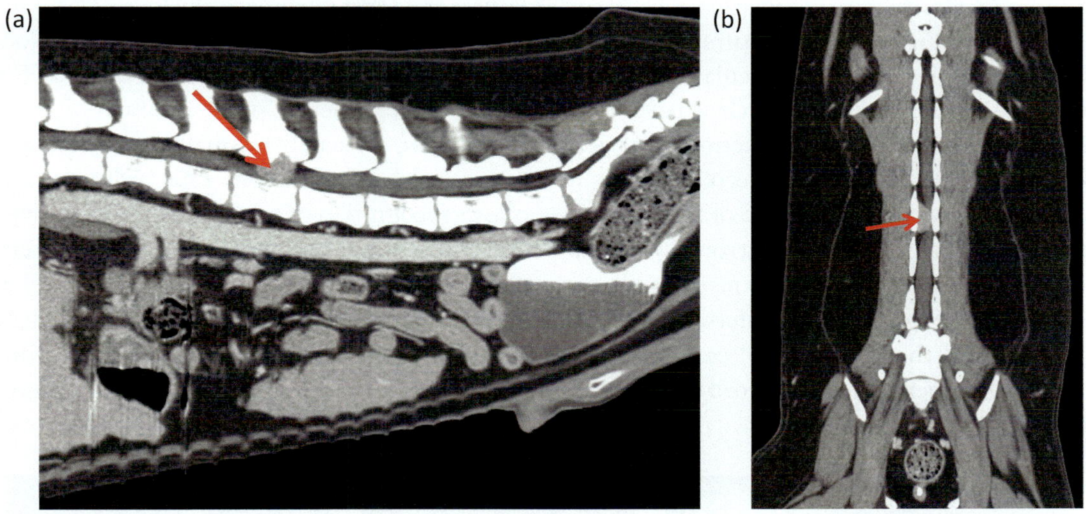

Figure 10.4 (a) Sagittal and (b) dorsal reformation of a spinal CT after the administration of IV contrast displaying an extradural, extramedullary mass at L3 (arrows).

of the whole scan region and then at the site of a lesion and be examined from the periphery to the centre, starting with the perivertebral musculature and adjacent tissue, vertebral body and disc, articular facets, spinal cord section and nerve roots, including peripheral patterns and accompanying vessels.

Perivertebral musculature and adjacent tissue should be examined for asymmetry, space-occupying tissues (such as cervical, thoracic or abdominal vessels; dorsal lung sections; lymph nodes; or sections of the large bowel or bladder), masses and tissue-density changes, following intravenous (IV) injection

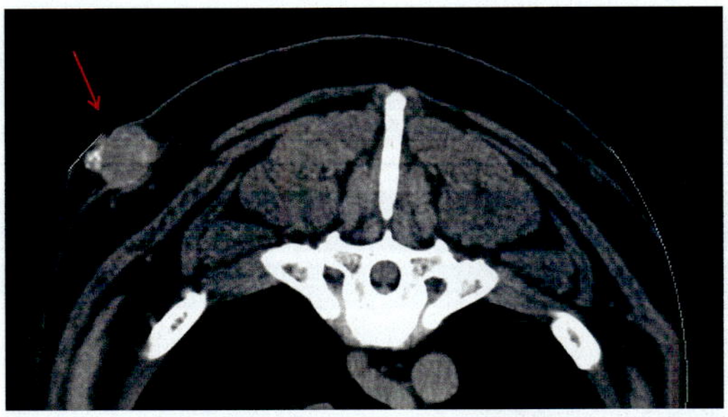

Figure 10.5 **Transverse CT image at the level of T7 in soft tissue window postcontrast (IV). A subcutaneous nodule is present; notice the moderate contrast enhancement (arrow).**

of the contrast medium (Figure 10.5). The vertebral section forming the vertebral canal, including the endplate, spongiosa, surrounding cortical bone, pediculi, laminae and spinal process, should be checked for loss of contour, changes in density or malformation such as spinal bending (Figure 10.6). Ensure that special attention is given to the articular facets and vertebral joints, including their shape, location, alignment, periosteal reactions and density (Figure 10.7). With the vertebral discs, it is important to check the width of the intervertebral space, as well as any narrowing, mineralisation and bulging, because protrusions can occur in varying shapes at different locations. Extrusions of the disc material can be present with displacement and spinal-cord compression (Figure 10.8).

Multiplanar reformation (MPR) is essential when reporting spinal studies. Sagittal imaging allows evaluation of alignment, wherein the width of the intervertebral disc spaces can be obtained and suspicion of sliding vertebra or traumatic displacement might be detected; furthermore, spinal cord displacement by any pathologic process (disc, mass, hematoma, …) and every destructive lesion is simpler to judge (also compare Figure 10.4).

The spinal cord's diameter is larger in the cervical and lumbar region than in the thoracic region. The spinal cord's section is always circular, with slightly ovoid enlargement in the cervical and lumbar areas, the intumescentia cervicalis and lumbalis, respectively. The cord's shape and position can be estimated from the amount of intracanalicular fat surrounding the dural sac. Myelo-CT is the only way to differentiate the exact medullary thickness of the spinal cord and its transition to the dural sac along the cauda equina (Figure 10.9). Myelo-CT is also helpful to detect small amounts of extradural mass, especially in small breed dogs with suspected disc herniation. This technique also allows recognition of pooling effects at the level of subarachnoid cysts or intramedullary remnants of CM due to high-velocity disc extrusion.

Intravenous injection of contrast medium helps to distinguish the segmental nerve roots from arteries and veins, as well as the ventral internal venous plexus, which can appear as a

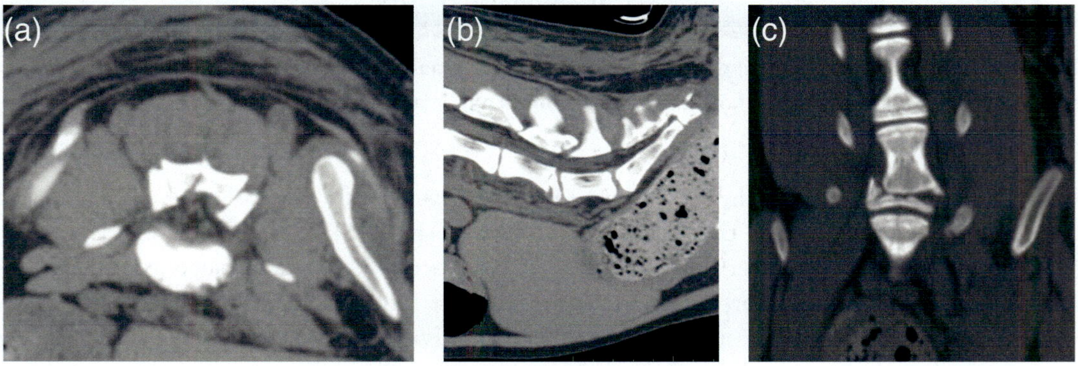

Figure 10.6 **(a) Transverse CT image at the level of L6: notice the fractures of the vertebral body and the vertebral roof. (b) and (c) Sagittal and dorsal reformatted images show clearly the fractured endplate of L6.**

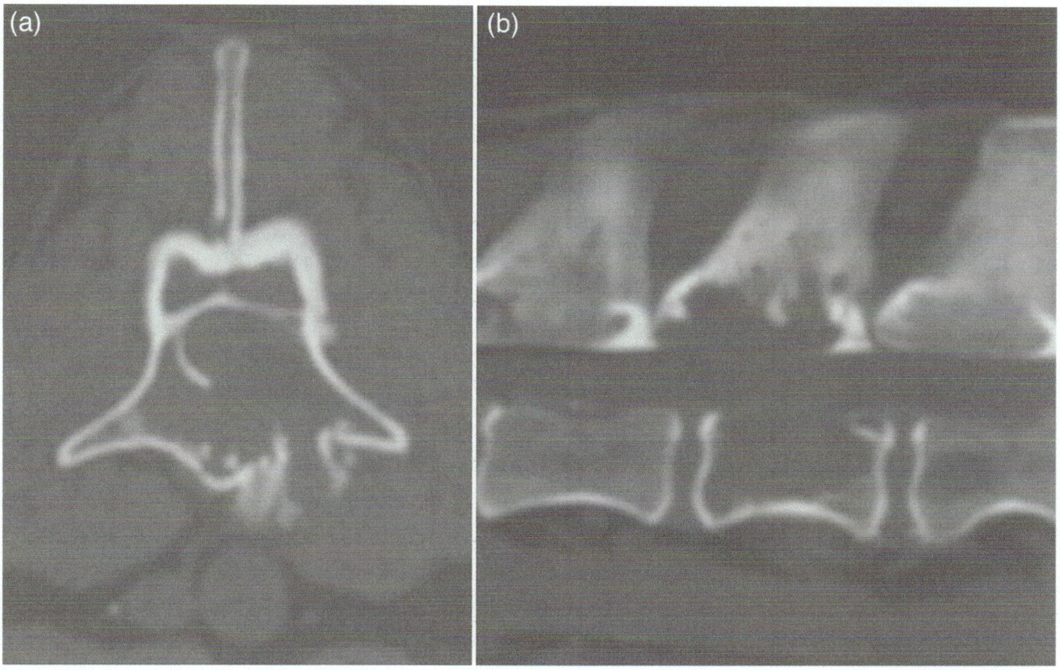

Figure 10.7 (a) Transverse and (b) sagittal reformatted CT images of L4 in the bone window. Severe osteolysis of the vertebral body corresponding to a neoplasia is visible.

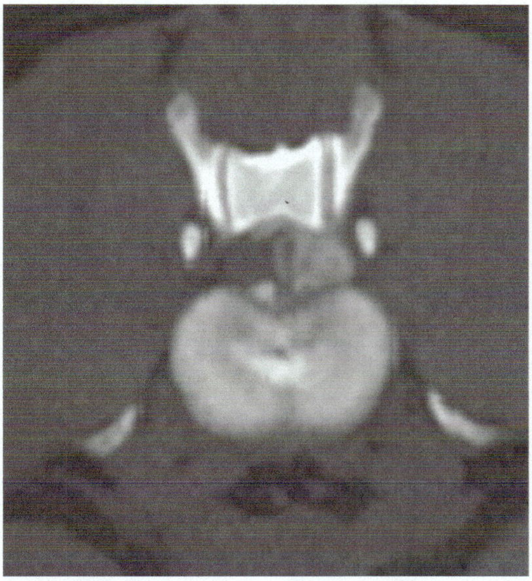

Figure 10.8 Transverse CT image at the level of T13–L1 in bone window. A large amount of mineralised disc material at the left side is present, causing marked extradural compression of the spinal cord.

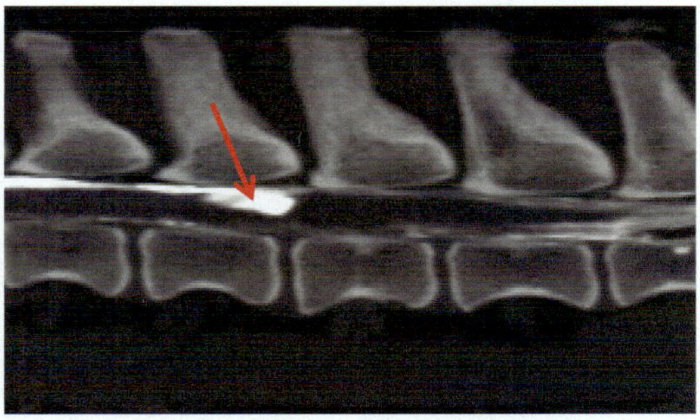

Figure 10.9 Sagittal reformatted CT myelogram of the lumbar spine in bone window: Contrast medium within the subarachnoid space is pooling dorsally. This finding is compatible with a subarachnoid diverticulum (arrow).

very large bilateral soft-tissue structure and mistaken for enlarged segmental nerve roots or protruding disc material (Figure 10.10). The segmental nerve roots can be traced from the vertebral canal near the middle of the vertebral body, where they move caudally and separate into the lateral recessus, whereas the peripheral nerve

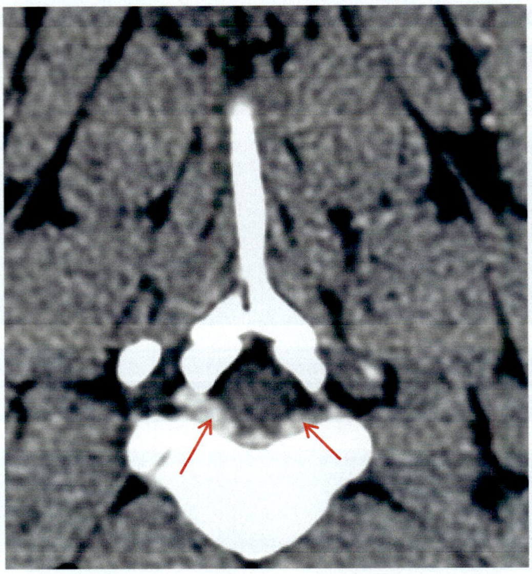

Figure 10.10 **Transverse CT image of the spinal cord in soft tissue window postcontrast (IV). The paired aspect and the contrast enhancement of the ventral venous sinuses are clearly visible (arrows).**

almost cuts into the bone at the caudal vertebral incisura, bends ventrolaterally and heads towards the ventral section of the next adjacent vertebra's transverse process. Comparison of the size and enhancement of nerves, which can be seen as roundish structures surrounded by a thin fatty layer within the perivertebral muscles on each side, is not difficult. The peripheral nerves forming the brachial or sciatic plexus can be traced ventral to the scalenus muscle or the sacral bone, (Figure 10.11), and their thickness shall be compared with respect to the right and left regions.

10.3 GENERAL INTERPRETATION OF CT IMAGES OF JOINTS

It is recommended to orientate the single series on the screen by a proper screen layout comparing the left and right joints at the same level of the section (see Section 10.1 – How to start!).

The whole joint should be documented by checking the beginning and the end of the image series and then comparing the diameter of both joints recognising joint distension and swelling or soft tissue atrophy. This usually indicates the lame side of the animal. Evaluate the bony contours and structures on plain images and then compare the IV or and

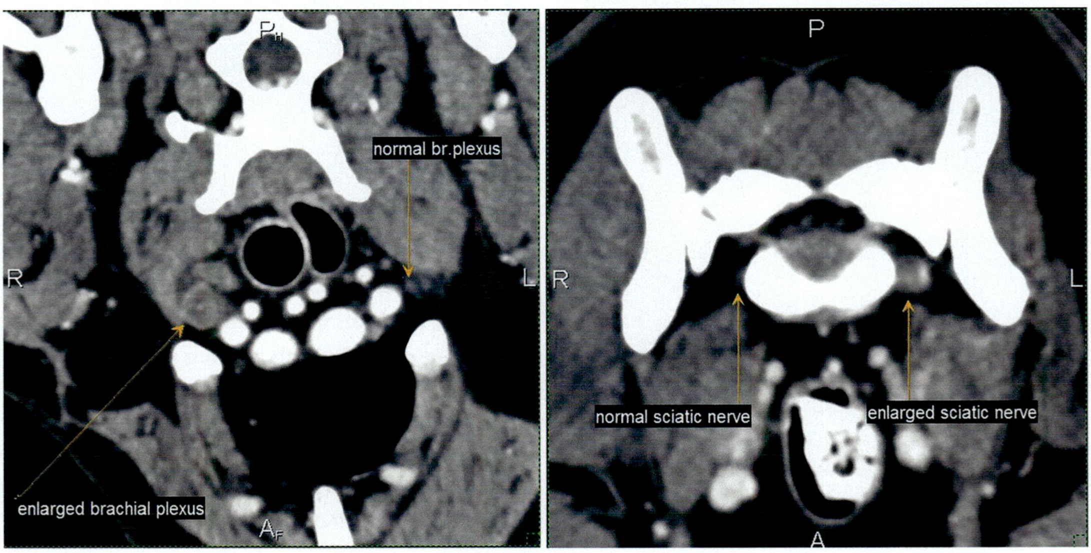

(a) (b)

Figure 10.11 **(a) Transverse CT image of the brachial plexus comparing right and left located ventrally to the scalenus muscle. (b) Transverse CT image of the sciatic plexus comparing right and left located ventrally to the sacral bone.**

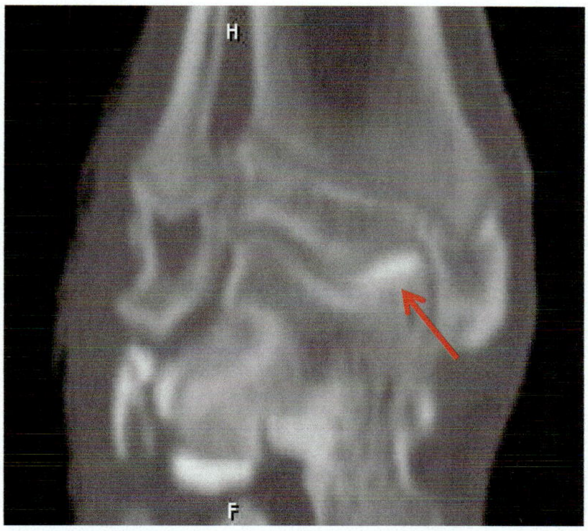

Figure 10.12 **Dorsal reformatted CT image of a tarsal joint after intra-articular contrast is administered. There is a loss of cartilage and subchondral bone at the medial talar ridge (arrow).**

iodinated (IA) contrast-enhanced images of both sides with the plain series (Figure 10.12). IV contrast is strongly recommended in orthopaedic cases, too, to detect various soft tissue lesions and active synovial-associated tissue.

The subchondral bone should be evaluated in a high-contrast window; data on shape, contour, possible periosteal proliferation and mineral density are obtained. Tiny bony fragments will be better visualised in a narrower soft-tissue window because of their less mineral content. Beam-hardening artefacts may mimic fissure lines, especially within the medial coronoid process; therefore, each joint shall be scanned without superimposition of the head/neck or the other foot. Visualisation of fluid in the joint(s) and recesses is possible after IV contrast, which allows the detection of the thickness of joint capsules and the size of their pouches as well as adjacent synovial structures, such as tendon sheaths (Figure 10.13). Contrast administration is also necessary to detect focal neoplastic or inflammatory conditions within periarticular tissues or the surrounding musculature.

10.4 POST-PROCESSING TOOLS AND 3D TECHNIQUES

Post-processing techniques and 3D imaging tools help translate a lesion's complexity into a format that can be read by surgeons, owners and students who are less familiar with pure axial/transverse or MPR images than radiologists. Nearly all the modern CT scanners offer diverse techniques, and the sophisticated tools can not only create MPR reformations but also produce

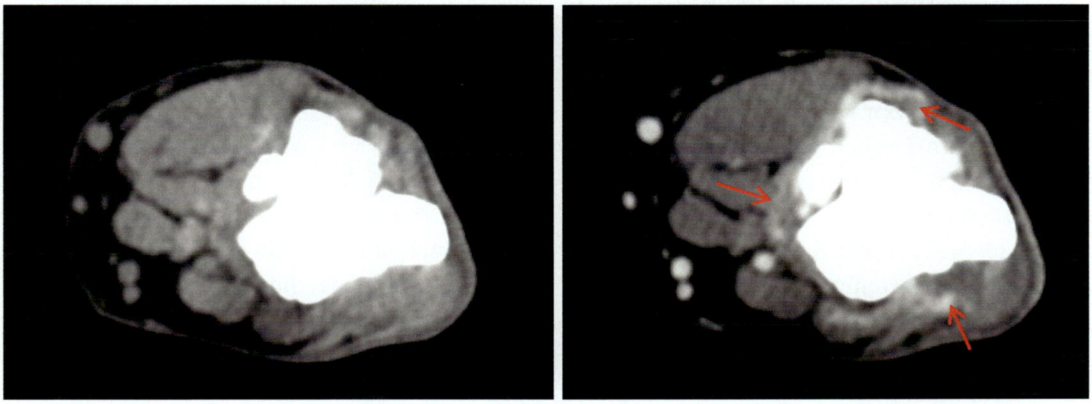

(a) (b)

Figure 10.13 **Transverse CT image of an elbow joint: (a) survey and (b) postcontrast (IV) images in soft tissue window. Thickening and contrast enhancement of the joint capsule are clearly demonstrated on the postcontrast image (arrows).**

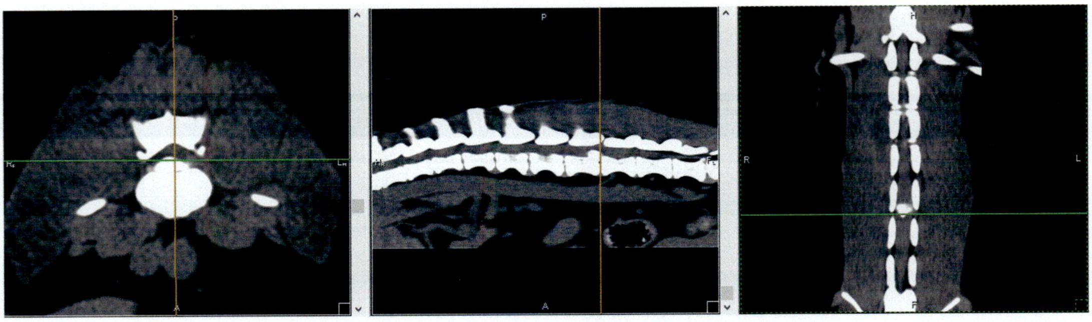

Figure 10.14 MPR view of a calcified disc extrusion L4/5 in selected planes simultaneously displayed (mixed-breed dog, mc, 8 years.).

maximum intensity projection (MIP), minimum intensity projection (MinIP), shaded surface display (SSD), volume rendering (VR), virtual endoscopy or subtraction technique images.

It is important to position the animal correctly during the CT examination, i.e., positioned straight and symmetrical to avoid oblique sections, especially when the head and spine are imaged. Most DICOM viewers allow MPR to be drawn along any plane on a corresponding transverse, dorsal or sagittal image; furthermore, advanced techniques now allow curvilinear or oblique reformations to follow the vertebral canal, vessels or bronchi. These reformations are highly advantageous because different views can be obtained at the same location of a suspected lesion and often simultaneously displayed (Figure 10.14). Lesions of unclear extension can be specified, or hindering partial volume artefacts can be ruled out.

Virtual endoscopy can also be used to explore tubular structures such as the vertebral canal, vessels, bronchi and the (large) bowel; a lesion (ulcus or mass) on the wall's surface can be tracked virtually and a decision made whether to perform endoscopic or surgical intervention (Figure 10.15).

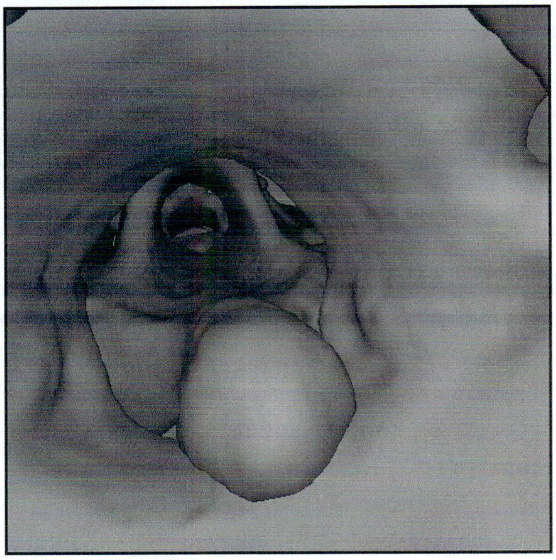

Figure 10.15 Virtual endoscopy of the cervical spinal canal presenting the normal atlantoaxial joint with a prominent dens axis on the fly through (courtesy Stephan Kaiser, Braunschweig, Germany).

MIP visualises the maximum number of voxels representing particularly hyperattenuating structures such as lung nodules and small metastases, slight calcifications, contrast uptake or the course of hyperattenuating vessels (shunt vessels). MIP is mainly used in CT angiography. The MinIP, on the other hand, visualises very low attenuating voxels in search of air or free gas along thickened soft tissue like peribronchial laceration or a perforating ulcus. These techniques show a lesion more clearly within an otherwise noisy image or that other structures may undergo disturbances due to partial volume artefacts (Figure 10.16).

When volume-rendering techniques and segmentation are combined, a 3D model of the region can be created, and advanced software can highlight the lesion and adjacent tissue

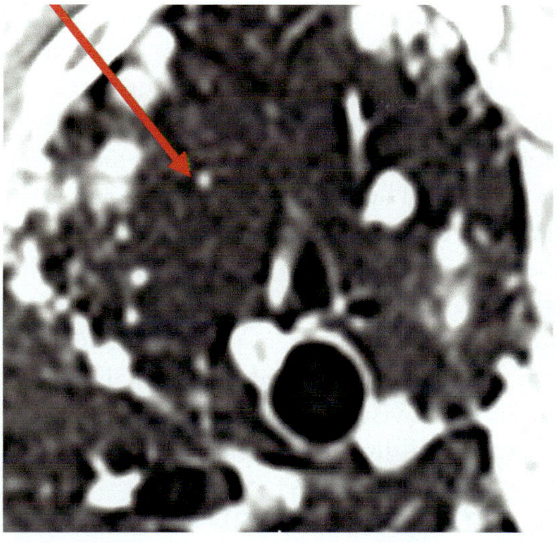

Figure 10.16 **Volume rendering technique MIP of a tiny lung nodule (arrow) besides obvious metastatic lung disease (Mixed Breed Dog, mc 14 years).**

in different colours (Figure 10.17). The subtraction technique can document areas where the blood supply to the entire organ, a specific part or a node, is increased or decreased. Precisely positioned pre- and post-contrast images (in breathhold-technique) are essential, especially for the caudal lung lobes, liver, stomach, spleen or pancreas that otherwise move around the diaphragm.

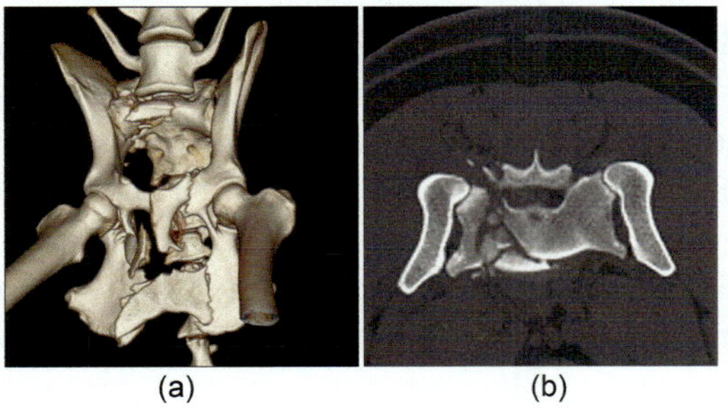

(a) (b)

Figure 10.17 **(a) Three-dimensional CT of sacral and pelvic fractures of a dog after trauma. (b) Corresponding transverse image of the traumatised sacrum. (Courtesy Stephan Kaiser, Braunschweig, Germany.)**

10.5 ADVANTAGES OF THE THREE-DIMENSIONAL SCANNER

Since 1972, when the first clinical CT scans were performed in London, CT has revolutionised medicine. One of the reasons is the technique's lack of superimposition, which gives it an edge over other methods, even though image interpretation sometimes may be difficult because of a lack of familiarity with normal transverse sectional anatomy. All CT scanners create three-dimensional (3D) images that closely resemble the anatomic structures during physical examination or surgery; therefore, the information that 3D CT images provide is often easier to interpret than that seen on transverse CT images. CT is now more readily available to veterinarians and more affordable for pet owners. Its clear, detailed imaging is crucial for making precise diagnoses and formulating treatment plans.

This chapter will cover the basic understanding of 3D CT, describe its most common uses in orthopaedic surgery and neurosurgery and focus on its advantages in joint fracture imaging.

10.5.1 Background

CT images can be processed directly by computer algorithms. Based on the original images, reconstructions can be generated in an arbitrary orientation in 2D or 3D displays to provide the required views. In principle, every image in this chapter—despite its appearance—is two-dimensional. The terms 2D and 3D refer to the content of the image, with 2D referring to displays of 2D distribution or slices and 3D to displays of complete distribution or volumes.

To perform 3D CT, the computer first combines all the digital information from a set of transverse CT images. The 3D displays, often called renderings, represent the scan volume in only one image. Next, the outside picture elements (pixels) of each transverse slice are selected to create a surface display of the volume image. This can only be performed successfully when a single structure (for example, the skeleton or vessels filled with contrast medium) needs to be displayed—in other words, only when structures with a high contrast in relation to their surroundings are involved. The original value of the CT information is lost in these cases, so different procedures that display, for example, only surfaces (SSD) or maximum CT values (MIP), as well as VR techniques that allow more flexible forms of volume displays, should be considered.

Although the images are actually shown in 2D, surface shading is used for depth perception. The volume image appears to be illuminated by a single source of light. Each tissue is assigned values for opacity and colour; for example, normal soft tissue can show high transparency, contrasted vessels can be slightly opaque and bone can be strongly opaque. Colour is often also documented to generate depth information and form a 3D impression because its intensity decreases as the observer moves further away. Modern workstations allow for fast, interactive work with near real-time performance, and operators are able to remove unwanted overlying structures from the image, display only selected structures, assign colours to selected structures and measure the volume of a specific region.

For MIPs, the point, or voxel, with the highest CT value along the search ray to the image plane is determined and displayed directly in the image, which preserves the minimal value of the CT information. This also implies that an MIP image is simply a 2D-projection image that can be rotated on any plane for a 3D evaluation. MIP is very useful for evaluating contrast-enhanced vascular and bone structures.

Volume rendering with user-friendly, preset evaluation protocols and intelligent editing tools may be considered the method of choice in 3D imaging; however, VR techniques will augment, but not replace, 2D displays. In most cases, a diagnosis is made on review of the original images or multi-planar displays.

Three-dimensional scans are limited because reconstruction artefacts may be seen, and a “stair-step” pattern can occur on the surface of the images. Artificial bone defects may occur in areas with low tissue density. The artefacts can be minimised by reducing the threshold for the range of bone densities to be displayed, while the stair–step effect can be minimised by using thinner axial slices and increasing the overlapping slices. In addition, a smoothing and shading algorithm that is used to create 3D images can limit the visualisation of subtle lesions such as small, non-displaced fractures.

10.5.2 The use of 3D CT in orthopaedic surgery and neurosurgery

Three-dimensional CT is most commonly used in orthopaedic surgery and neurosurgery to help with choosing surgical procedures and planning for surgery; evaluate growth deformities and limb alignment; to determine the volume and extent of involvement for tumour staging; check the outcomes of surgical procedures; prepare a custom-made prosthesis; and to facilitate client communication. Advanced computer programs can also perform virtual endoscopy, simulate surgery and plan radiation therapy.

10.5.2.1 Choosing surgical procedures and planning for surgery

In immature dysplastic dogs, for example, the decision to perform pelvic osteotomy must be made early to get the maximum benefits of the surgery. The prognosis for this procedure is most favourable for patients with radiographic evidence of hip subluxation, where osteoarthritis either does not exist or is minimal and where minimal cartilage damage is visualised on hip arthroscopy. Radiographs may not accurately reveal all the lesions on borderline cases, but using 3D CT will clarify the dysplastic hip’s exact geometry and anatomy and help the surgeon to decide whether patients with an abnormal acetabula (Figure 10.18) should be excluded. Most of

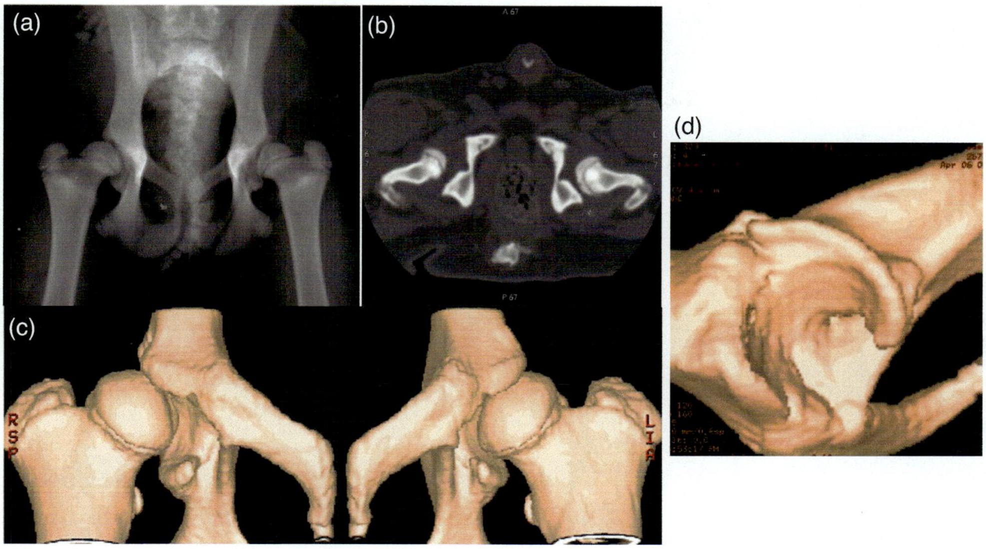

Figure 10.18 **Dorsoventral radiograph and 3-D CT of a 6-month-old French bulldog with bilateral hip dysplasia. (a) Ventrodorsal radiograph with severe subluxation without osteoarthritis. (b) Axial CT view with a bone filter demonstrating the dorsal subluxation as well as the flattening of the acetabulum. (c) 3-D CT showing a ventral view of the hips and pelvis and dorsal subluxation. (d) 3-D CT of the right acetabulum: note the remodelling and flattening of the dorsal rim of the acetabulum precluding the use of a triple pelvic osteotomy. The subtraction of the proximal femur was necessary for proper evaluation of all the subtle changes.**

this chapter's illustrations may be used as examples to plan and perform appropriate surgeries.

10.5.2.2 Evaluating growth deformities and limb alignment

A defined preoperative surgical plan is essential to correct angular deformities successfully, and such a plan will require a precise imaging modality. An angulated bone is a complete 3D form, but it cannot only be captured by radiographs in 2D representation. Typically, straight orthogonal films are sufficient for planning surgery on angular deformities in the frontal or sagittal planes, or both, but if the deformity includes torsion, it is more difficult to plan the surgical correction using only standard radiographs. Torsion of frontal plane deformities greater than 15 degrees can result in a miscalculation of more than 5° if the surgical plan is based on standard orthogonal radiographs. Such deformities can be assessed radiographically using a segmental technique, but a serious miscalculation may still occur, and surgical planning for such severe deformities may benefit from the additional diagnostics that a CT examination can provide.

The most common forms of angular limb deformities are antebrachial. Asynchronous growth may cause deformities, which in turn cause joint incongruence, lameness and degenerative joint disease. A subset of radial deformities, the bifocal or biapical radial deformities, have varus angulation of the proximal portion of the radius and valgus angulation of the distal portion. Because the affected antebrachii have both varus and valgus angulations, it is more challenging to assess and correct these deformities than uniapical ones. It is also challenging to accurately identify the components of biapical deformities on imaging because they can include varus, valgus, procurvatum, radial torsion, lateral subluxation of the radial head and distal humero-ulnar and carpal subluxations

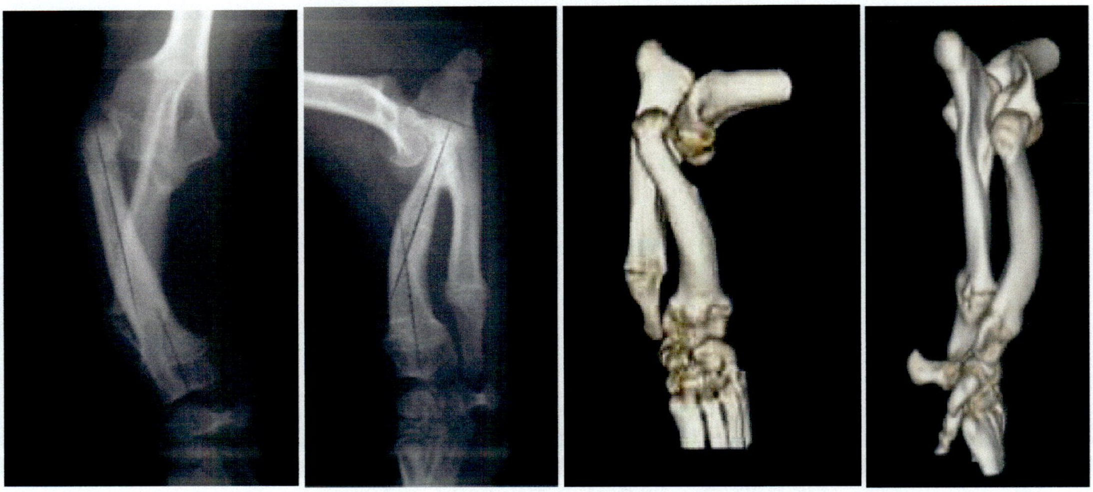

Figure 10.19 **Radiographs and CT of a 9-month-old Shi Tzu with congenital lateral subluxation of the radial head. The craniocaudal (a) and mediolateral (b) radiographs show the caudolateral luxation of the radial head, procurvatum and the lateral translation of the proximal radius related to the carpus. The radius appears straight on the frontal view on which the carpus is seen as a mediolateral view. (c) Lateral 3-D CT view of the antebrachium: the radial torsion is more apparent with a lateral view of the elbow and a frontal view of the carpus. (d) Caudal oblique 3-D CT view of the antebrachium, revealing the mediolateral curvature of the radius not seen on any other views.**

(Figure 10.19). Recent studies support the segmental radiographic technique associated with the use of 3D CT scans and stereolithographic models. CT appears to be the most accurate and reproducible method for assessing radial torsion, even in conjunction with severe procurvatum and valgus antebrachial angular limb deformities (Figure 10.19). Stereolithography, a form of prototyping technology, rapidly converts computer-aided design information from a CT scan to a stereolithographic machine that generates a 3D model. These methods provide supplemental information that the surgeon can use to diagnose the deformity and assess its severity. They can also increase the surgeon's knowledge, confidence and efficiency and potentially reduce the duration of surgery and anaesthesia.

Surgical planning and correction of rear-limb patellar luxation and limb alignment also benefit from modern imaging technology. As with antebrachial deformities, accurate radiographic assessment of limb alignment is difficult, but using a CT method that analyses individual slices or a 3D reconstruction of the skeletal elements into views analogous to radiography can quantify the deformities. The underlying cause of patellar luxation is not entirely understood. In large-breed dogs with medial or lateral patellar luxation, just as in small breeds, a variety of skeletal abnormalities that induce malalignment of the quadriceps mechanism may be present. During the dog's development, this malalignment may lead to genu varum associated with medial patellar luxation and distal femoral varus; hypoplasia of the medial femoral condyle, which accentuates the femoral varus deformity; external torsion of the distal femur with decreased femoral torsion (Figure 10.20); a shallow trochlear sulcus (Figure 10.20) with poorly developed or absent medial and/or lateral trochlear ridges; proximal tibial varus or valgus; internal tibial torsion; and medial displacement of the tibial tubercle. Internal tibial rotation may be observed at the

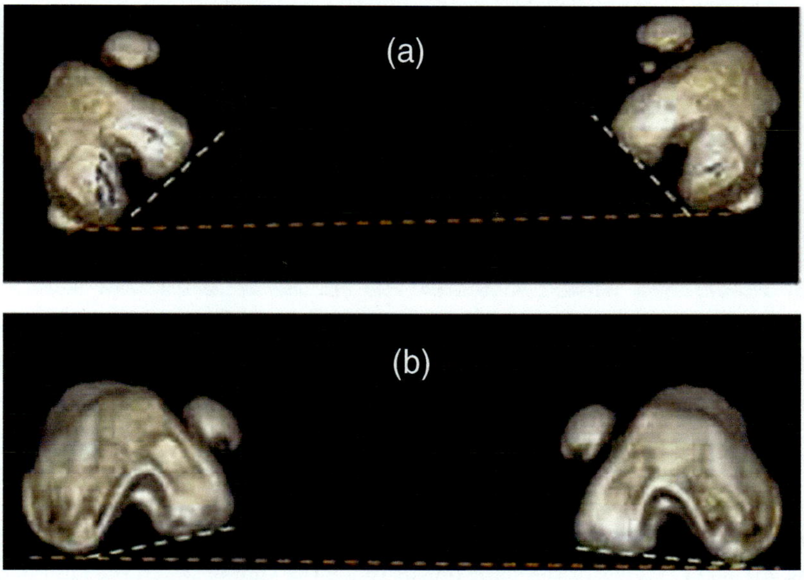

Figure 10.20 Axial 3-D scan of distal femurs. (a) Both femurs of this 9-month-old female English bulldog exhibit bilateral 45° external rotation of the femurs associated with bilateral medial patellar luxation and decreased femoral anteversion. (b) Same view as figure (a) in a 2-year-old Yorkshire with 9° external rotation of the femurs.

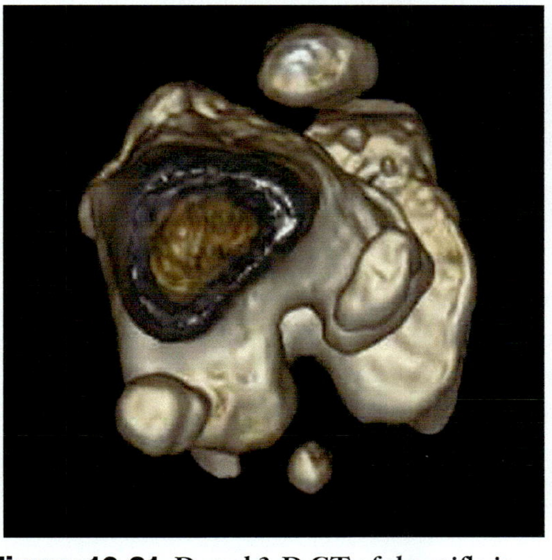

Figure 10.21 Dorsal 3-D CT of the stifle in the same dog as Figure 10.20a. Note the internal tibial rotation at the stifle joint and the medial patellar luxation.

stifle joint (Figure 10.21). These anomalies must be identified and corrected to prevent failure of surgical repair. Studies have shown that corrective femoral osteotomies have a very low rate of recurrent patellar luxation in cases with significant femoral varus and torsional deformity.

Medial coronoid disease and elbow joint incongruity are often associated with other intra-articular structure lesions. Elbow joint incongruity, which may occur alone or in combination with medial coronoid disease and cartilage disease in dogs with elbow dysplasia, is a complex issue because it has multiple manifestations and is therefore difficult to diagnose. CT has traditionally been considered the gold standard for identifying medial coronoid disease, but its inability to image cartilage lesions is an important disadvantage, as 29% of false-negative diagnoses based on CT of elbow joints with medial coronoid disease are confirmed by arthroscopy. Diagnosing elbow joint incongruity is the biggest challenge with elbow dysplasia. Many studies that focused on precise estimation of radio-ulnar incongruity (RUI), both *in vivo* and *in vitro*, have attested to the high accuracy and reliability of arthroscopy and the rendering of 3D images of the radio-ulnar articulation (Figure 10.22).

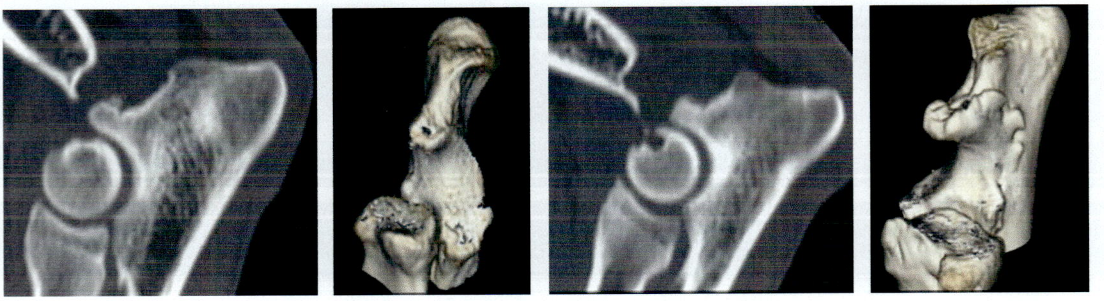

(a) (b) (c) (d)

Figure 10.22 Sagittal reformatted and 3-D CT views in two dysplastic elbows. (a) Sagittal reformatted view in a 10-month-old Labrador with moderate elbow dysplasia: note the increased humero-ulnar joint space and the osteophyte on the dorsal anconeal process. (b) 3-D CT view of the same dog as in (a): the humerus has been removed to demonstrate the moderate step between the lateral coronoid process and the radial head. (c) Sagittal reformatted view in an 11-month-old Labrador with severe elbow dysplasia: the step between the lateral coronoid processes and the radial head is more important. (d) 3-D CT view of the same dog as in (c); the step between the lateral and medial coronoid processes and the radial head is more severe.

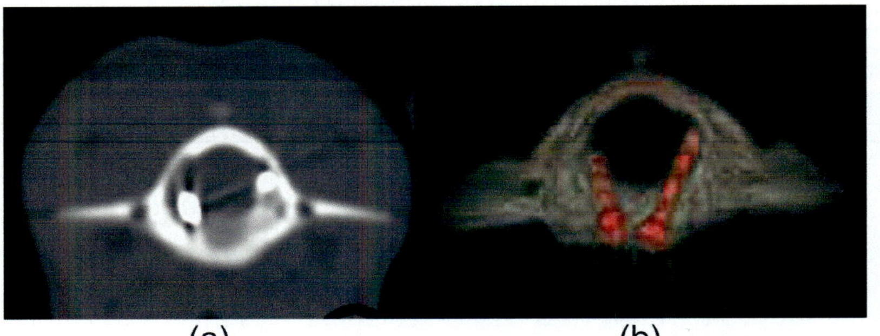

(a) (b)

Figure 10.23 Twenty-one month post-operative control CT views of a 20-month-old Yorkshire treated for atlanto-axial instability using a C1–C1 screw arthrodesis. Cervical pain developed 6 weeks before the CT. (a) CT C1–C2 axial view using a bone window demonstrating the left screw in the spinal canal. (b) 3-D CT with a special software for spinal implant: note the right screw in the spinal canal. The pain disappeared after screw removal.

Plain radiographs give mixed results and overall are considered unreliable when dealing with small (≤2 mm) RUI. Consequently, appreciation of joint congruence before or after geometric correction of the dysplastic elbow with RUI, such as dynamic proximal ulnar osteotomy (DPUO), should not rely on plain radiographs alone.

10.5.3 Checking the outcomes of surgical procedures

Most surgical procedures in orthopaedics and neurosurgery require post-operative radiographs. Conventional radiographs may be sufficient for the repair of long bone fractures, but intra-articular and spinal fractures, spinal decompression and arthrodesis may require post-operative CT scans for accurate evaluation (Figure 10.23).

10.5.4 Facilitating client communication

Three-dimensional reconstruction is beneficial when presenting information to clients because it allows them to visualise the lesions, and lay people can more easily understand what they can see. This includes not only orthopaedic and neurosurgical lesions but also soft-tissue masses and vascular anomalies.

10.5.5 The role of CT scans in articular fracture imaging

For decades, conventional radiography was the only diagnostic imaging modality available to assess fractures in animals. In most cases, plain radiography is still sufficient; however, some areas of the skeleton are difficult to study radiographically because the anatomy is complex, and the superimposition of important structures makes it difficult to perform a detailed examination of individual sections. Several radiographic projections are therefore necessary. CT is now considered the most accurate and useful diagnostic imaging procedure to assess complex articular structures. Suboptimal imaging may lead to a wrong diagnosis and the lack of or incorrect treatment, which could damage an animal's health, function and comfort.

The advantages of CT for imaging fractures have been demonstrated in several areas in veterinary patients: the pelvis, skull, tarsus, carpus and spine. CT may also be used to detect occult fractures, show the anatomy of fractures and the extent of bone involvement accurately, map soft-tissue changes and tissue damage, identify joint instability and to control post-operative reconstruction of joint surfaces.

10.5.6 Pelvic fractures

Human medicine studies found a 30%–65% error rate and a 9% misdiagnosis rate for pelvic fractures in adults with acute trauma. Those findings were supported by one veterinary study in which CT changed the fracture classification in 60% of cases, but case management was altered in only 29%. CT is the most useful method for evaluating sacral and acetabular fractures because it is more sensitive in detecting sacral fractures than plain radiographs. CT is also more sensitive in detecting the accurate configuration of the sacral fracture (Figure 10.24), even if it does not profoundly influence the prognosis. Both CT and plain radiography are recommended for acetabular fractures if a large degree of uncertainty exists about the fracture configuration. CT offers greater spatial relationship of any fracture fragments, eliminates the superimposition of faecal matter or colonic air and can be used to concurrently evaluate soft-tissue injuries. These fractures must be correctly classified to help the surgeon decide whether the fracture can be reconstructed.

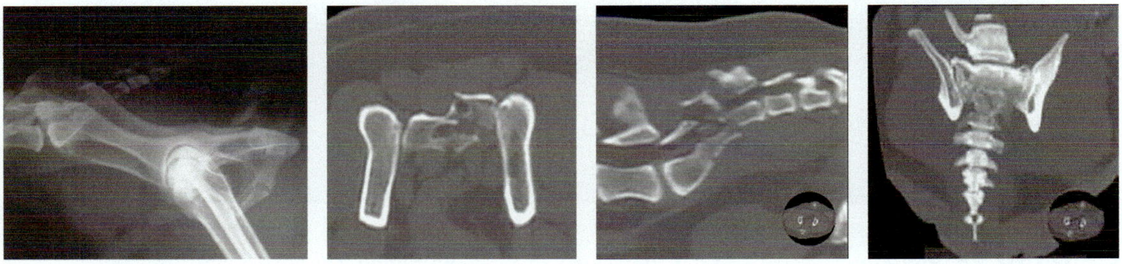

(a) (b) (c) (d)

Figure 10.24 Radiographs and 3-D CT views of a sacral fracture in a 1-year-old male mixed-breed dog. (a) Lateral radiographs of the pelvis: note the comminuted mid-body sacral fracture. (b) Axial transverse CT view through the fracture line showing the dorsal displacement of the right hemi-sacrum. (c) Sagittal reformatted CT view demonstrating the severity of the comminution. (d) MIP 3-D CT view: note the severity of the comminution of this sacral fracture.

10.5.7 Maxillofacial trauma

CT images are superior to conventional radiographs because they allow the identification of various anatomic components of the skull and the detection of maxillofacial trauma (MFT) injuries in dogs and cats. This study found 17 of 26 easily identifiable anatomic features on radiographs and six features that were difficult or impossible to identify, whereas CT allowed easy identification of every structure. CT demonstrated MFT injuries in 1.6 and twice as many cases in dogs and cats, respectively, than conventional radiographs. The average number of MFT injuries diagnosed with radiographs and CT scans was 3.8 and 7.6, respectively, in dogs and 3.8 and 7.7 in cats. Both radiographs and CT scans of the skulls of dogs and cats must be obtained under general anaesthesia, although CT significantly shortens the anaesthesia time compared with radiographs. The two most obvious traumas, symphyseal separation and mandibular body fractures, are associated with less obvious lesions in 36% of cases. However, skull radiographs are superior to CT images for visualising the mandibular body and dental occlusion.

10.5.8 Accurate visualisation of fractures and the extent of bone involvement

In complex and comminuted fractures, evaluating the size and position of the fragments and the degree of involvement of articular fractures may be difficult (Figure 10.25). Therefore, these fractures must be correctly classified to help the surgeon decide whether the fracture can be fixed, increase the accuracy of the diagnosis and to choose the surgical procedure and plan for surgery.

10.5.9 Detection of occult fractures

Identification of carpal or tarsal bone fractures or detection of small and avulsed fragments in all the joints can be challenging (Figure 10.26). Orthogonal radiographs may not show rare fractures, such as the fourth carpal bone. One human medicine study showed that wrist and ankle fractures were detected on radiographs in only 78% of cases compared with MIP CT views. Using conventional radiographs and CT for these areas, where possible, would benefit veterinary patients.

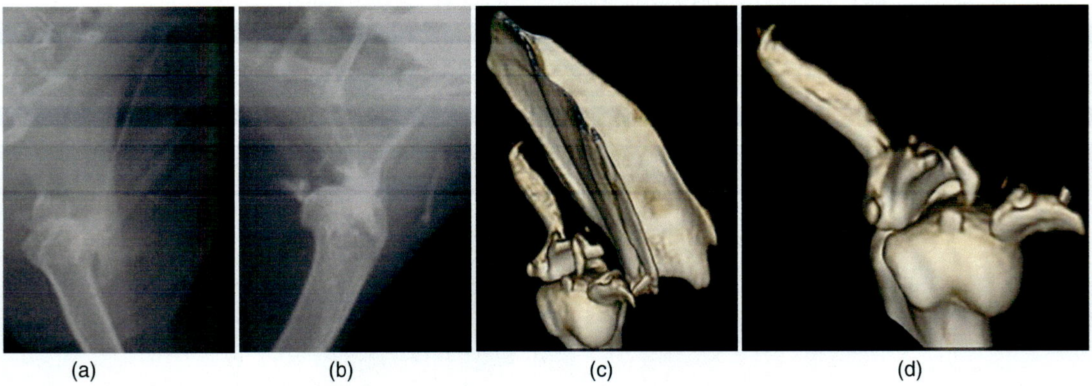

Figure 10.25 Radiographs and 3-D CT views of a 15-year-old female cat with a comminuted fracture of the right scapula. (a) Dorsoventral radiograph with medial displacement of the proximal humerus relative to the scapular body. (b) Mediolateral radiograph showing the caudal fragment of the glenoid articular surface and the longitudinal fracture line of the caudal scapula. (c) 3-D CT image showing a caudal view of the right shoulder with two main articular fragments. (d) 3-D CT caudal view when the scapular body has been removed.

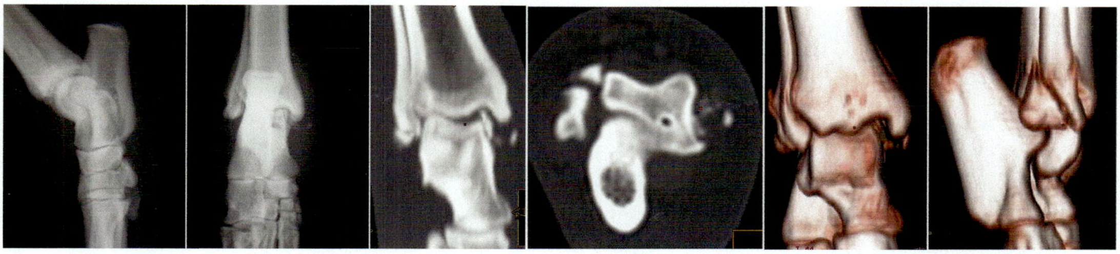

(a) (b) (c) (d) (e) (f)

Figure 10.26 Radiographs and CT views of a 15-month-old Labrador. (a) Normal mediolateral radiograph. (b) Avulsed medial collateral ligaments off the medial malleolus and slab fracture of the talus on this plantodorsal radiographs. (c) Reformatted frontal CT view with tarsocrural instability, slab fracture of the talus and distal lateral malleolar fracture. (d) Axial CT view of the distal tibia and malleoli illustrating the distal lateral and medial malleolar fractures. (e) Colour 3-D CT view of the right tarsus showing the lesions described in Figure c. (f) Colour 3-D CT lateral view of the right tarsus illustrating the distal lateral malleolar fracture not seen on the mediolateral radiographs.

Traumatic intercondylar humeral fractures may be observed occasionally. Craniocaudal radiographs are usually normal. CT scans of the distal humeral condyle demonstrate a complete or incompleted (fissure) fracture, which is usually seen across the caudal aspect of the condyle (Figure 10.27). The fracture or fissure may be associated with sclerosis if the animal has chronic lameness. Other occult fractures may be observed on the patella, sesamoid bones, medial

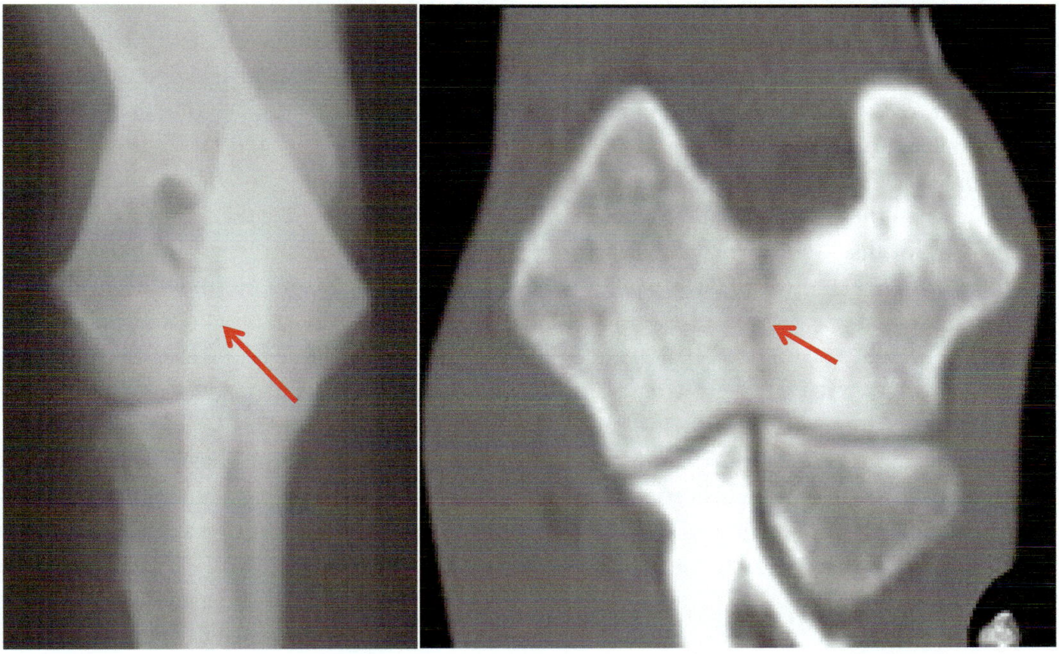

(a) (b)

Figure 10.27 Four-year-old hunting male Khortals with a 4-month-duration forelimb lameness. (a) Craniocaudal radiograph and (b) dorsal CT reconstruction showing a radiolucent line surrounded by a sclerotic rim representing an intercondylar humeral fissure (arrows).

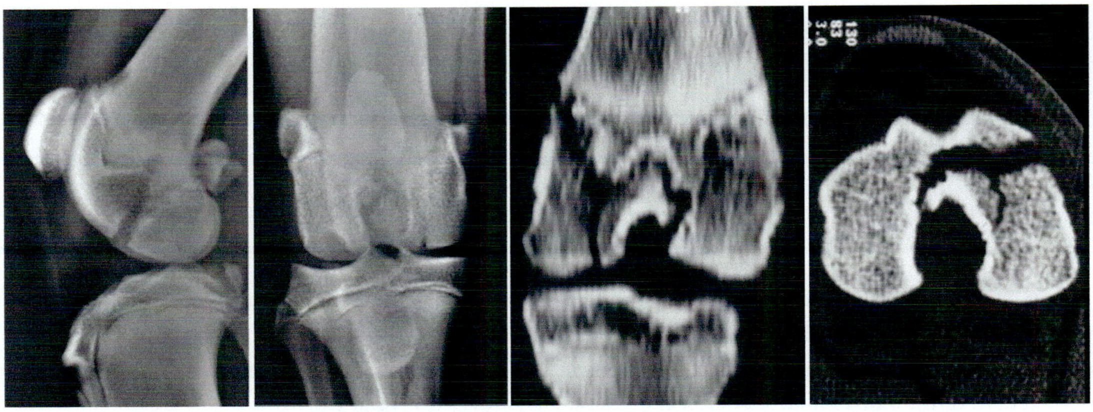

(a) (b) (c) (d)

Figure 10.28 Radiographs and CT views of a 5-month-old Labrador with a distal femoral fracture. (a) Femoral epiphyseal fracture perpendicular to the articular surface. (b) Femoral epiphyseal fracture of the medial condyle. (c) Reformatted frontal CT view with a comminuted fracture of the distal femoral epiphysis. (d) Axial CT view illustrating the anatomy of the fracture.

coronoid process, supraglenoid tubercle and avulsion of the cranial cruciate ligament of the distal femur.

10.5.10 Identification of soft-tissue damage and assessment of joint instability

The location of soft-tissue damage and the identification of avulsed fragments may not be obvious on initial standard radiographs (Figure 10.26); in these cases, a functional stress CT may help localise and quantify the lack of ligamentous support (Figure 10.26).

10.5.11 Correct evaluation of fractures in immature animals

Most fractures in immature animals are periarticular Salter–Harris type 1 or 2; the most common articular fractures are avulsion fractures. Because cancellous bone has less resistance, radiographs may not show clearly the severity of the fracture (Figure 10.28).

10.5.12 Postoperative control of joint reconstruction and implant positioning

Because anatomic reconstruction of joint surfaces is essential to prevent degenerative joint disease, CT is often useful for controlling not only the positioning of implants but also the postoperative outcome (Figure 10.21).

10.5.13 Conclusion

CT is now more readily available to veterinarians and more affordable for dog and cat owners. CT includes computer-processing techniques that can create three-dimensional (3D) images, which are most frequently used for choosing the correct surgical procedure and planning for surgery; evaluating growth deformities and limb alignment; checking the outcomes of surgical procedures; and facilitating client communication. Radiography is still commonly used as a primary imaging modality in acute joint trauma cases; however, many human medicine studies have shown that plain radiographs are moderately to poorly sensitive with high-energy, multi-trauma patients and complex fracture patterns. More frequent use of CT, when possible, therefore seems desirable because CT discloses the complex fracture morphology and clearly shows all the fracture components and their degree of displacement in these cases.

10.6 REACHING A DIAGNOSIS

Knowledge of anatomy and especially cross-section anatomy is essential. It is also important to remember that dogs, because of the breed and mixed-breed varieties, exhibit a wider range of "breed associated" normalities, whereas cats are all much the same.

Establishing a diagnosis requires thorough examination of all the images in multiplanar views. It has a high temporal benefit to display the images on the screen always in the same order and screen layout. The use of post-processing methods can assist to confirm suspected clinical changes or find additional or artificial pathologies that would influence or limit the treatment options.

CT-guided biopsy is the method of choice to confirm CT findings and support differential diagnoses.

Lesions are generally defined by their location, size, number, shape, borders/adjacent fatty spaces, structure and contrast enhancement—the more criteria, the more precisely a lesion can be defined. The animal's history, clinical appearance, treatment options and prognosis should also always be considered. The radiologist's responsibility is to provide documentation of a lesion with information that includes an accurate description and possible treatment options, considering distance to sound structures, blood supply, possibility for resection, invasiveness and TNM criteria. When a definitive diagnosis is not possible to be decided, a list of differential diagnoses should be provided.

FURTHER READING

Arzi B, Cissel DD, Verstrate FJM, et al. 2013. Computed tomography findings in dogs and cats with temporomandibular joint disorders: 58 cases (2006–2011). *Journal of the American Veterinary Medical Association* 242: 69–86.

Bar-Am Y, Pollard PE, Kass PH, Verstrate FJM. 2008. The diagnostic yield of conventional radiographs and computed tomography in dogs and cats with maxillofacial trauma. *Veterinary Surgery* 37:294–299.

Böttcher P, Bräuer S, Werner H. 2013. Estimation of joint incongruence in dysplastic canine elbows before and after dynamic proximal ulnar osteotomy. *Veterinary Surgery* 42: 371–376.

Crawford JT, Manley PA, Adams WM. 2008. Comparison of computed tomography, tangential view radiography, and conventional radiography in evaluation of canine pelvic trauma. *Veterinary Radiology & Ultrasound* 44: 619–628.

Crosse KR, Worth AJ. 2010. Computer-assisted surgical correction of an antebrachial deformity in a dog. *Veterinary and Comparative Orthopaedics and Traumatology* 23:354–361.

DeVasConCellos P, Balla VK, Bose S, et al. 2012. Patient specific implants for amputation prosthesis: Design, manufacture and analysis. *Veterinary and Comparative Orthopaedic Traumatology* 4:286–296.

Dismukes DI, Fox DB, Tomlinson JL. 2008. Use of radiographic measures and three-dimensional computed tomographic imaging in surgical correction of an antebrachial deformity in a dog. *Journal of the American Veterinary Medical Association* 232:68.

Draffan D, Clements M, Farrell M, et al. 2009. The role of computed tomography in the classification and management of pelvic fractures. *Veterinary and Comparative Orthopaedics and Traumatology* 3:190–197.

Dudley RM, Kowaleski MP, Drost WT, et al. 2006. Radiographic and computed tomographic determination of femoral varus and torsion in the dog. *Veterinary Radiology & Ultrasound* 47:546.

Eljack H, Werner H, Böttcher P. 2013. Sensitivity and specificity of 3D models of the radioulnar joint cup in combination with a sphere fitted to the ulnar trochlear notch for estimation of radioulnar incongruence. *Veterinary Surgery* 42:365–370.

Gemmill T, Mellor DJ, Clements DN, et al. 2005. Evaluation of elbow incongruency using reconstructed CT in dogs suffering fragmented coronoid process. *Journal of Small Animal Practice* 46:327.

Goldman LW. 2007. Principles of CT and CT technology. *Journal of Nuclear Medicine Technology* 35:115–130.

Griffon DJ. 2012. Surgical diseases of the elbow. In: Tobias KM, Johnston SA (eds) *Veterinary Surgery Small Animal*, Elsevier Saunders, St Louis, MO, pp. 724–751.

Haapamäki VV, Kiuru MJ, Mustonen AO, Koskinen SK. 2005. Multidetector computed tomography in acute joint fractures. *Acta Mechanica* 46(6):587–598.

Heiken JP, Brink, JA, Vannier MW. 1993. Spiral (helical) CT. *Radiology* 189:647–656.

Hofer M. 2007. Basic rules for reading CT examinations. In: Hofer M (ed) *CT Teaching Manual: A Systematic Approach to CT Reading*, Thieme Medical Publishers, Stuttgart, ch 2, pp. 14–25.

Jones JC. 1999. Advanced imaging concepts: A pictorial glossary of CT and MRI technology. *Clinical Techniques in Small Animal Practice* 14: 65–111.

Kalender WA. 2011. *Computed Tomographic. Fundamentals, System Technology, Image Quality, Applications*, Publicis, Erlangen, p. 230.

Kramer A, Holsworth IG, Wisner ER, et al. 2006. Computed tomographic evaluation of canine radioulnar incongruence in vivo. *Veterinary Surgery* 35:24.

Kowaleski MP. 2010. Computed tomographic planning of distal femoral osteotomy. In: *Proceedings of the 3rd World Orthopedic Veterinary Congress*, Bologna, Italy, September 15–18.

Kowaleski MP, Boudrieau RJ, Pozzi A. 2012. Stifle joint. In: Tobias KM, Johnston SA (eds), *Veterinary Surgery Small Animal*, Elsevier Saunders, St. Louis, pp. 982–989.

Kwan TW, Marcellin-Little DJ, Harrysson OL. 2014. Correction of biapical radial deformities by the use of bilevel hinged circular external fixation and distraction osteogenesis in 13 dogs. *Veterinary Surgery* 43:316–329.

Langley-Hobbs SJ. Fractures of the humerus. In: Tobias KM, Johnston SA (eds) *Veterinary Surgery Small Animal*, Elsevier Saunders, St. Louis, p. 722.

Mason DR, Schulz KS, Samii VF, et al. 2002. Sensitivity of radiographic evaluation of radio-ulnar incongruence in the dog in vitro. *Veterinary Surgery* 31:125.

Moores AP, Benigni L, Lamb CR. 2008. Computed tomography versus arthroscopy for detection of canine elbow dysplasia lesions. *Veterinary Surgery* 37:390.

Napoli A, Fleischman D, Chan FP, et al. 2004. Computed tomography angiography: state-of-the-art imaging using multidetector-row technology. *Journal of Computer Assisted Tomography* 28(Suppl 1):32–45.

O'Brian S, Darwish N. 2023. 3D visualisation of the spine. *Advances in Experimental Medicine and Biology* 1406:139–168.

Piras LA, Peirone B, Fox DB. 2010. Effects of antebrachial torsion on the measurement of frontal plane angulation: A cadaveric radiographic analysis. Radiographic measurements of the axes and joint angles of the canine radius and ulna. In: *Proceedings of the 37th Veterinary Orthopedic Society Symposium*, Breckinridge, CO, February 20–27, p. 36.

Reed N, Gosling M. 2021. Intra-thoracic gossypiboma (textiloma) in a 6-year-old cocker spaniel. *Australian Veterinary Journal* 99:6–10.

Rutherford S, Ness MG. 2012. Dorsal slab fracture of the fourth carpal bone in a racing Greyhound. *Veterinary Surgery* 41:944–947.

Rutty GN, Robinson C, Morgan B. 2008. Computed tomography of projectile injuries. *Clinical Radiology* 63:1160–1166.

Samii V, Gielen I, et al. 2011. Joints. In: Schwarz T, Saunders J (eds) *Veterinary Computed Tomography*, Wiley-Blackwell, Chichester, UK, p. 387.

Saunders J, Schwarz T. 2011. Principles of CT image interpretation. In: Saunders J, Schwarz T (eds.) *Veterinary Computed Tomography*, John Wiley & Sons Ltd, Chinchester, UK, ch 3, p. 29–34.

Wegener O. 1992. Image analysis. In: Wegener O (ed) *Whole Body Computed Tomography*, Blackwell Scientific Publications, Cambridge, MA, ch 3, pp. 81–95.

Wisner E, Zwingenberger A. 2015. *Atlas of Small Animal CT and MRI*. Wiley Blackwell, Chinchester, UK.

CHAPTER 11

DICOM VIEWING SYSTEMS AND DISPLAY EQUIPMENT

Eberhard Ludewig

Computed tomography (CT) images result from a stepwise process, comprising the following parts: data acquisition, data processing, and image presentation. The diagnostic utility of the system is limited by the performance of the weakest of those steps. Image presentation represents a key component within the imaging chain. “Image presentation” summarizes a number of hardware and software features. Due to ongoing rapid technical developments, it is a challenge to be up to date and understand upcoming new terms, interactions, and the background philosophy.

The most important “passive” parts of the reading process are the workstation and the physical environment. Both parts have a major impact on the radiologists’ performance: an adequate workstation setup (hardware, software, and network capabilities) and a well-designed reading environment can substantially improve the diagnostic accuracy and/or increase the productivity. This is of particular importance as image data volume has been increasing and becoming more complex. CT gives a good example of this development. In the past, CT studies comprised 100 to 150 images. Since the introduction of multislice CT, the number has increased and can easily exceed 1,000 images. This has contributed to substantial changes in image display, which began with the use of the frame mode and now involves the application of three-dimensional complex multiplanar reformation tools and advanced volumetric data renditions.

11.1 CHOOSING A WORKSTATION

The computational system that manages the acquisition, transmission, storage, distribution, display, and interpretation of medical images is called the Picture Archiving and Communication System (PACS). The PACS infrastructure includes imaging modalities, network and archive components, workstations, software, and interfaces with the hospital – or radiology – information system. When choosing a workstation, the workstation must be recognized as an integral part of the PACS.

The fundamental configuration of a workstation depends on the purpose for which it will be used and the modalities displayed. Consequently, when choosing a workstation, these two aspects have to be considered first. In particular, the first decision to be made is to define whether the workstation should be used for primary image reading (for detailed image analyses to make a diagnosis) or if it will be used for secondary review (for viewing images without the requirement for interpretation). Primary-read workstations have to fulfil higher technical requirements and are, therefore, more expensive than those used for subsequent viewing tasks. The second question is whether the workstation is going to be of a multimodality type or dedicated exclusively to CT imaging. A multimodality workstation must be capable of displaying the highest-resolution images as close to full resolution as possible. The highest-resolution requirements must be fulfilled when

DOI: 10.1201/9781315118130-12

the workstation is dedicated to the interpretation of radiographs. The resolution requirements for CT images are lower. Consequently, if the single-CT option is going to be used, the workstation could be equipped with monitors with lower resolution.

These two fundamental decisions will be followed by a number of activities that aim to find an optimal workstation configuration:

- When contacting vendors, it is advisable to request any – preferably independent – research papers on their workstation design and functionality.
- The experiences other users already had with the same or a similar product to that which you intend to purchase are often very helpful. Seeing how the system is functioning in clinical routine practice gives you a realistic impression of the potentials and shortcomings of the system.
- Ask the vendor of your favourite system for a test installation. Hands-on testing in your clinical environment gives you the opportunity to gain experience and generate open questions. It is important to include a large number of users in your institution to test the system. The results should be discussed and summarized. These are the basis for a subsequent conversation with the manufacturer. In-house testing is time-consuming, but this effort often prevents you from prospective trouble and therefore represents a good investment.
- Before purchasing a system, ask the manufacturer if he is able and interested in a co-operation to enhance the system according to your specific requirements.
- Once the workstation has been installed, an early evaluation of its functionality based on active working by several users is highly desirable to identify functional deficiencies. In the case of malfunction, the vendor should be obligated to provide full functionality.

The purchase of a workstation is a major and sustainable investment. Once the system is installed, it is unwise to exchange it within the next few years.

Soft and hardware are the two main components of a workstation. At the hardware level, a radiologist's performance can be greatly impacted by the monitor and graphic card. At the software level, the graphic user interfaces (GUIs) implemented by the manufacturer enable various functionalities necessary for image viewing. Such functionalities include image query and retrieval, display tools, and various image manipulation functions.

11.1.1 Hardware

Monitors are the key components of the workstation. Important considerations when assessing workstation monitors include the number of monitors necessary to meet the viewing requirements and the monitor quality, dependence on onscreen diameter, matrix, luminance, contrast ratio, bit depth, and greyscale versus colour capabilities.

A workstation for primary viewing should be equipped with two side-by-side monitors. Two monitors are required because it is necessary to compare multiple CT series or to handle images from other modalities simultaneously (Figure 11.1). It can be advantageous to have a third additional standard monitor for computer programs, next to the medical-grade monitors. When a single workstation is used for primary reading of images from different modalities, the specification of the monitors must match the highest level needed.

For human radiology, guidelines that define the minimum technical prerequisites for monitors and methods of their quality assurance exist. To the author's knowledge, comparable standards are not available for monitors used in veterinary radiology. Because of the underlying radiation safety issue, those medical standards should be adopted (Table 11.1).

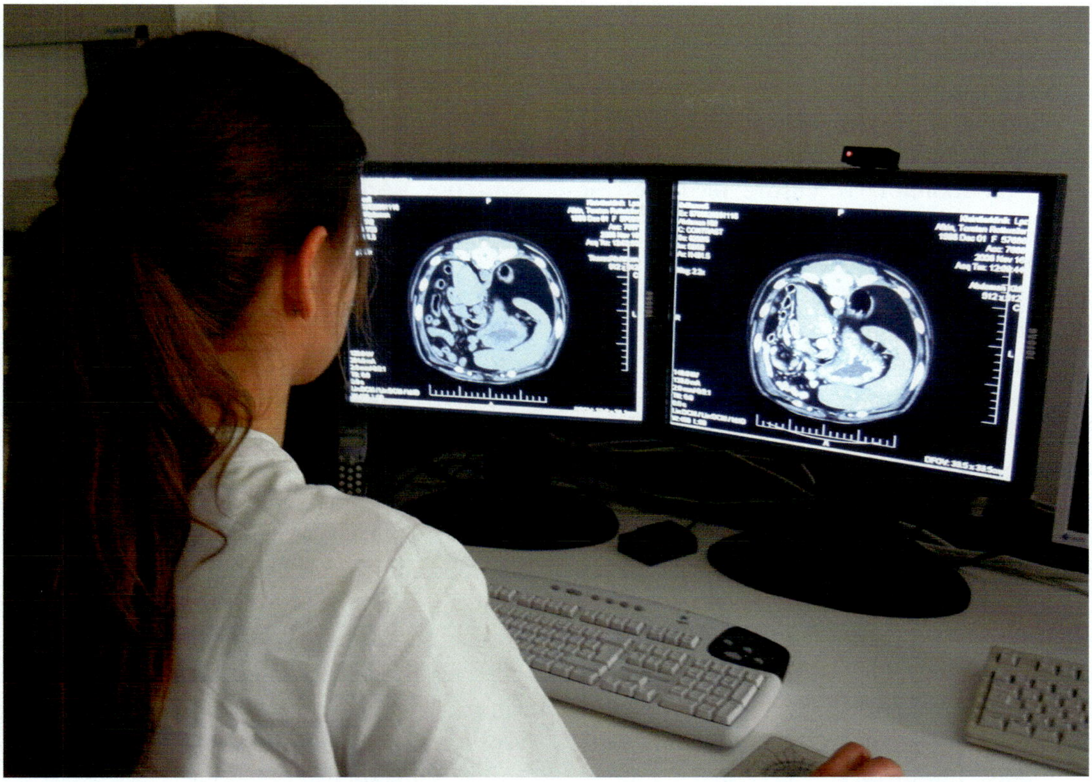

Figure 11.1 (Workstation). Multimodality workstation for primary image-reading. The system is equipped with two, landscape-mode, side by side greyscale monitors. Monitors and input devices such as mouse and keyboard are ergonomically positioned. A luminance meter positioned on the top of the right monitor displays when the ambient brightness exceeds a certain level.

Table 11.1 **Recommendations for monitor requirements for a multimodality veterinary workstation**

SPECIFICATION	PARAMETER
Number of monitors	≥2 greyscale monitors
Screen diameter	≥53 cm (≥21 inches)
Pixel pitch	≤0.2 mm
Pixel array (matrix)	≥2 megapixel (1,600 × 1,200 pixel)
Aspect ratio (width to height)	4:3 or 5:4
Maximum luminance	≥350 cd/m^2
Minimum luminance	1 cd/m^2
Contrast ratio	≥400:1
Greyscale bit depth	≥8 … 10 bits

Nowadays, grey-scale liquid crystal display (LCD) monitors represent the standard in medical imaging. Screen size and matrix (i.e., the number of pixels) determine the size of the pixel and consequently the spatial resolution of the monitor. To ensure adequate resolution, the matrix of the monitor should be as close as possible to the matrix of the recorded image data. In other words, "squeezing" a large detector matrix onto a small monitor matrix will result in loss of diagnostic information. For medical-grade monitors, it is recommended that the pixel pitch (the distance between the centres of a pixel) should be about 0.2 mm or less. Medical-grade monitors have 2–5 megapixel (MP), while consumer-grade monitors have 0.75 to 2 MP.

Contrast ratio (or dynamic range) is the ratio of luminance between the brightest and the darkest shade displayed. The higher the ratio, the better the monitor. The contrast ratio of medical-grade monitors ranges from 600:1 to 1,000:1. Luminance is a related parameter because a monitor with high brightness often has a wide dynamic range.

Standard computer monitors are generally colour LCDs with less luminance, lower contrast ratio, and larger pixels (smaller matrix). Often the graphic card is less sophisticated, and the monitor is not calibrated to the DICOM greyscale display function (GSDF) standard. The GSDF standard is a quality assurance metric. It ensures uniformity in image display between calibrated monitors even when the monitors degrade over time. This is realized by distributing the total contrast of the display across the entire greyscale. As a result, objects were presented with the same contrast regardless of whether they were located in dark or bright parts of the image.

Recent studies have shown that large screen standard computer colour LCD monitors with high brightness and resolution (screen diameter: ≥21 inch; maximum luminance: ≥200 cd/m^2; matrix: ≥2 MP) perform similar to greyscale monitors. Therefore, they could represent an alternative to expensive medical-grade monitors.

11.1.2 Software

Soft-copy workstations provide a dynamic and changeable presentation of the image data to the clinician. Therefore, the role of the radiologist is much more active in the soft-copy reading than it has been in analogue "lightbox" laser-camera printout image viewing. The interactive application of digital imaging tools enables the radiologist to enhance her/his performance either by improving the accuracy or increasing the productivity. Accuracy can be improved by using decision support and computational tools to deconstruct complex processes in image perception and diagnostic reasoning. Productivity is increased if the software provides features for work-flow optimization that automates time-consuming manual handling. Workstation capabilities important to enhance productivity are listed in Table 11.2.

There are many different versions of DICOM viewing software, which range from cost-free downloadable up to expensive programs. The GUI is a key feature. It must be user-friendly so that even with little or no training a user is able to work with the system. Until now, most workstation software work with pull-down menus, icon clicks, or keyboard commands for user input. However, there is evidence that eye movements away from the image are accompanied with reduced radiologist performance. In the light of anticipated increasing data volumes, it is becoming increasingly important to have advanced GUIs allowing total image manipulation with minimal or even completely lacking visual input (Figure 11.2).

Medical viewing programs are DICOM viewers capable of handling images from numerous digital imaging modalities. For optimized manipulation and evaluation of different modalities, the set of the displayed viewing tools can normally be adapted specifically.

Table 11.2 **Desired functionalities of workstations**

OPTION OR FUNCTION	PURPOSE
Hanging protocols	Specific information is incorporated into the DICOM header so that the software can recognize them and display images in a predefined order and orientation
Cropping tools	Automatic collimation detection to minimize white areas surrounding the exposed portion and reduction of background lighting
Image manipulation	Algorithms used to improve the accuracy by enhancing image features; presets allow for increased productivity
Decision support tools	Diagnostic aids to assist in soft-copy interpretation (e.g., 3D, segmentation and computer-aided detection tools), helping reduce the "human weaknesses" of bias, fatigue, and inconsistency
Prefetch algorithms	Automatic retrieval of previous examination results for comparison with the current study
Navigation tools	Interface for the navigation between studies of the same patient or different patients
Communication features	Features include the creation of a radiology report and allow real-time bidirectional collaboration between clinical partners

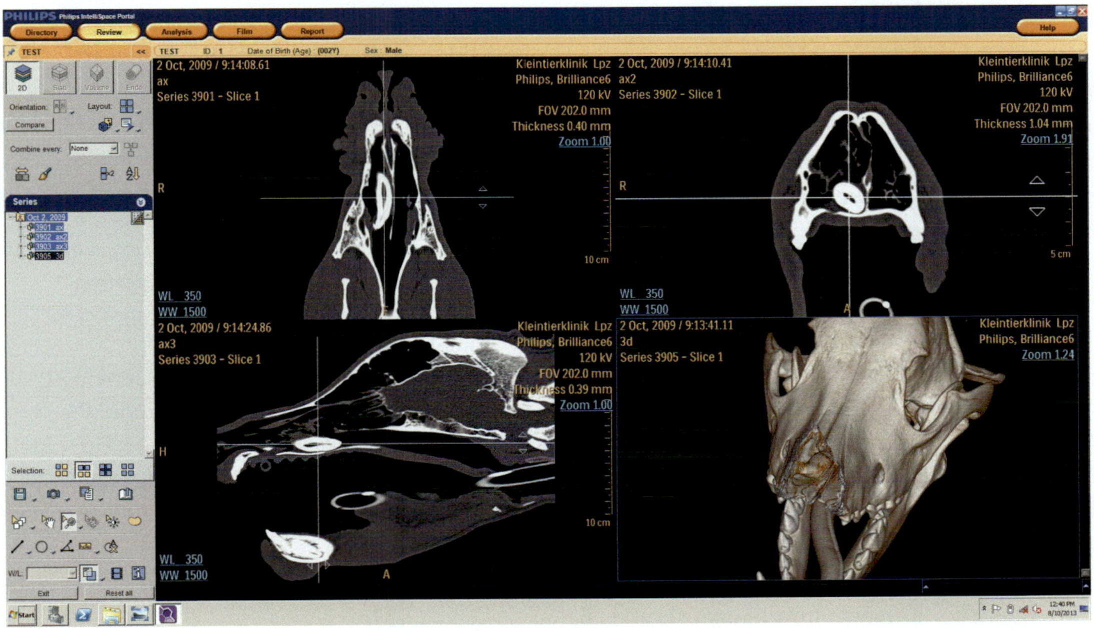

Figure 11.2 (Screenshot – GUI). Screenshot of an advanced workstation monitor. User-friendly customization includes a clearly structured graphic user interface (GUI), easy manual image-manipulation to meet personal visual preferences, and the option of easy restoration of default and set up values. (Photo: Philips.)

Programmes used for primary reading of CT series must exceed the minimum standard for DICOM viewer software, which includes simply controls for window and level (analogous to contrast and brightness), pan and zoom, flip, rotate, and measuring tools. Additional features, such as multiplanar reformation tools, volume rendering techniques, segmentation methods, and methods for quantitative analysis, are mandatory.

Because the functionality of GUIs differs between software programs, it is very helpful to explore programmes by hands-on testing to identify the optimum of functionality required to meet their own specific needs and to check user comfort.

11.2 VIEWING ENVIRONMENT

Human visual performance is influenced by a number of "environmental" factors, such as room layout, room climate (temperature, ventilation, and humidity), ambient lighting, ambient sound, and the ergonomics of the workplace design. Inadequate conditions result in increased eye fatigue and the degradation of diagnostic accuracy.

Among those factors, ambient lighting plays a decisive role. Parts of ambient lighting are reflected off the monitor's faceplate and the frame. This is added to the light emitted by the screen, and light reflection and veiling glare, therefore, substantially degrade the display quality by increasing the luminance in the dark regions. With increasing brightness of the ambient light, the ability for the identification of low-contrast structures decreases. Consequently, ambient luminance should be as low as possible and indirect. In principle, it should be significantly less bright than the darkest greyscale level on the screen. Standards recommend that the luminance measured at the monitor surface should not exceed 20–40 lux. A luminance meter positioned on the frame of the monitors facilitates control of the level of ambient brightness and therefore helps ensure that a certain level of ambient brightness is not exceeded.

Indirect lighting that reflects off the walls behind the reporting stations is advisable because it eliminates reflections on the screen. Individual workplaces should be equipped with dimmable light. The radiologist can adjust the lighting based on their specific requirements. Paperless working reduces the need for deploying additional reading lights.

FURTHER READING

Krupinski EA, Kallergi M. 2007. Choosing a radiology workstation: technical and clinical considerations. *Radiology* 242:671–682.

Puchalski SM. 2008. Image display. *Veterinary Radiology & Ultrasound* 20 49(Suppl 1):S9–S13.

Resolution 35. ACR-AAPM-SIIM technical standard for electronic practice of medical imaging. Revised 2012, https://gravitas.acr.org/PPTS/DownloadPreviewDocument?DocId=136

Robertson ID, Saveraid T. 2008. Hospital, radiology, and picture archiving and communication systems. *Veterinary Radiology & Ultrasound* 49 (Suppl 1):S19–S28.

Samei E, Seibert JA, Andriole K, et al. 2004. AAPM/RSNA tutorial on equipment selection: PACS equipment overview. *Radiographics* 24:313–234.

Siddiqui KM. 2009. Human factors and ergonomic issues in a radiology reading environment. *Applied Radiology* 38:18–21.

Siegel E, Reiner B. 2002. Radiology reading room design: the next generation. Applied *Radiology* 31:11–16.

Weiss D. 2008. The PACS workstation: a theory of evolution. *Applied Radiology* 37:24–29.

CHAPTER 12

PURCHASE

Stef Janssens, Henri van Bree, and Cees Verlooij
(From Helical/Spiral CT to Volume CT)

12.1 PURCHASE CT

Once the decision is made to install a CT equipment, various options to finance this purchase, such as renting, leasing, or buying the equipment, exist. In any case, a business plan needs to be worked out with the purpose of making it profitable or at least break even and offer a wider portfolio of services.

The reasonable number of 'patients' examined by the new equipment must be predicted. When this figure is defined, the budget needed for the project to succeed can be allocated. According to the budget size, there is the option to buy a new CT or a second-hand one from a vendor company, broker, or even directly from a hospital.

The business case needs to rely on a strong financial plan. Various models are possible, but the Total Cost of Ownership (TCO) model gives a realistic indication of the investment.

12.2 DEFINITION TOTAL COST OF OWNERSHIP (TCO)

TCO is a financial estimate to determine the direct and indirect costs of a project. TCO is the total net sum of the purchase price in addition to all other costs necessary to keep the CT up and running for a specific period (number of years).

12.2.1 Direct costs

12.2.1.1 Purchase price

First of all, and easy to understand, the price of the equipment has to be calculated for the TCO. This is the total price for CT gantry, table, post-processing console hardware, and software licenses, including VAT. Note that in some countries, when buying a second-hand CT from a hospital, they may not be under VAT-regulation, and VAT does not have to be taken into consideration for the buyer concerned with TCO calculation (Figure 12.1).

12.2.1.2 Warranty

The advantage of buying a new CT is that the machine has a warranty. The warranty period describes a firm defined time/period in year(s) or months (for example, with a second-hand system) and includes preventative maintenance in most of the contracts. Nearly always, the warranty also includes the X-ray tube, one of the most expensive parts of the CT. During the period, there is no need to pay for a service contract.

12.2.1.3 Service or maintenance

It is important to know that there are certain running costs associated with using a CT, to keep it in good shape, well-calibrated, and controlled, in order to avoid downtime. In case of buying second-hand equipment, there is mostly no warranty, which means the service cost will be considered right from using the CT. It is recommended to take a warranty, even for second hand devices. In some countries, there is also a need for annual check-up and certificates to be able to keep using the system. The service contract works like an insurance, with a price for services every year.

DOI: 10.1201/9781315118130-13

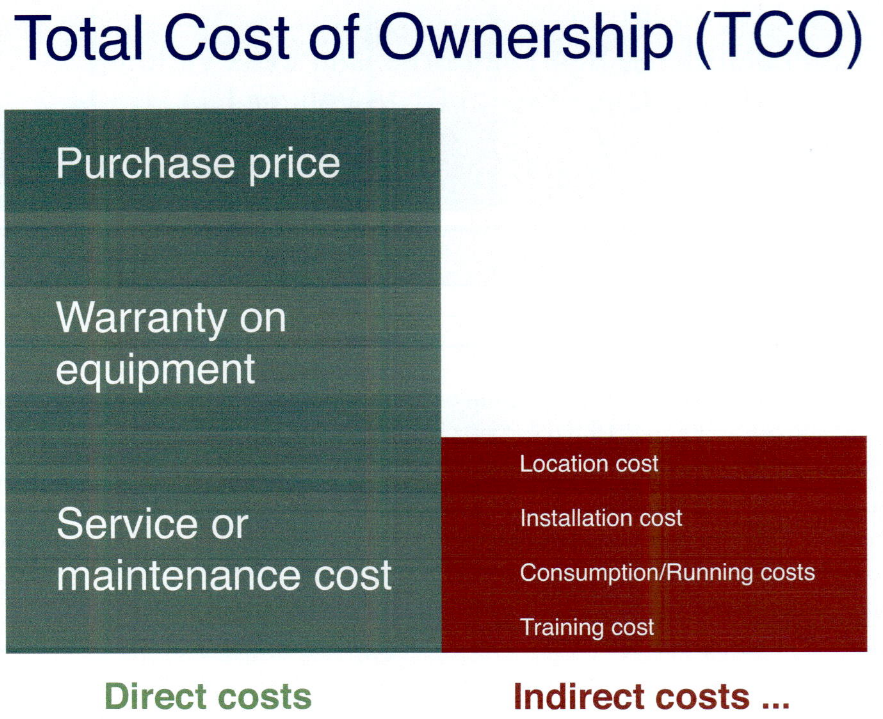

Figure 12.1 The TCO-calculation.

There are various types of service contracts:

- Preventative service contract: includes one or two service interventions every year, including working time and travel of field engineers (excluding any spare part costs)
- Standard service contract: includes one or two service interventions every year, working time and travel of the field engineers, and most of the spare parts, except the x-ray-tube (the expensive part of the CT). VPN monitoring of the equipment can be an option. Note the X-ray tube: there is still a certain unpredictable factor about the lifespan of every X-ray tube. It depends on being subjected to a good warming up every day, and it is better to use it on a regular basis at normal conditions rather than subjecting to extreme conditions, leading to higher damage to the tube itself
- Omnium service contract: this is the standard service contract and covers all parts, including the X-ray tubes. In this case, there are no big (expensive) surprises, and a real TCO can be calculated without a risk of budget increase.

12.2.2 Indirect costs

There are many indirect costs that can be taken into account to calculate the TCO. Only the most important costs are considered in this model: costs for location, installation, consumption/running, and training.

12.2.2.1 Location cost

This represents the total amount of hiring (cost per month/year) of building (acquiring and/or construction works) the room and, if necessary, the technical room. Sometimes, depending on the type of CT, the technical cabinet or installation can also be implemented in the same room as the CT gantry.

12.2.2.2 Installation cost

This is defined as the total costs for preparing the room where the CT and console are to be installed. An important part of this installation cost could be the lead protection and screen. A plan needs to be made in close collaboration with the local authorities with the needed lead protection in order to protect the personnel performing the CT exam. A stability study and costs for architects are necessary to determine which civil works are necessary.

12.2.2.3 Consumption/running costs

Here, many items could be listed. This often includes costs of electricity (estimation per exam/total per year) and yearly controls by official authorities to require certificates. Some CT companies provide cooling for CT by releasing the surplus heat of the CT gantry into the CT room. Check if additional air conditioning (and electricity consumption) is required. Some firms are cooling the CT on city water, and thus an invoice for water consumption has to be taken into account.

12.2.2.4 Training/application costs

The local staff needs to be trained to be able to scan and use the different system protocols. As the CT is often delivered with post-processing capabilities, the staff needs to be trained. With a new CT, this is included most of the time, but for a second-hand system, this has to be taken into consideration.

The most important costs are described above. To complete the TCO with more details, the following costs might be calculated:

- costs for future upgrades
- insurances for exploitation
- internal of external audits
- additional staff needed

etc. The TCO is always calculated over a certain number of years (7 years on average).

12.3 FROM HELICAL/SPIRAL CT TO VOLUME CT

12.3.1 General

CT scanners have developed from axial CT (CAT) to helical CT and recently to wide-area CT. Helical CT, which was invented in the 1980s, in comparison to axial sequential CT, made examinations possible. The commonly used terms helical and spiral represent the same principle; helical CT will be used in this chapter for consistency.

A helical CT scan, unlike sequential scanning, is not slice-by-slice, but it acquires a volume scan where thin slices are obtained to make three-dimensional (3D) images in a short amount of time, which is useful for visualizing blood vessels or fractures. Helical CT's main advantage, however, is its short scan time for longer anatomical ranges. Faster rotation, reconstruction methods, computer capacity, and tube technology are determining factors for further applications.

12.3.2 Acquisition

It is necessary to distinguish between acquisition and reconstruction. Typical acquisition items include Z-sampling (flying focal spot), detector rows, scan slice thickness, tube voltage, tube current, and rotation speed. With helical scanning, data are acquired by continuously rotating the tube and the detector and moving the couch through the gantry during exposure. The X-ray path resembles a helix. Transfer in the cranial direction can be seen as the tube moving in a caudal direction. Helical scanning (Figure 12.2) has three important parameters:

1. Number of detector rows
2. Scan thickness
3. Pitch

A system equipped with 80 detector rows can scan with dimensions of 0.5 mm×80, which means that in the Z- or longitudinal direction,

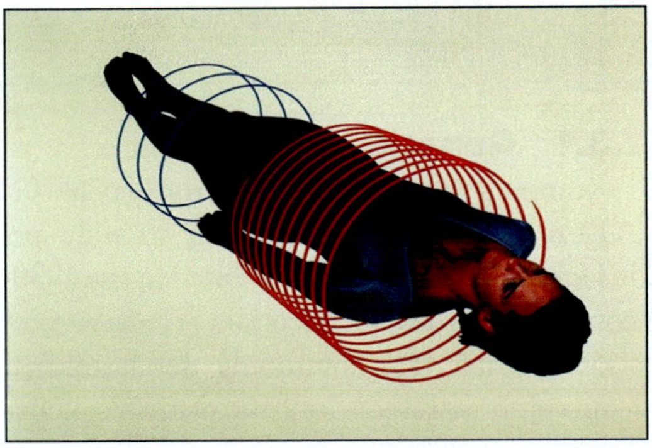

Figure 12.2 The concept of helical (spiral) scanning.

80 rows of 0.5 mm each are used, providing 40 mm of coverage.

Scan thickness—the thinner, the better—is very important for the spatial resolution of the reconstructed images. The thinnest CT scan slices currently available are 0.5 mm (Toshiba), 0.6 mm (Siemens), and 0.625 mm (Philips and GE). All slices are thin, but a 20% increase in thickness of 0.6 mm and 0.625 makes a difference when visualizing very small structures such as very small joints, bone structure, and nasal conchae. A thicker slice contains more photons and will have a larger signal-to-noise ratio (SNR); however, a thicker scan slice also suffers from partial volume averaging and leads to artefacts.

Couch speed, another determining parameter in acquisition, determines the coverage in the longitudinal direction (Z-coverage); this is shown as a ratio without entity (i.e., the pitch factor) in helical scanning.

The pitch factor is another important parameter in helical scanning because it influences image quality, effective slice thickness and dose, and determines couch speed. It represents the ratio between table movement per rotation and total beam width.

$$PF = \frac{\text{Couch Stroke}}{\text{Beam Width}}$$

Compare the pitch to the threads of a screw, where closer threads represent a small pitch and those far apart a large pitch (Figure 12.3).

With same parameters, a larger pitch (>1) results in a shorter scan time and lower radiation dose, but this may not be the best choice because the so-called over-ranging is also larger. Helical scanning requires over-ranging or Z-over scanning because the reconstruction algorithm needs additional data at the beginning and end of the planned scan. These extra rotations are used to reconstruct the scan's beginning and end. Every system faces this issue, but the latest models are equipped with active (or sliding) collimators that reduce the extra dose caused by Z-over scanning. A large pitch may lead to a thicker effective slice and higher partial volume effect that will degrade the image quality.

A smaller (overlapping) pitch leads to a higher dose, but it collects more information, which leads to better spatial resolution. Some systems work with effective mAs—those divided by the pitch factor. When the pitch is changed, the mA changes automatically. Keeping the pitch low and lowering the mA values, when possible, so that the effective mAs stay the same, are advised. The effective mAs can be a determining factor to keep the image quality equal between different scan protocols or different systems. CT systems that do not adjust the mAs automatically with a changing pitch can be distinguished from CT systems that do.

12.3.3 Detector rows and slices

The number of rows and slices causes much confusion. Several CT vendors employ the flying focal spot (FFS) technique to increase the number of slices, where moving the focal spot very quickly doubles the sampling frequency. This can be done in the Z-direction (longitudinal) equally

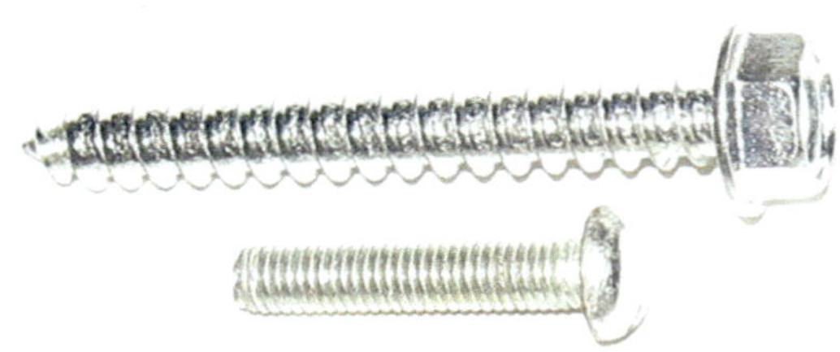

Figure 12.3 Screws representing a high pitch (above) and a low pitch (below).

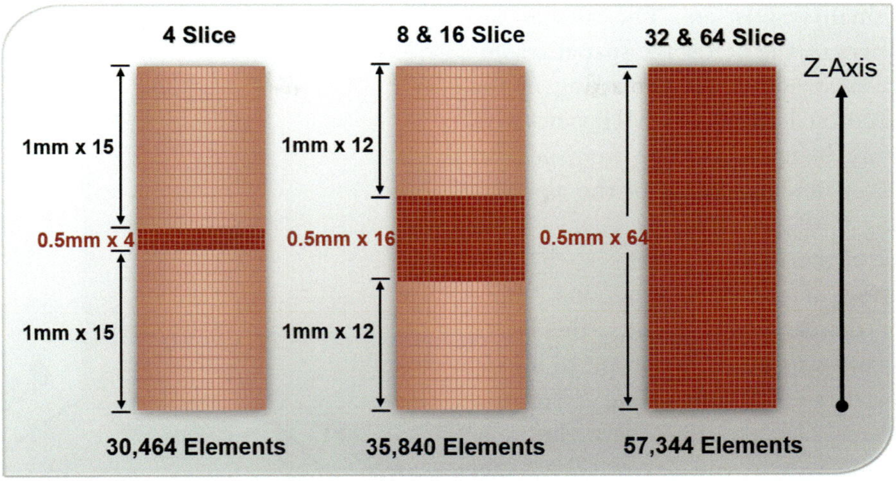

Figure 12.4 Detector: four row till 64 row. On the left: a four-row detector. If one wants to use 0.5 mm, the coverage is restricted to the middle of the detector: 0.5 mm × 4. The maximum coverage is 2.0 mm. To increase the coverage, one has to shift to 1.0 mm × 4. Two 0.5-mm detector elements are combined to 1.0 mm. When a faster scan is needed, more detector rows are combined to 2.0 mm × 4, 3.0 mm × 4 up to 8.0 mm × 4. In the middle: a 16-row detector is shown. Now more 0.5-mm elements are used, and the coverage is 0.5 mm × 16 = 8 mm. When a faster scan is desired, the coverage per rotation has to be increased. Detector elements are combined to 1.0 mm × 16 or even 2.0 mm × 16. The spatial resolution, however, will be decreased. On the right: the 64-row detector. No compromise in scanning because 0.5 mm × 64 can be selected always. Spatial resolution for every scan. (Image courtesy Toshiba Medical Systems.)

well as the *X* and *Y* directions. The sampling frequency determines the number of reconstruction errors, which improves image quality. The reconstruction errors referred to in this chapter are called windmill artefacts (Figure 12.4).

The FFS is also used to double the number of slices per rotation, that is, double compared with the number of detector rows. For example, a 64-row detector system with 0.6 mm using FFS can generate 128 slices of 0.6 mm; it behaves as a 64-row system concerning pitch, scan speed, and detector width. The *Z*-coverage remains 64 × 0.6 mm (38.4 mm). This scan technique increases the dose because a larger focal spot is used and the penumbra also is larger. A dose increase of 6%–10% is assumed. Consider that slices double only in axial (sequential) scans. Helical acquisition scans volumes, not slices, and the slice doubling plays no role.

Another vendor uses reconstruction to double the slices, and this method is explained in the following section.

12.3.4 Reconstruction

Dedicated reconstruction techniques are required to obtain axial images from the raw data, and both filtered back-projection (FBP) Fourier and Iterative reconstruction are used. The latter method, which is used on the latest CT systems, provides a lower dose and improved image quality. The larger the cone beam becomes, the more issues with artefacts are encountered. Also, 320-detector row CT scanners need special reconstruction algorithms. These 3D-reconstructions of cone beam projections are based on Feldkamp, Davis, and Kress (FDK).

The determining factor for visualizing small structures is spatial resolution, and the detector width that can be used routinely influences the quality achieved when examining small joints, inner ear, small vessels, etc. The smaller the detector elements, the better the spatial resolution. 3D reconstructions, reformatting in the sagittal and coronal planes, and MIP views with CT angio are being used more frequently to obtain a better understanding of the anatomy and pathology.

Detector elements also determine the spatial resolution of the reconstruction. Again, the smaller the detector elements, the better the spatial resolution. It is important to always use isotropic datasets in daily practice. Isotropic voxels (Figures 12.5 and 12.6) have the same dimensions in the *X*, *Y*, and *Z*-axes. The *X* and *Y* plane dimensions are the result of the reconstructed Field of View divided by the matrix, whereas the *Z*-axis is the result of the reconstructed slice thickness.

When considering the purchase of a CT scanner, the big question is whether it fits a veterinary practice. Many systems are available—single slice, dual slice, or four slice—that perform sufficiently, but they cannot scan routinely with the smallest detector elements. Therefore, to maintain a practical scan duration, keep the tube heating within limits, the generator capacity workable, and the dose low, a non-isotropic dataset is scanned. The thicker slices complicate diagnosing the pathology of small structures, so using the smallest detector elements for routine scanning allows high-quality 3D reconstructions, reformats in the sagittal and coronal planes, and curved MPR to be made, which makes the diagnosis easier and more reliable.

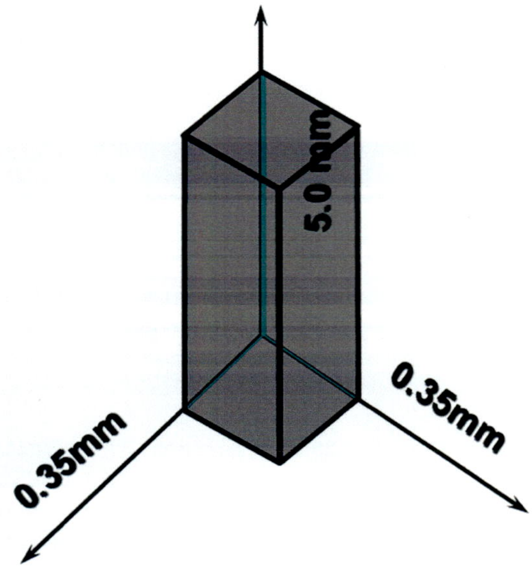

Figure 12.5 Non-isotropic voxel (0.35 × 0.35 × 5.0 mm).

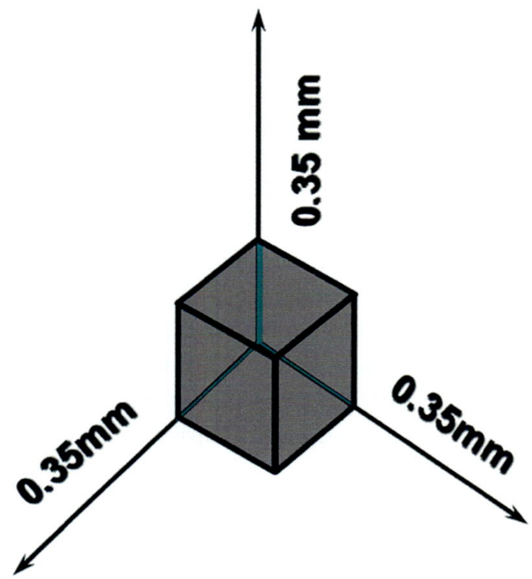

Figure 12.6 Isotropic voxel (0.35 × 0.35 × 0.35 mm).

12.3.5 Reconstruction interval

Reconstruction index, interval, or increment all are used to refer to the amount of applied overlapped reconstruction. Any desired *Z*-position can be selected to reconstruct a slice within the scanned volume. The 0.5 mm/0.3 mm indication means that a 0.5-mm slice is reconstructed every 0.3 mm, giving a 0.2-mm overlap. The reconstruction interval, which is a reconstruction parameter (independent of pitch, which is an acquisition parameter) can be chosen in tenths of millimetres and will have no effect on the given dose. Overlapped reconstruction has the benefit of achieving a higher detection rate and optimal lesion detection. The quality of 3D, MPR, MIP, etc. will also be much improved. Reconstruction and the transfer to PACS is time-consuming and more storage is needed, but this is not a problem with modern equipment. A 50% overlap is ideal, but 30% overall is sufficient.

The reconstruction interval is sometimes related to the number of slices that can be generated. Calculate the maximum number of slices in a helical scan using the following formula: $N_p=(\Delta b/RI_s)\times N_d$ or the number of slices=(detector element: smallest possible interval) multiplied by the number of detector rows. An example: The scan is made with 0.5 mm acquisition on an 80-row scanner, and a 0.1-mm interval is desired. The number of slices in one helical rotation will be $(0.5/0.1)\times 80=400$.

12.3.6 Detector technology

The detector forms the core of a CT scanner, and much effort has been put into developing better detectors for the following reasons:

- Higher spatial resolution (high contrast resolution)
- Improved response time (speed to process detected photons and be ready for the next readout)
- Good contrast resolution (low contrast detectability)
- Lower dose (sensitivity for incoming radiation).

The *Z*-coverage of detectors has been discussed earlier in this chapter, but development of wide area detectors (e.g., Toshiba's Aquilion ONE that has a 320-row detector of 0.5 mm elements) is crucial. A maximum coverage in the *Z*-axis of 16 cm can be achieved in one rotation, with no couch movement or pitch compared with a helical scan. A volume is made in one rotation, and slices are reconstructed from that volume. This technique is inseparable in relation to acquiring and reconstructing with a very large cone angle (over 15°), which is extremely difficult and requires a dedicated reconstruction algorithm. Toshiba Medical System developed the ConeXact™ dedicated algorithm that is employed on other systems to double the number of slices acquired with one rotation. On the Aquilion ONE, this algorithm will generate 640 slices from one rotation, whereas an Aquilion 32 generates 64 slices, an Aquilion 64 generates 128 slices, and an Aquilion PRIME 160 slices. This double-slice model is the counterpart of the FFS technique, whose goal—improving the spatial resolution in the *Z*-axis—is the same. The FFS technique halves the number of photons per slice sampling and leads to higher dose, whereas the ConeXact algorithm does not need a higher dose, improves spatial resolution, and preserves low-contrast resolution.

12.3.7 Wide-area CT

Wide-area CT, or cone beam CT, is one of the latest exciting developments. In 2007, Toshiba Medical Systems presented its Aquilion ONE system that has 16-cm coverage in the *Z*-direction in the rotation center. Philips has a system with 8-cm coverage, and GE recently introduced a 16-cm coverage system. These systems have the great advantage of enabling scanning of the (human) heart and brain in one rotation, eliminating the need for helical scanning. They can capture an entire organ and produce high-quality 3D, MPR, etc. Very fast scan times (the scan time is the rotation time) result in decreased contrast and patient dose. A wide-area CT with whole-organ coverage makes it possible to not only produce morphologic images but also physiological ones. Four-dimensional imaging, where time is the fourth dimension, is now a reality, and dynamic series or moving joints, incoming arterial contrast, washing out in the veins, and perfusion are enabling completely new applications.

FURTHER READING

Blobel J, de Vries H, Irwan R, Mews J, Ogawa Y. 2013. 640 Multislice Reconstruction with Dynamic Volume CT, https://www.visions-journal.com/index.php?id=1&lang=gb&ausgabe=38

Feldkamp LA, Davis L, Kress J. 1984. Practical cone-beam algorithm, *Journal of the Optical Society of America* 1:612–619.

Galanski M, Hidajat N, Maier W, Nagel H, Schmidt T. 2002. In *Radiation Exposure in Computed Tomography*, 4th edition. Editor: H.D. Nagel, CTB Publications, Hamburg.

Kalender WA. 1994. Spiral or helical CT: right or wrong, *Radiology* 193(2):583.

Katsevich A, Taguchi K, Zamyatin A. 2006. Formulation of four Katsevich algorithms in Native geometry, *IEEE Transactions on Medical Imaging* 25:855–868.

Mori I. 1986. Computerized Tomographic Apparatus Utilizing a Radiation Source. US Patent No. 3.866.047, admitted in 1973, granted in 1986.

Prokop M, Galanski M. 2003. *Spiral and Multislice Computed Tomography of the Body*, Thieme Verlag, Leipzig.

Radon J. 1917. Uber die bestimmung von funktionen durch ihre Integralwerte langs gewisser Mannigfaltigkeiten, Saechsische Akademie der Wissenschaffen. *Leipzig: Berichte uber die Verhandlungen* 69:262–277.

CHAPTER 13

ADDITIONAL CT PROCEDURES, IV AND IA CONTRAST APPLICATIONS, AND CT GUIDED BIOPSIES

Ingrid Gielen, Wolfgang Henninger (IV contrast), and Henri van Bree

13.1 CONTRAST APPLICATION

13.1.1 Intravenous and intraarticular (CT arthrography)

13.1.1.1 Intravenous

A computed tomography (CT) examination is not complete if no intravenous (IV) contrast medium is administered.

13.1.1.1.1 Advantages of iodine contrast medium administration

CT images are reconstructed from data acquired from a 360° data set generated by the penetration and absorption of X-rays; radiographic imaging has shown that high contrast tissue such as bone and gas/air provides the best imaging resolution, whereas soft tissue resolution is low. The density of two tissues must differ at least 10 HU to be visible on a CT image. The administration of IV contrast medium will provide an increase of 40–75 HU, resulting in a natural difference of attenuation between tissues and making them become visibly different in the image. Therefore, contrast examinations are well-established in radiography and CT. Dosages used regularly with radiography are too high for CT, so they usually need to be diluted within cavities, or they will be diluted intravascularly by the plasma, as the cross-sectional computerized imaging modality is much more sensitive.

Water-soluble, non-ionic iodine contrast medium of low osmolarity with an iodine content of about 300 mg iodine per ml solution is recommended for IV administration. Many products of equal quality are available for veterinary use.

13.1.1.1.2 Contrast medium application and contrast phases

After a plain serie(s), an IV contrast medium is administered, and the region(s) of interest are again scanned, obtaining different phases of contrast enhancement according to the clinical situation. Iodine is administered either by an injection pump at a dose of 500–800 mg/kg body weight (about 1–2 mL of the contrast medium solution) and a speed of 0.5–4 mL/s, or manually, again depending on the clinical situation. The contrast medium administered should always be closer to body temperature to keep the animal safe during anaesthesia and to flush the otherwise very viscous solution.

Any accessible vein can be used for IV access, although the cranial cephalic vein on the front leg or the lateral saphenous vein on the hind leg are preferable. A more central vein, such as the jugular vein, can also be used. Streak artefacts caused by high concentrations of iodinated contrast material extend throughout a relatively large part of the image and may obscure structures. Preferably, the IV access is at the opposite site of the one where the pathology is expected. It is more efficient for a single bolus injected in a vein close to the heart rather than at a location with a longer distance, e.g., the hind legs, which poses a higher risk of venous dilution. The intravascular distribution of the contrast medium during the cardiac cycles shows an aortal peak and an arterial phase over 8–15 seconds, followed by immediate recirculation

DOI: 10.1201/9781315118130-14

(approximately 20 seconds) through the capillary bed, into the venous circulation, and back to the heart. There, the contrast medium is redistributed via the arterial system and recirculated via the venous system; these cycles continue while excretion via both kidneys starts simultaneously. After a few minutes, the high attenuation between the arterial and venous vessels equalizes, and the organs and the muscles are contrast-enhanced to their maximum.

The speed at which the blood and the contrast medium mix depends first on the injection rate and then on the heart rate or ejection fraction. Injecting the contrast medium at a higher rate leads to a rather high aortal/arterial peak, followed by a quick decline and rapid recirculation, so not only the tissue with delayed enhancement (e.g., the spleen) but also the musculoskeletal system needs more time to be contrast-enhanced than better-perfused organs, which will not be enhanced by the contrast medium during the first run. In multi-slice helical scanning, enhancing the foci should not be missed as the scanning time is short, and so the contrast may still not be visible within the area of interest at that moment. A bolus-tracking system to ensure the scan begins at the correct time is a standard requirement in most multi-slice scanners.

13.1.1.1.3 Dealing with contrast phases with single- and multi-slice scanners

Single- and multi-slice scanners differ in the way they deal with contrast phases. With single-slice scanners, an expensive injection pump with one plunger is more useful to obtain a homogeneous, reliable enhancement in reasonable time, although manual administration is still possible. The thorax and abdomen/pelvis are imaged in a helical mode, using pitch 2 for a higher table speed; reconstruction intervals can be kept as short as possible. All other body parts are better imaged in an incremental mode, where the often prolonged inter-scan delays obtaining images of several contrast phases. A suspected lesion from the plain scan should be imaged first in the (arterial) contrast phase, even if the plain series was taken from the other direction.

With multi-slice scanners, a more expensive injection pump with a modern two-way plunger system is recommended because it allows a specific application, depending on the scan level, using a defined bolus of the contrast medium accompanied by a saline flush before and after the bolus administration. Using this kind of pump arterial, portovenous and venous contrast phases can be performed. All examinations using multi-slice scanners are performed in the helical mode because of the higher resolution.

Patients with neurologic diseases or musculoskeletal disorders should be examined in an arterial or early venous phase, although a late-phase enhancement may be necessary in some cases. Usually the scanning starts about 1–3 minutes after the contrast injection. A delayed-phase protocol (6 minutes post injection) may also be needed for some patients with musculoskeletal problems.

13.1.1.2 CT arthrography

13.1.1.2.1 Introduction

Arthrography is a diagnostic imaging technique used for decades in human medicine to enhance the visualization of intra-articular structures that cannot be seen clearly on plain radiographs. As ultrasonography (US) and magnetic resonance imaging (MRI) became more available, the use of arthrography declined. In veterinary medicine, particularly for small animals, arthrography has been used mostly for imaging of the shoulder joint. US and MRI are the current modalities of choice for imaging soft-tissue structures; however, both techniques have their advantages and disadvantages.

Computed tomographic arthrography (CTA) is an imaging technique whereby intra-articular administration of the contrast medium is used to improve the visualization of the intra-synovial and peri-synovial soft-tissue structures. The contrast medium diffuses within the joint and these structures, which appear as filling defects outlined in the joint by the contrast and can be clearly visualized (Figure 13.1). CTA is used to identify abnormalities in the articular surfaces, intra-synovial soft-tissue structures, synovial surface of the joint capsule, and peri-articular structures that have a close relationship with the joint compartments. Abnormalities in the size, shape, or margins of the intra-synovial soft-tissue structures can be identified by an ill-defined contrast pattern seen when the contrast enters any tears or defects resulting from injury to the soft-tissue structures or articular surfaces.

The cartilage layers in canine joints are much thinner than those in humans and horses, which makes evaluation difficult—perhaps impossible. CTA has some advantages over MRI or MR-arthrography (MRA), and human medicine studies have shown that CTA is equal to or better than MRI or MRA for diagnosing various entities. One of CTA's important advantages is the sub-millimetre resolution capability of current generation multi-detector scanners; another is that higher contrast can be seen between bone, soft tissues such as hyaline cartilage, and the iodinated contrast material, which make it a superior technique for defining morphologic cartilage defects. CTA also defines calcified structures more clearly.

CTA is usually performed fast because one or two volumes of data provide adequate multiplanar reformatted images tailored to the clinical problem. Also, individual CT-scans are much faster than MRI sequences, which reduces motion artefacts.

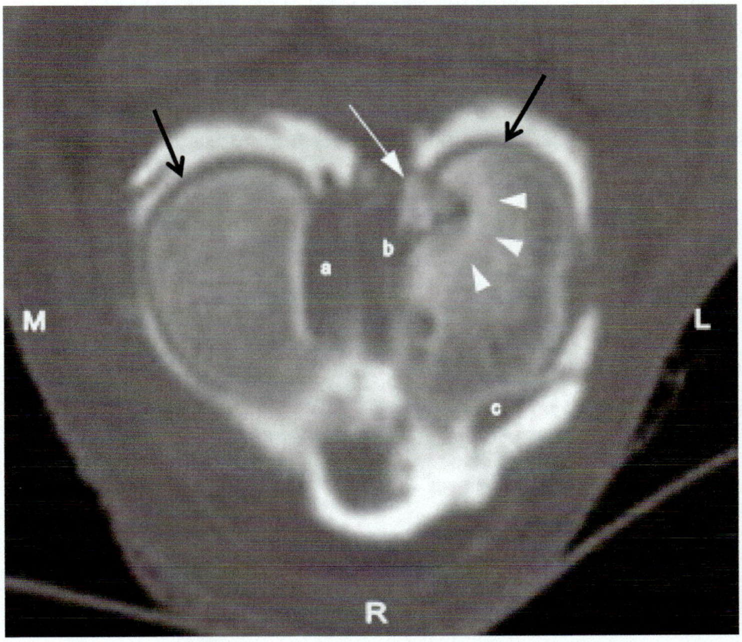

Figure 13.1 **Transverse CTA image of a right stifle joint at the level of the femoral condyles. There is a diffuse sclerosis of the lateral femoral condyle (arrow heads). An osseous avulsion fragment is present at the proximal attachment of the cranial cruciate ligament (white arrow). The caudal cruciate ligament (a), the cranial cruciate ligament (b), and the tendon of the extensor digitorum longus muscle (c) can be appreciated. The joint cartilage is seen as a hypodense line between the subchondral bone and the contrast medium (black arrows). Medially (M) and laterally (L) is indicated for orientation.**

13.1.1.2.2 Indications for CTA

- Evaluation of intra-articular ligaments and structures;
- Evaluation of calcified structures within the joint, in addition to internal derangement;
- Examination of the stability of osteochondral lesions of the articular surfaces;
- Examination of the stability of avulsion fractures of intra-articular ligaments;
- Detection of intra-articular fragments ("loose bodies");
- Evaluation of hyaline cartilage defects (Figure 13.2)

13.1.1.2.3 Patient preparation and injection

Before performing the arthrography study, a plain CT study has to be performed in both bone and soft-tissue algorithms. General anaesthesia should be maintained during the examination with either injectable or gas anaesthetics. The patient is placed on the CT table in lateral recumbency on the non-affected joint; the upper leg that will be injected must be extended in line with the table. The area for arthrocentesis should be clipped and aseptically prepared before the injection.

13.1.1.2.4 Arthrography technique

Firstly, remove any excess joint fluid, if possible. Insert a 22-gauge, 1-inch needle into the joint, and the synovial fluid is aspirated to ensure correct needle placement. After the contrast medium has been injected, flex and extend the joint repeatedly to facilitate the distribution of the contrast medium throughout the joint. Then, the limb is repositioned on the CT table, as before, and the CT acquisition is repeated.

Arthrography requires the use of a non-ionic iodinated contrast medium, which the joint tolerates better than an ionic iodinated contrast. Many products are available; they are all similar, except for the concentration of iodine per millilitre. Whichever product is used, the contrast medium should be diluted with sterile 0.9% physiologic saline solution (NaCl) to obtain a contrast medium concentration of approximately 80 mg of iodine per millilitre. If the contrast medium is too dense, it will surround the joint structures and impede their delineation. The amount of the contrast medium injected into the joint differs between patients and depends on the size of the joint to be examined.

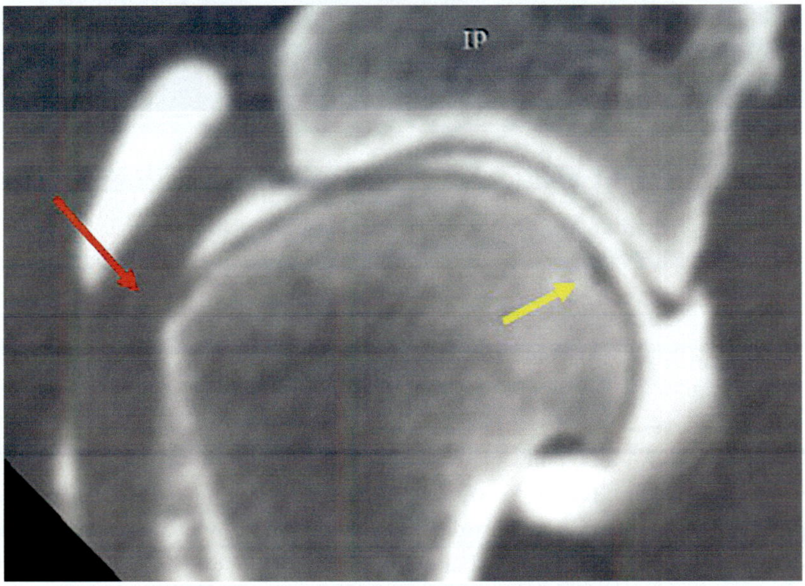

Figure 13.2 CT arthrography, sagittal reconstruction, of a shoulder joint of a non-clinical OC lesion. Thickened cartilage is covering the small subchondral defect (yellow arrow). There is no contrast medium visible underneath the cartilage. The biceps tendon is nicely delineated (red arrow).

13.1.1.2.5 *Pitfalls*

- Poor contrast-medium distribution within some joints, which can be avoided by repeatedly flexing and extending the joint;
- Insufficient density of the contrast medium, the result of dilution in the joints from a massive joint effusion, which can be avoided by removing the excess joint fluid;
- Diffuse delineation of the intra-articular structures, the result of rapid resorption of the contrast medium by the inflamed synovia in the diseased joints, which can be avoided by admixture of 0.2 mL of epinephrine; and
- An interpreter's incorrect diagnoses because of his or her lack of experience in assessing CT arthrograms.

The CT scanning protocol will vary, depending on the equipment and the size of the dog. Bone and soft-tissue algorithms are ideal for evaluating bone and soft-tissue structures.

Assess pre-arthrography and post-arthrography CT images with a bone and a soft-tissue window to allow proper evaluation and comparison of all the structures comprising the joint. MPR, an indispensable software tool available in the CT workstation and with the commercially available DICOM viewing software programs, allows the images to be analysed in all planes. Also, accurate interpretation is dependent upon detailed knowledge of the internal anatomy of the joint to be examined.

The articular surface is visible when the contrast medium is in contact with the articular cartilage, which is seen as a smooth, uniform hypo-attenuating line between the contrast medium and the subchondral bone. The joint must be fully distended, and the contrast medium must completely cover the superficial margin of the articular cartilage. Ligaments are difficult to evaluate on pre-contrast images, but on CTA, normal ligaments can be seen as uniform, hypo-attenuating bands of tissue surrounded by the contrast medium.

13.2 PERCUTANEOUS CT-GUIDED BIOPSY OF MUSCULOSKELETAL LESIONS AND SPINAL OSSEOUS LESIONS

13.2.1 Introduction

Medical imaging is an essential technique for the diagnostic workup of many soft tissue, bone, and spinal lesions in dogs and cats; however, the available modalities do not always result in a differentiation between inflammatory or infectious conditions from neoplastic disorders. Medical imaging is essentially morphologic; its low specificity only allows for aetiological diagnosis in a few patients. A CT-guided biopsy is used for body tissue sampling and is considered to be a safe and accurate procedure. CT examinations are performed to obtain images, which can be used to guide a biopsy device through the skin to the right place inside the body. Closed or percutaneous biopsy can be performed with a fine needle (soft tissue), a trocar, or a True-Cut needle (bone). A small piece of tissue is collected, which is sent to the laboratory to be analysed. Indications for musculoskeletal biopsy are primary or secondary tumours and infections. Absolute contraindications for musculoskeletal biopsy are coagulopathy and circumstances where biopsies do not provide additional information or do not change patients' management.

The patient should be evaluated for incidence of haemostatic disorders when there is a significant risk of haemorrhage. The general condition of the animal should be checked as well because for this procedure, general anaesthesia is required. Animals are positioned to allow the easiest access to the lesion, the hair is clipped in the area of interest, and the site is surgically prepared. For fine needle aspiration (FNA), 21–25 G needles are used, mostly injection needles (3 cm in length) or spinal needles (9 cm). Tissue core biopsy (TCB) are taken with 14 or 18 G needles. It has been reported that using a small needle decreases the risks of complications, but with a normal coagulation profile, better diagnostic samples are obtained with larger needles.

A pre- and post-contrast CT examination is performed to confirm the lesion, to assess its extent and vascularity, and to identify possible metastases. A target plane is chosen at a location that shows significant changes so that viable tissue is likely to be sampled. Necrotic areas and regions containing large blood vessels should be avoided. For TCB of bone or spinal lesions, the selected area should provide enough cortical support to hold the needle, and the complete bone thickness must be sampled to avoid loss of bone material along the needle tract. In one study on CT-guided FNA and TCB of bone lesions in 21 dogs and two cats, 17/17 TCBs and 5/6 FNAs were diagnostic, and no major complications were described.

The CT table is moved to the target plane, and the site of insertion of the needle is chosen and marked by inserting a small syringe needle into the skin (Figures 13.3 and 13.4). The CT slice thicknesses and intervals, usually 3 mm, should be related to the location and size of the lesion. A few additional CT slices are acquired in the area of the marker to measure the distance between the skin and the lesion to be biopsied, as well as to define the insertion angle of the needle. The CT table is moved out of the gantry so that the needle and the metallic guide or the bone needle (for TCB) can be inserted and advanced to the pre-set distance and angle after a small skin incision (number 11 blade) (Figure 13.5). The position of the needle tip is evaluated with additional images, and the needle placement can be corrected if necessary (Figure 13.6). Precise localization of the needle tip during CT-guided percutaneous biopsy is a key element of a successful procedure (Figure 13.7).

Multiple tissue samplings should be harvested depending on the size, location, and suspected diagnosis of the lesion. After obtaining the samples, additional images in the area of the lesion can be taken to check for complications. The overall time of the biopsy and sampling, including the pre-biopsy CT examination time, varies from 15 to 30 minutes. The diagnostic accuracy of the procedure is related to the diameter of the biopsy needle or trocar; the site, size, histology of the lesions, and the number

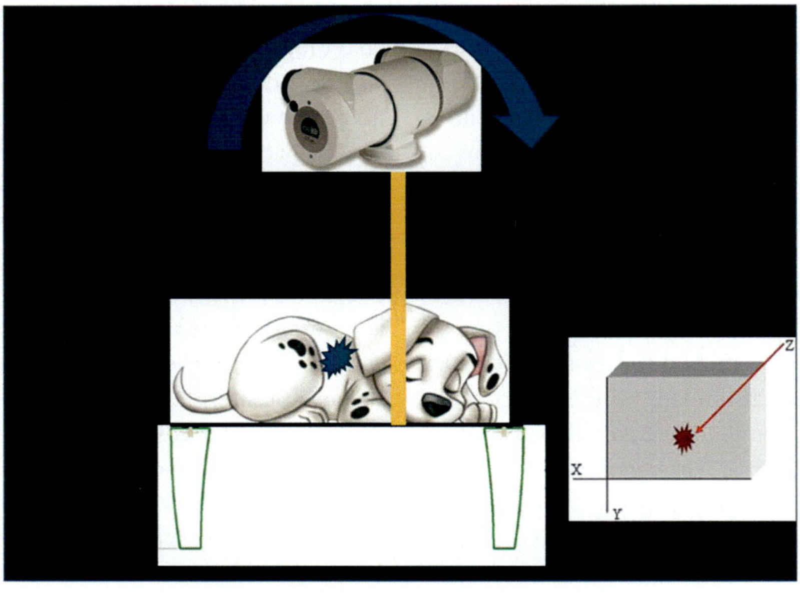

Figure 13.3 The patient is scanned, and the target region is visualized.

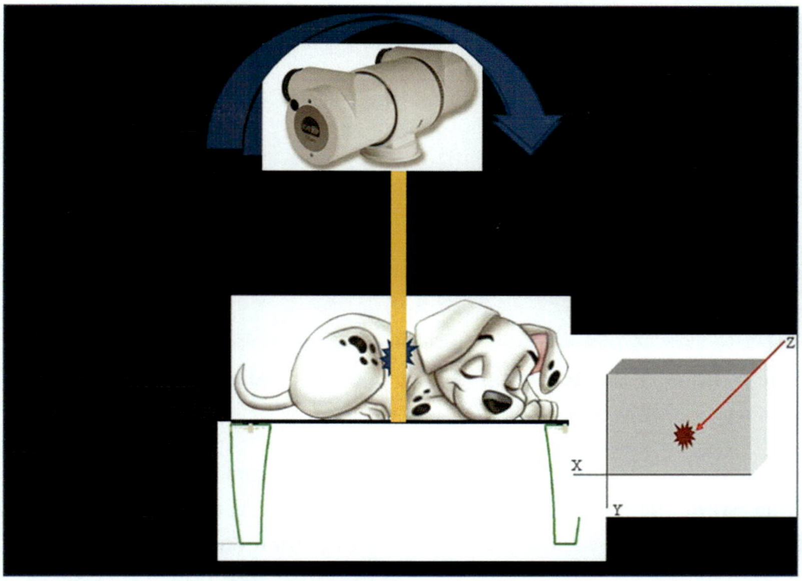

Figure 13.4 The table is moved to the target lesion.

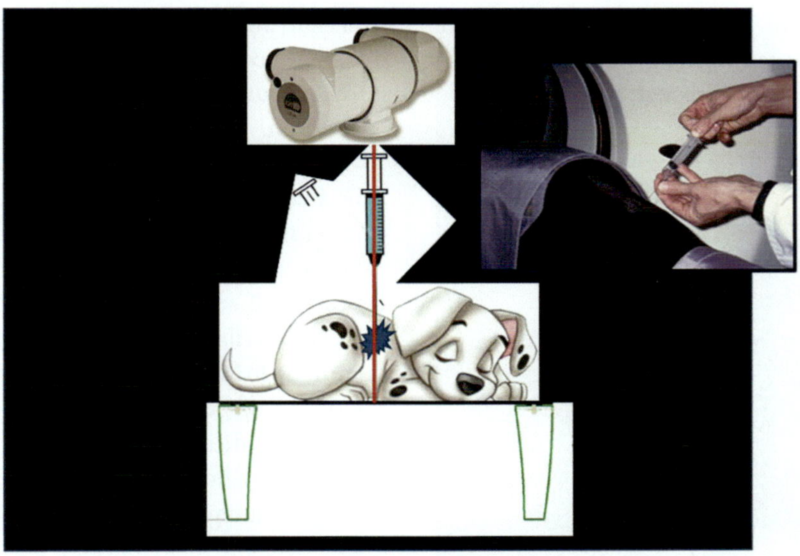

Figure 13.5 Multiple CT-guided samples are taken at the level of the target lesion.

of tissue samples obtained. CT-guided biopsies should be repeated for non-diagnostic initial workups.

Possible complications associated with this procedure include haemorrhage, neurovascular damage, procedure-related bone fracture, and non-diagnostic tissue sampling, although the overall reported complication rate in humans as well as in animals is low.

13.3 LOADING CONDITIONS AND SUBCHONDRAL BONE DENSITY

Joint loading is important in the (patho)physiology of both the joint cartilage and the whole joint, but, unfortunately, it is very difficult to evaluate *in vivo*. Different techniques have been used to evaluate joint loading under specific conditions, from *in vitro* testing to implanting

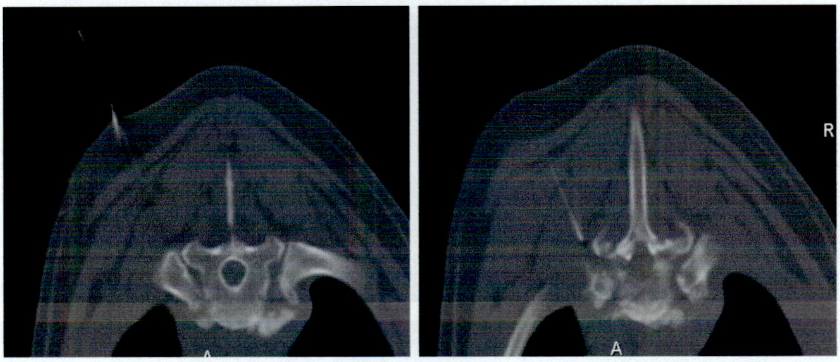

Figure 13.6 The position of the needle tip is evaluated with additional images, and the needle placement can be corrected if necessary.

pressure-sensitive devices either *in vitro* or *in vivo*, all which include some degree of dissection and invasiveness and may lead to altered joint biomechanics.

Bones generally provide valuable information about locomotion and loading because they continually adapt to the loading conditions, in orientation and magnitude, using mechanical stimuli through a process called mechanotransduction to withstand the loading conditions. At the level of the whole bone, adaptation includes trabecular orientation, bone curvature, and bone cross-sections. At the level of the joints, the subchondral bone plate (SBP) is continually adapted to fulfil its load-bearing function and to withstand mechanical stress during locomotion. This life-long process is best summarized as mechano-biological adaptation.

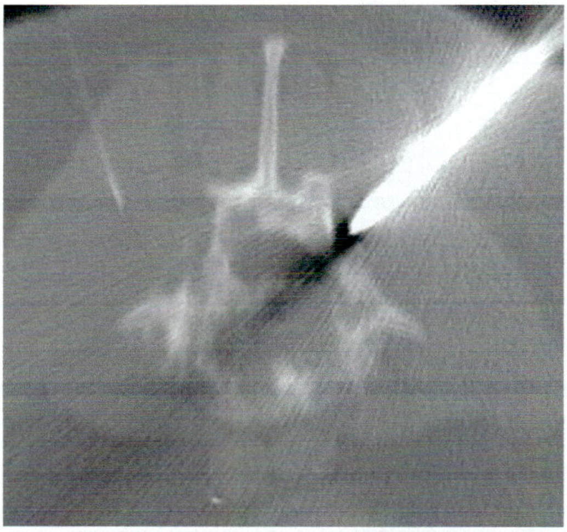

Figure 13.7 Transverse CT image in bone window at the level of L5 showing the track of the bone biopsy needle. White streaks, metal artefacts from the biopsy needle, obscure the image. Histologically, an osteosarcoma was confirmed.

The articular cartilage and the underlying SBP that supports the cartilage form a functional unit. The SBP is composed of a layer of calcified cartilage directly beneath the articular cartilage and a layer of lamellar bone. The density distribution is adapted not only to the dominant loads acting on the joint but also to the joint shape, which explains why similar loading conditions can give rise to different density distributions, depending on the joint shape. In human joints, wider joint sockets have a monocentric distribution, and deeper joint sockets have a bicentric distribution. The subchondral bone density distribution correlates to the joint's mechanical loading.

Examination of the SBP's density distribution allows the long-term joint loading, i.e., the loading history, to be evaluated. Imaging techniques are crucial for *in vivo* SBP studies because experimental studies always result in joint disruption.

13.3.1 Imaging techniques in bone density evaluation

The attenuation of X-rays varies with tissue density, allowing bone density evaluation. Different techniques have been developed, but dual-energy X-ray absorptiometry (DEXA) is the most frequently used and widely recognized in the medical imaging field, including veterinary applications. DEXA, however, is limited to planar views and does not allow evaluation of the bone density in specific areas within the SBP, so quantitative computed tomography (QCT) is often used to evaluate bone density, mostly for osteoporosis research and fracture risk management. QCT can also be used to evaluate trabecular and cortical bone separately and can provide a three-dimensional orientation. QCT results describe a mean value for the region of interest (ROI) that the operator defines. Using multiple ROIs allows evaluation of the SBP's regional bone density differences, but it is very labour-intensive and prone to inaccuracies because of ROI placement. Computer tomographic osteoabsorptiometry (CTOAM), which was developed to display relative differences in the SBP's density, can provide absolute values when a density calibration phantom is used.

13.3.2 Computer tomographic osteoabsorptiometry (CTOAM)

This reliable technique that has been validated for use *in vivo* can evaluate and quantify the subchondral bone density. First, the Hounsfield Units (HUs), based on the tissue-dependent attenuation of the X-rays, are divided into different density ranges. Then, lines are drawn to connect points of equal density and form isodensities, and these isodensity lines will provide an image of the subchondral bone's density distribution. Colour is applied for visualization. The image is a two-dimensional representation of the articulating surface that is created using a maximum intensity projection (MIP). The thickness of the SBP, which depends on the joint type and the species, is displayed in columns of voxels perpendicular to the joint surface, and the voxel with the highest HU is projected onto the surface.

Computed Tomographic Mapping of Subchondral Bone Density (CT-TOMASD), a depth-specific technique that is more sensitive to trabecular bone changes, has also been described. Because an MIP is used with CTOAM, this technique primarily displays the cortical bone, which has more overall density compared with trabecular bone. In *in vivo* experiments, the first step in the CTOAM workflow is the segmentation of the bone of interest. Next, based on the segmented images, a three-dimensional image that will be used to delineate the articular surface is reconstructed. If CTOAM is applied to isolated bones, three-dimensional images can be reconstructed directly. Once the articular surface has been segmented separately, an MIP projection is generated, and the false colour scheme added (Figure 13.8).

13.3.3 Applications of CTOAM

At the SBP level, joint loading alterations can lead to physiologic (for example, during growth or exercise) or pathologic modelling or remodelling. As a result of this structure–function relationship, the subchondral bone density distribution provides valuable information on both physiologic and pathologic joint (loading) conditions (Figure 13.9).

When evaluating subchondral bone density, the different influencing factors, including age and (orthopaedic) pathology, must be taken into account. Morphologic changes in subchondral bone density in humans are largely age-dependent. In dogs, the tarsal and elbow joints have shown a significant increase in density with increasing age. Depending on the orthopaedic pathology, the altered joint loading can be a cause or an effect. In addition, because the SBP and the joint cartilage are so closely connected, anatomically and physiologically, pathologic

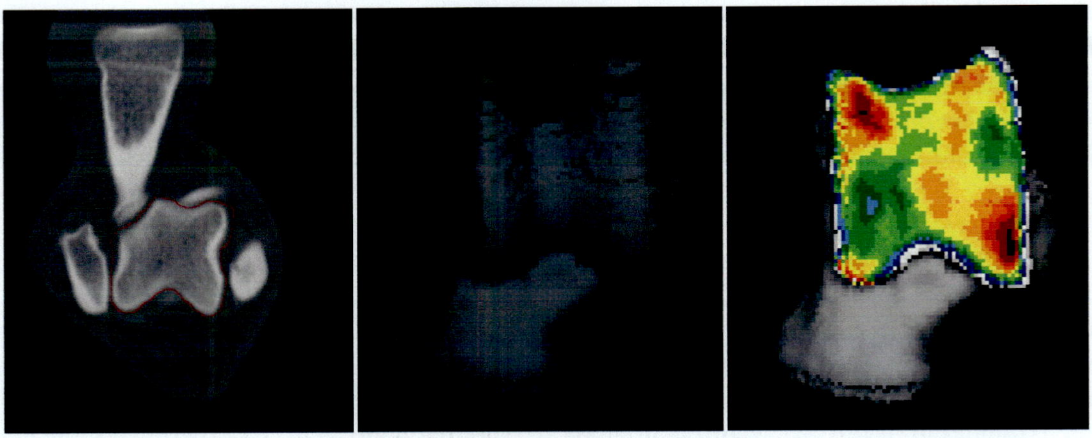

Figure 13.8 The general workflow of the CTOAM procedure. Left: Segmentation of the bone of interest (in this case the talus). Middle: Three-dimensional reconstruction. Right: Isolation of the articular surface and applying a false colour map to the MIP image.

conditions such as osteochondrosis (OC) cannot be seen as a condition of just the cartilage. Altered joint loading leads to altered bone mineral density in focal areas of the subchondral bone.

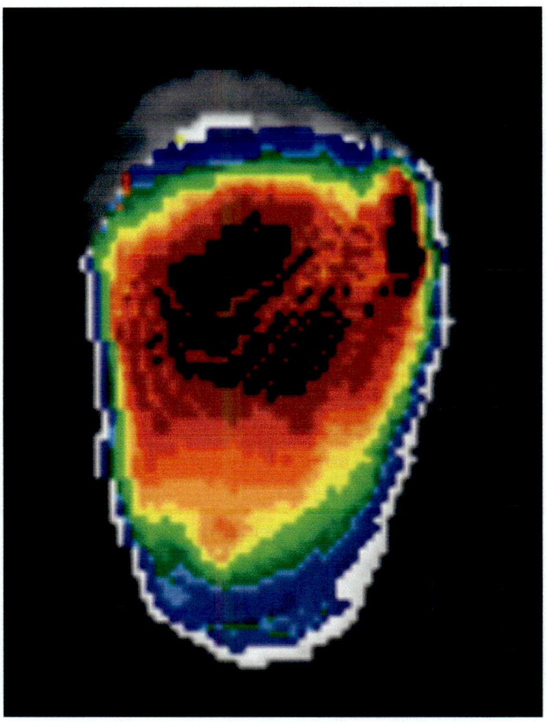

Figure 13.9 Densitogram of the right patella of a dog. The density maximum is located centrally on the articulating surface.

The overall subchondral bone density and the location of the density maxima can be evaluated using CTOAM, which, therefore, can be used to evaluate the effects of orthopaedic surgery or joint angulation and habitual posture comparisons (Figure 13.10).

In a rabbit meniscectomy model, bone mineral density was seen to decrease as early as 2 weeks after surgery, and changes in the location of the density maxima were visible 4 weeks after surgery. In human medicine, the canine anterior cruciate ligament transsection (ACLX) model is often used to study subchondral bone changes and osteoarthritis (OA) progression, with changes seen as early as 4 weeks after surgery. In dogs, surgical treatment of cranial cruciate ligament (CCL) tears is often aimed at restoring the knee's cranio-caudal stability. The goal of tibial tuberositas advancement (TTA) and tibial plateau leveling osteotomy (TPLO) procedures is to change the joint biomechanics to redistribute joint loading. In addition to the clinical results, the effects on joint loading can be evaluated using CTOAM, making it a specifically interesting research method. The same principle applies to other orthopaedic surgical approaches that aim to restore normal joint biomechanics (varus and valgus deformities, for example).

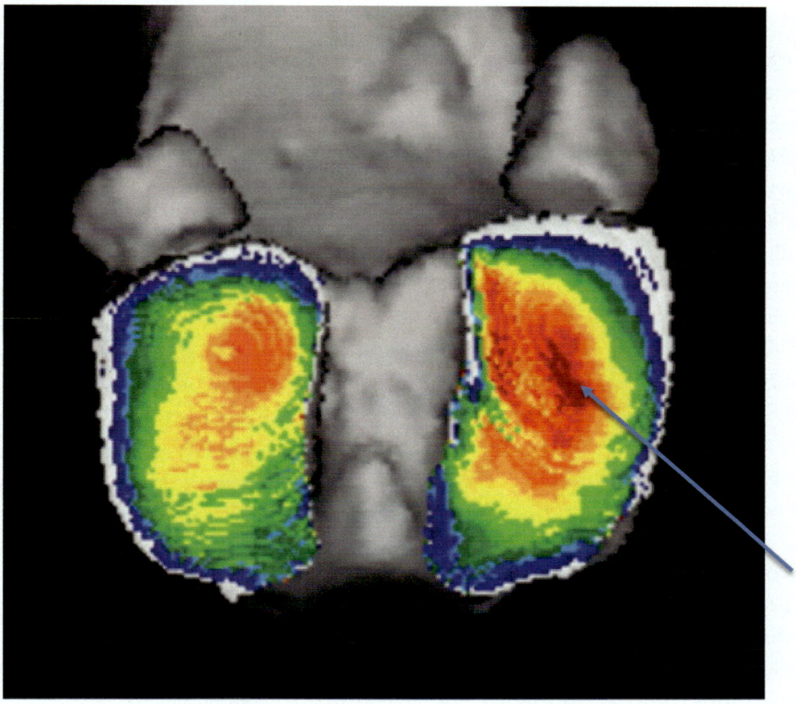

Figure 13.10 **Densitogram of the right distal femur of a dog. The lateral femoral condyle shows a higher density compared to the medial femoral condyle (arrow).**

CTOAM is a non-invasive technique and can be used not only *in vivo* but also on osteological specimens from research and museum collections without any bone destruction. Using CTOAM to evaluate habitual postures and taxonomy provides very interesting data; for example, primate locomotor behaviour (i.e., bipedal, suspensory, and quadrupedal) can be differentiated based on the subchondral bone density distribution in the radiocarpal joint. Sloths and anteaters, which display similar locomotor behaviour, have revealed similar results.

CTOAM's strength in biomechanical research is its ability to show the joint's morphologic changes, combined with other changes over time. Studies of different species and locomotor behaviours also suggest an important structure–function relationship that is very useful in functional morphology. CT examinations now are being performed in veterinary medicine more frequently, so CTOAM is a valuable technique for orthopaedic research. Pre- and post-operative evaluations of SBP density patterns can provide additional information about the biomechanical efficacy of specific surgical approaches, as well as the opportunity for further refinement.

FURTHER READING

Adams JE. 2009. Quantitative computed tomography. *European Journal of Radiology* 71(3):415–424.

Coughlin KM, Peura GD, Fleming BC, Hallock S, Beynnon BD. 2005. In vivo loads in the medial compartment of the rabbit knee. *Clinical Biomechanics* 20(9):1007–1009.

De Rycke L, van Bree H, et al. 2015. Epinephrine enhanced computed tomographic arthrography of the canine shoulder. *Research in Veterinary Science* 102:15–21.

Dickomeit MJ, Bottcher P, Hecht S, Liebich HG, Maierl J. 2011. Topographic and age-dependent distribution of subchondral bone density in the elbow joints of clinically normal dogs. *The American Journal of Veterinary Research* 72(4): 491–499.

Dingemanse W, Müller-Gerbl M, Jonkers I, Vander Sloten J, van Bree H, Gielen I. 2016. Subchondral bone density distribution of the talus in clinically normal Labrador Retrievers. *BMC Veterinary Research* 12(1):56. https://doi.org/10.1186/s12917-016-0678-8. (Q1, IF 1.643).

Dingemanse W, Müller-Gerbl M, Jonkers I, Vander Sloten J, van Bree H, Gielen I. 2017. A prospective follow up of age related changes in the subchondral bone density of the talus of healthy Labrador Retrievers. *BMC Veterinary Research* 13:57. https://doi.org/10.1186/s12917-017-0974-y.

Dingemanse W, Gielen I, van Bree H, Muller-Gerbl M, Krstić N, Mitrović M, Ćirović D, Lazarević Macanović M. 2019. Spatial subchondral bone density reflecting joint loading of the talus in different canidae. *Veterinary and Comparative Orthopaedics and Traumatology* 32(3):207–214.

Eckstein F, Jacobs CR, Merz BR. 1997. Mechanobiological adaptation of subchondral bone as a function of joint incongruity and loading. *Medical Engineerng & Physics* 19(8): 720–728.

Eckstein F, Merz B, Schon M, Jacobs CR, Putz R. 1999. Tension and bending, but not compression alone determine the functional adaptation of subchondral bone in incongruous joints. *Anatical Embryology* 199(1):85–97.

Eivers CR, Corzo-Menéndez N, Austwick SH, Thomson DG, Gibson SM, Handel I, Schwarz T. 2018. Computed tomographic arthrography is a useful adjunct to survey computed tomography and arthroscopic evaluation of the canine shoulder joint. *Veterinary Radiology and Ultrasound* 59(5):535–544.

Espinosa LA, Jamadar DA, Jacobson JA, DeMaeseneer MO, Ebrahim FS, Sabb BJ, Kretschmer MT, Biermann JS, Kim S-M. 2008. CT-guided biopsy of bone: A radiologist's perspective. *Merican Journal of Radiology* 190:W283–W289.

Gefen A, Seliktar R. 2004. Comparison of the trabecular architecture and the isostatic stress in the human calcaneus. *Medical Engineering & Physics* 26(2):119–129.

Gendler A, Keuler NS, Schaefer SL. 2014. Computed tomographic arthrography of the normal canine elbow. *Veterinary Radiology & Ultrasound* 56(2):144–152.

Gielen I, Saunders J, Van Ryssen B, van Bree H. 2010. Computed tomography of the stifle. In *Advances in the Canine Cranial Cruciate Ligament*, ed. Peter Muir, pp. 123–134. Ames, IA: Wiley-Blackwell.

Gielen I, De Rycke L, Dingemanse W, Kromhout K, van Bree H. 2015. Computed tomographic and low-field magnetic resonance arthrography: A comparison of techniques for visualizing the intra-articular structures of the normal canine shoulder. *Veterinary Surgery* 44(6):704–712.

Hofer M. 2007. *CT Teaching Manual: A Systematic Approach to CT Reading*. 3th ed, pp. 20–25. Stuttgart: Thieme Medical Publishers. Administration of contrast.

Johnston JD, Kontulainen SA, Masri BA, Wilson DR. 2010. A comparison of conventional maximum intensity projection with a new depth-specific topographic mapping technique in the CT analysis of proximal tibial subchondral bone density. *Skeletal Radiology* 39(9):867–876.

Johnston JD, Masri BA, Wilson DR. 2009. Computed tomography topographic mapping of subchondral density (CT-TOMASD) in osteoarthritic and normal knees: Methodological development and preliminary findings. *Osteoarthritis Cartilage* 17(10):1319–1326.

Laborda-Vidal P, Martín M, Orts-Porcar M, Vilalta L, Melendez-Lazo A, García de Carellán A, Ros C. 2022. Computed tomography-guided fine needle biopsies of vertebral and paravertebral lesions in small animals. *Animals* 12:1688.

Madry H, van Dijk CN, Mueller-Gerbl M. 2010. The basic science of the subchondral bone. *Knee Surgery, Sports Traumatology, Arthroscopy* 18(4):419–433.

Manhire A, Charig M, Clelland C, Gleeson F, Miller R, Moss H, Pointon K, Richardson C, Sawicka E. 2003. Guidelines for radiologically guided lung biopsy. *Thorax* 58:920–936.

Mason DR, Schulz KS, Fujita Y, Kass PH, Stover SM. 2008. Measurement of humeroradial and humeroulnar transarticular joint forces in the canine elbow joint after humeral wedge and humeral slide osteotomies. *Veterniary Surgery* 37(1):63–70.

Müller-Gerbl M, Putz R, Hodapp N, Schulte E, Wimmer B. 1990. Computed-tomography-osteoabsorptiometry: A method of assessing the mechanical condition of major joint in a living subject. *Clinical Biomechanics* 5(4):193–198.

Müller-Gerbl M, Putz R, Kenn R. 1992. Demonstration of subchondral bone density patterns by three-dimensional CT osteoabsorptiometry as a noninvasive method for in vivo assessment of individual long-term stresses in joints. *Journal of Bone and Mineral Research* (7):411–418.

Polk JD, Blumenfeld J, Ahluwalia D. 2008. Knee posture predicted from subchondral apparent density in the distal femur: An experimental validation. *The Anatomical Record* 291(3):293–302.

Reis Silva H, Uosyte R, Clements DN, Bergkvist GT, Schwarz T. 2013. Computed tomography and positive contrast computed tomographic arthrography of the canine shoulder: Normal anatomy and effects of limb position on visibility of soft tissue structures. *Veterinary Radiology & Ultrasound* 54(5):470–477.

Romans L. 2010. *Computed Tomography for Technologists: A Comprehensive Text.* Philadelphia, PA: Lippincott Williams and Wilkins. Chapter 12 Contrast agents, pp. 120–141 and Chapter 13 Injection techniques, pp. 142–164.

Samii VF, Dyce J. 2004. Computed tomographic arthrography of the normal canine stifle. *Veterinary Radiology & Ultrasound* 45:402–406.

Samii V, Dyce J, Pozzi A, Drost WT, et al. 2009. Computed tomographic arthrography of the stifle for detection of cranial and caudal cruciate ligament and meniscal tears in dogs. *Veterinary Radiology & Ultrasound* 50:144–150.

Schiavon LH, Jeng Tyng CJ, Travesso DJ, Rocha RD, Schiavon ACSA, Vieira Bitencourt AGV. 2018. Computed tomography-guided percutaneous biopsy of abdominal lesions: Indications, techniques, results, and complications. *Radiologia Brasileira* 51(3):141–146.

Shahabpour M, et al. 2008. The effectiveness of diagnostic imaging methods for the assessment of soft tissue and articular disorders of the shoulder and elbow. *European Journal of Radiology* 65:194–200.

Sungyoung H, Haengbok C, Hangmyo C, Juhyung K, Ji-Houn K, Mhan-Pyo Y, Youngwon L, Heechun L, Chang D. 2008. Evaluation of partial cranial cruciate ligament rupture with positive contrast computed tomographic arthrography in dogs. *Journal of Veterinary Science* 9(4):395–400.

Tidwell A, Johnson KL. 1994. Computed tomography-guided percutaneous biopsy in the dog and cat: Description of technique and preliminary evaluation in 14 patients. *Veterinary Radiology and Ultrasound*. https://doi.org/10.1111/j.1740-8261.1994.tb02070.x

Tivers MS, Mahoney P, Corr SA. 2008. Canine stifle positive contrast computed tomography arthrography for assessment of caudal horn meniscal injury: A cadaver study. *Veterinary Surgery* 37:269–277.

van Bree H, Van Ryssen B. 1995. Positive contrast shoulder arthrography with iopromide and diatrizoate in dogs with osteochondrosis. *Veterinary Radiology & Ultrasound* 36(3):203–206.

Vandevelde B, Van Ryssen B, Saunders J, Kramer M, van Bree H. 2006. Comparison of the ultrasonographic appearance of osteochondrosis lesions in the canine shoulder with radiography, arthrography, and arthroscopy. *Veterinary Radiology & Ultrasound* 47(2):174–184.

Vignoli M, Tamburro R, Felici A, Del Signore F, Dettori A, Di Tommaso M, Ghiraldelli A, Terragni R, Simeoni F, Falerno I, Miglio A. 2021.

Clinical value of CT-guided fine needle aspiration and tissue-core biopsy of thoracic masses in the dog and cat. *Animals (Basel)* 11(3):883.

Vignoli M, Ohlerth S, Rossi F, Pozzi L, Terragni R, Corlazzoli D, Kaser-Hotz B. 2004. Computed tomography-guided fine-needle aspiration and tissue-core biopsy of bone lesions in small animals. *Veterinary Radiology and Ultrasound* 45(2):125–130.

Zurcher K, Sugi MD, Naidu SG, Oklu R, Knuttinen G, Alzubaidi S, Eversman W, Dahiya N, Young SW, Kriegshauser JS. 2020. Multimodality imaging techniques for performing challenging core biopsies. *Radiographics* 40(3):910–911.

CHAPTER 14

IS THERE A PLACE FOR CONE BEAM CT IN ORTHOPAEDIC AND NEUROIMAGING?

Ingrid Gielen and Henri van Bree

As computed tomography (CT) utilization increases in veterinary medicine, there has been an increase in the imaging options offered on the veterinary market. Cone beam CT (CBCT) has become more and more popular in veterinary medicine because it offers some advantages over traditional CT (Figure 14.1).

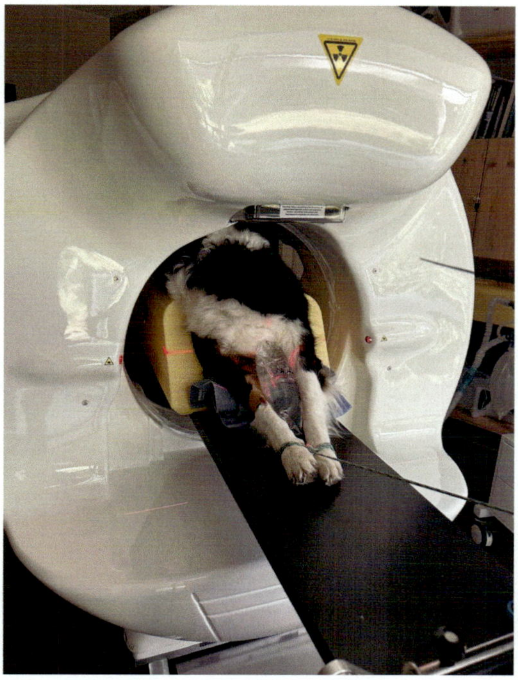

Figure 14.1 **Operational zone of a cone beam computed tomography system. The X-ray generator and a flat collecting plate are located on opposite sides of the patient. The device completes one full spin around the patient to collect the information.**

Because these systems operate in very different ways, differences in image quality and radiation emission are rather stark. Traditional CT, also known as fan beam CT, utilizes a "fan-like" X-ray beam and detectors which collect data from a small layer, "slice", of the anatomy. Multiple beam rotations are required to collect enough data for the computer to recreate the 3D structure. This results in a longer acquisition time and more radiation.

14.1 ARTEFACT

Traditional CT is less affected by motion artefacts. This is because the slice thickness is 1–2 mm. A single instance of motion will only affect that portion of the scan, which could be only a few slices pertaining to the multi-slice study, while other kinds of exposures remain unaffected. Motion artefacts are more likely to occur in cone beam CT. Hereby, a single motion artefact can degrade the image quality, making the study diagnostically unusable, thus requiring a second examination.

14.2 DIAGNOSTIC QUALITY

Cone beam CT emits a cone-shaped X-ray beam onto a more broad detector. This takes less time and emits less radiation because the X-rays are emitted via a broad "cone", producing more "scattered" radiation, which is also captured by the broad detector. The radiation scatter

DOI: 10.1201/9781315118130-15

deviating from the beam path reduces image contrast, ultimately reducing the image quality and making CBCT more prone to artefacts.

14.3 WHAT IS CONE BEAM COMPUTED TOMOGRAPHY?

Cone beam computed tomography (or CBCT, also referred to as C-arm CT, cone beam volume CT, flat panel CT or digital volume tomography) is a type of 3-D imaging that uses radiation, similar to traditional CT. However, with traditional CT, the X-ray beam is fan-shaped, and linear detectors are used, and with CBCT, the cone-shaped beam falls on a flat panel detector. In CBCT, the X-ray tube and the detector synchronously rotate 360° around the patient. CBCT uses the cone-shaped beam to encompass the area of interest, and the X-ray source makes only one pass around the patient to acquire all the data needed for the image. Because these systems operate in very different ways, differences in the image quality and radiation emission are rather stark. Traditional CT, also known as fan beam CT, utilizes a "fan-like" X-ray beam and detectors which collect data from a single "slice" of the anatomy. This requires multiple beam rotations to collect enough data for the computer to recreate the 3D structure. Because traditional CT requires more rotations, it also emits more radiation and takes more time to complete the scan, although now, CT devices have several rows of detectors, resulting in a little difference in scanning duration between CBCT and CT (Figure 14.2).

In man CBCT scans are used in dentistry. The high-resolution and detailed images are well-suited for dental applications, allowing

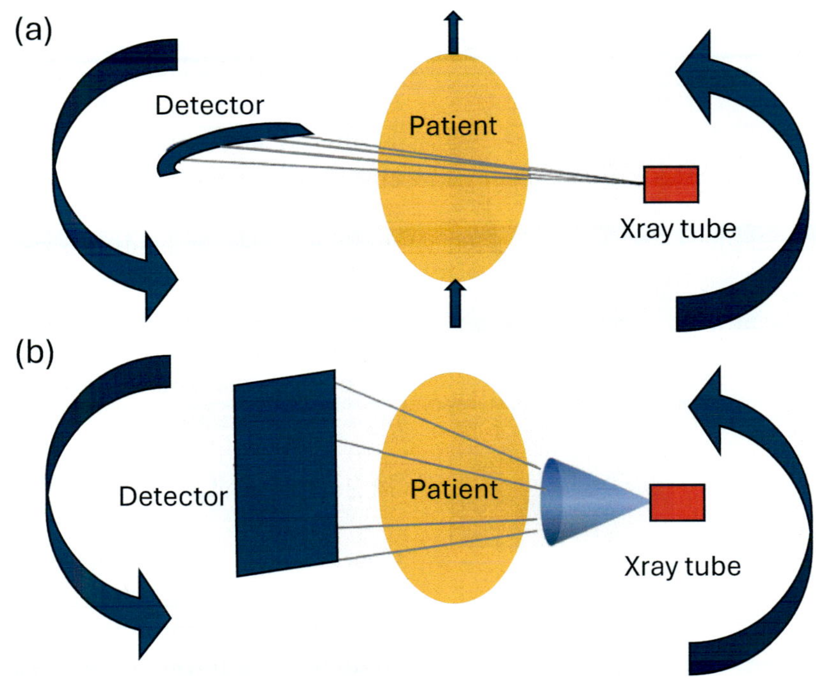

Figure 14.2 **In the functioning of a conventional CT machine, the emitted fan-shaped X-ray beam is captured by detector elements positioned in an arc that rotates in an opposite direction of the X-ray source around the patient positioned on the movable patient table (a). In contrast, CBCT utilizes a cone-shaped beam for image capturing, which is detected by a flat-panel detector (b).**

dentists to examine oral and maxillofacial structures better. Its use has also been described in radiation therapy, angiography, and intra-operative imaging procedures. Recent advances in system design, artefact correction, and 3D image reconstruction methods open the possibility for improved low-contrast resolution suitable for brain neuroimaging.

CBCT has found widespread applications in veterinary medicine, with its advantages extending beyond general diagnostics to specialized fields such as dentistry. Furthermore, reports exist on its use in large animal head problems.

14.4 ADVANTAGES AND DISADVANTAGES OF CBCT

14.4.1 Advantages

- *Lower radiation dose* (a factor which may be insignificant in veterinary scanning). The lower radiation dose of CBCT results from the cone-shaped beam and the need for only one pass around the patient, thereby exposing the patient to a substantially lower dose of radiation, approximately 10% less than conventional or spiral CT of the same area.
- *High spatial resolution images of bone architecture*: CBCT systems can also be used in orthopaedics, e.g., to display smallest joint structures. CBCT is also suitable for head- and even whole body examinations of exotics or small pets like rabbits, guinea pigs, or even of the size of a small rodent. In addition, the resolution of CBCT images tends to be higher than in traditional CT images. CBCT slice thickness is smaller, allowing detailed imaging of small structures. Most traditional CT scanners obtain images in the range of 1–3 mm slice thickness, and the CBCT slice thickness is 0.1–0.4 mm.
- *High spatial resolution images of cartilage surface* after intra-articular contrast injection: CBCT offers high anatomical detail of articular cartilage, which may be an advantage to routine MRI in evaluation of osteochondral lesions in joints with thin cartilage, such as the tarsal joint.
- Less artefacts with metallic objects
- *Relative low cost of equipment*: its low cost in purchase, maintenance, and energy costs. Traditional CT units must be housed in a dedicated lead-lined room for radiation safety.

14.4.2 Disadvantages

- *Limited soft tissue contrast resolution* (poor soft tissue resolution). The downside of CBCT imaging is its decreased ability to differentiate soft tissue structures and decreased contrast enhancement, which results from the design of the system. Traditional CT systems have a collimated beam producing little scattered radiation. The X-rays from CBCT are not collimated. The X-rays are forming a divergent, uncollimated cone, so more scattered radiation is produced, leading to a reduced soft tissue contrast resolution, especially in the thorax and abdominal region (Figure 14.3). The larger the body part images, the greater the scatter. In a large dog's abdomen or thorax, the contrast is lower than for a small pet or limb. However, some CBCT manufacturers are refining the soft tissue imaging quality; therefore, future improvements in this area are likely.
- *Scatter & higher noise*: In an X-ray system with a large area detector, such as CBCT, the scatter-to-primary ratio (SPR) can be as high as 8 in certain areas of projection images. The traditional method to deal with noise problems is to increase the dose. The dose needs to be increased by a factor of more than 10 in order to make the noise level of the scatter-corrected

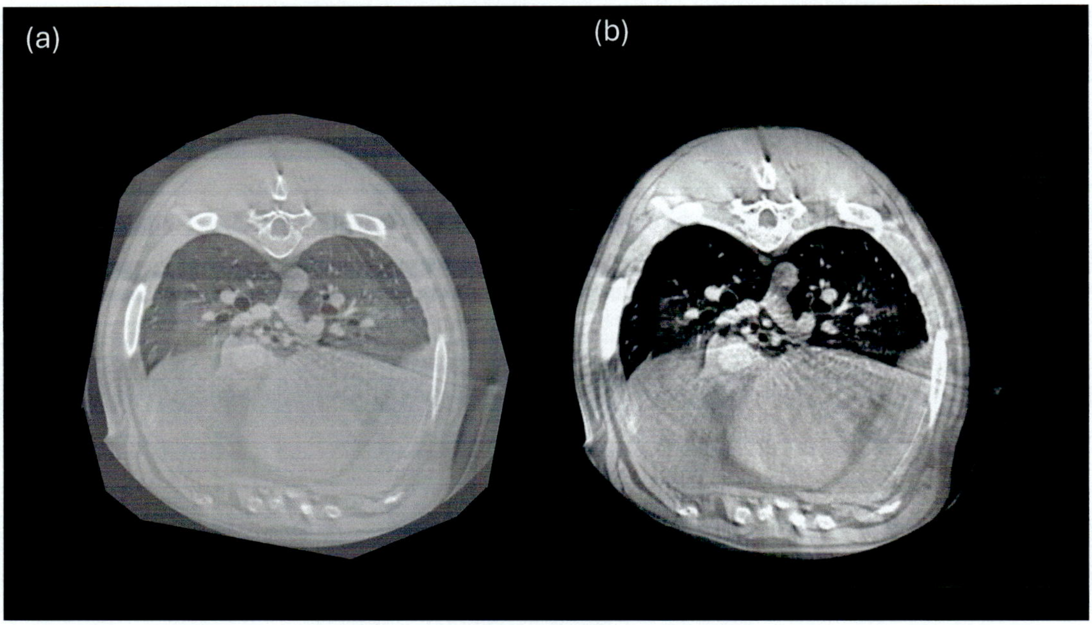

Figure 14.3 **Transverse CBCT images in bone window (a) and soft tissue window (b) setting at the level of the thorax and abdomen showing motion artefacts, blurring, and long-range streaks.**

image comparable to that without scatter correction. This excessive dose increase is hardly acceptable in clinical practice. Other techniques are investigated.

- *Prone to streaking and motion artefacts*: In CBCT, the images are acquired in only one pass around the patient: even a single motion artefact can reduce the overall image quality, leading to investigating again (Figure 14.3). Traditional CT produces numerous image slices, each of which requires a separate scan. Therefore, only the motion slices show artefacts, while the other slice exposures are not affected. Traditional CT offers the required high contrast and less susceptibility for motion artefacts, e.g., due to breathing.
- *Limited scan range and field of view*: CBCT is not suitable for imaging large-body areas or large joints. It is often impossible to image both sides of joints simultaneously within the same scan for comparison. As lesions are often bilateral, a second scan may be needed, which is more time-consuming than traditional CT, where both sides can be simultaneously scanned.
- *Hounsfield units (HU)*: It's essential to know that the grey values in CBCT do not directly correspond to HUs, which are utilized in traditional CT scans for measuring the tissue density. In CBCT, the accuracy of the provided HU is diminished.
- *Contrast administration?*: Although it has been claimed that the contrast dose required for CBCT is twice that of spiral/helical/ traditional CT, still research is going on to determine the optimal dose requirements.

A recently performed pilot study in dogs suffering from elbow pathology pointed out that although in many studies the image quality was moderate, the studies were considered diagnostic in the majority of the cases (Figure 14.4). In most of the cases, this was related to motion artefacts and those having a streak-like appearance (Figure 14.5). The overall conclusion of this study was that CBCT can be applied as a diagnostic technique in elbow disease, although

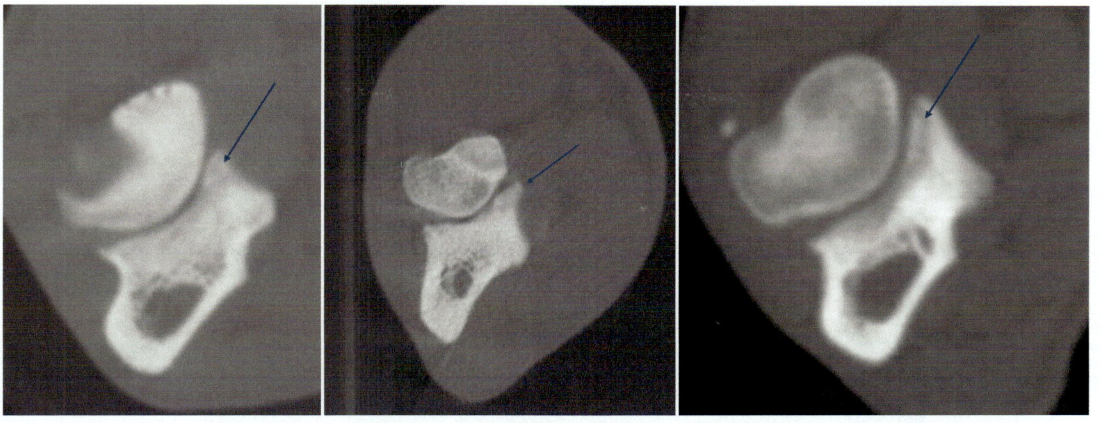

Figure 14.4 Good-quality transverse images of elbow dysplasia, obtained from different CBCT devices of different dogs. A nondisplaced fragment at the medial coronoid process is observed (blue arrows).

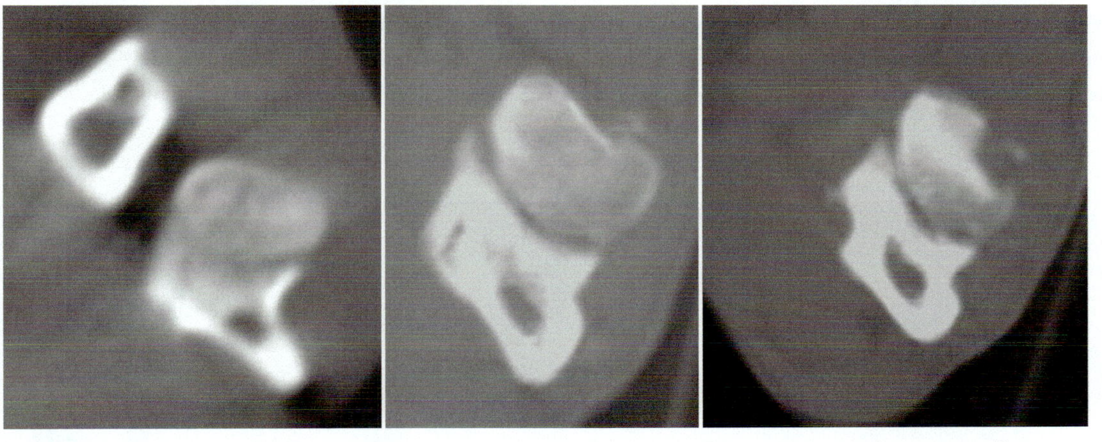

Figure 14.5 Transverse elbow CT images of poor to limited quality, which are diagnostically unusable.

subtle lesions may be overlooked. Due to poor soft tissue resolution, the evaluation of muscular and tendinous pathology like flexor tendinopathy may be challenging (Figure 14.6). Traditional CT still is considered the imaging modality of choice for elbow screening and more particular in appeal cases where a definitive decision needs to be made.

In neuroimaging, there are many limitations. Beam hardening artefacts especially on reconstructed images hinder the interpretation of discrete disc pathology (Figure 14.7).

Calcified protrusions and extrusions can be easily seen, but recognition of acute disc herniation may be challenging (Figure 14.8).

CBCT is able to identify large space occupying brain lesions, but more subtle lesions can easily be overlooked (Figure 14.9).

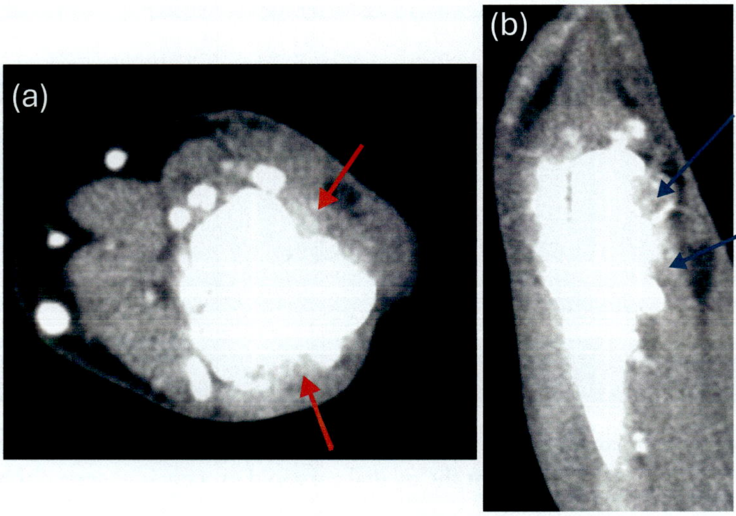

Figure 14.6 Transverse (a) and dorsal reconstruction (b) of an elbow after administering IV contrast. There is inferior contrast resolution present. General diffuse contrast enhancement of the joint capsule (red arrows) and of the attachment of the flexor tendons (blue arrows) is noticed.

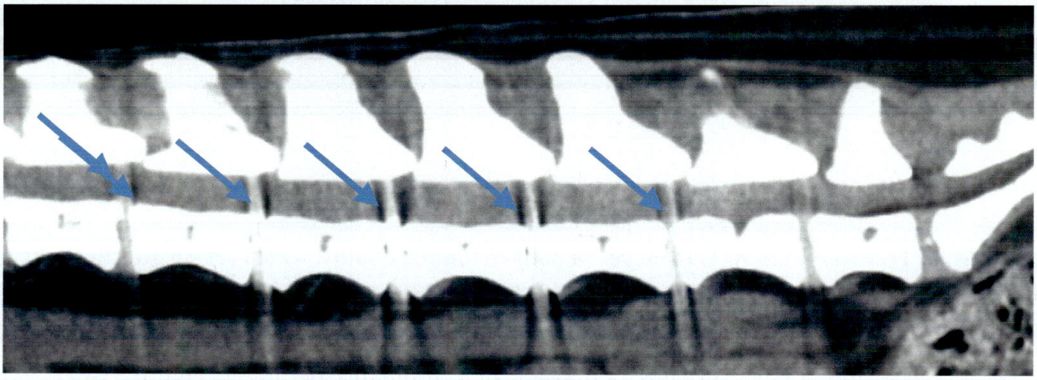

Figure 14.7 Sagittal reconstructed image of a spine. Beam hardening artefacts are present as streaks (blue arrows) caused by dense bones, hindering the interpretation of discrete disc pathology.

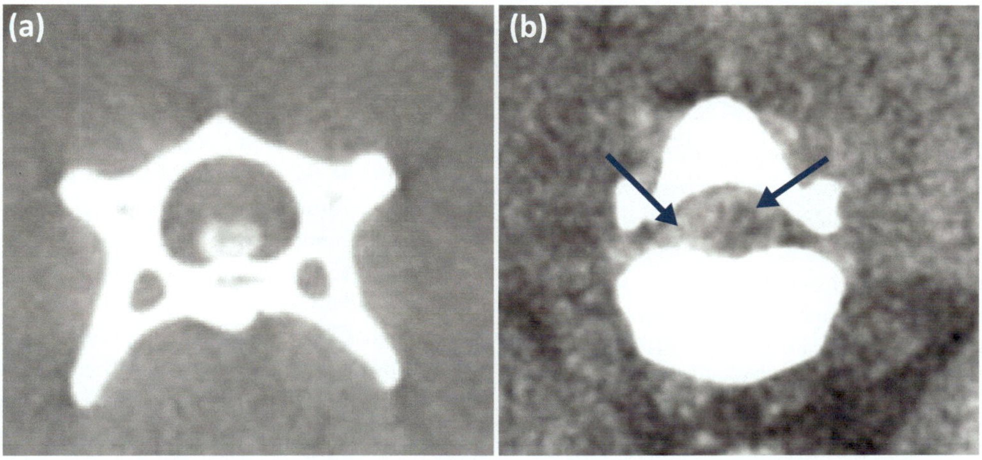

Figure 14.8 Transverse slices at C3 (a) and at Th12–Th13 (b) showing disc pathology. The heavy mineralized disc extrusion at C3 is readily seen. Due to the poor soft tissue resolution, the disc extrusion with extensive haemorrhage at Th12–Th13 is not clearly visible (blue arrows).

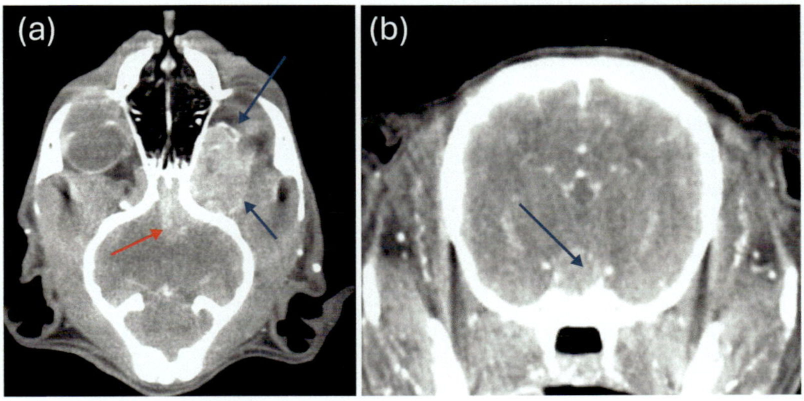

Figure 14.9 Dorsal reconstructed image of the brain of a cat (a) and transverse image of the brain of a dog (b) after IV contrast administration. Both images show a space occupying mildly contrast-enhancing mass. The image of the cat presents a retrobulbar mass (blue arrow). The mass extends through the optic canal contiguous with an extra-axial mass, suspicious of a meningioma (red arrow). The one of the dog is showing a macroadenoma of the hypophysis (arrow).

ADDITIONAL READING

J. Bailey, M. Solan, E. Moore. Cone-beam computed tomography in orthopaedics. *Orthopaedics and Trauma* (2022), 36:4.

K. Bannon. Cone beam computed tomography in veterinary dentistry. *Journal of Veterinary Dentistry & Oral Surgery* (2022), 2:70–77.

M. D. Bregger, C. Koch, R. Zimmermann, D. Sangiorgio, D. Schweizer-Gorgas. Cone-beam computed tomography of the head in standing equids. *BMC Veterinary Research* (2019), 15:289.

J. Carrino, A. Al Muhit, W. Zbijewski, G. Thawait, J. Stayman, N. Packard, R. Senn, D. Yang, D. Foos, J. Yorkston, J. Siewerdsen. Dedicated cone-beam CT system for extremity lmaging. *Radiology* (2014), 270:3.

M. Doan, J. Long, E. Verhey, A. Wyse, K. Patel, J. Flug. Cone-beam CT of the extremities in clinical practice. *Home RadioGraphics* (2024), 44:3.

E. Duckworth, G. Chintalapani, K. Camstra, P. Kan. Optimizing contrast-enhanced cone-beam CT protocol to facilitate simultaneous visualization of neurovascular pathologies and surrounding structures of interest. *Interventional Neuroradiology* (2019), 25(1):102–110.

S. Kumar, K. Sidappa. Cone beam computed tomography – know its secrets. *Journal of International Oral Health* (2015), 7(2):64–68.

R. Mackie, J. Hayes. Fan beam CT versus cone beam CT for veterinary medicine. Asto CT (2021): 1–5.

M. Posadzy, J. Desimpel, F. Vanhoenacker. Cone beam CT of the musculoskeletal system: clinical applications. *Insights Imaging* (2018), 9:35–45.

U. Pyakurel, R. Sabounchi, M. Eldib, F. Bayat, H. Phan, C. Altunbas. Evaluation of a compact cone beam CT concept with high image fidelity for point-of-care brain imaging. *Scientific Reports* (2024), 14:28286.

G. E. Rupp, C. L. Farnsworth, J. L. Hughes, V. V. Upasani, A. T. Pennock, J. D. Bomar, E. W. Edmonds. A comparison of cone beam computed tomography, standard computed tomography and plain radiographs in the evaluation of medial epicondyle humerus fractures. *Journal of Pediatric Orthopaedics B* (2024), 33(3):308.

R. Schulze, U. Heil, D. Grob, D. Bruellmann, E. Dranischnikow, U. Schwanecke, E. Schoemer. Artefacts in CBCT: a review. *Dentomaxillofacial Radiology* (2011) 40:265–273.

B. Van Thielen, F. Siguenza, B. Hassan. Cone beam computed tomography in veterinary dentistry. *Journal of Veterinary Dentistry* (2021), 29:1.

L. Zhu, J. Wang, L. Xing. Noise suppression in scatter correction for cone-beam CT. *Medical Physics* (2009), 36:3.

PART II

CT Procedures

CHAPTER 15

CT OF THE SHOULDER

Ingrid Gielen, Annemie Van Caelenberg, and Henri van Bree

15.1 INTRODUCTION

The shoulder joint is the most mobile of all main limb joints. It is a ball-and-socket joint, which allows movement in essentially every direction. Although most of its movement takes place in a direct cranial-to-caudal plane, resulting in mainly extension and flexion, it also undergoes significant abduction and adduction and internal and external rotation. Its stability is ensured by the joint capsule and by the medial and lateral gleno-humeral ligaments, which are in fact reinforcements of the capsule. Also, large tendons located inside (e.g., tendon of origin of the biceps brachii muscle) or immediately outside the joint (e.g., supraspinatus, infraspinatus, teres minor, and subscapularis) ensure its stability. Shoulder lameness, although a common condition in the dog, can be difficult to localise. Even when the lameness has been associated with a shoulder problem, identifying the real source of the problem can be unrewarding and frustrating. Frequently, a poor correlation exists between the radiographic findings and clinical signs. There may be extensive alterations, such as degenerative joint disease, and yet no lameness present.

15.2 INDICATIONS FOR USING CT

Radiography is still the first-line imaging modality being used when evaluating small animals with shoulder lameness. Besides plain radiography, arthrography, ultrasound (US), and arthroscopy (AS) are very useful imaging modalities in diagnosing shoulder problems. Scintigraphy localises lameness as being a shoulder problem and evaluates the significance of equivocal radiographic findings. In some conditions, the use of CT can be justified, especially to avoid superimposition of surrounding (bony) structures, to evaluate the extent of processes, or to determine the source of calcifications or fragments, seen on plain radiographs. New bone formation in the bicipital groove can be easily seen. For complex fractures in the shoulder area, CT can be indicated for an exact evaluation of the nature of the fracture and for better treatment planning. Also, to evaluate decalcification, CT can be justified as it will detect loss of bone mineral content in an earlier phase than radiology. In order to differentiate a shoulder problem from a plexus brachialis problem, intravenous (IV) CT contrast can be indicated. To evaluate intra-articular structures, ligaments, articular cartilage, and biceps tendon, intra-articular contrast (arthro-CT) can be indicated.

15.3 TECHNICAL ASPECTS

15.3.1 Positioning

The anaesthetised patient is positioned in dorsal or ventral recumbency, with the front legs extended cranially and positioned as symmetrical as possible in order to be able to compare both sides (Figure 15.1). The scans are performed from the acromion till the proximal third of the humerus and are performed parallel to the joint space. In case of a necessity to examine the scapulum, the extension of the scanning region needs to be enlarged.

DOI: 10.1201/9781315118130-17

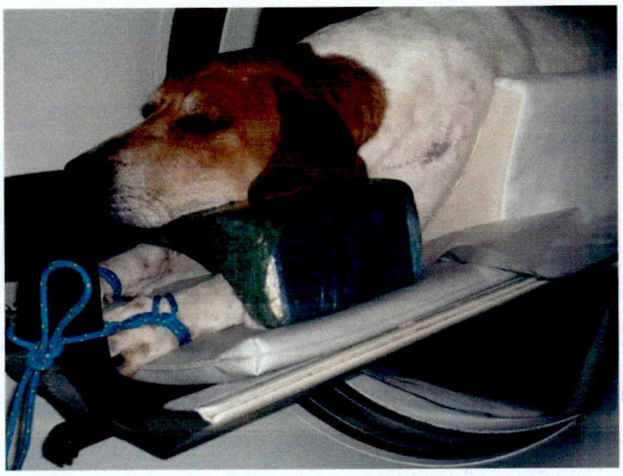

Figure 15.1 The anaesthetised patient is positioned in ventral recumbency with the front legs extended cranially and positioned as symmetrical as possible in order to be able to compare both sides.

15.3.2 Technical parameters

A spiral acquisition with slice thickness of 1–2 mm with 50% overlap and pitch of 1 with reconstruction algorithm in bone and soft tissue is appropriate. Both shoulder joints are included within the field of view. Settings for the CT imaging technique are as follows: 120 kV and 120–140 mAs. Individual images are reviewed using a bone setting (WW = 3,000 HU; WL = 500 HU) and a soft tissue setting (WW = 350 HU; WL = 90 HU).

15.3.3 Contrast

CT arthrography provides additional information to survey radiographs, such as the extent of the cartilage flap and delineation of the cartilage and biceps tendon, and small subchondral bone cysts can be evaluated. This contrast CT technique can be performed after an intra-articular injection of 1–4 mL of a 50–80 mg I/mL solution of a preferably non-ionic contrast medium. In case of soft tissue involvement and to detect intratendinous abnormalities of the shoulder region, IV contrast can be applied with a dose of 200 mg I/kg bodyweight.

15.4 CT ANATOMY AND NORMAL VARIANTS

In dogs and cats, however, most of the movement of this ball-and-socket joint takes place in a direct cranial-to-caudal plane, resulting in mainly extension and flexion of the joint. Because of the thickness of the musculature surrounding the shoulder joint, (sub)luxations are relatively rare in small animals.

Bones contributing to the formation of the shoulder joint are proximally the scapula (by the glenoid cavity) and distally the humerus (by its head). The main difference between dogs and cats is the presence of the clavicula in the cat, although a small remnant may be present in dogs (Figure 15.2). There is also a difference in the trabecular bone structure between these species and of the shape of mainly the scapula.

The joint capsule consists of a single cavity that extends between the scapula and humerus. A subdivision of this capsule surrounds the

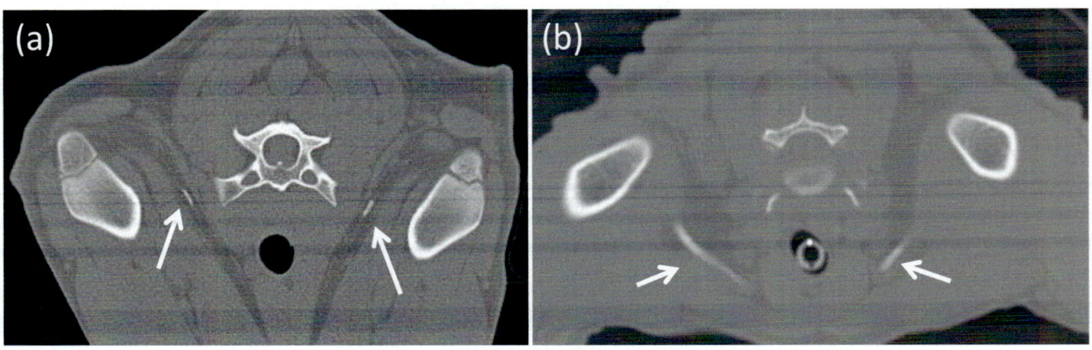

Figure 15.2 Transverse CT slices at the proximal part of the humeri in a dog (a) and a cat (b). Note the clavicula in the cat and remnants in a young dog (arrows).

proximal tendon of the biceps and follows it in the intertubercular groove as the tendon's synovial sheath. The bicipital tendon and its sheath are retained within the intertubercular groove by the transverse humeral ligament, which is a thick band of ligamentous tissue bridging the groove from the tuberculum major to the tuberculum minor (Figures 15.3–15.5).

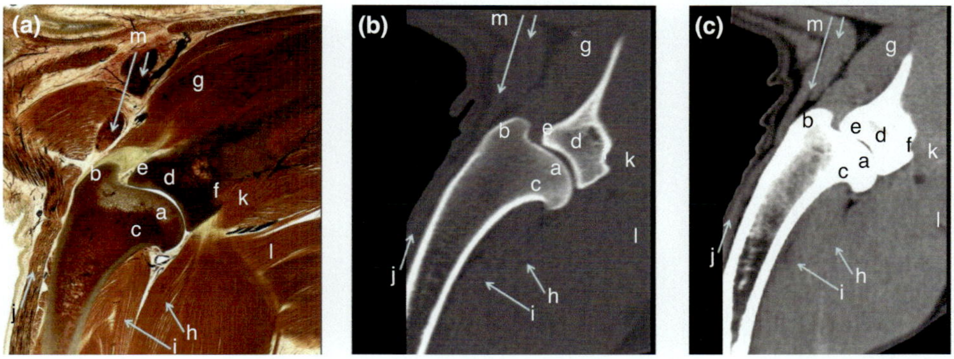

Figure 15.3 Normal anatomy: Sagittal plane of a normal shoulder joint; (a) anatomic specimen; (b) bone window; (c) soft tissue window. a. head of the humerus; b. greater tubercle; c. neck of the humerus; d. glenoid cavity of the scapula; e. supraglenoid tubercle; f. infraglenoid tubercle; g. supraspinatus muscle; h. long head of the triceps brachii muscle; i. medial head of the triceps brachii muscle; j. cleidobrachialis muscle; k. teres major muscle; l. latissimus dorsi muscle; m. superficial cervical lymph nodes.

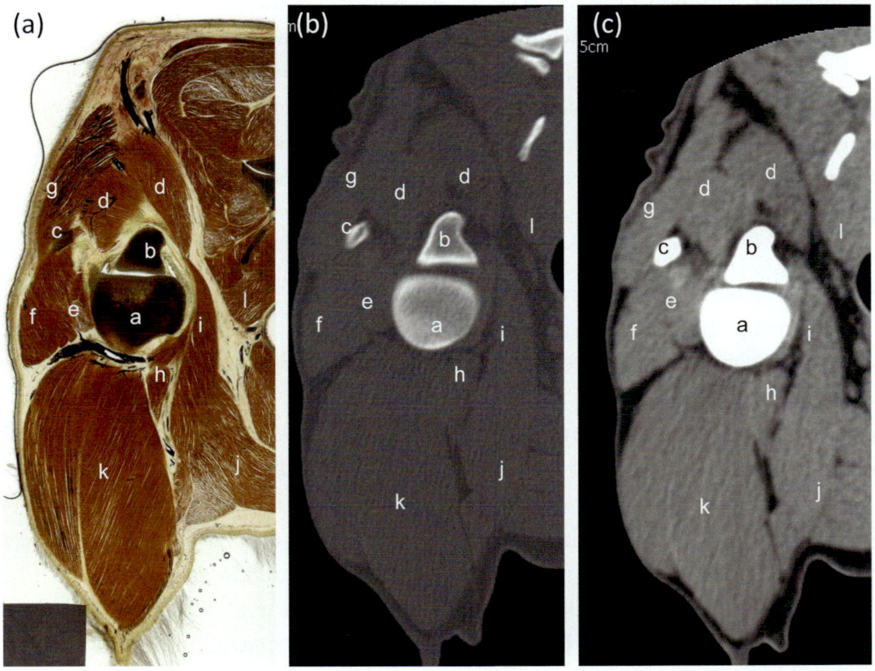

Figure 15.4 Normal anatomy: Transverse slices of a normal shoulder at the level of the supraglenoid tubercle: (a) Anatomic section. (b) Bone window. (c) Soft tissue window. a. head of the humerus; b. supraglenoid tubercle; c. acromion; d. supraspinatus muscle; e. infraspinatus muscle; f. deltoideus muscle; g. omotransversarius muscle; h. coracobrachialis muscle; i. deep pectoralis muscle; j. superficial pectoralis muscle; k. triceps brachii muscle; l. longus capitis muscle.

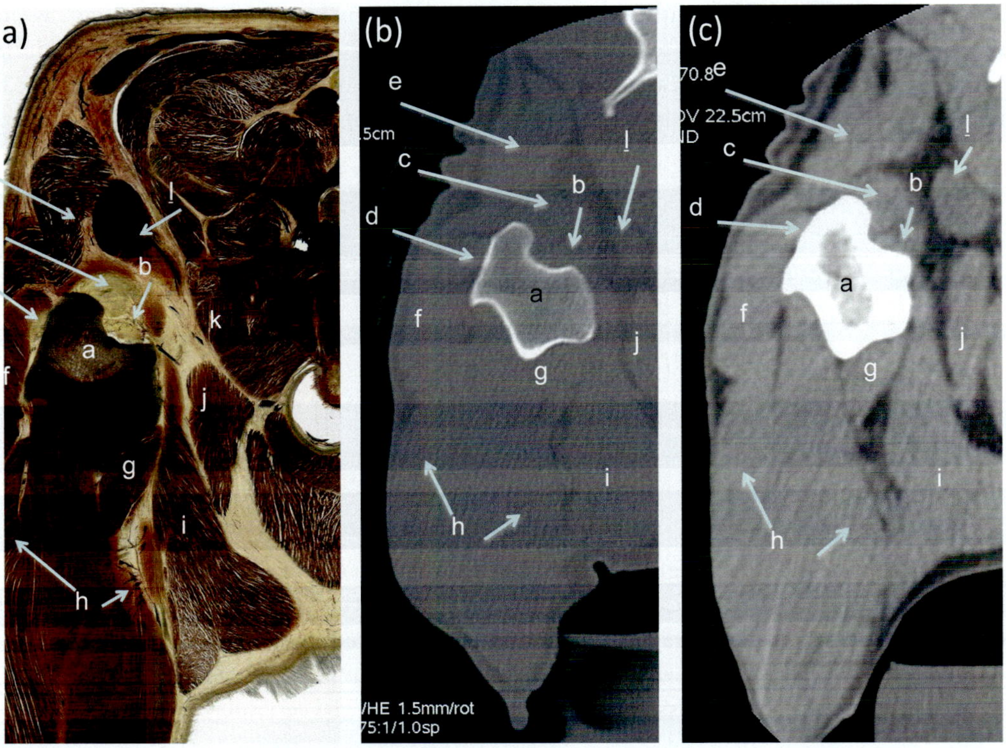

Figure 15.5 Normal anatomy: Transverse slices of a normal shoulder at the level of the bicipital groove: (a) Anatomic section. (b) Bone window. (c) Soft tissue window. a. humerus; b. biceps brachii muscle tendon; c. supraspinatus muscle, tendon of attachment; d. infraspinatus muscle, tendon of attachment; e. omotransversarius muscle; f. deltoideus muscle; g. medial head of triceps brachii muscle; h. long and lateral head of triceps brachii muscle; i. pectoral muscles; j. sternocephalicus muscle; k. longus capitis muscle; l. superficial cervical lymph nodes.

The medial and lateral glenohumeral ligaments are thickenings of the respective surfaces of the joint capsule and are no real ligaments. They are rather weak; it is the strength of the musculature surrounding the shoulder joint that maintains its integrity. The glenohumeral ligaments are visible on arthro-CT as filling defects. The delineation of the ligaments can be visualised and evaluated on artho-CT images (Figure 15.6). The shoulder joint muscles responsible for movement in the shoulder attach to the scapula and the humerus.

15.5 PATHOLOGIC FINDINGS

15.5.1 Developmental/congenital

15.5.1.1 Osteochondritis dissecans (OCD) shoulder

Transverse images in the area of the caudal humeral head demonstrate multiple radiolucencies surrounded by sclerotic rims. These features are typical for all OCD lesions in all types of joints. Reformatted images display a radiolucent subchondral defect of variable size involving the caudal area of the humeral head (Figure 15.7). Using CT arthrography, the condition of the

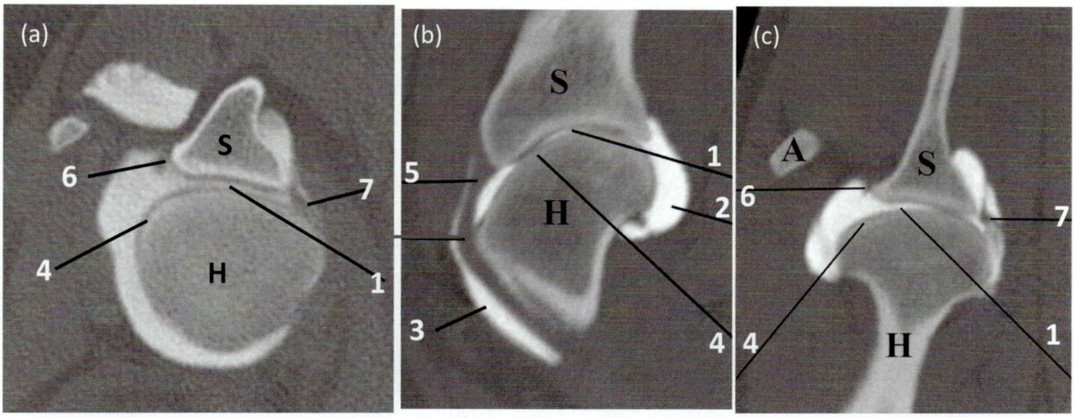

Figure 15.6 Arthro-CT of a normal shoulder joint: (a) Transverse slice; (b) sagittal reformatted image; (c) dorsal reformatted image; S: scapula; H: humerus; A: acromion; 1 = joint space; 2 = caudal pouch; 3 = biceps tendon sheath; 4 = joint cartilage; 5 = biceps tendon/silhouette; 6 = lateral glenohumeral ligament; 7 = medial glenohumeral ligament.

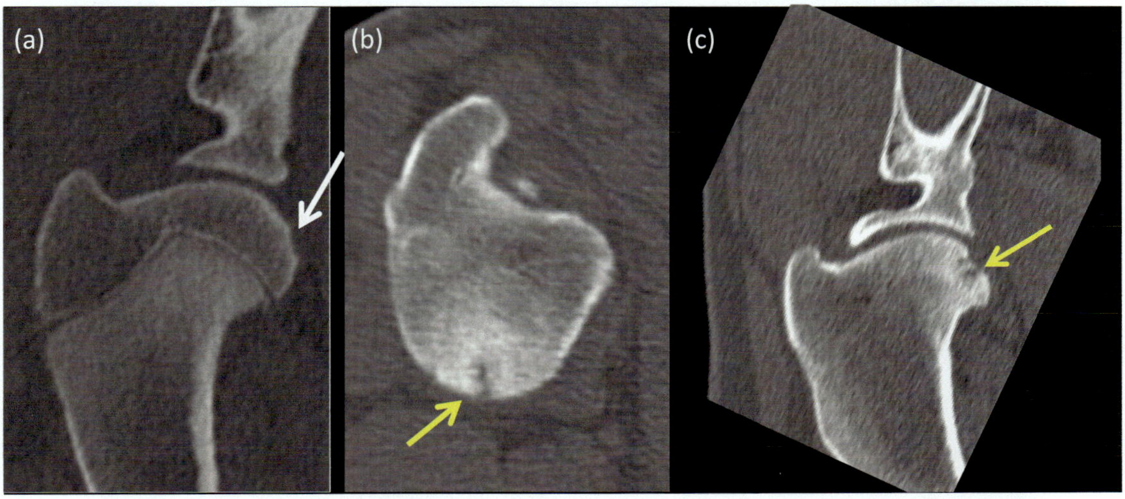

Figure 15.7 Sagittal reconstructed (a) image of a 9-month-old dog, transverse (b) and sagittal reconstructed (c) images of a 13-month-old dog, all in bone algorithm. A small flattened area at the caudal aspect of the humeral head (white arrow) is noticed under which the bone density did not increase. These findings, confirmed by arthroscopy, corresponded with osteochondrosis. The image of the older dog shows rounded radiolucencies surrounded by sclerosis, typical for osteochondrosis lesions, with irregular delineation of the caudal humeral head (yellow arrows). This lesion was confirmed as osteochondritis dissecans on arthroscopy.

joint cartilage that covers the subchondral defect can be assessed and classified to determine the possible treatment modality. With a high accuracy rate, radiographic arthrography can expose humeral cartilage lesions and detect the presence of a nonmineralised cartilage flap (Figure 15.8a), which aids in deciding whether a dog should undergo conservative or surgical treatment. In osteochondrosis, thick cartilage covering the subchondral defect is detected by the overlying contrast medium (Figure 15.8b). Calcified joint mice are seen on survey images (Figure 15.9). Non-calcified flaps may be visible

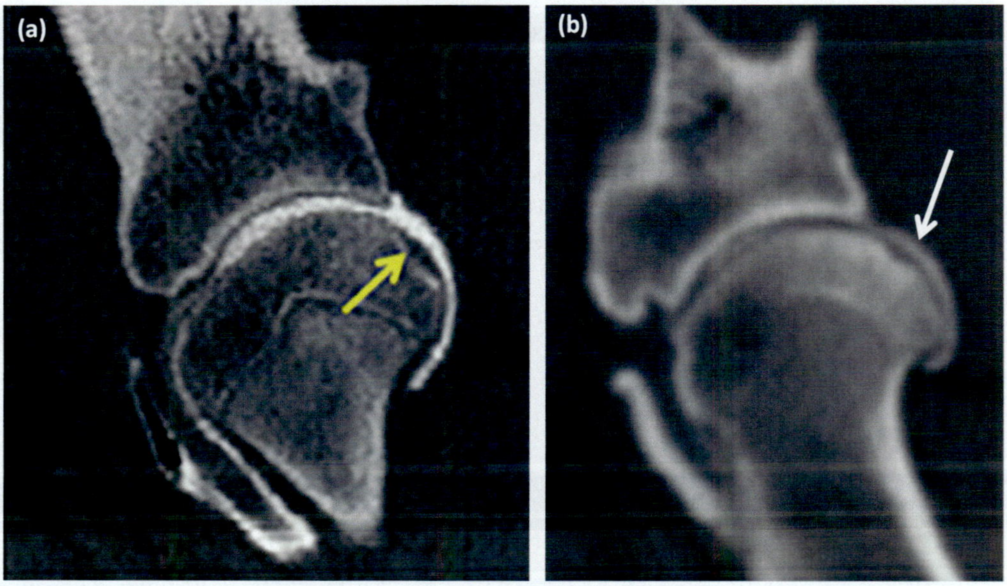

Figure 15.8 Sagittal reconstructed arthro-CT images of an OCD lesion (a) and an OC lesion (b). Thickened cartilage with a white line underneath is typical for a partially detached cartilage flap (yellow arrow). The present thickened cartilage with no contrast medium underneath, meaning the cartilage is thickened but not detached (white arrow).

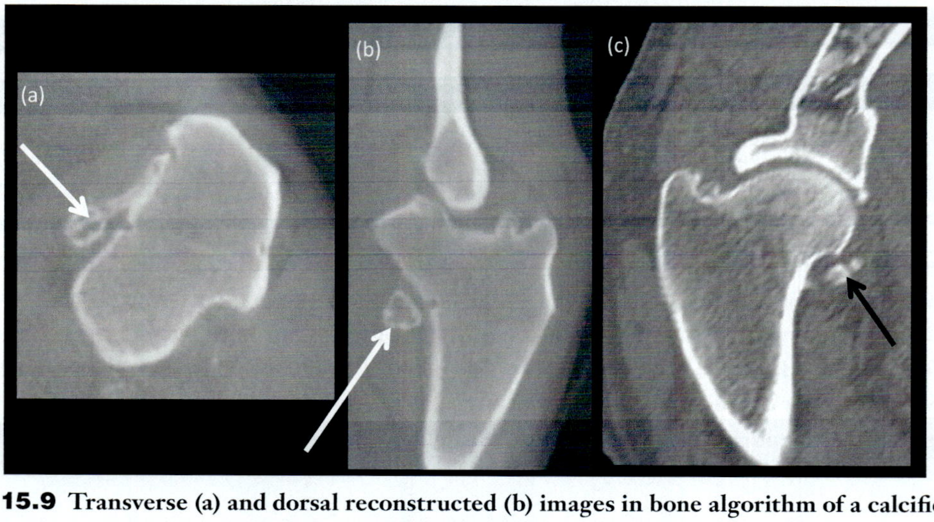

Figure 15.9 Transverse (a) and dorsal reconstructed (b) images in bone algorithm of a calcified joint mouse present within the biceps tendon sheath (white arrows). There is prominent new bone formation observed in the bicipital groove. Sagittal reconstructed image; (c) shows joint mice present within the caudal pouch (black arrow).

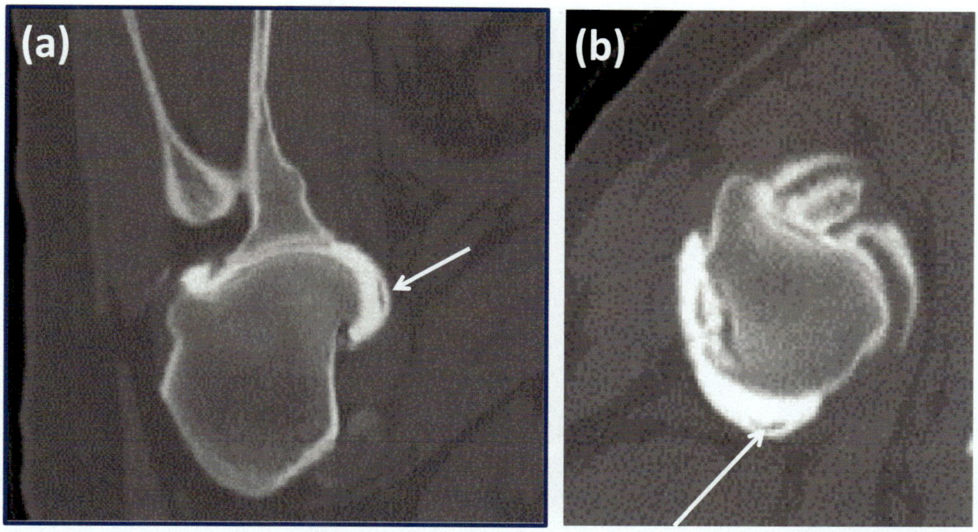

Figure 15.10 Sagittal reconstructed (a) and transverse (b) arthro-CT images in bone algorithm of the non-calcified flap seen as a filling defect within the contrast medium (arrows).

as a filling defect at the caudal pouch or at the biceps tendon sheath (Figure 15.10).

Osteochondrosis of the glenoid cavity is presented as an osseous cyst-like lesion of the glenoid cavity (Figure 15.11).

Thickened cartilage with a white line underneath is typical for a partially detached cartilage flap (yellow arrow). The present thickened cartilage has no contrast medium underneath, meaning the cartilage is thickened but not detached (white arrow).

15.5.1.2 Congenital shoulder luxation

Both congenital and acquired shoulder luxation or dislocation may occur (Figure 15.12). Dogs with congenital shoulder luxation frequently have glenoid and humeral head deformities. Traumatic luxation is rare in both cats and dogs, but when it does happen, medial luxation is more common. Large breed dogs, on the other hand, frequently exhibit lateral shoulder luxation after a trauma.

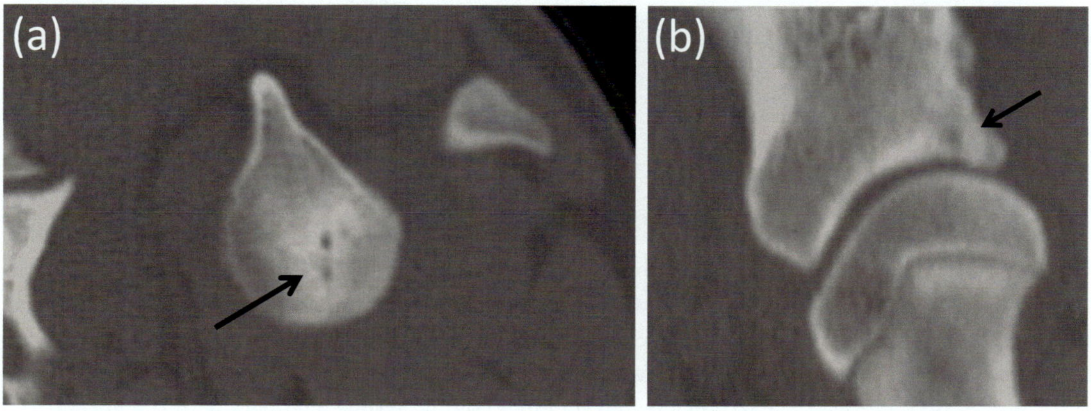

Figure 15.11 Transverse image through the caudal glenoid (a) and the corresponding sagittal reconstruction in bone algorithm (b), showing rounded lucencies surrounded by sclerosis, typical for an OC/OCD lesion, here of the glenoid (black arrows).

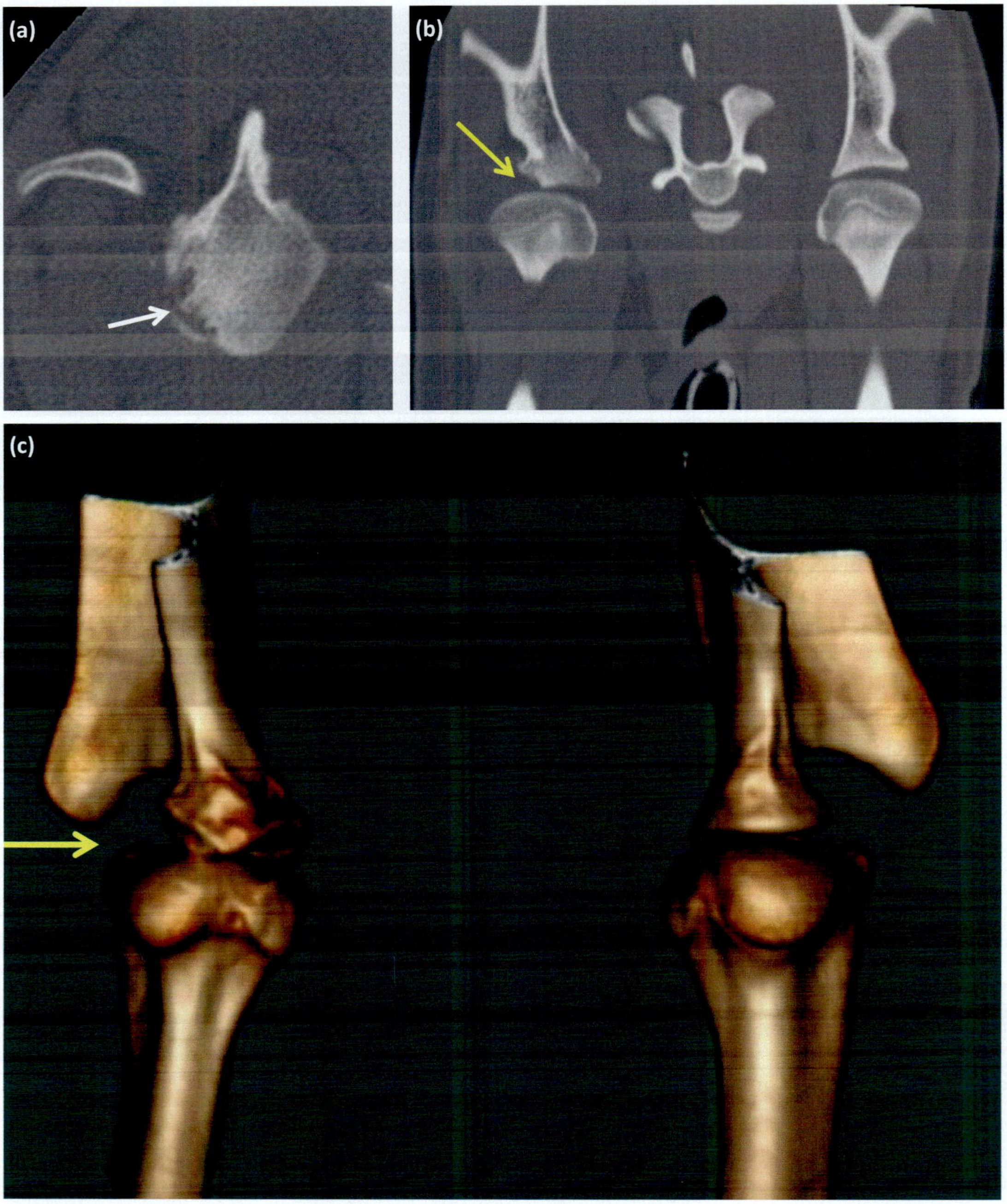

Figure 15.12 CT images in bone algorithm; transverse (a) and dorsal reconstruction (b) of a congenital shoulder subluxation in a 7-month-old springer spaniel. The caudo-lateral aspect of the glenoid is severely irregular outlined and sclerotic (white arrow). The glenoid cavity is flattened, and the humeral head is slightly displaced to the lateral side in respect to the glenoid (yellow arrow). (c) 3D reconstruction. Notice the partially luxated shoulder (yellow arrow) (compare with the normal contralateral side).

15.5.1.3 Fragments or separate ossification centres in the caudal aspect of the glenoid

Fragments or so called ossification centres in the caudal area of the glenoid can be found as an incidental finding. However, sometimes, they cause clinical symptoms. To define the fragment's location and structure, CT can be a very helpful tool. If they are mobile, they can cause impingement on the glenohumeral joint. In these cases, they are more elongated than nicely rounded. In clinical cases, arthro-CT shows impingement on the joint capsule, with the fragment demonstrated as a filling defect (Figure 15.13).

15.5.2 Trauma

15.5.2.1 Intra-articular fractures

In case of intra-articular fractures, subtle or complex fractures, CT will show more fragments than conventional radiography, allowing the surgeon to better plan appropriate surgery (Figure 15.14). Sagittal and dorsal reconstructions can be very useful in these cases (also refer to Section 22.4.2).

15.5.2.2 Avulsion fractures

An avulsion fracture of bone at the attachment site of tendon, ligament, or joint capsule is generally caused by physical trauma. The avulsions near the shoulder joint (infraspinatus and supraspinatus) and supraglenoid tuberosity are described in Sections 15.5.3.3.2 and 15.5.3.3.4.

15.5.3 Degenerative

15.5.3.1 Degenerative joint disease (DJD)

Like in any other joint on CT images, new bone formation can be detected at an earlier stage than on conventional radiographs. The vacuum phenomenon is detected on radiographs, CT, and magnetic resonance imaging (Figure 15.15). Although this phenomenon is most commonly associated with degenerative joint diseases, it is observed in other conditions.

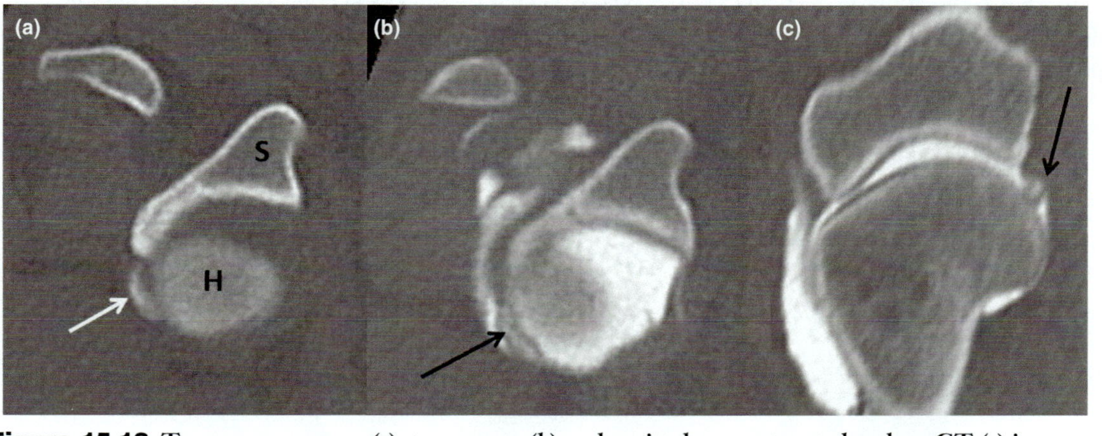

Figure 15.13 Transverse, survey (a); transverse (b) and sagittal reconstructed arthro-CT (c) images in bone algorithm of a fragment in the caudal aspect of the glenoid. Notice the elongated, irregular shape of the fragment (white arrow). S, scapulum; H, caudal humeral head. There is impingement of the fragment on the joint capsule, with the fragment demonstrated as a filling defect (black arrows).

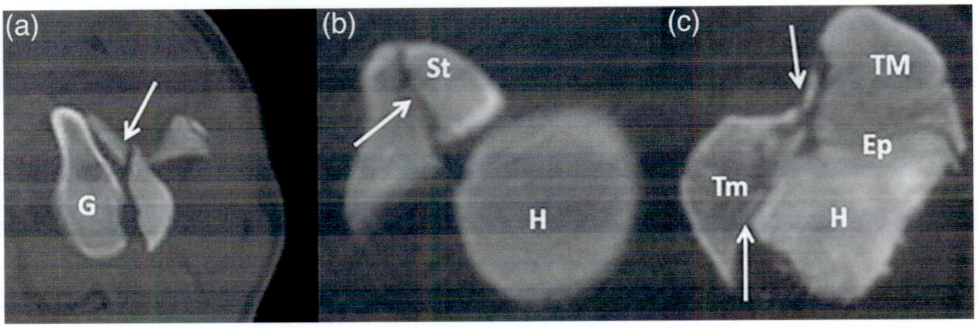

Figure 15.14 Transverse CT images in bone algorithm of the glenoid (a), the supraglenoid tubercle (b), and proximal humerus (c) of a young dog showing intra-articular fractures and epiphyseal fracture (white arrows). The trochanter minor, the glenoid, and the supraglenoid tubercle are fractured, and multiple fragments are present (white arrows). Ep, normal epiphysis; H, humerus; TM, trochanter major; Tm, trochanter minor; G, glenoid; St, supraglenoid tubercle.

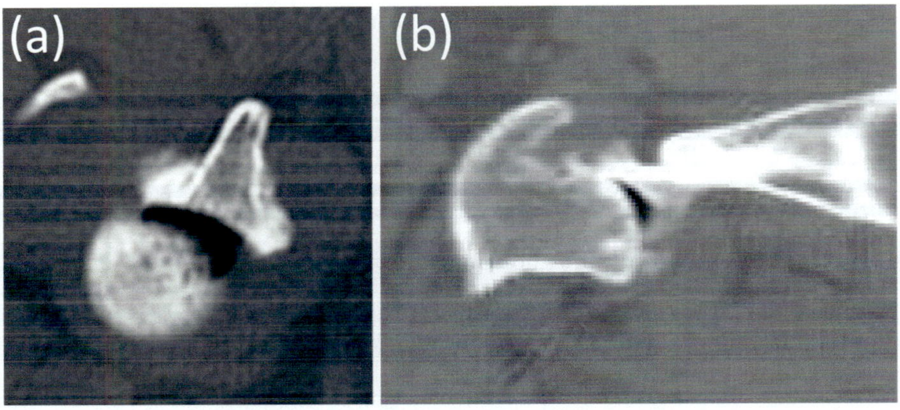

Figure 15.15 Transverse (a) and sagittal reconstructed (b) images in bone algorithm. There is degeneration with prominent arthrotic changes at the scapula and humerus. Intra-articular gas is present within the shoulder joint, vacuum phenomenon.

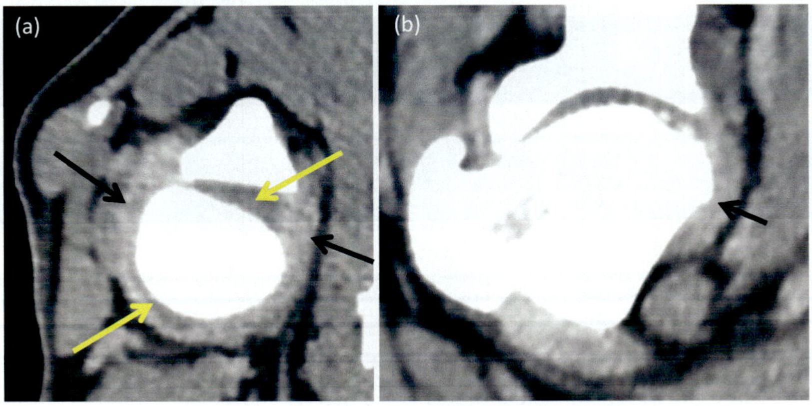

Figure 15.16 Transverse (a) and sagittal reconstructed (b) images after IV administration of contrast, of exhibiting synovitis. Thickening and contrast enhancement of the joint capsule (black arrows), with joint effusion (yellow arrows), are seen.

There is degeneration with prominent arthritic changes at the scapula and humerus.

Intra-articular gas is present within the shoulder joint, known as vacuum phenomenon.

15.5.3.2 Synovitis

Synovitis is characterised by thickened, irregular, synovium that is enhanced following contrast administration (Figure 15.16). The best indicator of inflammatory activity is early enhancement.

15.5.3.3 Biceps lesions

CT examination can be useful in cases of biceps lesions. In cases of bicipital pathology, arthro-CT may demonstrate changes in the contour of the bicipital groove and biceps tendon, incomplete filling of the synovial sheath, filling defects, and irregularities. Smaller lesions within the biceps tendon cannot be identified. Several biceps lesions can occur.

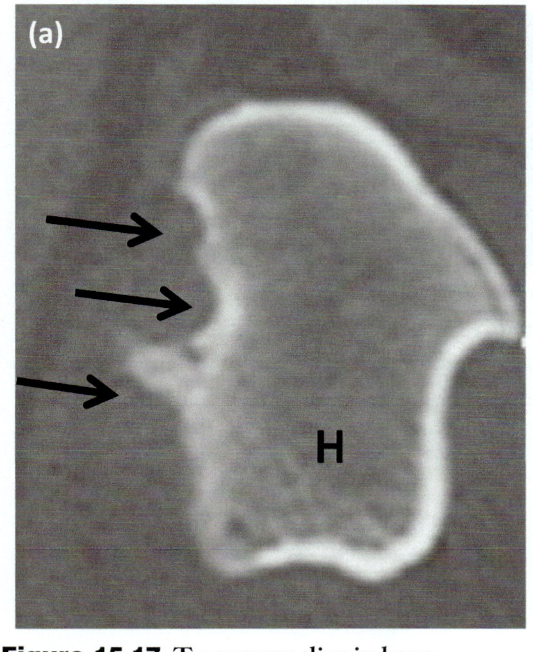

Figure 15.17 Transverse slice in bone algorithm at the level of the humeral head and bicipital groove. New bone formation is noticed at the sulcus of the biceps tendon (black arrows). These bony spikes within the groove hinder the mechanical action of the tendon and cause inflammation of the tendon.

15.5.3.3.1 Inflammation of the biceps tendon

In these cases, new bone formation inside the bicipital groove can be detected in an earlier stage than on radiographs. Some of these bony spikes are present in the region of the biceps tendon, the intertubercular groove, and the supraglenoid tuberosity, hindering the mechanical action of the tendon (Figure 15.17). Arthro-CT shows the biceps brachii muscle as a filling defect within the contrast medium column and is used to detect tendon thickening and irregularities of the tendon. Bicipital tenosynovitis is characterised by irregular or incomplete filling of the synovial bicipital tendon sheath (Figure 15.18). These alterations indicate adhesion formation between the tendon and its sheath as well as synovial proliferation.

15.5.3.3.2 Avulsion fractures of the insertion of the biceps tendon

Avulsion fragments of the supraglenoid tubercle of the scapula or irregular sclerotic supraglenoid tubercle with a bone defect can easily been picked up on plain CT (Figure 15.19a and b). Small bony fragments may be present within the bicipital groove. Often, there is new bone formation visible at the edges of the bicipital groove. Arthro-CT may show synovial effusion with thickening of the proximal end of the biceps, which suddenly ends (Figures 15.19 and 15.20). A contrast filling defect may be present distally to the supraglenoid tubercle (Figure 15.19c and d).In these cases, small enthesophytes may be present at the insertion of the supraspinatus tendon (Figure 15.6).

15.5.3.3.3 Partial/complete ruptures of the biceps tendon

With arthro-CT, the integrity of the biceps tendon can be evaluated. On a normal CT arthrogram, the biceps tendon is identified as a well-delineated tubular filling defect surrounded by the contrast medium. In case of rupture, this filling defect disappears and the contrast medium is visible within the biceps tendon sheath (Figure 15.21). Pooling of the contrast medium is observed close to the origin of the tendon.

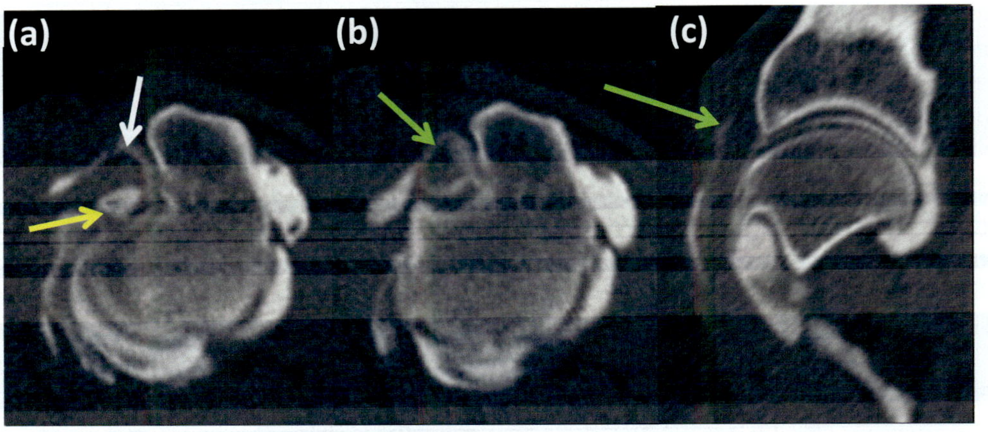

Figure 15.18 Transverse (a) at the level of the joint space and supraglenoid tubercle, transverse image (b) more distally, and sagittal reconstructed (c) arthro-CT images of a shoulder joint suffering from inflammation of the biceps tendon. The biceps tendon seems to be swollen at its proximal end (white arrow). There is irregular and poor filling of the tendon sheath, with unclear delineation of the biceps brachii tendon and its sheath (green arrows). These findings were arthroscopically confirmed. The yellow arrow points to the supraglenoid tubercle.

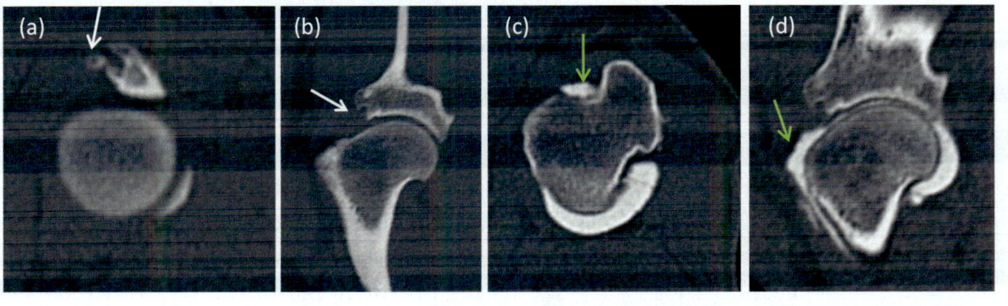

Figure 15.19 Transverse (a) at the level of the proximal humerus, sagittal reconstructed (b), survey images, transverse (c), and sagittal reconstructed (d) arthro-CT images of a dog with avulsion of the biceps brachii tendon. An avulsed fragment is present at the supraglenoid tubercle, which is irregularly delineated (white arrows). The tendon sheath is completely filled with the contrast medium (green arrows). The contours of the biceps tendon are not seen. The tendon is moved distally. There is distension of the joint capsule.

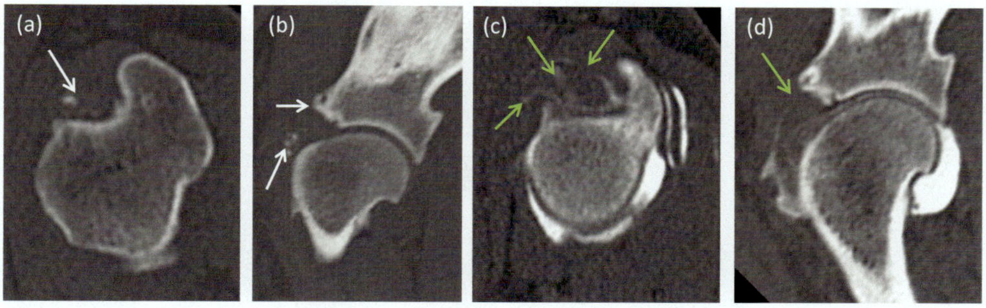

Figure 15.20 Transverse (a) at the level of the proximal humerus, sagittal reconstructed (b), survey images, transverse (c), and sagittal reconstructed (d) arthro-CT images of a dog with avulsion of the biceps brachii tendon. The supraglenoid tubercle shows irregular delineation, with multiple small mineralisations at the groove of the biceps tendon (white arrows). There is contrast leakage with filling defects at the origine of the biceps tendon, which is not well delineated by contrast (green arrows). There is insufficient filling of contrast medium of the tendon sheath. There is distension of the caudal pouch.

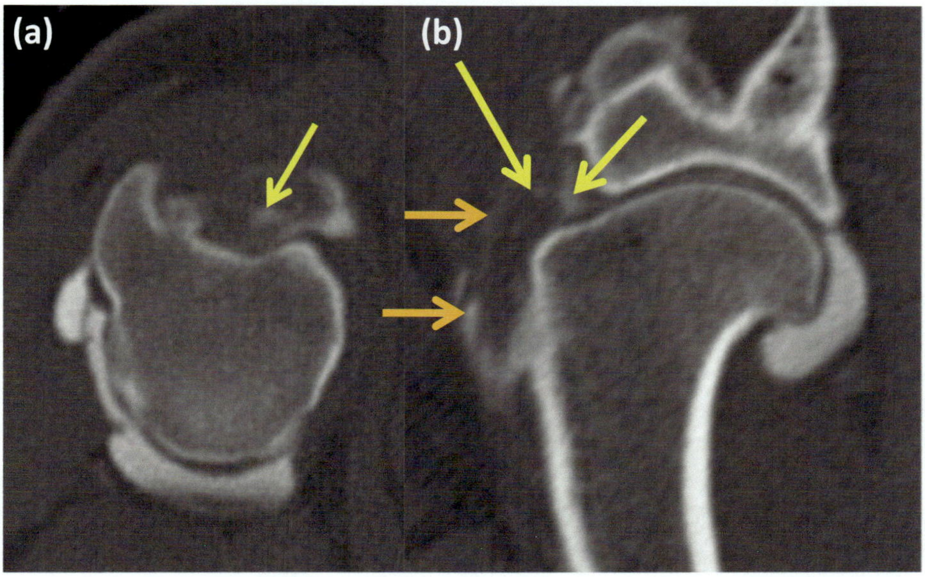

Figure 15.21 Transverse (a) at the level of the proximal humerus and sagittal reconstructed (b), arthro-CT images in bone algorithm of a dog with inflammation and a partial tear of the biceps brachii. There are no cartilage lesions. Several filling defects are visible within the biceps tendon sheath (orange arrows) and at the attachment of biceps tendon (yellow arrows), probably inflammation and partial tear. The caudal pouch is distended.

These findings were arthroscopically confirmed.

15.5.3.4 Affections of the tendons and ligaments

Contrast-enhanced CT is an additional imaging procedure for assessing dogs who may suffer a muscle or tendon injury. Both mineralised and non-mineralised supraspinatus tendinopathy are described. Mineralised foci may be an incidental finding or may cause impingement (Figures 15.22 and 15.23). Lameness may occur from inflammation of peritendinous tissues and impingement of the supraspinatus tendon, with

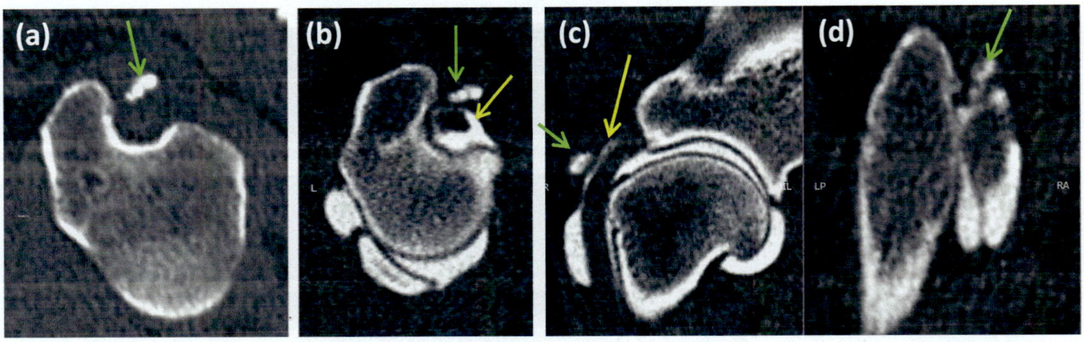

Figure 15.22 Transverse images, in bone algorithm, (a) survey, (b) arthro-CT, sagittal (c), and dorsal (d) reconstructed arthro-CT images showing well-defined mineralisations (green arrows), within the supraspinatus tendon. There is a sharp delineated aspect of the biceps tendon, (yellow arrows), seen as a filling defect, on the arthro-CT images. There is no impingement noticed of the opacities on the biceps tendon.

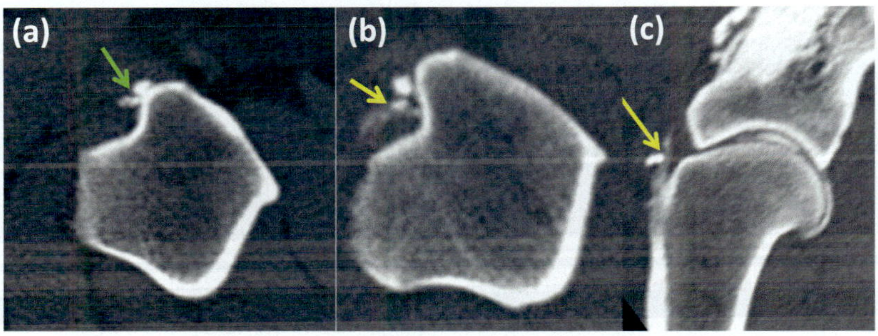

Figure 15.23 Transverse images, in bone algorithm, survey (a), transverse (b), and sagittal reconstructed (c) arthro-CT images of mineralisations within the supraspinatus tendon (green arrow). There is an irregular and unsharp contrast delineation of the biceps tendon, on the arthro-CT images. There is impingement of the opacities on the biceps tendon (yellow arrows).

eventual dystrophic mineralisation, compressing the biceps brachii tendon sheath (Figures 15.23 and 15.24).

The CT features of chronic infraspinatus tendinopathy or tendovaginitis are sclerosis and irregular bone delineation at the insertion of the muscle on the lateral side of the greater tubercle of the humerus. IV contrast may show contrast enhancement with fluid pocket (Figure 15.25).

Supra- and infraspinatus tendinous insertion avulsions are traumatic or degenerative. In these cases, CT clearly demonstrates the presence of a radiolucent defect that corresponds to the location on the proximal humerus, where the infraspinatus or supraspinatus tendons are attached (Figure 15.26).

Glenohumeral ligaments are visible as filling defects, sharply delineated on arthro-CT images of a sound shoulder joint (Figure 15.27).

Infraspinatus contracture is most commonly seen in hunting and working dogs and should be a differential diagnosis for dogs with shoulder lameness (Figure 15.28).

Disruption or fraying of the tendon of the subscapularis, arthroscopically seen, might show an abnormal appearance on CT (Figure 15.29).

15.5.4 Neoplastic

15.5.4.1 Osteochondromatosis of the shoulder joint

This is a benign metaplasia and neoplastic disease originating from the synovial membrane,

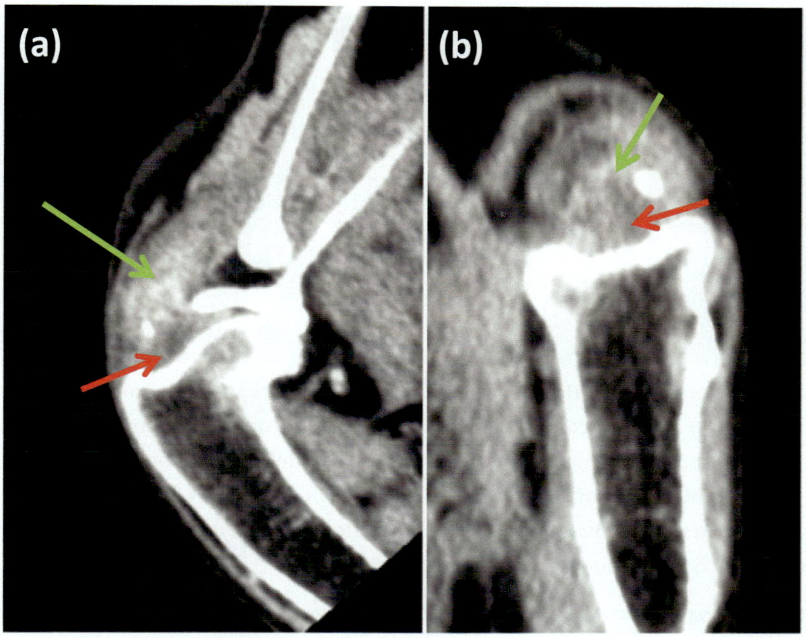

Figure 15.24 IV contrast images, sagittal (a) and dorsal (b) reconstructed, in soft tissue algorithm of an inflamed supraspinatus tendon. The supraspinatus tendon shows an increased volume and diffuse contrast enhancement at its insertion on the greater tubercle of the humerus (green arrows). It reveals compression of the biceps brachii tendon sheath (red arrows). There is a mineralization seen in the supraspinatus tendon.

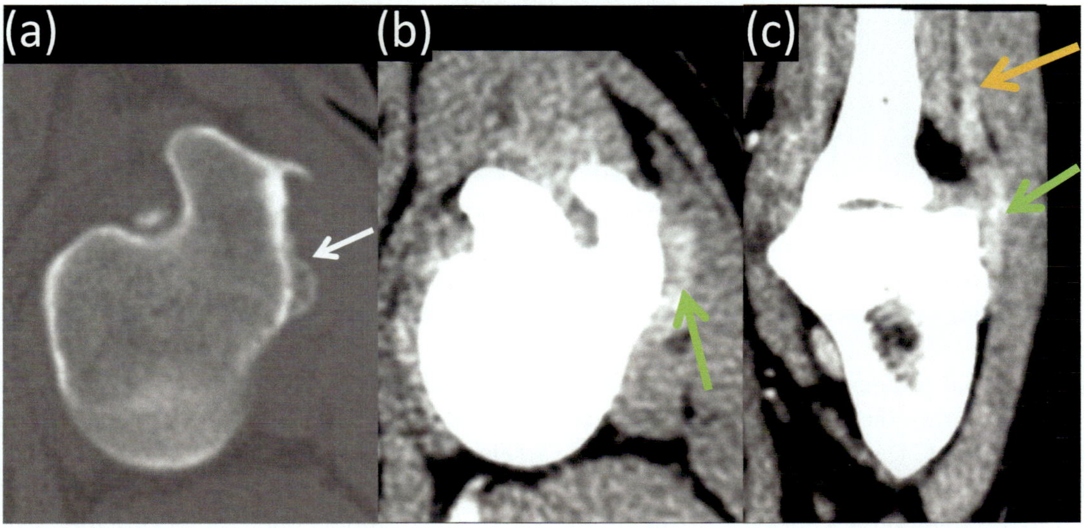

Figure 15.25 Transverse (a) survey in bone algorithm, transverse (b) and dorsal reconstruction (c) after IV contrast administration in soft tissue algorithm of infraspinatus tendinopathy. Enthesophytes are present at the level of the insertion of the infraspinatus (white arrow). There is contrast enhancement with some fluid at the attachment of the infraspinatus tendon (green arrows). The contrast enhancement is noticed within the whole course of the tendon (orange arrow).

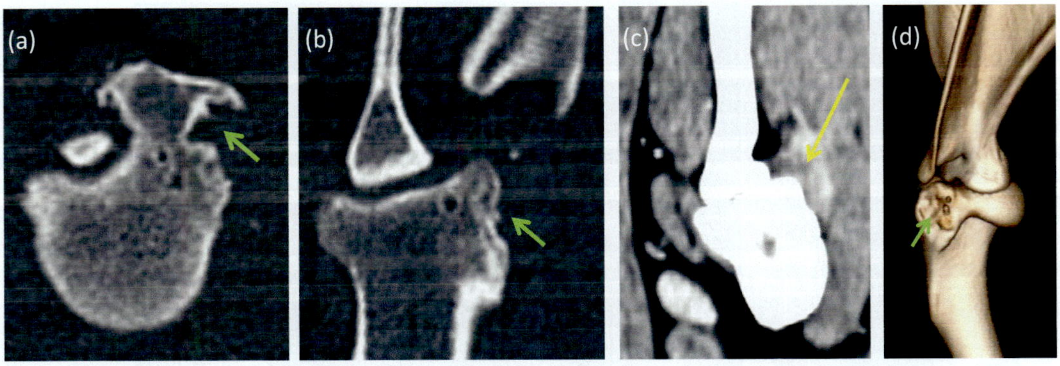

Figure 15.26 Transverse (a) and dorsal reconstructed (b) and transverse images in bone algorithm. Dorsal reconstruction (c) after IV contrast in soft tissue algorithm and 3D image (d) of an avulsion of the infraspinatus tendon. The greater tubercle shows clearly delineated concave defects with irregular borders at the attachment of the infraspinatus tendon (green arrows). A heterogenous aspect of the tendon with mixed-contrast enhancement is present (yellow arrow). More proximally is a hypodense centre which might correspond to necrotic tissue.

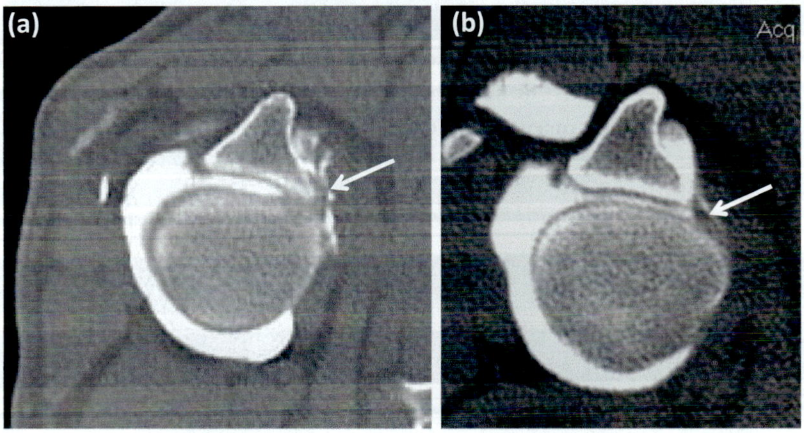

Figure 15.27 Arthro-CT transverse image (a) at the level of the medial glenohumeral ligament of a generally inflamed shoulder joint with fibrillation of the medial glenohumeral ligament. There is a flaccid and irregular aspect of this ligament (arrow). As comparison, an arthro-CT, (b), of a normal shoulder joint with a normal well delineated medial glenohumeral ligament (arrow).

with synovial proliferation producing multiple joint mice within the joint space, causing severe lameness. On CT, severe arthropathic changes and multiple calcified particles can be defined within the joint cavity, while on arthro-CT, filling defects are visible (Figure 15.30).

15.5.4.2 Synovial cell sarcoma

Aggressive bone lysis and periosteal reactions with thickening of the neoplastic synovial membrane is appreciated on CT. IV contrast shows enhancement of the neoplastic process and allows an increased demarcation of the tumour (Figure 15.31).

15.5.4.3 Primary bone tumours of the proximal humerus

On conventional radiographs, a loss of bone content of more than 30% can be appreciated, whereas on CT images, this can be seen in an earlier phase. CT is the best modality to visualise small calcifications, tumour mineralisation,

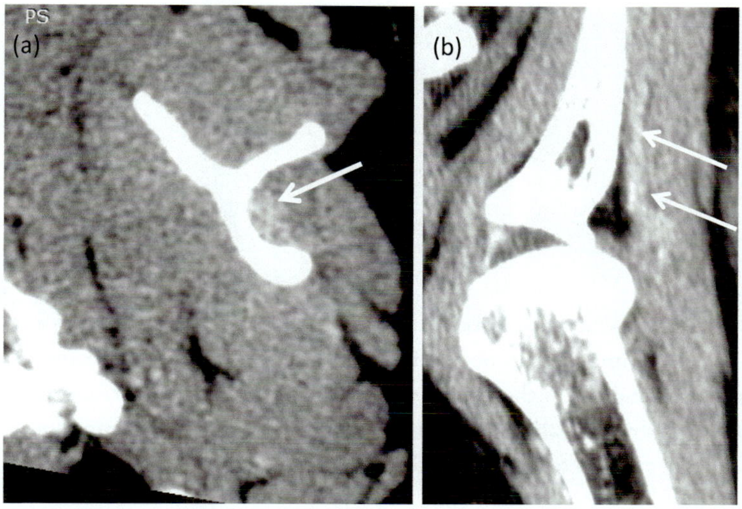

Figure 15.28 Transverse (a) and dorsal reconstruction (b) in soft tissue algorithm after IV contrast of a dog presenting with clinical signs of infraspinatus contracture. There is a noticeable reduction in infraspinatus muscle volume. The contrast enhancement indicates ongoing inflammation (white arrows).

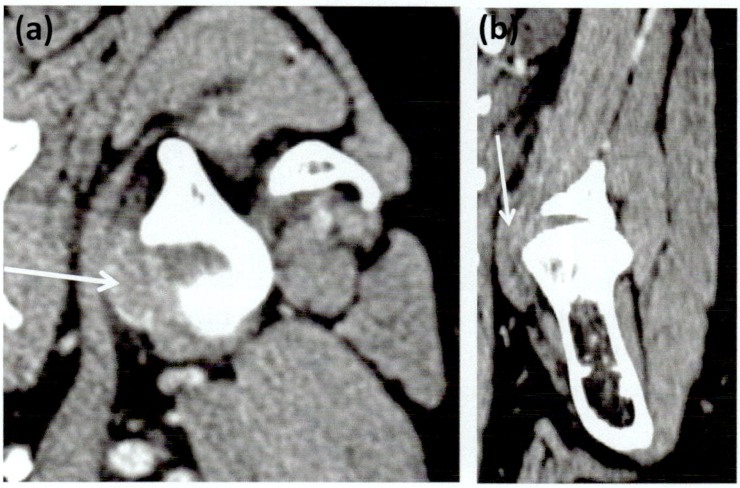

Figure 15.29 Transverse (a) and dorsal reconstruction (b) in soft tissue algorithm after IV contrast of a shoulder joint with subscapularis tendinopathy. There is diffuse thickening of the subscapularis tendon with heterogenous contrast enhancement (white arrows).

cortical changes, periosteal reactions, and minor bone abnormalities (Figures 15.32 and 15.33). CT is helpful in defining the full extent and precise anatomical delineation of a primary or metastatic bone tumour (Figure 15.34). Using CT guidance, an accurate bone biopsy can be performed, allowing a precise diagnosis and better treatment planning. Multiplanar and/or 3D reconstruction provides precise measurements, which may be of special value in these patients.

15.5.4.4 Soft tissue tumours

In addition to staging and the evaluation of the local tumour behaviour, advanced imaging should also be considered, particularly for tumours in difficult areas where surgical removal is a challenge. Both CT and MRI are useful for distinguishing

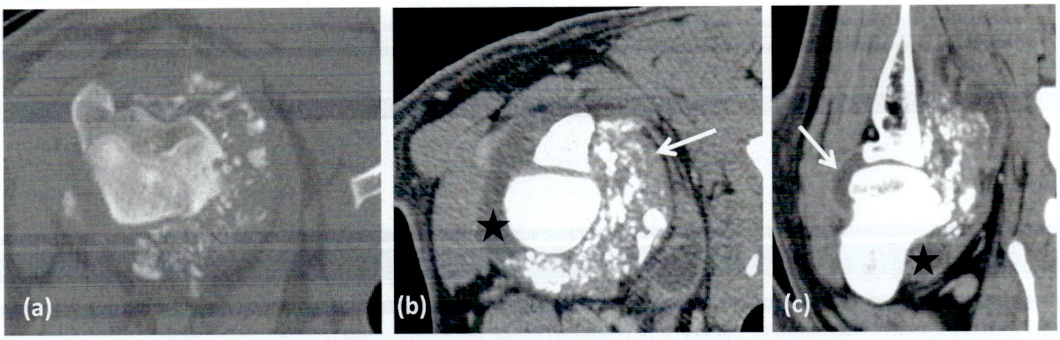

Figure 15.30 Transverse image (a) in bone window of synovial osteochondromatosis. Multiple calcifications are present, mainly on the medial side of the joint. (b) and (c) transverse and dorsal reconstruction after the administration of IV contrast in soft tissue algorithm. Notice the enhancement of the joint capsule (arrows) and the fluid pockets (aster).

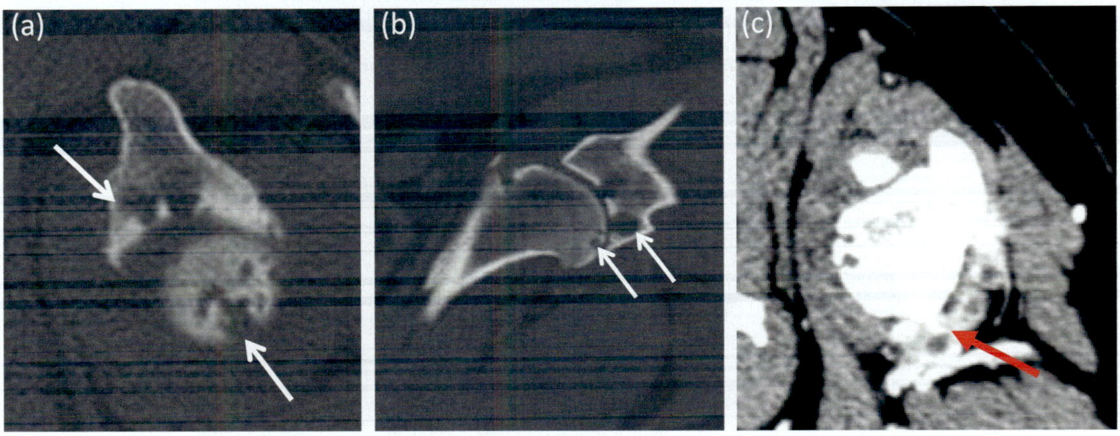

Figure 15.31 Transverse (a) and sagittal reconstruction (b) of a synovial cell sarcoma of the shoulder joint. Fingerprint like radiolucencies in the humerus and scapulum are visible (white arrows). Transverse image after IV contrast administration, (c) shows contrast enhancement of the neoplastic synovia (red arrow).

cystic tissue from solid tissue, demonstrating the vascularity of a mass, and detecting tumour necrosis (Figures 15.35 and 15.36).

15.5.5 Miscellaneous

15.5.5.1 Infection

CT with IV contrast medium can be used to assess soft tissue and the extent of the infection (Figure 15.37), which in turn can serve as a guide for aspiration or surgery.

15.5.5.2 Osteomyelitis

In addition to providing information about soft tissue changes, CT offers an ideal characterisation of cortical bone destruction and periosteal reaction (Figure 15.38). It is the most effective method for identifying zones of cortical erosion. CT may be useful in identifying the infection site and directing interventional procedures (biopsies and aspirations). Abnormalities in soft tissues can be detected with the use of post-contrast CT.

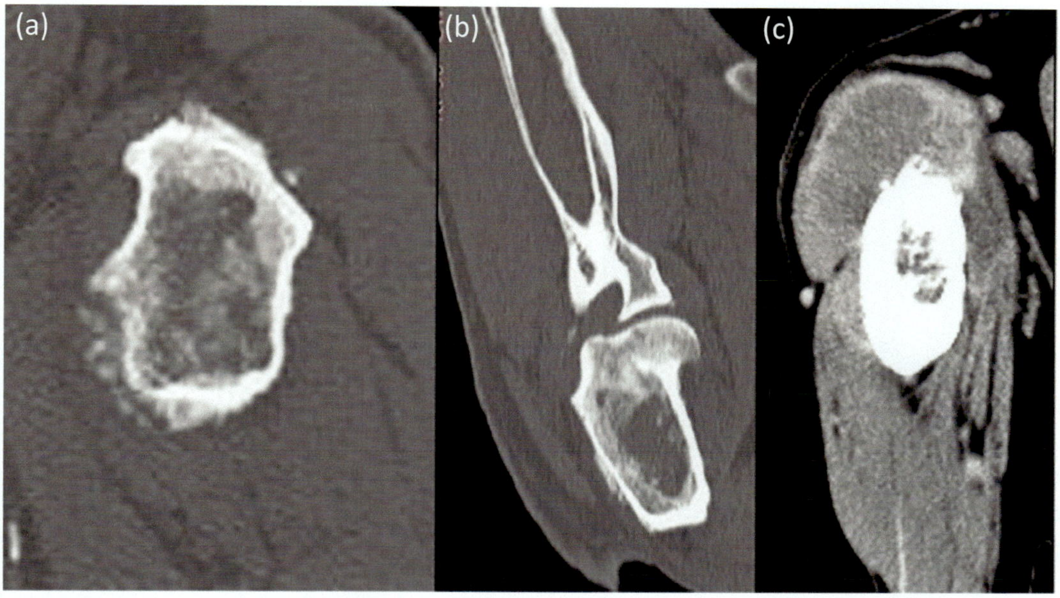

Figure 15.32 Transverse (a) and sagittal reconstruction (b) of an osteosarcoma of the proximal humerus. There is a mixture of osteolysis, and new bone formation is seen. IV contrast administration, transverse image (c) in soft tissue algorithm, shows the extension and the enhancement of the neoplastic mass.

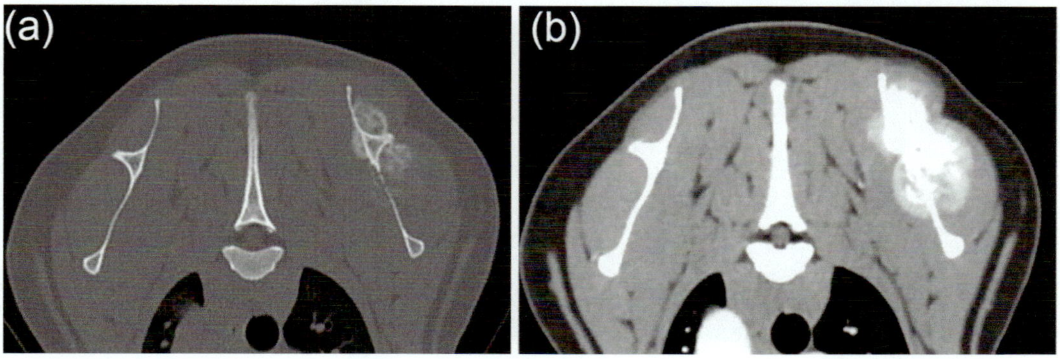

Figure 15.33 Transverse images, in bone algorithm (a) and soft tissue algorithm after IV contrast administration (b), of a 2-year-old Australian shepherd with a primary bone tumour of the right scapula, strongly suggestive of an osteosarcoma. The right shoulder musculature is distended. There is substantial lysis and new bone formation of the scapula, proximally, with marked contrast enhancement of the surrounding soft tissue and muscle.

15.5.5.3 Inflammation

Joint effusion, a sign of joint inflammation, and periarticular swelling are easiest identified on US, CT (Figure 15.39), and MRI.

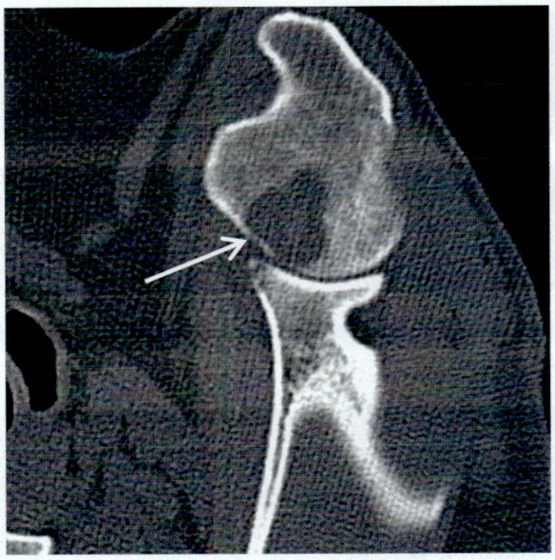

Figure 15.34 Transverse image in bone window of the proximal humerus, with predominantly osteolytic metastasis originating from a primary carcinoma. Note the osteolytic lesion breaking through the cortex (arrow).

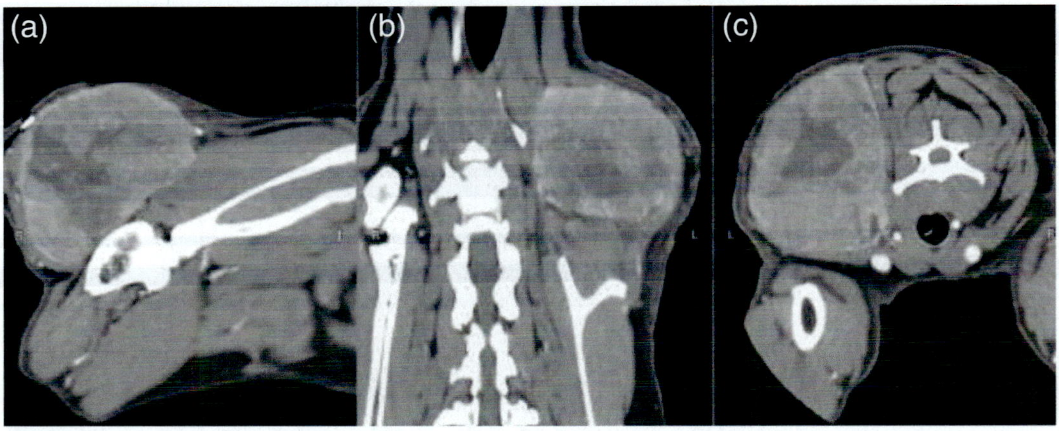

Figure 15.35 Sagittal (a) and dorsal (b) reconstructions, transverse image (c), in soft tissue algorithm after IV contrast administration, of the shoulder area. A large round/oval tumour with well-defined margins is present cranio-laterally of the joint. Fat stranding is seen in the subcutaneous area. The mass shows heterogeneous contrast enhancement after IV contrast administration with hypodense areas, indicating necrosis. There is a 'mass' effect on the surrounding muscles but no real infiltration. FNA revealed malignancy.

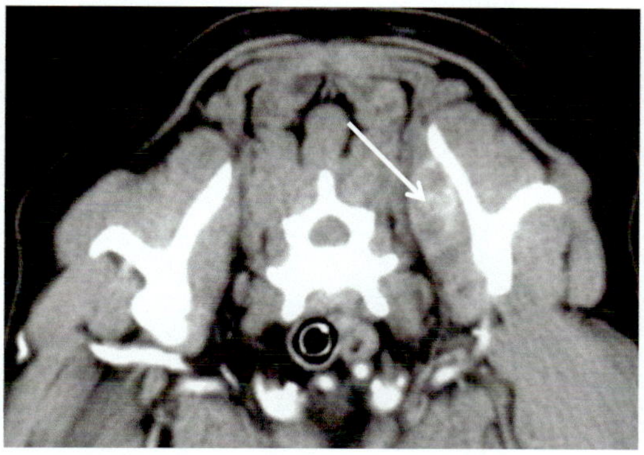

Figure 15.36 Transverse image, soft tissue algorithm after IV contrast administration, at the level of the scapula of a 10-year-old French bulldog. This dog suffered from a haemangiosarcoma of the spleen with metastasis in the lungs. The contrast-enhanced mass at the subscapularis muscle highly indicates metastasis (white arrow).

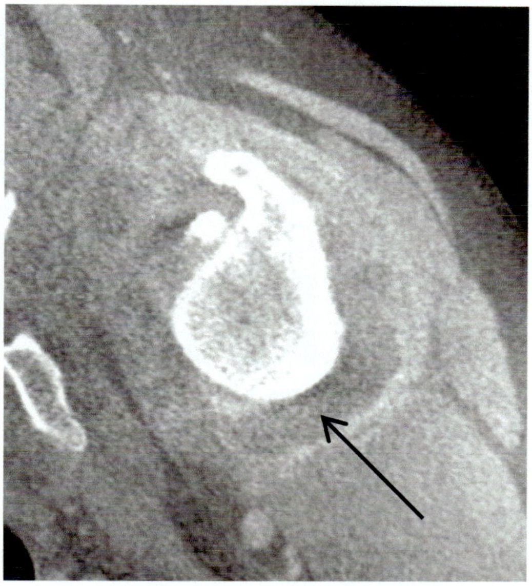

Figure 15.37 Transverse image, cone beam CT, in soft tissue algorithm after IV contrast administration at the level of the shoulder joint space. Around the joint, there is massive soft tissue swelling with several fluid pockets inside (black arrow). These images (massive new bone formation and soft tissue swelling) are suggestive for a severe joint infection. *Staphylococcus* infection could be confirmed.

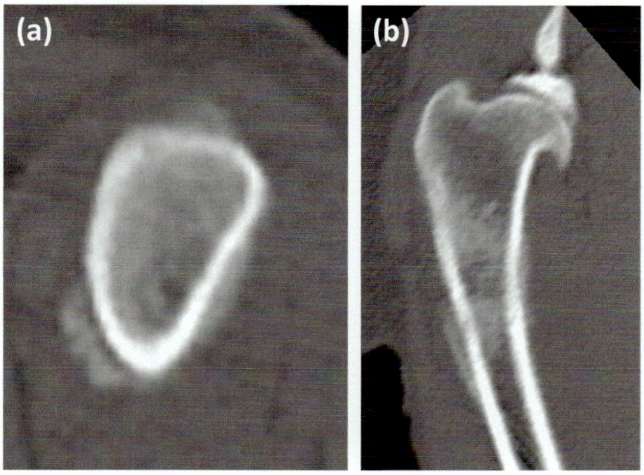

Figure 15.38 Transverse (a) and sagittal reconstruction (b) in bone algorithm of the humerus of an 8-year-old Labrador retriever with polyostotic involvement of the humeri. The humerus shows medullary sclerosis associated with smooth-to-irregular periosteal reactions on the caudal, lateral, and medial aspects of the proximal humeral diaphysis. In these cases, multifocal; osteomyelitis, primary bone neoplasia (osteosarcoma, fibrosarcoma, lymphoma, and histiocytosis), osteomyelitis, and fungal disease are possible differential diagnoses. Biopsies of the described lesions are advised.

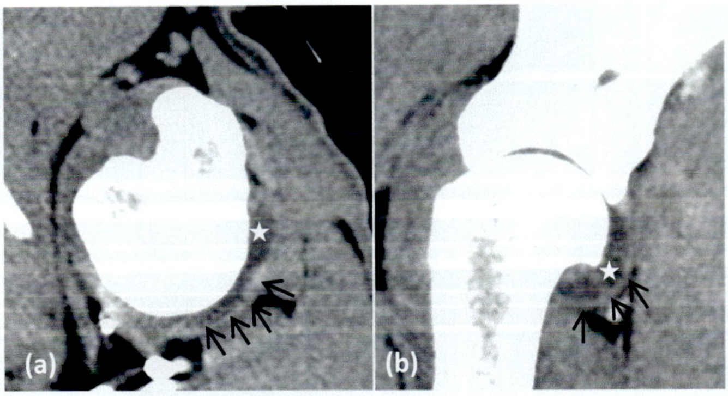

Figure 15.39 **(a) Transverse- and (b) sagittal reconstruction images after administration of IV contrast medium in soft tissue algorithm of joint inflammation. There is joint effusion (aster) and enhancement of the inflamed joint capsule (arrows).**

FURTHER READING

De Rycke LM, Gielen IM, Dingemanse W, Kromhout K, van Bree H. Computed tomographic and low-field magnetic resonance arthrography: A comparison of techniques for observing intra-articular structures of the normal canine shoulder. *Vet Surg*. 2015 Aug;44(6):704–12.

De Rycke L, van Bree H, Van Caelenberg A, Polis I, Duchateau L, Gielen I. Epinephrine-enhanced computed tomographic arthrography of the canine shoulder. Res Vet Sci. 2015 Oct;102:15–21.

De Simone A, Gernone F, Ricciardi M. Imaging diagnosis-bilateral abnormal ossification of the supraglenoid tubercle and cranial glenoid cavity in an English Setter. *Vet Radiol Ultrasound*. 2013 Mar–Apr;54(2):159–63.

Lande R, Reese SL, Cuddy LC, Berry CR, Pozzi A. Prevalence of computed tomographic subchondral bone lesions in the scapulohumeral joint of 32 immature dogs with thoracic limb lameness. *Vet Radiol Ultrasound*. 2014 Jan–Feb;55(1):23–8.

Maddox TW, May C, Keeley BJ, McConnell JF. Comparison between shoulder computed tomography and clinical findings in 89 dogs presented for thoracic limb lameness. *Vet Radiol Ultrasound*. 2013 Jul–Aug;54(4):358–64.

Mikola K, Piras A, Hakala L. Isolated avulsion of the tendon of insertion of the infraspinatus and supraspinatus muscles in five juvenile labrador retrievers. *Vet Comp Orthop Traumatol*. 2018;31:285–290.

Pappa E, Tomlinson A, Ferreire A, Petitt R, Mortier J. Ultrasonographic and CT arthrography findings in two dogs with biceps brachii tendon avulsion. *Vet Record Case Rep*. 2020,8:e000989.

Reis Silva H, Uosyte R, Clements DN, Bergkvist GT, Schwarz T. Computed tomography and positive contrast computed tomographic arthrography of the canine shoulder: normal anatomy and effects of limb position on visibility of soft tissue structures. *Vet Radiol Ultrasound*. 2013 Sep–Oct;54(5):470–7.

CHAPTER 16

CT OF THE ELBOW

Ingrid Gielen, Annemie Van Caelenberg, and Henri van Bree

16.1 INTRODUCTION

Elbow lameness, a common problem in small animal practice, can be caused by several disorders. One of the major problems is attributed to canine elbow dysplasia (ED), a hereditary developmental problem commonly diagnosed in young large-breed dogs, such as the Bernese Mountain dog, Labrador and Golden Retriever, Rottweiler, and German Shepherd dog. Elbow dysplasia is a complex of joint disorders and includes medial coronoid disease (MCD), osteochondrosis of the medial humeral condyle (OC), ununited anconeal process (UAP), and elbow incongruity (EI). A combination of different ED lesions within the same joint is possible. Although MCD is commonly reported as a developmental disorder affecting young large-breed dogs, it can occur in adult and older dogs and in atypical breeds. Medial compartment erosions (medial compartment disease) without macroscopic fragmentation of the medial coronoid process are frequently seen as a typical lesion in 'old' dogs. CT and arthro-CT offer a possibility to evaluate the associated cartilage damage and to differentiate this disease from other pathologies.

Flexor enthesopathy has been described as a pathologic condition of the flexor muscles and their attachment to the medial humeral epicondyle in dogs. It is considered a primary elbow disorder when underlying pathology of the elbow is absent. However, in many cases, flexor enthesopathy occurs concurrently with elbow dysplasia. In this disease,intravenous (IV) contrast may offer the possibility to distinguish clinical from non-clinical lesions.

Diagnosis of elbow disorders including ED is traditionally made by clinical examination followed by radiographic examination. Computed tomography (CT) and arthroscopy (AS) are useful to confirm the diagnosis. Resent results confirm that CT is very reliable to diagnose MCD in patients and is practically as accurate as arthroscopy in discerning coronoid lesions. CT also offers interesting options for objective measurement of subtrochlear sclerosis.

The elbow joint is a complex synovial hinge joint formed by the distal part of the humerus and the proximal part of the radius and the ulna. Three separate subjoints are thus identified, the humero-ulnar joint, the humero-radial joint, and the proximal radio-ulnar joint. Each of these plays an essential role in the normal functioning of the joint.

The joint is supported with strong collateral ligaments, and several muscles originate from it and insert on it. Its main motion is flexion–extension, while a small gliding motion is also present. Little rotation is also possible, allowing supination of the paw mostly pronounced in the cat.

16.2 INDICATIONS FOR USING CT

a. Definite diagnosis of medial coronoid disease and evaluation of the size, number, and location of the fragments.
b. Definite diagnosis of OC of the medial humeral condyle and differential diagnosis with contact lesions concomitant with coronoid fragments.

DOI: 10.1201/9781315118130-18

c. In cases of UAP to check a concomitant fragmented coronoid process and possible malalignment or displacement of the UAP fragment.
d. More accurate evaluation of elbow incongruity.
e. Detection and conformation of medial compartment (syndrome) disease (arthro-CT helpful).
f. Evaluation of flexor enthesopathy (IV contrast helpful).
g. Early detection of degenerative changes.
h. Conformation of severe synovitis (soft tissue window+IV contrast helpful)
i. Detection of neoplastic changes including synovial cell sarcoma, primary bone tumours, and metastases.
j. Differentiation between synovial cell sarcoma and severe degenerative changes.
k. Miscellaneous bony- and soft-tissue lesions in the elbow area like evaluation of intra-articular fractures. (Cfr Rovesti chapter intra-articular fractures).
l. Incomplete ossification of the humeral condyle (IOHC), also known as humeral intracondylar fissure (HIF),

16.3 TECHNICAL ASPECTS

16.3.1 Positioning

While generally anaesthetized, the dogs are positioned in lateral or ventral recumbency on the CT scanning table with the elbow joints parallel and extended cranially. An angle of less than 120° in the elbow joint (flexion) should be avoided. It is essential to prevent beam hardening artefacts due to the patient's head. This requires adjusting the patient's head laterally or caudally to prevent interference with the elbows, regardless of the patient's orientation. A recommended position is pulling the head of the dog caudally to the lateral side in order to scan only both elbow joints in the gantry (Figure 16.1). Wedge sponges are placed between both elbow joints in order to avoid rotation. This position allows a perfect symmetry and comparison of both elbows at the same level; and the obtained image quality is superior to the quality acquired with positioning the head and neck of the dog within the CT gantry (Figure 16.2). Especially in discrete coronoid lesions, like fissures or non-displaced fragments, this can be of importance. This protocol of avoiding superimposition of the head is also required by the International Elbow Working Group. It is not recommended to scan the elbows separately. As this way of scanning means longer anaesthesia, a high burden for

Figure 16.1 Positioning: the animal is positioned on left lateral recumbency with the front-legs extended and the head withdrawn out of the gantry. This positioning produces less artefacts. A sponge wedge is placed is placed between both elbows.

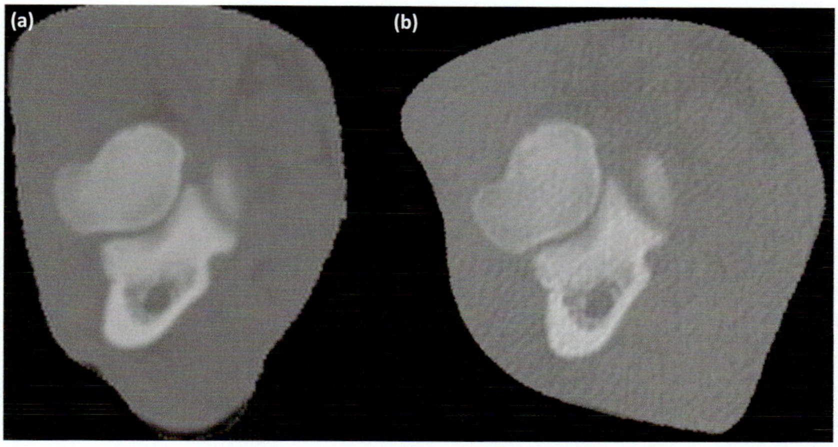

Figure 16.2 **Influence of positioning on image quality: CT images of the coronoid process of the same dog in bone window, positioned in left lateral recumbency (a) and in dorsal recumbency (b). Notice on image b, streak artefacts as a result of beam hardening due to the positioning of the head within the gantry.**

the tube and higher radiation dose are expected. Hyperdense material (positioning aids and anaesthetic equipment) needs to be outside the gantry.

16.3.2 Technical parameters

Lateral and cranio-caudal survey views are made to confirm correct positioning. These images are used to plan the scans, from the top of the olecranon till 3 cm distally to the cubital joint and parallel to the humero-radial joint space. Scans of 0.6–1 mm slice thickness, helical mode, with an overlap of 50%, in the bone algorithm (bone kernel) are performed from the top of the olecranon to 3 cm distal to the cubital joint and parallel to the humeroradial joint space. Two- and 3-mm slices are less effective and occasionally inconclusive when compared to 1-mm slices. Additional soft-tissue algorithm (standard kernel) images of 1.5–2 mm slice thickness, helical mode, with an overlap of 50% are reconstructed. Thinner slices in this algorithm need to be avoided as the images will be too grainy and show too many artefacts. Settings for the CT image technique are 120 kV and 140 mAs. The field of view (FOV) is adapted around both elbow joints, enhancing the resolution of the obtained image. Individual images are reviewed using a bone setting (window width=3,000–3,500 Hounsfield units; window level=500 Hounsfield units) and a soft-tissue setting (window width=350 Hounsfield units; window level=90 Hounsfield units). Reconstructions in dorsal and sagittal planes are made. Sagittal images are reconstructed in the lateral compartment and through the keel of the trochlear notch. A comprehensive assessment of the trochlear notch's shape and congruity of the three joint spaces can be conducted using these images.

16.3.3 Contrast (IV and IA)

In case of soft-tissue pathology (sever synovitis, neoplasia, and flexor enthesopathy), IV contrast can be very helpful in the differential diagnosis (Figure 16.3a and b).

Intra-articular contrast application (arthro-CT, CTA) is unrewarding to check for fragmented coronoid process (coronoid disease). It provides non-invasive visualization of the articular cartilage. In case of suspected cartilage damage (medial compartment syndrome), CTA can help determine the damage and cartilage erosions (Figure 16.3c).

16.4 CT ANATOMY & NORMAL VARIANTS

Besides full anatomic detail of the bony structures, some soft tissues such as tendons and large

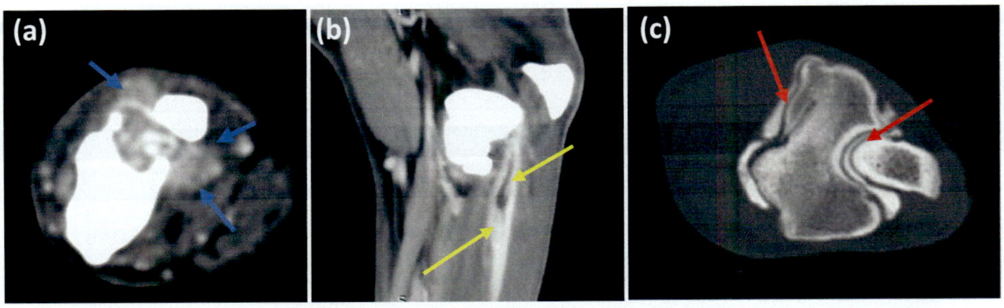

Figure 16.3 Transverse (a) and sagittal reconstruction (b) in soft tissue algorithm after IV application of contrast medium. Transverse CT image in bone algorithm after intraarticular contrast (c). In cases of tumors, mesenchymal tumor, IV contrast media provides information about vascularization and allows the estimation of demarcation from the surrounding tissue (blue arrows). In case of flexor enthesopathy, IV contrast provides enhancement at the level of the flexor muscles (yellow arrows). Intraarticular contrast delineates articular cartilage (black lines) of the normal canine elbow (red arrows).

blood vessels can also be evaluated, by adjusting the window and using the varying grey-scale formats (Figure 16.4). However, it is not possible to identify nerves and smaller blood vessels. The normal medial coronoid process is smoothly delineated and the joint space between the radius and the ulna is narrow and nicely parallel. On the lateral side of the radius, often a sesamoid is visible (sesamoid in the tendon of origin of the supinator muscle), which is a normal finding and should not be mistaken with a fragment (Figure 16.5a). Sometimes in the area of the attachment of the medial collateral, an ossicle can be seen as well (Figure 16.5b). In some young Golden Retrievers, the medial coronoid process looks as a separate bony structure on some slices (Figure 16.5c) (should be checked on previous and following slices). In very young dogs, the growth plates can be puzzling and should not be mistaken for fracture lines (Figure 16.5b). After obtaining transverse images, CT-reconstructions of the elbow joint in a sagittal and dorsal plane can be made. The entire elbow joint surface and congruity can be evaluated on these reconstructed images, without superimposition of any surrounding (bony) structure. (Figure 16.6).

16.5 PATHOLOGIC FINDINGS

16.5.1 Developmental

Elbow dysplasia (ED) is characterized by varying degrees of elbow incongruity, bony fragments (bone chips), and ultimately, severe arthrotic change. The term was introduced to describe generalized osteoarthrosis (arthrosis) of the elbow joint in which the anconeal process may be ununited (UAP), the medial coronoid of the ulna may be fragmented (FCP), and osteochondrosis of the humeral condyle (OCD) may be present. These three aetiologies resulting in ED can be present individually, but it is important to note that there can be considerable overlapping in their presence.

Abnormalities in the area of the medial coronoid process include the following: fragmentation (displaced or nondisplaced), fissure, abnormal shape, sclerosis, osteophytes, and lucencies. In the area of the medial humeral condyle, sclerosis, lucency, and/or flattening can be evaluated, and a differential diagnosis between kissing lesions and real OCD lesions can be made. All these abnormalities can be diagnosed on the transverse and reconstructed images. CT findings, like fissures or non-displaced fragments at the medial coronoid process and subchondral lucencies at the

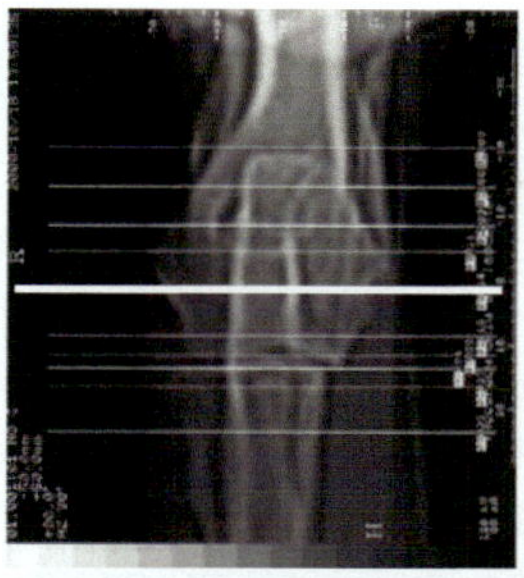

14: Pronator teres,
29: Lat collateral lig,
28: Joint capsule

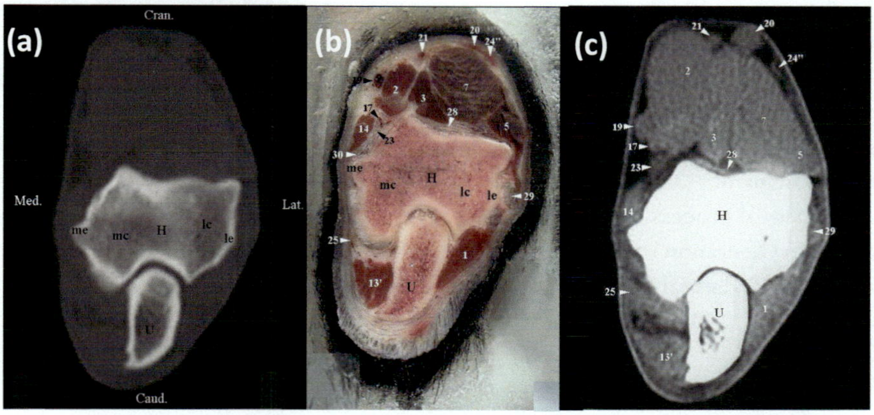

(a)

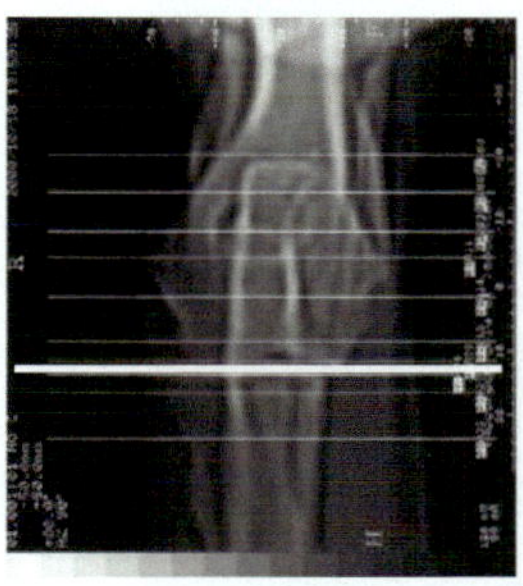

12: Flex carpi radialis,
30: Med collateral ligament

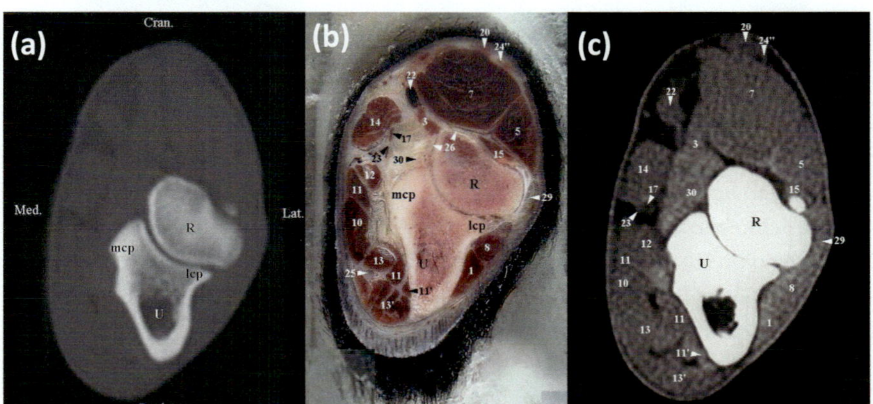

(b)

Figure 16.4 Normal CT anatomy of the elbow at the level of the humeral condyles (a) and at the level of the coronoid process (b). Transverse images in bone (a)- and soft tissue window

(Continued)

Figure 16.4 (*Continued*) (c) compared with the anatomic specimen (b) in the area of the proximal part. In the soft-tissue window, several soft tissue structures can be recognized. Caud=Caudal. Med=Medial. Lat=Lateral. Cran=Cranial. 1=Anconeus muscle. 2=Biceps brachii muscle. 3=Brachialis muscle. 5=Extensor digitorum communis muscle. 6=Extensor digitorum lateralis muscle. 7=Extensor carpi radialis muscle. 8=Extensor carpi ulnaris muscle. 9=Abductor digitialis muscle. 10=Flexor digitorum superficialis muscle. 11=Flexor digitorum profundus muscle (11=caput humerale, 11'=caput ulnare). 12=Flexor carpi radialis muscle. 13=Flexor carpi ulnaris muscle (13=caput humerale, 13'=caput ulnare). 14=Pronator teres muscle. 15=Supinator muscle. 17=Brachial artery. 18=Median artery. 19=Brachial vein. 20=Cephalic vein. 22=Median vein. 23=Median nerve. 24=Radial nerve (24'=medial superficial branch, 24''=lateral superficial branch). 25=Ulnar nerve. 26=Anular ligament of radius. 27=Antebrachial interosseous membrane. 28=Elbow joint capsule (fibrous membrane). 29=Lateral collateral ligament. 30=Medial collateral ligament. H=Humerus. lc=Lateral part of humeral condyle. le=Lateral epicondyle. mc=Medial part of humeral condyle. me=Medial epicondyle. of=Olecranon fossa. rf=Radial fossa. sf=Supratrochlear foramen. R=Radius. U=Ulna. AP=Anconeal process. lCP=Lateral coronoid process. MCP=Medial coronoid process. o=Olecranon. OT=Olecranon tuber. RN=Radial notch. TN=Trochlear notch (CR=central ridge, MF=medial flange, LF =lateral flange). ut=Ulnar tuberosity.

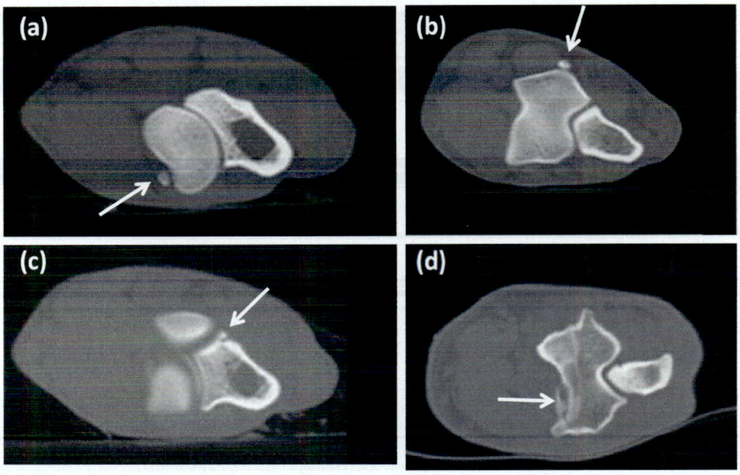

Figure 16.5 (a) Anatomic variations, transverse images at different levels, all in the bone algorithm: (a) A sesamoid on the lateral side of the radial head (arrow) (sesamoid in the tendon of the origin of the supinator muscle) not to be mistaken for a fragment. (a) The medial coronoid process (MCP) is normal. Notice the smooth delineation of the MCP and the narrow parallel joint space between the ulna and the radius in this normal joint. (b) Ossicle in the area of the medial collateral ligament (arrow). (c) The medial coronoid process often looks like a separate ossification centre (arrow) in some slices often in large-breed dogs like Golden Retrievers. (d) The attachment of the joint capsule may look irregular (arrow).

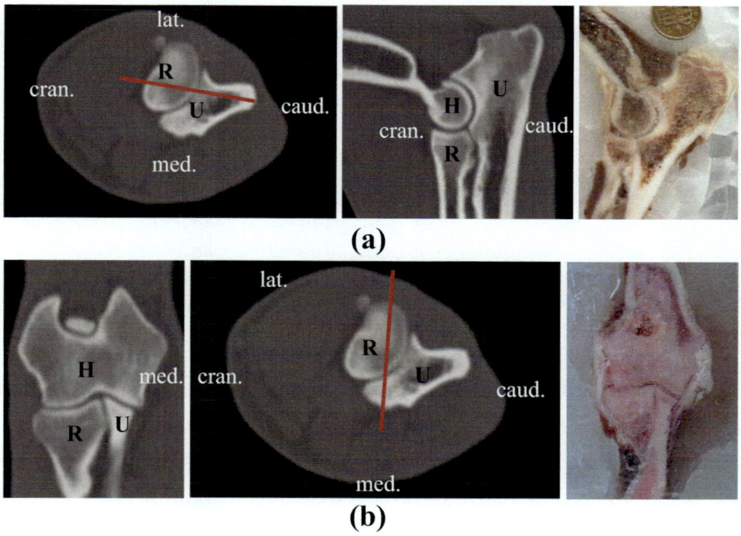

Figure 16.6 **Sagittal (a) and dorsal (b) CT reconstructions, in bone algorithm, of normal CT anatomy compared with an anatomic specimen: The entire elbow joint surface can be evaluated without superimposition of any surrounding structure. H=humerus; R=radius; U=ulna.**

medial humeral condyle, are useful for decision-making in the arthroscopic treatment of these lesions.

16.5.1.1 Medial coronoid disease

In addition to being the most common heritable orthopaedic disease in dogs, medial coronoid disease is the primary cause of elbow lameness in large and giant-breed dogs. It is a component of elbow dysplasia complex.

16.5.1.1.1 Chondromalacia

In these cases, older dogs, there is distension with irregular and soft cartilage present. Chondromalacia is best noticed on transverse CT images as a mixture of hypo- and hyperdense bone present at the medial coronoid process. There is no fragmentation of the medial coronoid process in these cases (Figure 16.7).

16.5.1.1.2 Fragmented medial coronoid process (FCP): non-displaced fragments/fissures versus displaced fragments

On transverse slices, the shape of the medial coronoid process and the incidence of one (Figure 16.8a) or multiple fragments (Figure 16.8b and c) can be evaluated and recognized. Fragments may be displaced to the cranial

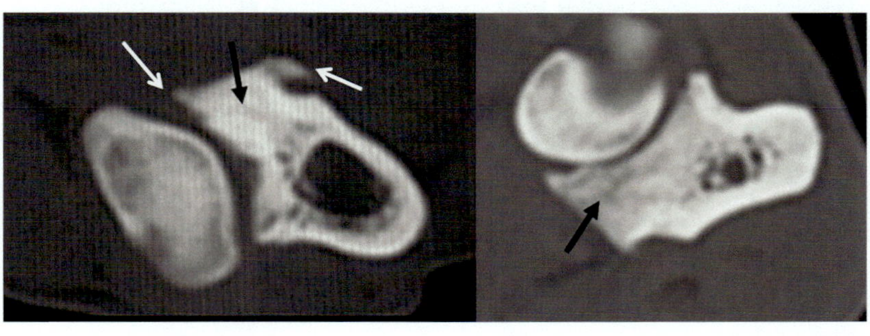

Figure 16.7 **Transverse slices in bone window of two different dogs. Both hypo- and hyperdense subchondral bone are present at both medial coronoid processes (black arrows). Osteophytes can be appreciated as well (white arrows). Notice the irregular joint space between the radius and ulna on the left image.**

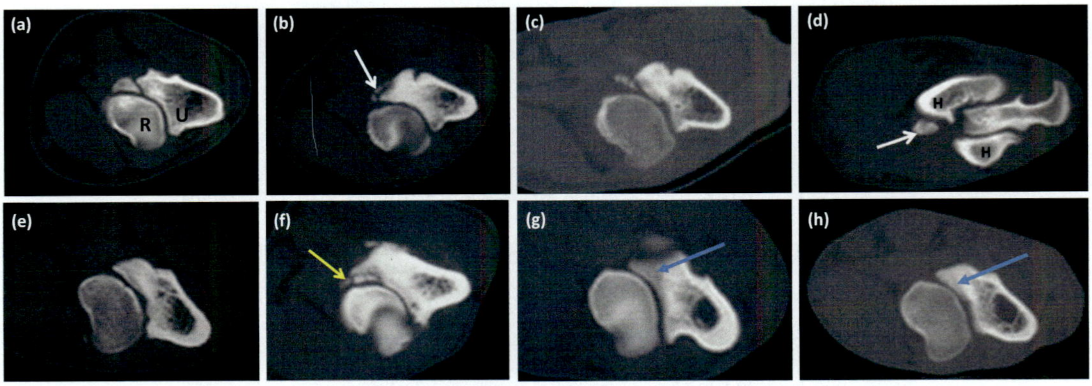

Figure 16.8 Several examples of medial coronoid disease, in bone algorithm, transverse slices in bone algorithm: (a) a single, slightly displaced fragment; (b) multiple slightly displaced fragments, there is subtrochlear sclerosis seen (arrow); (c) several displaced fragments; (d) a displaced fragment migrated towards the cranial part of the joint (arrow). Notice in all these lesions the degenerative changes and sclerosis within the ulna. (e) A single fragment in between the radius and ulna; (f) multiple fragments in between the radius and ulna; (g and h) discrete fissure lines. U=ulna; R=radius; H=humerus. Compare with Figure 16.4b showing a normal medial coronoid process.

part of the joint (Figure 16.8d). Fragments may be non-displaced and squeezed in between the radius and ulna (Figure 16.8e and f). The structures have to be examined in the correct bone window to accurately evaluate the subchondral bony structures and more specific to visualize fissures as their appearance can be very discrete (Figure 16.8g and h). In most pathologic cases, the joint space between the radius and ulna is irregular, not parallel, and shows lucencies. Often, sclerosis of the ulna can be appreciated. If hypoattenuation is present within the medial coronoid process after ossification by the age of 24 weeks, this is considered abnormal. In doubtful cases, reconstructions can be helpful to establish a definite diagnosis (Figure 16.9). High-quality images and correct image window/level are required to remark discrete abnormalities and non-displaced fragments.

MCD and FCP should be considered a cause of thoracic limb lameness in toy and small breed dogs as well. Also, in aged dogs, distinct fragments may be found often without any signs of DJD (Figure 16.10).

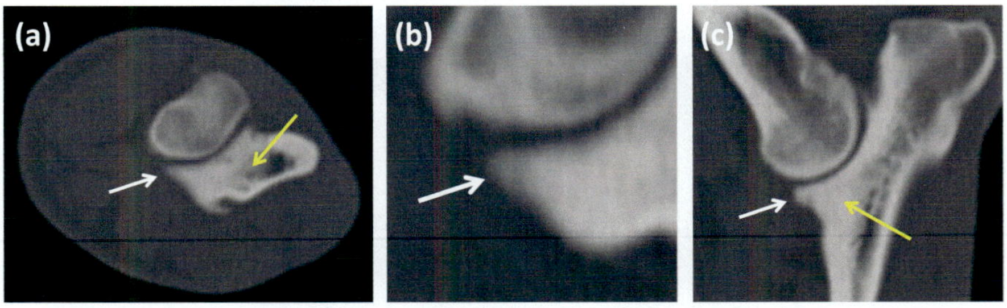

Figure 16.9 Transverse- (a, b) and sagittal-reconstructed (c) images in bone algorithm of a Golden retriever of 10 months. The medial coronoid process shows a hypoattenuated aspect; this is better visible on the magnification. (b) A discrete fissure line is appreciated on the reconstructed image (white arrows). Arthroscopy confirmed the incidence of a fissure and non-displaced fragment. Subtrochlear sclerosis is clearly visible in this case (yellow arrows).

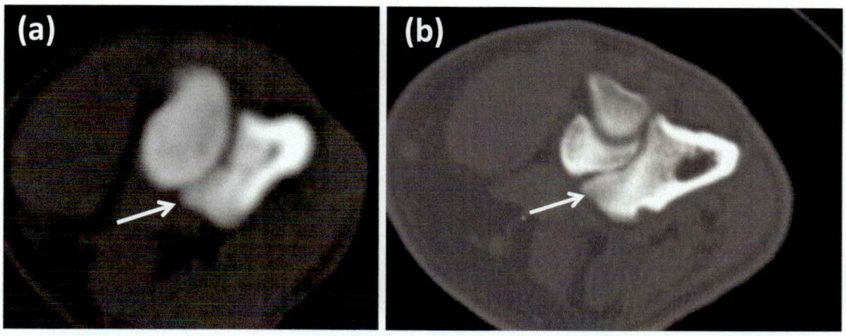

Figure 16.10 Transverse images in bone algorithm, of a 6-year-old Pomeranian (a). Due to the small size of the dog, this image looks blurry; however, the abnormal density and aspect of the medial coronoid process can be noticed (white arrow). Transverse image of a 7-year-old Newfoundlander (b) shows an abnormal appearance of the medial coronoid process, which was confirmed as a distinct fragment on arthroscopy (white arrow).

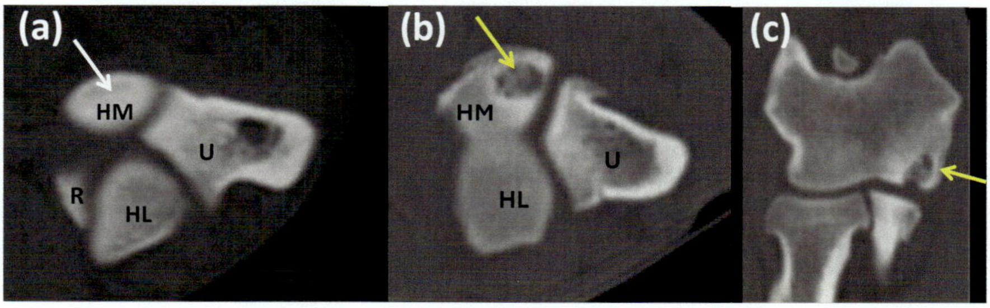

Figure 16.11 (a) Normal aspect of the medial condyle: notice the homogeneous subchondral bone structure (white arrow). (b) A typical OCD lesion: notice the radiolucent circular lesion surrounded by sclerosis. (c) Shows the corresponding dorsal reconstruction, with the large subchondral defect of the medial humeral condyle surrounded by sclerosis (yellow arrows). The flattening of the condyle is noticeable. U=ulna; R=radius; HL=humerus lateral condyle; HM=humerus medial condyle.

16.5.1.2 Osteochondrosis (OC) or osteochondritis dissecans (OCD) of the medial humeral condyle

OC/OCD appears on CT as a subchondral defect, radiolucent, surrounded by a sclerotic rim in the area of the medial humeral condyle. On reformatted images, flattening of the medial humeral condyle can be appreciated. These features are typical for all OCD lesions in all types of joints (Figure 16.11). Their aspect is often more extended on CT in comparison with radiography. Using CT, OC lesions can be detected before flap formation (OCD), and if such joints are arthroscopically inspected, no cartilage cleft will be detected (Figure 16.12). If touched with a probe, the eventually loose cartilage will detach.

These OC/OCD joints should be checked for the presence of an FCP, which is quite common.

16.5.1.3 Kissing lesions on the medial humeral condyle

Kissing lesions, on radiographs showing the same aspect as OC/OCD, display a completely different CT aspect. They are presented as a stripe-like form in most cases associated with significant new bone formation (Figure 16.13). When kissing lesions are present, FCP may be expected as well.

16.5.1.4 Ununited anconeal process (UAP)

UAP is noticed on transverse slices at the level of the humeral condyles and anconeal process

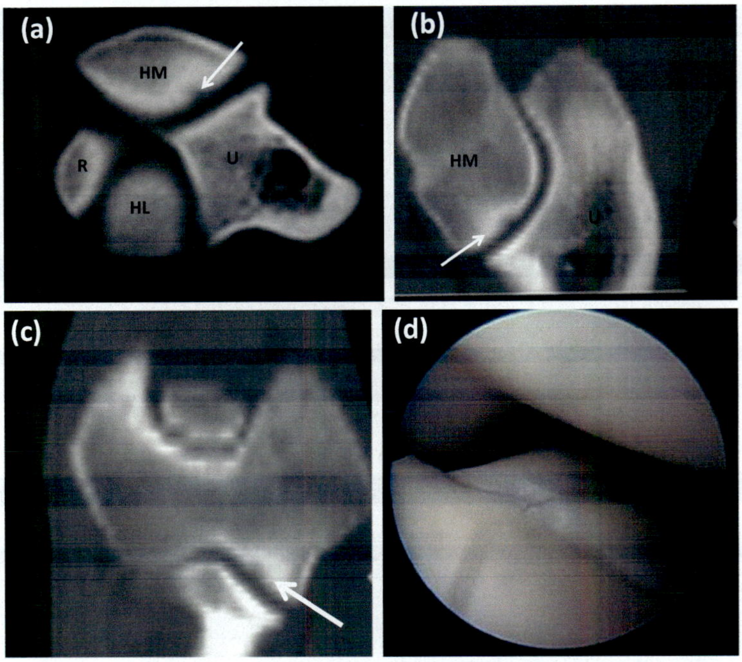

Figure 16.12 Transverse- (a), sagittal- (b), and dorsal (c)-reconstructed images of a 5-month-old Bernese mountain dog with OC. There is flattening of the medial humeral condyle present with a discrete OC lesion surrounded by sclerosis (white arrows). (d) On arthroscopy, no detached flap was visible. U=ulna; R=radius; HL=humerus lateral condyle; HM=humerus medial condyle.

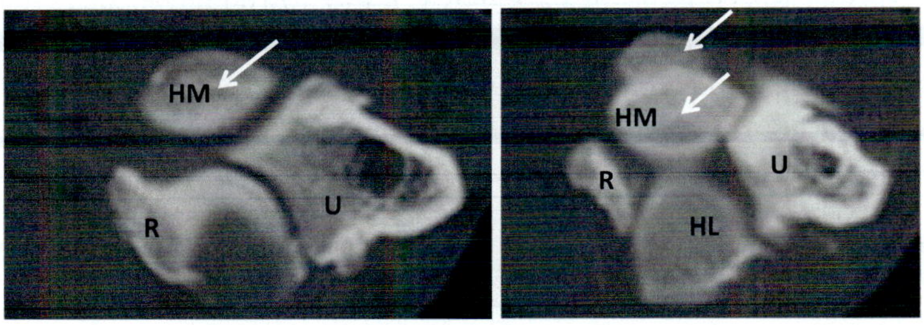

Figure 16.13 Transverse images in bone algorithm showing the typical aspect of kissing lesions: notice the stripe-like form lesion of the medial humeral condyle (arrows) and extensive new bone formation. U=ulna; R=radius; HL=humerus lateral condyle; HM=humerus medial condyle.

as an irregular gap. The position and alignment of the fragment in relation to the olecranon can best be evaluated on the sagittal reconstructions through the centre of the notch (Figure 16.14). This information is important for the surgeon to determine the surgical strategy and to decide whether an ulnectomy could be useful. In patients with UAP, it is indicated to check for an FCP as both lesions are often present.

16.5.1.5 Elbow incongruity

An incongruent elbow joint is often seen in combination with a fragmented coronoid process and in combination with some cases of the ununited anconeal process. The radiographic evaluation of elbow incongruity provides a low sensitivity, especially in discrete cases. Incongruity can be better appreciated on CT images avoiding superimposition of the different bony structures. CT demonstrates

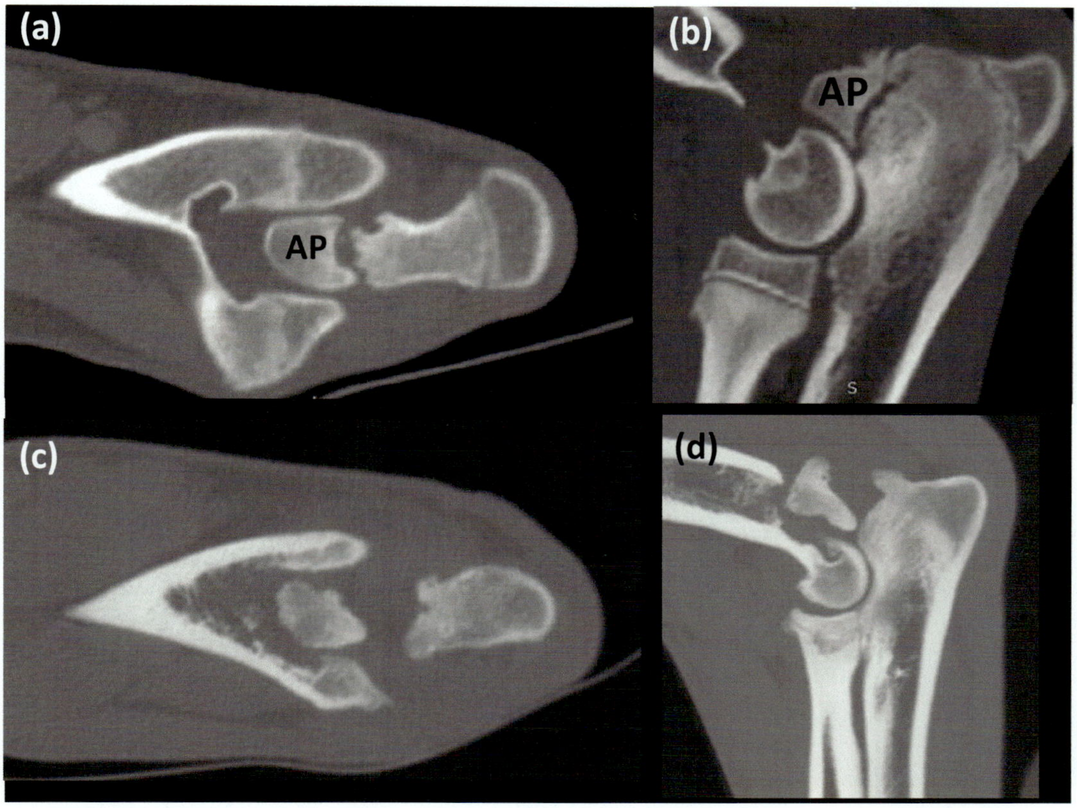

Figure 16.14 Ununited anconeal process (UAP), transverse, and sagittal reconstructions in bone algorithm: (a and c): transverse images of an UAP: the irregular gap can be appreciated in both cases but in (d) the MPR image, the fragment seems to be displaced, which is confirmed in the sagittal reconstruction, where it can be seen that the fragment is not only displaced but also tilted. (b): sagittal reconstruction of case A, a 5-month-old Bouvier with a non-displaced anconeal process, aligned with the olecranon. Ulnectomy may be considered in this case. U =ulna; R=radius; H=humerus. AP=anconeal process.

joint widening in both humero-ulnar and humero-radial joint spaces, radioulnar step, cranial displacement of the humeral condyle, a comma-shaped humero-ulnar joint space, and an elliptical shape of the trochlear notch in cases of severe elbow incongruity. On transverse slices at the level of the trochlear notch and the anconeus, a widened and diverging joint space can be recognized. In normal elbow joints, this joint space is narrow and parallel. On sagittal and dorsal reconstructions, a step between the radius and ulna can be appreciated, as well as an abnormal shape of the trochlear notch (Figure 16.15).

16.5.1.6 Incomplete ossification of the humeral condyle (IOHC) (Intercondylar fissures of the distal humerus)

On transverse slices at the level of the distal part of the humerus, a saw-toothed intercondylar complete or incomplete hypoattenuating defect with hyperattenuating margins is demonstrated (Figure 16.16a and b). Dorsal reconstructions are useful to evaluate the complete extent of the lesion (Figure 16.16c). In these cases, evidence of concurrent medial coronoid disease (about 26%) and degenerative joint disease (about 79%) has been reported. Co-existing incongruence in the elbow joint can be present.

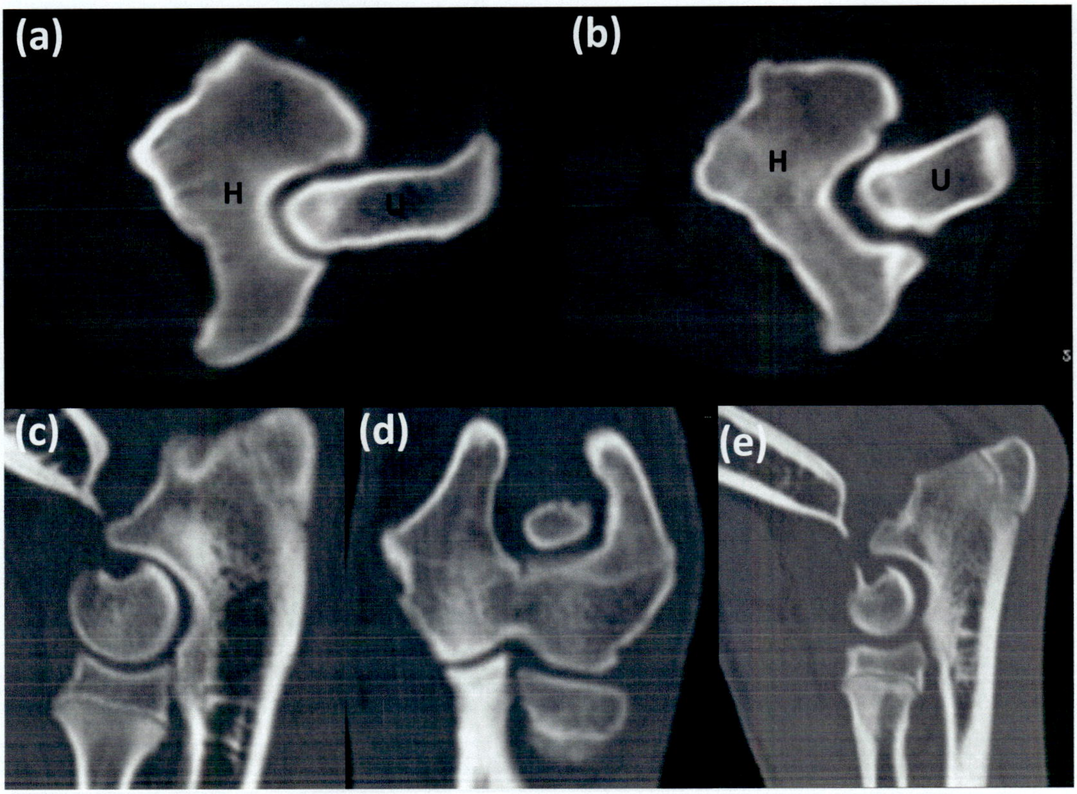

Figure 16.15 Incongruency, transverse, and sagittal reconstructions in bone algorithm: (a) transverse image at the level of the trochlear notch of a congruent elbow. The joint space between the ulna and humerus is narrow and parallel. (b) Transverse image through the trochlear notch of an incongruent elbow. The wide and divergent joint space between the ulna and humerus is obvious. Sagittal (c) and corresponding dorsal (d) reconstructions of an incongruent elbow. Notice the elliptic form of the trochlear notch and the step between the radius and ulna. Sagittal (e) reconstruction of a severe incongruent joint. U=ulna; R=radius; H=humerus.

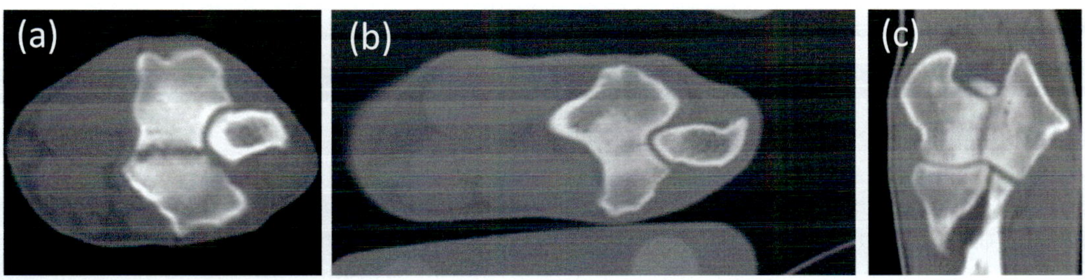

Figure 16.16 Incomplete ossification of the humeral condyle (IOHC): (a) transverse image showing an irregular saw-toothed intercondylar defect surrounded by extensive sclerosis. (b) Transverse image of an incomplete intercondylar defect. (c) Dorsal reconstruction on which the gap between the humeral condyles, surrounded by sclerosis is clearly visible.

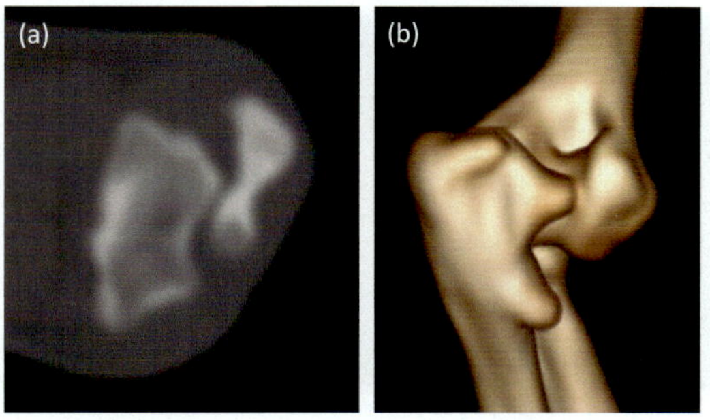

Figure 16.17 Transverse image in bone algorithm (a) and 3D (b) of the elbow of a 9-month-old Terrier, showing fore limb problems since birth. Congenital elbow luxation, type 1 is present. There is a subluxation of the ulna, and the ulna is rotated laterally. The radius is not luxated.

16.5.1.7 Congenital elbow luxation

This is a congenital developmental abnormality. MPR and 3D reconstruction might be helpful in determining the exact diagnosis and in treatment planning and follow-up evaluation (Figure 16.17).

16.5.2 Trauma

CT is commonly utilized to assess elbow fractures in order to provide a comprehensive evaluation beyond radiography and achieve precise fracture classification. By offering a meticulous depiction of all fracture fragments, CT enables a thorough examination of the injury, revealing its intricacy and severity. CT displays more fragments than conventional radiography, giving more accurate information to the surgeon, better for appropriate surgery planning. Sagittal and dorsal reconstructions can be very useful in these cases. In most cases, a subluxation of the anconeal process can be appreciated. Three-dimensional reconstructions are useful for evaluation and treatment planning of fractures. They can not only show the three-dimensional forms of lesions but also observe the fractures that are not shown in cross-sectional images (Figures 16.18–16.21).

16.5.3 Degenerative

16.5.3.1 Degenerative joint disease (DJD)

Like in any other joint, new bone formation can be detected in an earlier stage on CT images compared with conventional radiographs. The appearance of degeneration in CT is often more pronounced. DJD tends to occur at typical locations (Figure 16.22).

16.5.3.2 Synovitis

On a soft-tissue window, thickening of the joint capsule can be appreciated with fluid pockets within the swollen synovia. IV contrast

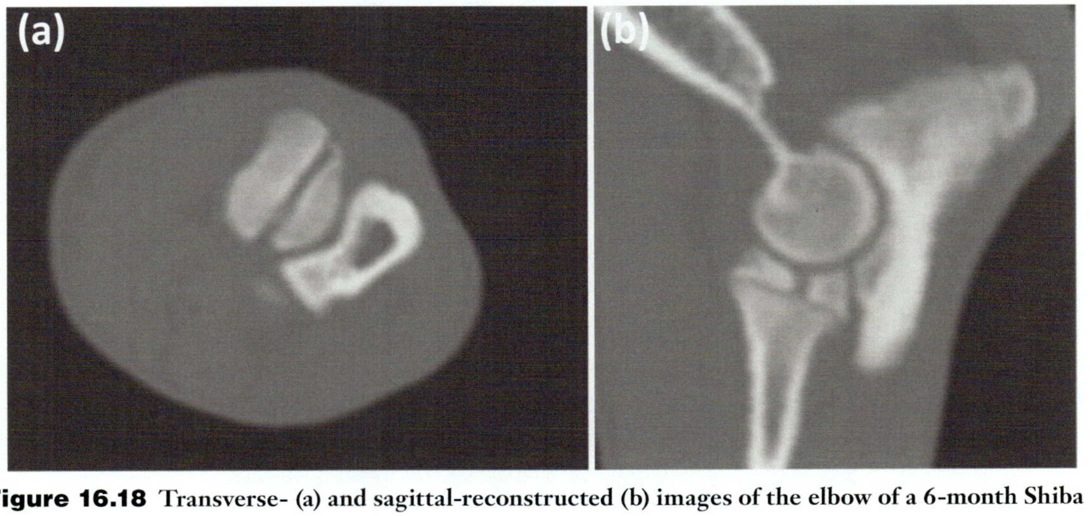

Figure 16.18 Transverse- (a) and sagittal-reconstructed (b) images of the elbow of a 6-month Shiba Inu. The proximal radial growth plate is widened. At the proximal part of the radial head, a wide longitudinal fracture line extending until the growth plate is present.

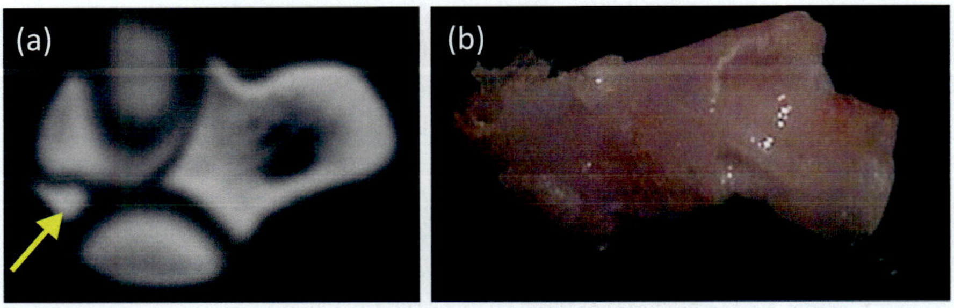

Figure 16.19 Transverse image (a) in bone algorithm at the level of the joint space of a 10-month-old Rottweiler who showed acute lameness was diagnosed with traumatic fracture of the medial coronoid process. A large cranially displaced fragment with a sharply delineated edge is seen (yellow arrow). The macroscopic image (b) of the removed fragment showing a sharp edge.

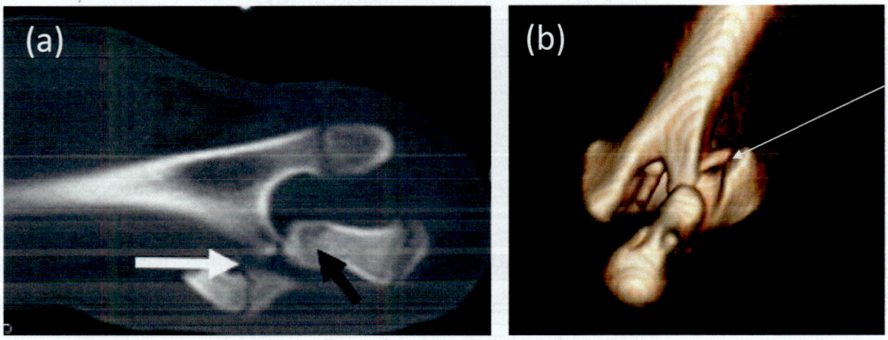

Figure 16.20 Transverse (a) in bone algorithm and 3D image (b) of an intra-articular fracture. The lateral column is fractured and displaced (white arrows). A subluxation of the anconeal process to the lateral side can be appreciated (black arrow). Salter Harris type IV fracture.

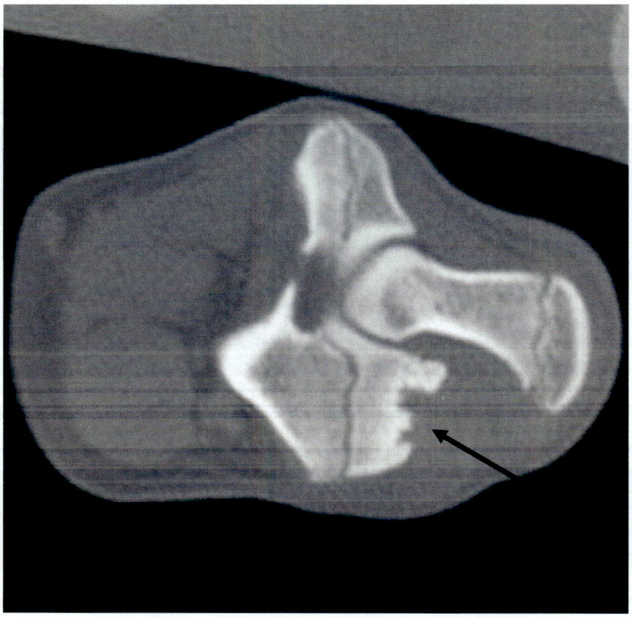

Figure 16.21 Transverse image of the elbow of a Labrador Retriever puppy, female, 16 weeks old, with an avulsion of the medial epicondyle of the humerus. Note the crater in the distal part of the medial epicondyle (black arrow).

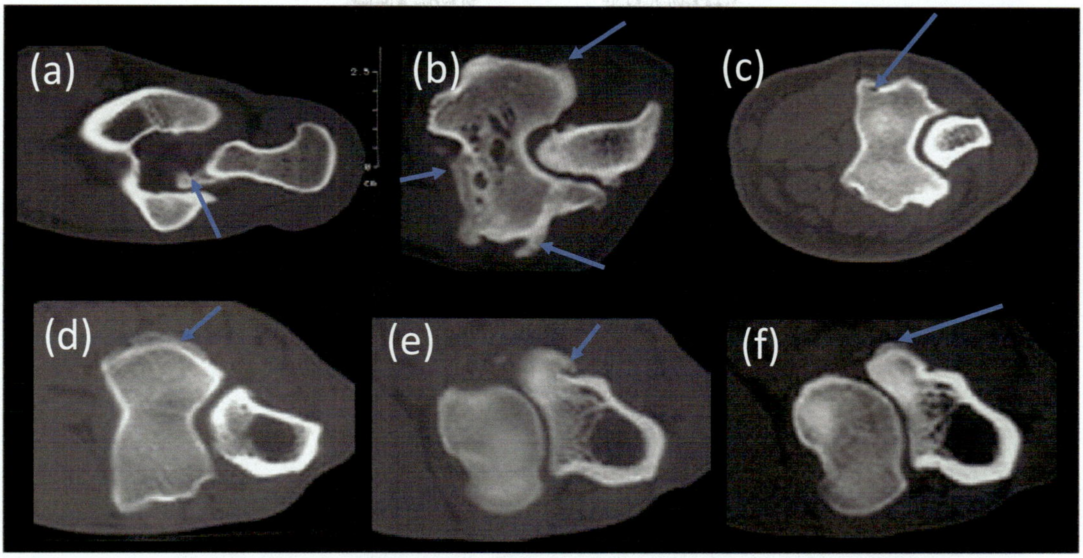

Figure 16.22 **Transverse images of different levels of the elbow in bone algorithm showing periarticular osteophyte production associated with the anconeal process (a), the humeral condyles (b–d), and the medial coronoid process (e, f) (arrows). Subchondral lucencies may be prominent (b).**

application shows enhancement of the joint capsule (Figure 16.23).

16.5.3.3 Affections of the insertion of the flexor tendons

CT is a sensitive technique for detecting flexor enthesopathy. It may be present as the only sign in a joint with primary flexor enthesopathy or concomitant with other elbow pathology.

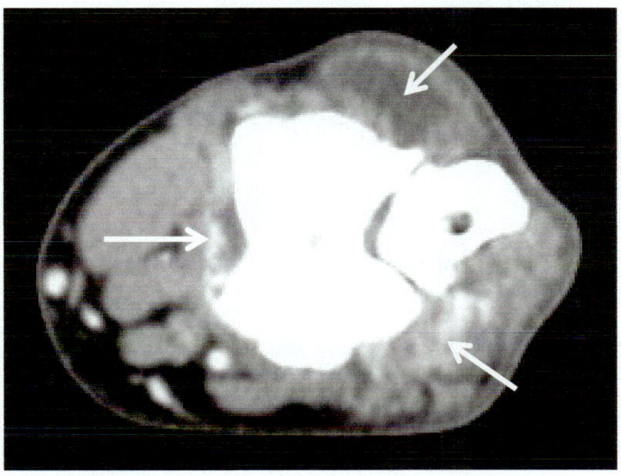

Figure 16.23 **Transverse image in soft-tissue window (a) of an inflamed elbow joint after IV contrast administration. Notice the contrast enhancement and obvious swelling of the joint capsule with fluid pockets and effusion (white arrows).**

Modifications at the medial epicondyle attributed to the skeletal reaction to elevated tensile forces within a tendon, like bony alterations at the flexor attachment to the medial epicondyle, sclerosis, and thickening of the cortex of the medial humeral epicondyle, are noticed. Calcifications within the soft tissues in the area of the medial humeral condyle can be easily visualized (Figure 16.24). In clinical cases, IV contrast administration will demonstrate enhancement around these calcified bodies (Figure 16.25).

Cats may suffer of medial humeral epicondylitis with chronic degeneration, mineralization, and metaplastic bone formation in damaged fibrillar matrix involving the origin of the humeral head of the flexor carpi ulnaris muscle (Figure 16.26).

16.5.3.4 Medial compartment syndrome

Medial compartment erosion (MCE) is an advanced stage of medial coronoid disease, an important cause of elbow lameness in dogs, referring to full-thickness cartilage loss of the medial part of the humeral condyle and the

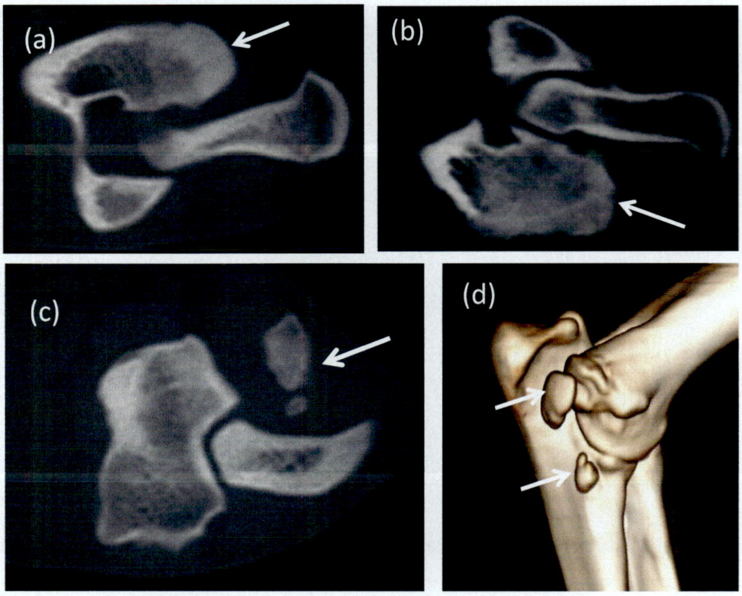

Figure 16.24 Flexor enthesopathy: Different stages of appearance of the attachment of the flexor tendons in the area of the medial condyle: (a) Thickened cortex with smooth delineation (arrow). (b) Irregular attachment (arrow). (c) Transverse and (d) 3D reconstructed image with calcified bodies within the flexor muscles (white arrows).

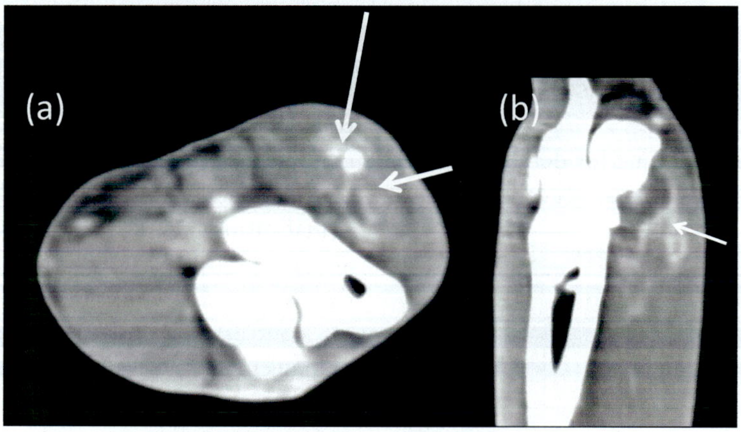

Figure 16.25 Transverse (a) and dorsal reconstruction (b) after administration of IV contrast of a dog suffering from flexor enthesopathy. There is heterogenous contrast enhancement with thickened, hyperattenuating appearance of the flexor muscles (arrows). The hypodense centre represents fluid.

corresponding ulnar contact area, with narrowing of the joint space in the medial compartment of the joint. The most prevalent signs on CT are periarticular osteophytes, an abnormal shape and or fragmentation of the medial coronoid process, a subchondral defect of the medial part of the humeral condyle, sclerosis of the medial part of the humeral condyle, ulna and medial coronoid process, and flattening of the medial part of the humeral condyle (Figure 16.27a–c). An interruption of the continuous delineation of the articular cartilage can be noticed on arthro-CT images (Figure 16.27d).

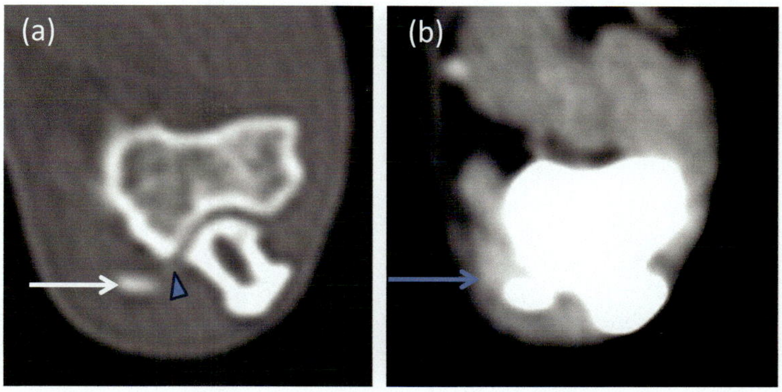

Figure 16.26 Transverse image in bone algorithm (a) and soft tissue algorithm after IV contrast administration (b) showing epicondylitis in a domestic short-haired cat. There is a mineralized intra-articular free joint body close to the medial joint capsule (white arrow). New bone formation (arrow head) is in the area of the origin of the humeral head of the flexor carpi ulnaris muscle. The thickening of the flexor tendons and contrast enhancement is noticed (blue arrow).

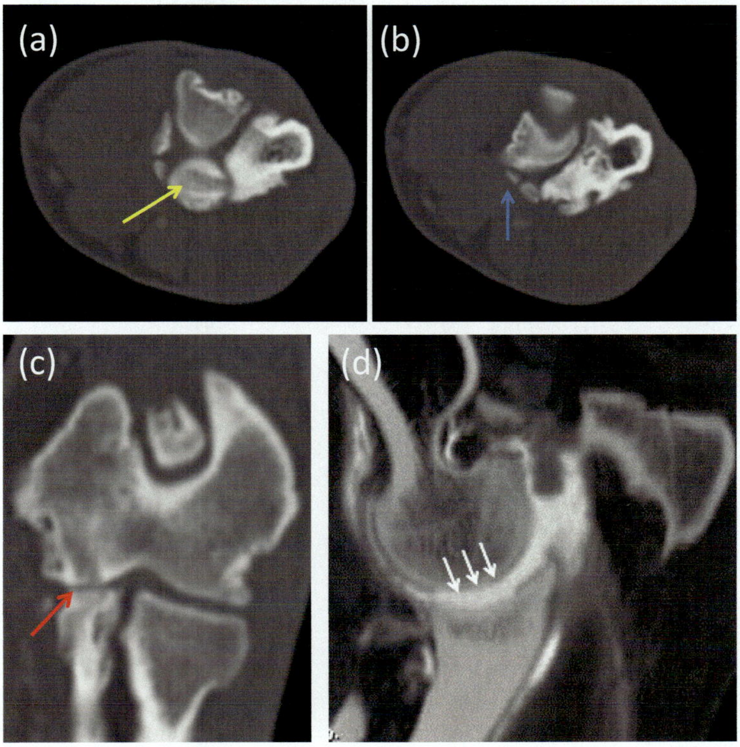

Figure 16.27 Transverse images at the level of the distal humeral condyles (a), at the level of the medial coronoid process (b), dorsal (c)- and sagittal-reconstructed (d) images, all in bone algorithm. There is severe DJD and severe sclerosis present with obvious kissing lesions at the medial humeral condyle (yellow arrow). There are two displaced fragments at the medial coronoid process (blue arrow). The joint space at the medial compartment is narrowed (red arrow). After performing arthro-CT (d), the absence of joint cartilage at the medial compartment of the joint is seen (white arrows).

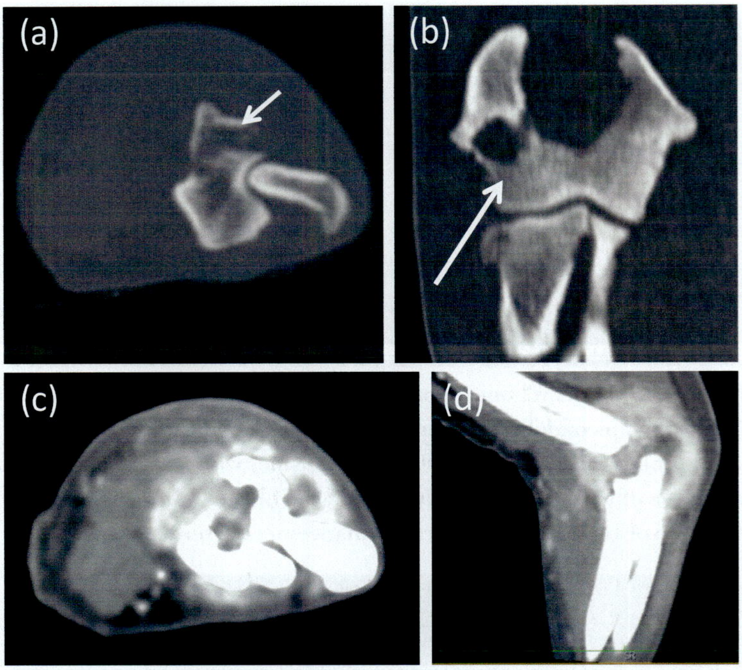

Figure 16.28 Transverse image (a) and dorsal reconstruction (b) both in bone window of a synovial cell sarcoma. Fingerprint-like osteolytic areas are noticed within the humerus (arrows). (c) Transverse and (d) sagittal reconstruction after IV contrast. There is substantial soft tissue swelling with massive contrast enhancement of the neoplastic synovial membrane.

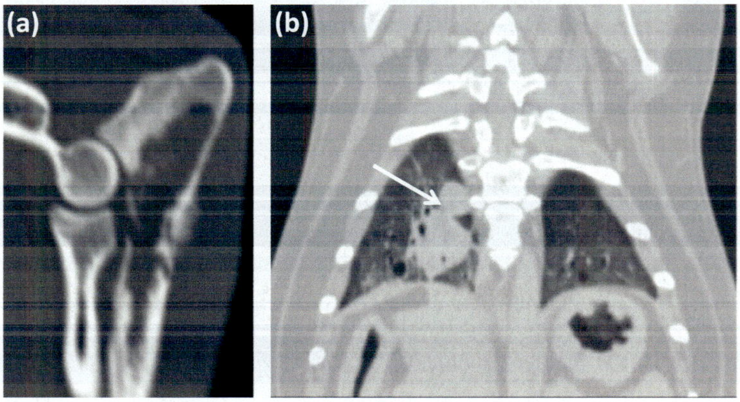

Figure 16.29 Sagittal-reconstructed image in bone algorithm (a) of an osteosarcoma of the proximal ulna of a 13-year-old mongrel dog and dorsal reconstruction in lung algorithm of the thorax (b). The destruction of bone and irregular new bone formation can be appreciated. Pulmonary metastasis is present (arrow).

16.5.4 Neoplasia

16.5.4.1 Synovial cell sarcoma

Like in other joints, bony lysis and thickening of the neoplastic synovial membrane can be appreciated. The process should be evaluated in bone- and soft-tissue window IV contrast; enhancement of the neoplastic process allows a more detailed demarcation of the tumour (Figure 16.28).

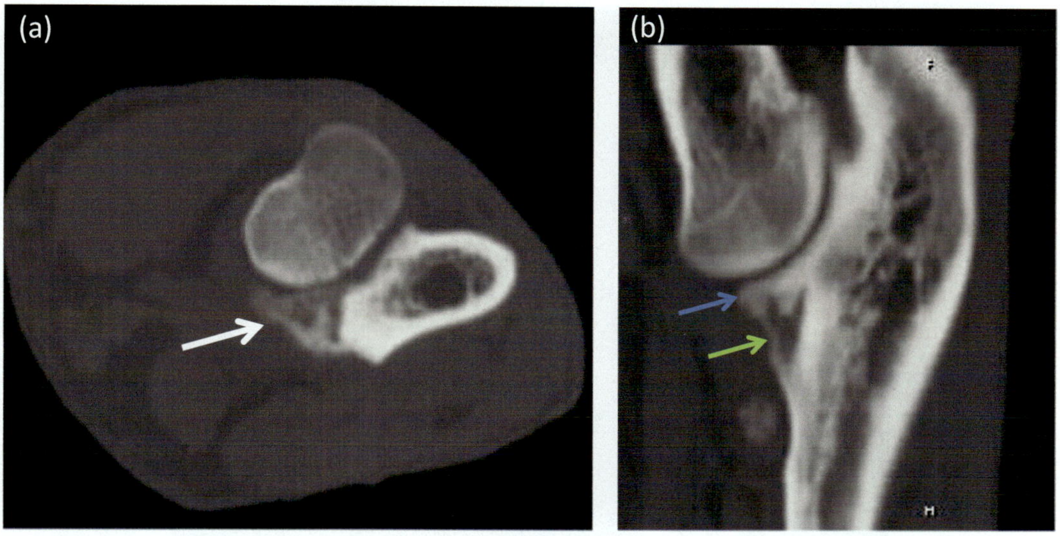

Figure 16.30 Transverse image at the level of the medial coronoid process (MCP) (a) and sagittal reconstruction (b). Abnormal delineation with demineralization of the MCP is present (white arrow). On the sagittal reformatted image, the surface of the MCP appears to be normal (blue arrow). More distally lysis and heterogeneous aspect of the MCP is noticed (green arrow). Calcified opacities are visible more distally. This is a confirmed chondroblastic osteosarcoma of the coronoid process, mimicking a fragmented coronoid process.

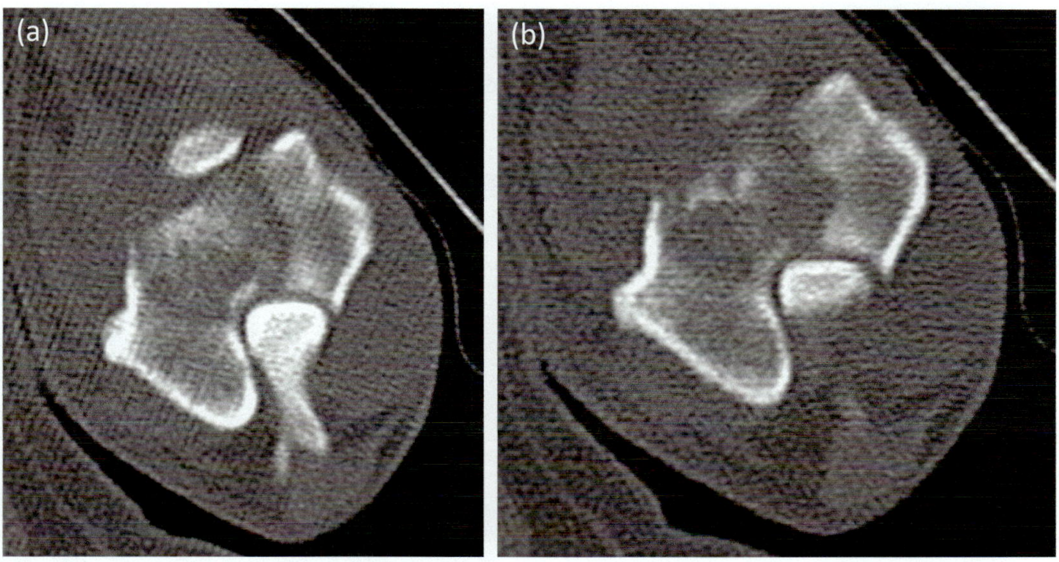

Figure 16.31 Metastases: (a, b) two consecutive transverse images in bone window of the distal humerus of predominantly osteolytic metastasis originating from a primary carcinoma.

16.5.4.2 Primary bone tumour

Primary bone tumours in the elbow area are rare. Bone lysis and bone formation are the common features. The process should be evaluated in bone- and soft-tissue windows (Figures 16.29 and 16.30). A biopsy under CT guidance can be performed.

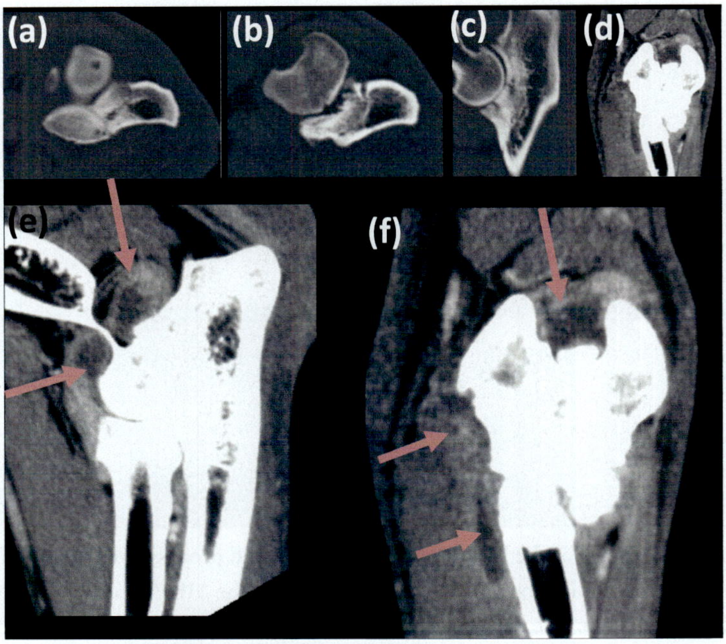

Figure 16.32 Transverse (a) at the level of the distal humeral condyles; (b) at the level of the medial coronoid process and sagittal reconstruction (c) in bone algorithm of an infected elbow joint with osteomyelitis. The humeral condyles, ulna, radius, and trochlear notch of the ulna show cortical defects, erosions, and lytic foci. IV contrast-reconstructed images (d, e, f) demonstrate joint effusion causing capsular distension with contrast enhancement and associated inflammatory changes with periarticular inflammatory soft tissue changes and fluid collections (arrows). The incidence of *Staphylococcus aureus* was confirmed.

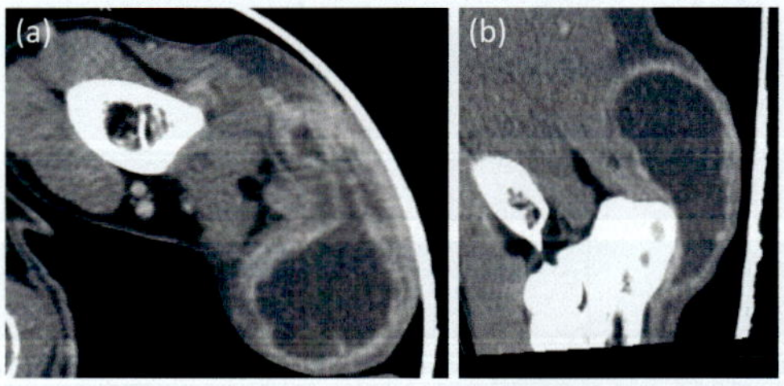

Figure 16.33 Transverse (a) and sagittal reconstructed image (b) in soft tissue algorithm after IV contrast administration of an abscess, which initially started with bursitis. There is contrast enhancement of the swelling on the olecranon demarcating an area with an obvious capsule. There is a mass effect on the triceps, but no involvement of the musculature itself. These images are suggestive for an abscess surrounding the olecranon. Bacterial culture was initiated, and cytology confirmed an infection.

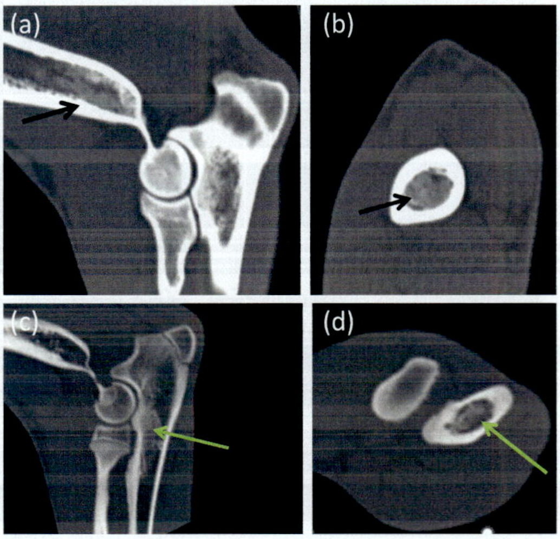

Figure 16.34 Sagittal reconstructed (a, c) and transverse (b: humerus, d: radius and ulna) images in bone algorithm of two young dogs. The medullary canal shows a patchy appearance with cloudiness (black arrows: humerus; green arrows: ulna).

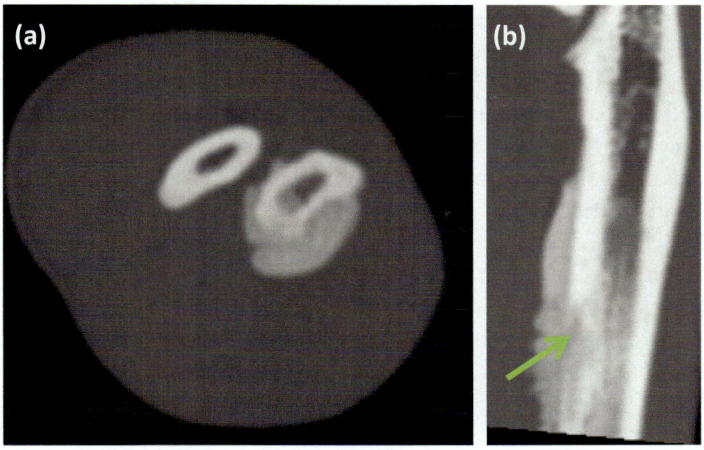

Figure 16.35 Transverse (a) and sagittal reconstruction (b) in bone algorithm. There is smooth periosteal reaction along the cortical margin of the ulna with extensive new endosteal new bone formation. The diaphysis shows an osteolytic lesion with cortical bone loss (arrow). There is the suspicion of ischaemic necrosis with possible osteomyelitis.

16.5.4.3 Metastases

The goal of diagnostic imaging is to detect skeletal metastases early, for which purpose CT is more sensitive than radiographs. Often, bone metastases have both lytic and blastic features (Figure 16.31).

16.5.5 Pathologic findings – miscellaneous

16.5.5.1 Infection

Both bacterial and fungal infections can cause osteomyelitis, which is a crucial differential diagnosis of osteosarcoma. Both diseases show a proliferative, mixed, or lytic appearance on imaging (Figure 16.32). Inflammation over the olecranon bursa; bursitis of hygroma, a pseudocyst, is a fluid-filled cavity surrounded by a dense fibrous connective tissue. This can lead to infection and abscess formation (Figure 16.33).

16.5.5.2 Panosteitis

Panosteitis is a painful inflammation of the outer surface or shaft of one or more long bones of the legs. It is sometimes called "growing pains." Panosteitis may occur in more than one bone at a time or may move around from area to area, cause a "shifting" lameness that goes from one bone to another. Imaging modalities show an increased density of the medullary canal in a patchy or mottled pattern with cloudiness and sclerosis, most frequently starting near the nutrient foramen (Figure 16.34).

16.5.5.3 Radioulnar ischemic necrosis

Osteonecrosis of the radius and/or ulna in dogs resulting from an unknown aetiology of ischaemic injury is known as radioulnar ischaemic necrosis, or RUIN. Upon imaging, it appears as a permeative bone destruction that spreads into the radius, ulna, or both bones from the interosseous ligament region. The lesion is usually located where the vein or artery enters the proximal third of the ulna and radius bones (Figure 16.35).

SUGGESTED READING

Alves-Pimenta S, Ginja MM, Colaço B. Role of elbow incongruity in canine elbow dysplasia: advances in diagnostics and biomechanics. *Vet Comp Orthop Traumatol.* 2019;32(2):87–96.

Alves-Pimenta S, Ginja MM, Fernandes AM, Ferreira AJ, Melo-Pinto P, Colaço B. Computed tomography and radiographic assessment of congruity between the ulnar trochlear notch and humeral trochlea in large breed dogs. *Vet Comp Orthop Traumatol.* 2017;30(1):8–14.

Anderson OJ, Langley-Hobbs SJ, Parsons KJ. Humeral condylar fractures and fissures in the French bulldog. *Vet Surg.* 2023;52(1):134–145.

Baud K, Griffin S, Martinez-Taboada F, Burton NJ. CT evaluation of elbow congruity in dogs: radial incisure versus apical medial coronoid process fragmentation. *J Small Anim Pract.* 2020;61(4):224–229.

Burton NJ, Warren-Smith CM, Roper DP, Parsons KJ. CT assessment of the influence of dynamic loading on physiological incongruency of the canine elbow. *J Small Anim Pract.* 2013;54(6):291–298.

Burton NJ, Meakin L, Hosworth A, Parsons KJ. Reliability of CT measurement of induced radioulnar step in dogs using a circle superimposition technique. *J Small Anim Pract.* 2018;59(2):92–97.

Carrera I. Computed tomographic features of incomplete ossification of the canine humeral condyle. *Vet Surg.* 2008;37(3):226–231.

Carpenter L, Schwarz P, Lowry J. Comparison of radiologic imaging techniques for diagnosis of fragmented medial coronoid process of the cubital joint in dogs. *J Am Vet Med Assoc.*1993;203:78–83.

Coppieters E, Gielen I, Verhoeven G, Van Vynckt D, Van Ryssen B. Erosion of the medial compartment of the canine elbow: occurrence, diagnosis and currently available treatment options. *Vet Comp Orthop Traumatol.* 2015;28(1):9–18.

Danielski A, Yeadon R. Humero-anconeal elbow incongruity in spaniel breed dogs with humeral intracondylar fissure: arthroscopic findings. *Vet Surg.* 2022;51(1):117–124.

Danner AL, Leonard K, Chandler J, Gendler A. Computed tomography-guided cannulated lag screw placement for treatment of ununited anconeal process: surgical technique, clinical outcome, and radiographic healing in 7 dogs. *J Am Vet Med Assoc.* 2024;14;262(5):1–7.

Davenport A, Clements D, Dancer S. Humeral intracondylar fissures and intracondylar sclerosis are common CT findings in the limb contralateral to a humeral condylar fracture in French bulldogs and spaniel breeds. *Vet Radiol Ultrasound.* 2023;64(4):686–693.

de Bakker E, Gielen I, van Caelenberg A, van Bree H, van Ryssen B. Computed tomography of canine elbow joints affected by primary and concomitant flexor enthesopathy. *Vet Radiol Ultrasound.* 2014; 55(1):45–55.

de Bakker E, Gielen I, Saunders JH, Polis I, Vermeire S, Peremans K, Dewulf J, van Bree H, Van Ryssen B. Primary and concomitant flexor enthesopathy of the canine elbow. *Vet Comp Orthop Traumatol.* 2013;26(6):425–434.

De Rycke LM, Gielen IM, van Bree H. Computed tomography of the elbow joint in clinically normal dogs. *Am J Vet Res.* 2002;63:1400–1407.

Ellis LF, House AK, Shimura K, Frances M. Comparison of hounsfield units within the humeral trochlea and medial coronoid process in a population of labrador X golden retriever guide dogs and border collies. *Vet Comp Orthop Traumatol.* 2024;37(5):223–229. doi: 10.1055/s-0043-1778631.

Farrell M, Trevail T, Marshall W, Yeadon R, Carmichael S. Computed tomographic documentation of the natural progression of humeral intracondylar fissure in a cocker spaniel. *Vet Surg.* 2011;40(8):966–971.

Franklin C, Herrtage M, Harris K, Genain MA. Prevalence of incidental humeral intracondylar fissures in brachycephalic breed dogs in CT studies. *Vet Radiol Ultrasound.* 2023;64(2): 194–200.

Gander Soares D, Allen MJ, Burton NJ. Bone density of the humeral condyle in Labrador retrievers with medial coronoid process disease. *J Small Anim Pract.* 2022;63(11):821–828.

Gasch EG, Labruyère JJ, Bardet JF. Computed tomography of ununited anconeal process in the dog. *Vet Comp Orthop Traumatol.* 2012;25(6):498–505.

Gemmill T. Completing the picture: use of CT to investigate elbow dysplasia. *J Small Anim Pract.* 2004;45(8):429–430.

Gemmill TJ, Hammond G, Mellor D, Sullivan M. Use of reconstructed computed tomography for the assessment of joint spaces in the canine elbow. *J Small Anim Pract.* 2006;47:66–74.

Gemmill TJ, Mellor DJ, Clements DN, Clarke SP, Farrell M, Bennett D, Carmichael S. Evaluation of elbow incongruency using reconstructed CT in dogs suffering fragmented coronoid process. *J Small Anim Pract*. 2005;46:327–333.

Gendler A, Keuler N, Schaefer S. Computed tomographic arthrography of the normal canine elbow. *Vet Radiol Ultrasound*. 2015;56:144–152.

Gielen I, Van Ryssen B, Buijtels J. Canine elbow incongruity evaluated with computerised tomography (CT), radiography and arthroscopy. *7th Annual Conference, Euopean Association for Veterinary Diagnostic Imaging*, July 18–21, 2001, Paris, France, p 22.

Glück T, Zablotski Y, Meyer-Lindenberg A. Microcomputed tomographic analysis of the ulnar trochlear notch in medium- and large-breed canine cadavers with and without medial coronoid disease. *Res Vet Sci*. 2024;166:105108.

Griffon DJ, Mostafa AA, Blond L, Schaeffer DJ. Radiographic, computed tomographic, and arthroscopic diagnosis of radioulnar incongruence in dogs with medial coronoid disease. *Vet Surg*. 2018;47(3):333–342.

Groth AM, Benigni L, Moores AP, Lamb CR. Spectrum of computed tomographic findings in 58 canine elbows with fragmentation of the medial coronoid process. *J Small Anim Pract*. 2009;50(1):15–22.

Hebel M, Panek WK, Ruszkowski JJ, Nabzdyk M, Niedzielski D, Pituch KC, Jackson AM, Kiełbowicz M, Pomorska-Mól M. Computed tomography findings in a cohort of 169 dogs with elbow dysplasia — a retrospective study. *BMC Vet Res*. 2021;17(1):296.

Hersh-Boyle RA, Chou PY, Kapatkin AS, Spriet M, Filliquist B, Garcia TC, Marcellin-Little DJ. Comparison of needle arthroscopy, traditional arthroscopy, and computed tomography for the evaluation of medial coronoid disease in the canine elbow. *Vet Surg*. 2021;50(Suppl 1):O116–O127.

Holsworth IG, Wisner ER, Scherrer WE, Filipowitz D, Kass PH, Pooya H, Larson RF, Schulz KS. Accuracy of computerized tomographic evaluation of canine radio-ulnar incongruence in vitro. *Vet Surg*. 2005;34:108–113.

House MR, Marino DJ, Lesser ML. Effect of limb position on elbow congruity with CT evaluation. *Vet Surg*. 2009;38(2):154–160.

Humphreys WJ, Wilder K, Pettitt R, Comerford EJ, Maddox TW. CT attenuation of the medial coronoid process is reduced in dogs with medial coronoid disease but independent of arthroscopic disease severity. *Am J Vet Res*. 2022;83(9):1–8.

Klumpp S, Ondreka N, Amort K, Zwick M, Gesierich K, Engert M, Tacke S, Failing K, Kramer M. Diagnostic value of CT and MRI for the diagnosis of coronoid pathology in the dog. *Tierarztl Prax Ausg K Kleintiere Heimtiere*. 2010;38(1):7–14.

Kramer A, Holsworth IG, Wisner ER. Computed tomographic evaluation of canine radioulnar incongruence in vivo. *Vet Surg*. 2006;35:24–29.

Kroner K, Cooley K, Hoey S, Hetzel SJ, Bleedorn JA. Assessment of radial torsion using computed tomography in dogs with and without antebrachial limb deformity. *Vet Surg*. 2017;46(1):24–31.

Krotscheck U, Böttcher PB, Thompson MS, Todhunter RJ, Mohammed HO. Cubital subchondral joint space width and CT osteoabsorptiometry in dogs with and without fragmented medial coronoid process. *Vet Surg*. 2014;43(3):330–338.

Kubiak-Nowak D, Kiełbowicz Z, Hebel M, Przadka P, Agnieszka Antonczyk A and Jankowski M. Assessment of Elbow Joint Incongruity as a Primary Cause of Canine Elbow Dysplasia: Comparative Imaging Analysis Using CT and Radiography in 108 Dogs—A Retrospective Study. *Life*. 2025; 15(1673): 1–18, https://doi.org/10.3390/life15111673

Kunst CM, Pease AP, Nelson NC, Habing G, Ballegeer EA. Computed tomographic identification of dysplasia and progression of osteoarthritis in dog elbows previously assigned OFA grades 0 and 1. *Vet Radiol Ultrasound*. 2014;55(5):511–520.

Labrador J, Witte PG, Holdsworth A. Use of an extended CT scale reconstruction to diagnose metal implant failure in a canine elbow. *Vet Radiol Ultrasound*. 2021;62(6):e63–e66.

Lappalainen AK, Mölsä S, Liman A, Laitinen-Vapaavuori O, Snellman M. Radiographic and computed tomography findings in Belgian shepherd dogs with mild elbow dysplasia. *Vet Radiol Ultrasound*. 2009;50(4): 364–369.

Lau SF, Theyse LF, Voorhout G, Hazewinkel HA. Radiographic, computed tomographic, and arthroscopic findings in labrador retrievers with medial coronoid disease. *Vet Surg*. 2015;44(4):511–520.

Mella S, Dirrig H, Meeson RL. Computed tomographic features of non-lame French bulldog elbows. *Vet Comp Orthop Traumatol*. 2022;35(3):175–183.

Moores AP, Benigni L, Lamb CR. Computed tomography versus arthroscopy for detection of canine elbow dysplasia lesions. *Vet Surg*. 2008;37(4):390–398.

Pardo M, Morris AP, Frazzica F, Barnes DM. Accuracy of medial-to-lateral transcondylar screw placement using an aiming device and preoperative computer tomography planning for the treatment of humeral intracondylar fissure. *Vet Surg*. 2021;50(6):1218–1226.

Payne DJL, Sparks TH, Smith MAJ, Macdonald NJ. Computed tomography topographical analysis of incomplete humeral intracondylar fissures in English Springer Spaniel dogs. *Vet Comp Orthop Traumatol*. 2024;37(2):64–73.

Punke JP, Hulse DA, Kerwin SC, Peycke LE, Budsberg SC. Arthroscopic documentation of elbow cartilage pathology in dogs with clinical lameness without changes on standard radiographic projections. *Vet Surg*. 2009;38(2):209–212.

Rau FC, Wigger A, Tellhelm B, Zwick M, Klumpp S, Neumann A, Oltersdorf B, Amort K, Failing K, Kramer M. Observer variability and sensitivity of radiographic diagnosis of canine medial coronoid disease. *Tierarztl Prax Ausg K Kleintiere Heimtiere*. 2011;39(5):313–22.

Reichle JK, Park RD, Bahr AM. Computed tomographic findings of dogs with cubital joint lameness. *Vet Radiol Ultrasound*. 2000;41:125–130.

Rohr B, Reese S, Zöllner M, Meyer-Lindenberg A. Diagnostic value of contrast-enhanced CT for elbow joint disorders in dogs. *Front. Vet*. Sci. 2025; 12:1626472.

Samii VF, Les Clifford M, Schulz KS, Keyak JH, Stover SM. Computed tomographic osteoabsorptiometry of the elbow joint in clinically normal dogs. *Am J Vet Res*. 2002;63(8):1159–1166.

Samoy Y, Van Ryssen B, Gielen I. Review of the literature: elbow incongruity in the dog. *Vet Comp Orthop Traumatol*. 2006;19(1):1–8.

Samoy Y, Gielen I, Van Caelenberg A, van Bree H, Duchateau L, Van Ryssen B. Computed tomography findings in 32 joints affected with severe elbow incongruity and fragmented medial coronoid process. *Vet Surg*. 2012;41(4):486–494.

Samoy Y, Gielen I, Saunders J, van Bree H, Van Ryssen B. Sensitivity and specificity of radiography for detection of elbow incongruity in clinical patients. *Vet Radiol Ultrasound*. 2012;53(3):236–244.

Seidler S, Rhode M, Volk H, Harms O. Comparative evaluation of radiographic and computed tomographic findings in dogs with bilateral medial coronoid disease (MCD) presenting with unilateral forelimb lameness. *PLoS One*. 2023;18(4):e0282656.

Shimizu N, Warren-Smith CM, Langley-Hobbs SJ, Burton NJ, Kulendra E, Bradley K, Bowen E, Holdsworth A, Parsons KJ. Inter- and intraobserver agreement in interpretation of CT features of medial coronoid process disease. *J Small Anim Pract*. 2015;56(12):707–713.

Shubert MP, Filliquist B, Chou PY, Kapatkin AS, Spriet M, Kim SY, Garcia TC, Marcellin-Little DJ. Results of using multiplanar reconstructed CT images for assessing elbow joint osteoarthritis in dogs are consistent with results of radiographic assessment. *Am J Vet Res*. 2022;83(10). doi: 10.2460/ajvr.22.04.0066

Skinner OT, Warren-Smith CMR, Burton NJ, Parsons KJ. Computed tomographic evaluation of elbow congruity during arthroscopy in a canine cadaveric model. *Vet Comp Orthop Traumatol*. 2015;28:1.

Tromblee TC, Jones JC, Bahr AM, Shires PK, Aref S. Effect of computed tomography display window and image plane on diagnostic certainty for characteristics of dysplastic elbow joints in dogs. *Am J Vet Res*. 2007;68(8):858–871.

Vali Y, Villamonte-Chevalier A, Dorny M, Ludewig E, van Bree H, Gielen I. Radiographic and computed tomographic evaluation of supinator sesamoid bones in dogs. *BMC Vet Res*. 2022;18(1):377.

van Bree H, Van Ryssen B. Diagnostic imaging of the canine elbow including radiology, arthroscopy and computed tomography (CT). Oral Abstracts, 10th IRVA Meeting. *Vet Radiol Ultrasound*. 1994;35:248.

van Bree H, Gielen I, Van Ryssen B. Comparative joint imaging in small animals. *Eur J Compan Anim Pract*. 2002;12:25–36.

Van Ryssen B, de Bakker E, Beaumlin Y, Samoy YC, Van Vynckt D, Gielen I, Ducatelle R, van Bree H. Primary flexor enthesopathy of the canine elbow: imaging and arthroscopic findings in eight dogs with discrete radiographic changes. *Vet Comp Orthop Traumatol*. 2012;25(3):239–245.

Vermote K, Bergenhuyzen A, Gielen I, van Bree H, Duchateau L, Van Ryssen B. Elbow lameness in dogs of six years and older: arthroscopic and imaging findings of medial coronoid disease in 51 dogs. *Vet Comp Orthop Traumatol*. 2010;23:43–50.

Villamonte-Chevalier A, van Bree H, Broeckx B, Dingemanse W, Soler M, Van Ryssen B, Gielen I. Assessment of medial coronoid disease in 180 canine lame elbow joints: a sensitivity and specificity comparison of radiographic, computed tomographic and arthroscopic findings. *BMC Vet Res*. 2015;11:243.

Wagner K, Griffon DJ, Thomas MW, Schaeffer DJ, Schulz K, Samii VF, Necas A. Radiographic, computed tomographic, and arthroscopic evaluation of experimental radio-ulnar incongruence in the dog. *Vet Surg*. 2007;36(7):691–698.

Wennemuth J, Tellhelm B, Eley N, von Pückler K. Computed tomography enhances diagnostic accuracy in challenging medial coronoid disease cases: an imaging study in dog breeding appeal cases. *Vet Comp Orthop Traumatol*. 2020;33(5):356–362.

Zweifel RT, DiDonato P, Hartmann A, Kramer M, von Pückler KH. Improved computed tomography accuracy with a 1-mm versus 2- or 3-mm slice thickness for the detection of medial coronoid disease in dogs. *Vet Comp Orthop Traumatol*. 2020;33(1):45–50.

CHAPTER 17

CT OF THE CARPUS

Ingmar Kiefer, Eberhard Ludewig, Amalia Agut, and R. Latorre (Anatomy)

17.1 BRIEF INTRODUCTION TO CARPAL JOINT ANATOMY AND PATHOLOGY

The carpus consists of two rows of small bones. It represents a complex set of joints. In many cases, a lesion of an individual structure has an impact on the other components of the joint and its neighbourhood. Therefore, a lesion often results in a significant loss of function. CT is able to identify even subtle structural bony changes and helps categorize findings and guide biopsies if required. However, its ability to demonstrate soft-tissue abnormalities is limited.

17.2 INDICATIONS FOR CARPAL CT

Normally, radiography is the preferred imaging method when evaluating the carpus. CT in particular is helpful in the following cases:

- when radiography fails to identify a lesion,
- further information is required for treatment planning or assess prognosis.

Indications for a CT-study of the carpus are as follows:

- Joint-related lameness and/or pain
- Joint laxity
- Limb deformities with suspected joint involvement
- Mass-like lesion or soft-tissue swelling
- Systemic disease with suspected joint involvement or involvement of the adjacent radius, ulna, and the bones of the digits.

17.3 TECHNICAL ASPECTS

17.3.1 Positioning

Positioning in ventral or dorsal recumbency allows investigation of both limbs within a single scan (Figure 17.1). Especially in cases where subtle lesions are expected, the comparison of the suspected lesion with corresponding structures of the contralateral limb is helpful. Alternatively, the limbs can be scanned with the animal in lateral recumbency, with the affected limb kept upmost. To investigate the opposite

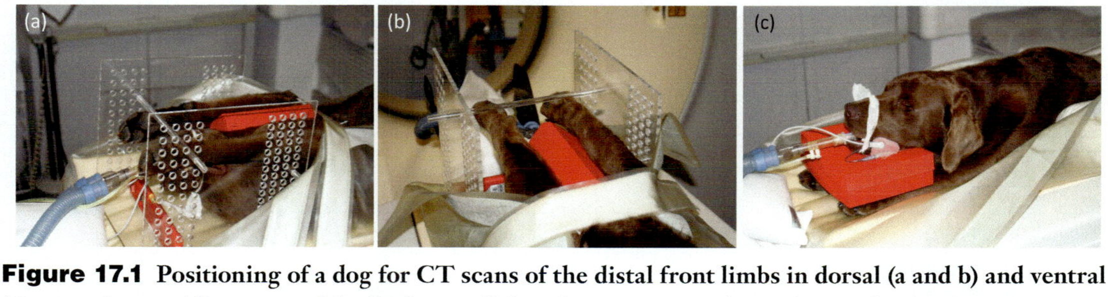

Figure 17.1 Positioning of a dog for CT scans of the distal front limbs in dorsal (a and b) and ventral (c) recumbency. Alignment of the limbs parallel to the scan axis reduces the needs of compensatory manual post-processing and helps avoid misinterpretations due to positioning differences.

DOI: 10.1201/9781315118130-19

limb, the position of the animal must be changed. The advantage of this technique is that the Scan-Field of View is much smaller compared to scanning both limbs in a single scan. Care should be taken to position the antebrachium along the scan-axis and that in the case of parallel scanning of both joints, the joints further should be placed at the same level relative to the scan plane. This prevents time-consuming subsequent image-manipulation.

Additional functional studies ("Stress-CT") help clarify reasons for instability that could be demonstrated in the clinical examination. Useful stress manoeuvres in the carpus are shear and wedge stress forces. Sometimes, it is important to compare both joints when evaluating stress views. Helpful positioning devices are bandages, tapes, and radiolucent foam material.

17.3.2 Technical parameters

Due to the small dimensions of the target structures, detail rendition is one of the most important image quality parameters. Therefore, special effort is necessary to achieve small voxel sizes and ensure a high signal-to-noise ratio (Table 17.1).

The protocol is a recommendation for the majority of indications in medium-sized and large-breed dogs. The investigation of small structures such as the carpus of toy breed dogs, cats, and young animals requires further specification to achieve adequate detail visibility.

17.3.3 Contrast studies

Contrast studies (angiography, arthrography, fistulography, and lymphography) are methods to improve the differentiation of soft-tissue structures. Protocols of such studies follow general principles described in another chapter.

Table 17.1 **CT imaging protocol**

PARAMETER	SETTINGS
Motion control	Not required
Scan margin – proximal[a]	Middle of the antebrachium
Scan margin – distal[a]	Distal to the digits
Scan direction	Disto-proximal or proximo-distal
Voltage	100 … 140 kVp
Current	80 … 140 mAs
Rotation time	0.5 … 2 seconds
Slice width (collimated or binned)	<1 mm
Axial slice interval	0.5 mm
Detector pitch	≤1
Helical image reconstruction interval	<1 mm
Kernel frequeny	High
Window – bone	
Window level (WL)	200 … 500 HU
Window with (WW)	2,500 … 3,000 HU
Window – soft tissue	
Window level (WL)	50 … 100 HU
Window with (WW)	280 … 500 HU

[a]To study the axis of the leg (e.g. to plan corrective surgery), the scan margins must be extended.

17.4 NORMAL CT ANATOMY AND/OR ANATOMIC VARIATION(S)

The seven bones of the carpus are arranged in two rows. The proximal row consists of the radial carpal bone (Rc) (formed by fusion of three separate ossification centres), the ulnar carpal bone (Uc), and the accessory carpal bone (Acc). The distal row is composed of four carpal bones (C1–C4). A sesamoid bone medial to the middle carpal joint is located in the tendon of the abductor pollicis longus muscle.

There are three joint levels: the antebrachiocarpal joint, the middle carpal joint, and the carpometacarpal joint. They are responsible for approximately 80%, 15%, and 5% of carpal motion, respectively. The joints between bones within the rows (intercarpal joints) are of minor importance with respect to the range of motion. Short intraarticular ligaments link the bones and hold them in place. None of the intraarticular ligaments are spanning all three joint levels. Most are crossing only one joint.

The joint capsule of the carpus is composed of two non-communicating departments. The proximal part covers the antebrachiocarpal articulation, the distal radioulnar articulation, and the articulation of the accessory carpal bone. The distal part encloses the articulations of the middle carpal, the carpometacarpal, and the intermetacarpal junctions. The joint capsule blends with the palmar carpal fibrocartilage and ligaments.

Ligamentous support comprises medial and lateral collateral ligaments. The accessory carpal bone is stabilized by short ligaments to the ulnar carpal bone and long ligaments to the base of metacarpal bones IV and V. Muscles and the tendons of the flexor and extensor muscles are located around the carpus. Some of these soft-tissue dense structures can be identified in CT. However, due to the lack of fat delineation, it is often difficult. Smaller structures, such as tendon sheaths, bursae, fasciae, and nerves, usually cannot be identified (Figures 17.2–17.5).

Appearance of the ossification centres and physeal closure varies between species and breeds (Table 17.2). In toy breed dogs, ossification centres mineralize earlier and physeal closure occurs earlier than in large-breed dogs. In gonadectomized cats, physes close later in comparison to intact individuals.

CT studies in normal dogs have shown that the subchondral bone density varies depending on the location of the measurement and age of the dog. Areas of high density are seen in the subchondral bones of the dorsomedial aspects of the antebrachiocarpal joint. In both the midcarpal and carpometacarpal joint, there is a more homogenous distribution. The subchondral bone density increases with increasing age, but the distribution patterns do not change.

17.5 PATHOLOGIC FINDINGS

17.5.1 Developmental

A variety of carpal anomalies are seen in dogs and cats. Malformation relates to a diverging number of bones, disturbances in ossification, premature closure and other disturbances of the physes, and deformities of the distal limb.

17.5.1.1 Diverging number of bones

Aplasia and supernumerary bones can be found sporadically in dogs and cats.

In some of the subtypes of the subsequently listed malformations, not only the digits but also the carpus is involved. Depending on their degree, the structural changes of the musculature can vary in a wide range.

Polydactyly is a rare hereditary disease in which supernumerary toes are present. The condition is reported to be more common in cats than in dogs. In cats, various combinations ranging from four to seven toes per foot can be found. The number of toes is typically the same in each limb. Polydactylism affects the front legs more frequently than the hind legs. Usually, the hind legs are involved in addition to having polydactyl front paws. The malformation is an autosomal-dominant trait (Figure 17.6).

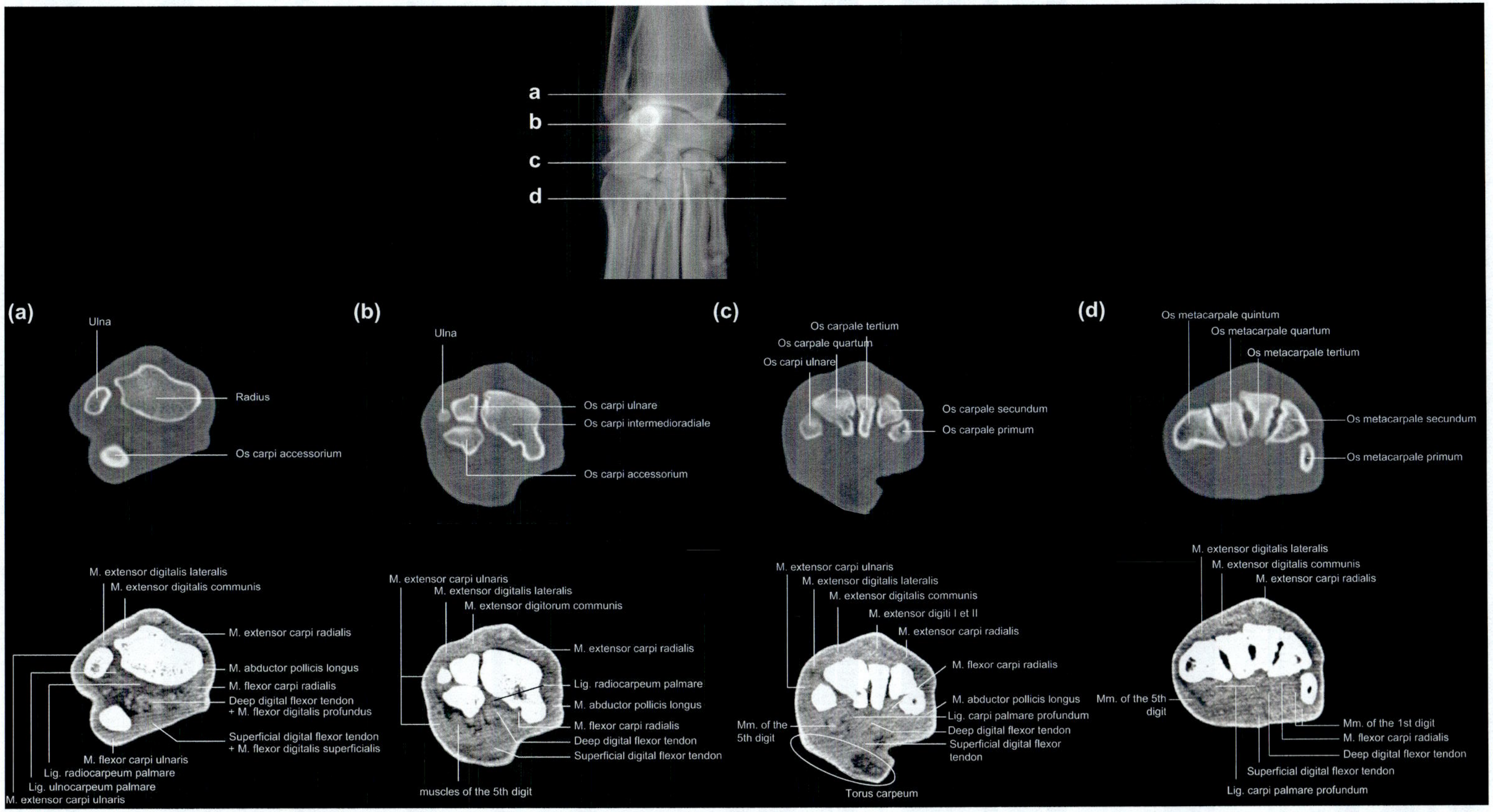

Figure 17.2 CT-anatomy of the canine carpus: (a) Transverse CT slice at the level of the distal radius and ulna: 1 bone window; 2 soft-tissue window. (b) Transverse CT at the level of the intermedioradial carpal bone: 1 bone window; 2 soft-tissue window. (c) Transverse slice of the carpus at the level of distal row intercarpal bones: 1 bone window; 2 soft-tissue window. (d) Transverse slice of the metacarpal bones: 1 bone window; 2 soft-tissue window.

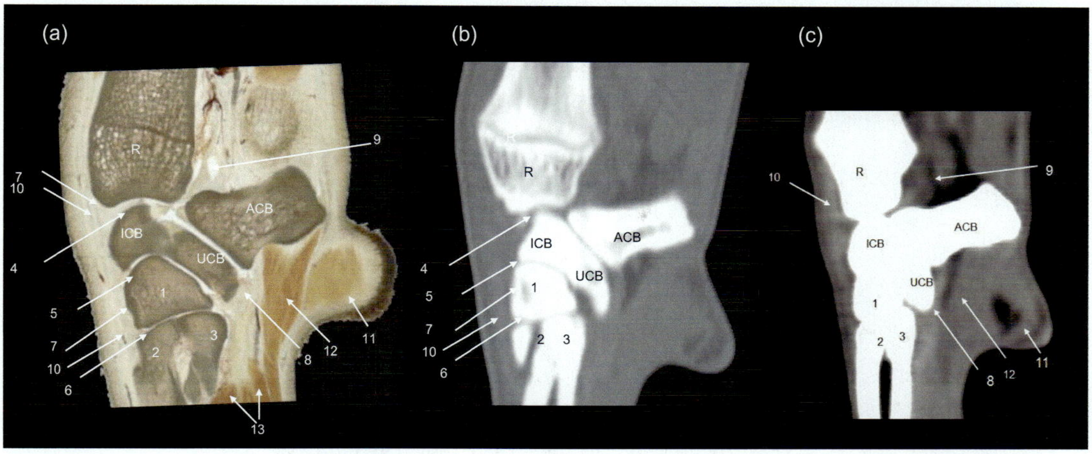

Figure 17.3 Sagittal plane of the carpus at the level of the accessory carpal bone. (a) Anatomic specimen; (b) CT reconstruction in the bone window; (c) CT reconstruction in the soft-tissue window. (R: radius, ACB: accessory carpal bone, UCB: ulnar carpal bone, ICB: intermedioradial carpal bone. 1 = four carpal bone; 2 = metacarpal bones; 3 = metacarpal bones; 4 = antebrachiocarpal joint; 5 = middle carpal joint; 6 = carpometacarpal joint; 7 = dorsal aspect of the capsule of the carpal joint; 8 = palmar ligaments of the carpal joint; 9 = palmar synovial pouch; 10 = common digital extensor tendon; 11 = carpal pad; 12 = short digital muscles of the fifth digit; 13 = interosseous muscles).

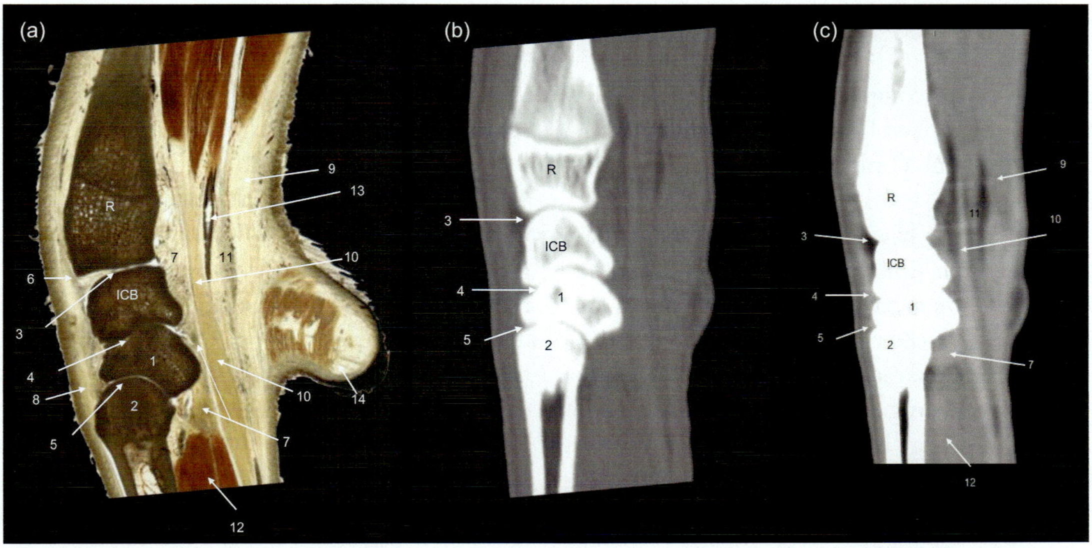

Figure 17.4 Sagittal plane of the carpus at the level of the intermedioradial carpal bone. (a) Anatomic specimen. (b) CT reconstruction in the bone window. (c) CT reconstruction in the soft-tissue window. (radius R, intermedioradial carpal bone ICB. 1 = third carpal bone III; 2 = metacarpal bone III; 3 = antebrachiocarpal joint (ACJ); 4 = middle carpal joint (MCJ); 5 = carpometacarpal joint (CMJ); 6 = dorsal aspect of the capsule of the carpal joint (CJ); 7 = palmar ligaments of the carpal joint (PL); 8 = extensor carpi radialis tendon (ECR); 9 = superficial digital flexor tendon (SDF); 10 = deep digital flexor tendon (DDF); 11 = flexor retinaculum (FR); 12 = interosseous muscles (IM); 13 = median artery MA; 14 = carpal pad (CP)).

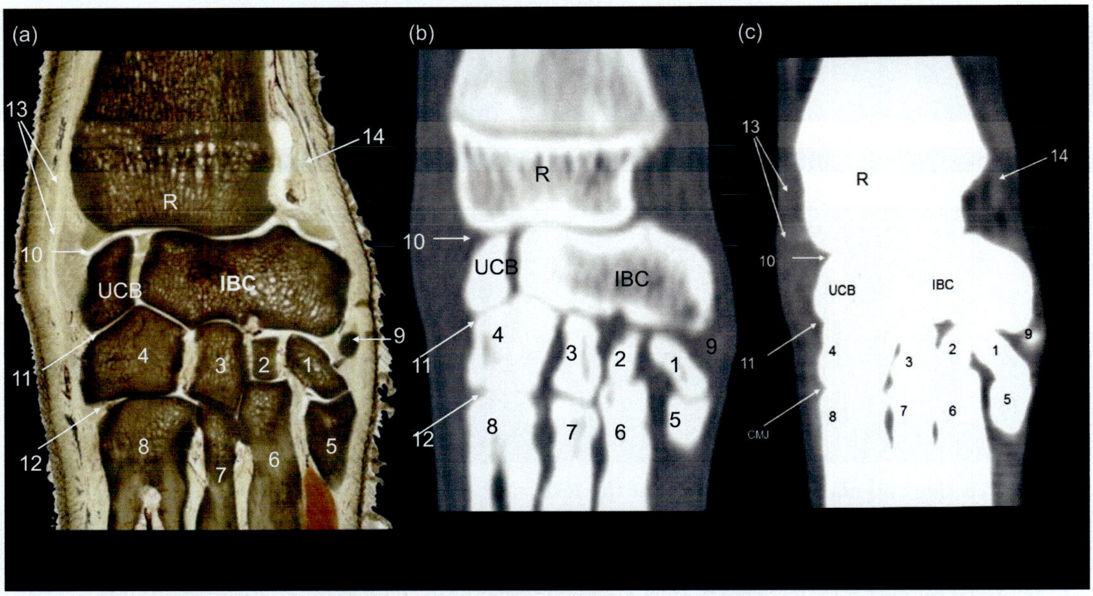

Figure 17.5 **Dorsal plane of the carpus. (a) Anatomic specimen. (b) CT reconstruction in bone window. (c) CT reconstruction in soft-tissue window. Dorsal section of the carpus (radius R, ulnar carpal bone UCB, intermedioradial carpal bone ICB. 1 = first carpal bone I; 2 = second carpal bone II; 3 = third carpal bone III; 4 = four carpal bone IV; 5 = metacarpal bone I; 6. metacarpal bone II; 7 = metacarpal bone III; 8 = metacarpal bone IV; 9 = sesamoid bone (SB); 10 = antebrachiocarpal joint (ACJ); 11 = middle carpal joint (MCJ); 12 = carpometacarpal joint (CMJ); 13 = lateral colateral ligament (LCL); 14 = medial colateral ligament (MCL)).**

Dimelia is a congenital musculoskeletal disorder characterized by duplication of the whole or part of a limb ("mirror hand"). It is reported in dogs.

In ectrodactyly, occurring in dogs and cats, the distal front limb is split. A part of the phalanges, the metacarpal and carpal bones are associated with the radius or the ulna.

Disorders characterized by reduced number of bones range from absence of a single part of the distal limb to partial or complete absence of the limb. The term hemimelia describes conditions with missing limb segments. Several subtypes are reported in dogs and cats. Transverse hemimelia refers to an anomaly characterized by a complete absence of part of the limb across the full width (congenital amputation). In paraxial hemimelia, only cranial (pre-) or caudal (post-)axial limb structures are absent. The term terminal hemimelia is used for malformations where there is a partial or complete defect in the distal limb portion.

17.5.1.2 Disturbances in ossification

Ossification centres are less observed in congenital disorders of endochondral ossification (e.g., chondrodystrophy) and congenital diseases influencing bone development (e.g., hypopituitarism and hypothyroidism). These disturbances cause osseous malformation.

17.5.1.3 Premature closure of physes

Premature closure and retarded development of the radial and ulnar physes caused by chondrodystrophies or physeal trauma during growth result in malarticulation of the antebrachiocarpal joint and change in the ligamentous support. The degree of limb deformity is dependent on the age of the animal when trauma or other initiating incident occurs and the time left to the actual physeal closure.

Table 17.2 **Radiographic appearance of the ossification centres and physeal closure**

SITE	DOG (GSD)		CAT (DSH)	
	FIRST APPEARANCE OF THE OSSIFICATION CENTRES (IN DAYS)	FUSION OF THE APO- OR EPIPHYSEAL LINES (IN MONTHS)	FIRST APPEARANCE OF THE OSSIFICATION CENTRES (IN DAYS)	FUSION OF THE APO- OR EPIPHYSEAL LINES (IN MONTHS)
Radius—distal	49–65	9–11	20–24	19–25
Ulna—distal	14–29	9–11	21–31	19–22
Os carpi radiale	28–29		29–43	
Os carpi intermedium	16–22	3–4	22–36	3–7
Os carpi centrale	28–36		20–24	
Os carpi ulnare	28–36		29–42	
Os carpi accessorium	14–16		20–28	
Apophysis ossis carpi accessorium	49–72	4–5	43–59	6–7
Os carpale I	21–29		22–35	
Os carpale II	28–36		20–28	
Os carpale III	28–36			
Os carpale IV	21–29		20–29	
Metacarpalia - proximal	21–35	5–6	20–42	9–12
Os sesamoideum m. abductor pollicis longus	120		99–183	

Source: Data abstracted from Schröder (1978) and Horvath (1983).

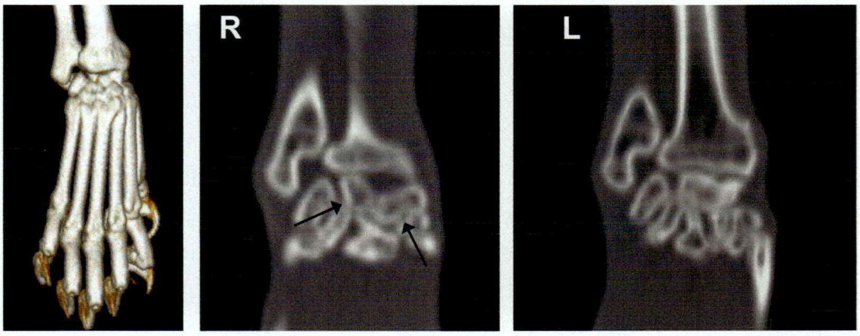

Figure 17.6 **Polydactyly with six digits in each foot in a 4-year-old male cat. Besides the supernumerary bones distal to the right carpus, there are abnormalities of the carpal bones with respect to number, shape, and position. Instead of the radial carpal bone (Os intermedioradiale), two separate bones can be identified (arrows). For comparison, the dorsal view of the left carpus is mirrored.**

17.5.1.4 Deformities of the distal limb

Carpal deformities related to growth are seen in rapidly growing medium and large puppy breeds. Several forms of deformities are described, such as carpal flexural deformity, carpal hyperflexion, and carpal hyperextension. The conditions are characterized by excessive hyperextension or hyperflexion at the carpus, which frequently is related with a varus or valgus deformity. Aetiologies are unclear. It has been suggested that poor muscle tone or weakness between the extensor and flexor muscle groups or imbalances in growth rates between the bones and supporting soft-tissue structures, such as tendons, might cause the deformity. The bones of the limb exhibit no abnormalities.

17.5.2 Traumatic

The carpus represents a site often affected in patients with the history of trauma. Trauma of the distal front limb can be classified into several injury patterns. The larger and more athletic the animal, the more severe are injuries in this region. CT is an important imaging modality to ensure the selection of adequate treatment options. The aims of CT studies are as follows:

- to detect fractures
  The identification of fractures especially in carpal bones of small animals and of small fragments is challenging. CT better than radiography is able to visualize fractured bones.
- to describe the degree of bone involvement
  CT improves the knowledge about the size and position of the fragments and the degree of the involvement of articular surfaces. This information is important for treatment planning.
- to map soft tissue changes and avulsed fragments
  Traumatic injury of bones is accompanied by soft tissue swelling due to damage of these tissues. The normal delineation of soft-tissue structures is getting lost. Nevertheless, the location of the soft-tissue damage and the identification of avulsed fragments can help differentiate injury patterns.
- to identify sources of instability and assess the degree of functional loss
  Functional investigations are helpful to localize and to quantify the lack of ligamentous support.
- to control postoperative outcomes
  A complete reconstruction of bony surfaces is important to reduce the risk of the development of secondary degenerative changes. Despite artefacts encountered after surgery using metal implants, the question of integrity of surfaces often can be answered reliably. Furthermore, CT can answer other questions related to post-operative outcomes; e.g., CT can support decision-making in cases of arthrodesis or osteosynthesis when the removal of implants is considered.

17.5.2.1 Fractures

Fractures of the carpal and adjacent bones can take many forms. While fractures of the radius and ulna, the radial carpal bone, and metacarpal bones are frequent, fractures of the ulnar carpal bone and the numbered carpal bones are rare (Figure 17.7).

The majority of fractures of the carpal bones are manifested as chips or slabs off the articular surface. These fragments result from combined jamming or compressive and shear forces.

There is little tendency for these fragments to heal spontaneously. The fragments cause inflammatory reactions with subsequent degenerative changes. Therefore, early diagnosis and adequate treatment are required to achieve a good prognosis.

Several types of fractures of the radial carpal bone can be seen. When load is directed axially, the bone is typically crushed. Chip fractures of the dorsal border are seen when the joint is in hyperextension during injury. Transverse fractures are rare.

Both fractures of the ulnar carpal bone and fractures of the numbered carpal bones occur only rarely. Most of the fractures of the numbered carpal bones are seen in racing or working dogs.

Fractures of the accessory carpal bone are avulsion fractures predominantly seen in racing greyhounds. They were categorized into the following:

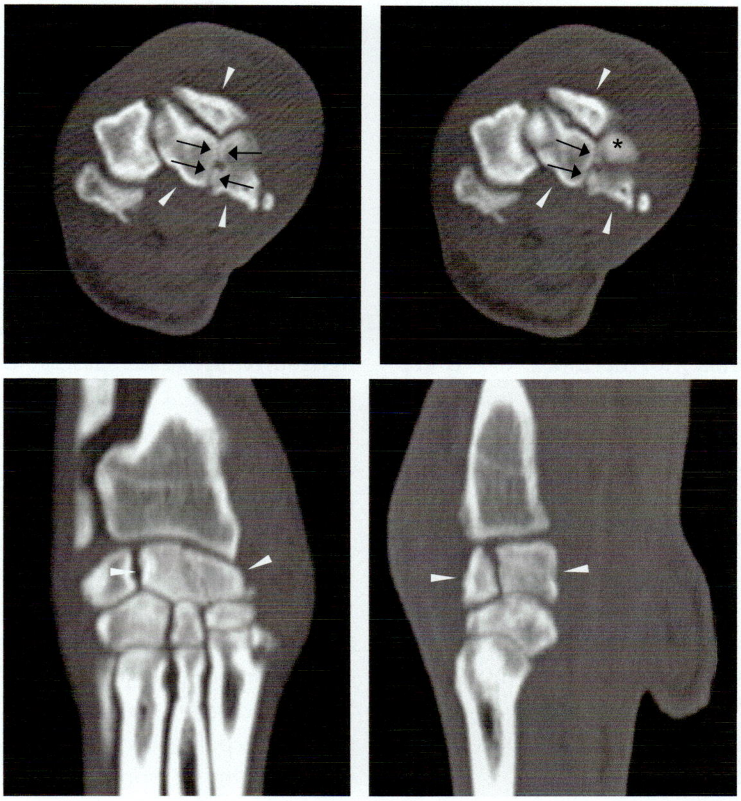

Figure 17.7 **Left carpus of an 18-month-old Boxer after assumed trauma. The dog showed acute lameness since about 4 weeks. The radial carpal bone is fractured into three fragments (white triangles). Better than radiography, CT demonstrates the degree of dislocation of fragments and the incongruities of the joint surfaces. The sclerotic rims bordering the fragments (black arrows) might be a sign of incomplete ossification of the radial carpal bone. If being so, the bone is prone to fracture due to minor forces. The radial carpal bone on the right side was normal. (asterix: second carpal bone.)**

- Type I: distal-basilar fracture
- Type II: proximal-basilar fracture
- Type III: distal-apical fracture
- Type IV: proximal-apical fracture
- Type V: comminuted fracture of the body

Whereas Type I and II fractures are intraarticular, Type III and IV fractures are extraarticular. Type V fractures are intra- and extraarticular. In racing greyhounds, Type I fractures account for 67% of the injuries. Type II injuries rarely occur alone. Usually, they are found concurrently with Type I fractures. Type III fractures are the least common. Prognosis depends on the type of fracture. Type IV and V fractures are reported to be the most frequently seen in non-racing dogs.

Salter–Harris fractures can be seen in the distal radius and ulna. Salter–Harris type III and IV fractures of the radius involve the antebrachiocarpal joint. Those types of fractures as well as joint involving fractures in adult animals (e.g., styloid process of the radius and ulnar styloid) give rise to joint instability. Type V fractures could be non-displaced. Their detection in CT is found to be -improved on comparison of both extremities. The proximal physes of the metacarpal bones are already closed at birth. Consequently, they are not prone to fracture.

Soft tissue damage associated with carpal fractures is usually confined to the local area. Injuries of tendons, supporting ligaments, the cephalic vein, or the branches of the superficial radial nerve are uncommon.

CT features of fractures depend on several factors (e.g., size and composition of the fragment, location of the fracture, degree of dislocation, and time of fracture relative to the time of CT investigation). CT is very helpful to identify and localize small fragments.

In fractures involving large fragments, signs of healing can be observed. Early follow-ups might show the appearance of zone of bone resorption, followed by progressive mineralization of the callus. Furthermore, CT is able to record the degree of soft-tissue injury better than radiography. Inadequate load on the joint cartilage (e.g., due to malpositioned fragments or limb deformities) results in subsequent degenerative joint disease. Therefore, it is important to control the completeness of fragment repositioning by means of appropriate imaging techniques in cases where osteosynthesis is performed.

17.5.2.2 Radial carpal bone fractures without known trauma

Fractures of the radial carpal bone can occur without a known episode of violent trauma. The radial carpal bone has three separate centres of ossification which eventually fuse; their planes of fusion correspond approximately to the two main fracture lines. These areas of fusion could be weak points within the radial carpal bone. This condition can be bilateral, and the majority of dogs are boxers. These dogs showed varying degrees of lameness, which tend to be exacerbated by exercise. The radiographic diagnosis can be challenging as there is no displacement of the fragments and the fracture line is not always recognized depending on the angulation of the primary x-ray beam. On CT, the fracture line can be easily appreciated (Figure 17.8).

17.5.2.3 Luxations and subluxations

Complete and incomplete tears or sprain of supportive ligaments result in deficiencies of joint stability. All carpal joints can be affected. Soft-issue involvement is evident. In acute cases, haemorrhage, oedema, or joint effusion can be seen. In chronic cases, soft tissue swelling is caused by fibrous tissue replacing the initial changes. Over time, mineralization of the scar tissue and the formation of enthesiopathy at the site of origin of the ligaments can be observed.

It must be considered that in animals with a pre-existing joint disease, such as rheumatoid arthritis or systemic lupus erythematodes, minor trauma might be sufficient to cause severe damage. Physical examination can demonstrate instability. In many cases, stress radiography and/or stress-CT are able to localize the site of the instability, to identify damaged structures, and to differentiate pre-existing joint disease if present.

Commonly, subluxations or luxations of carpal joints are palmar. The injury occurs with the carpus in an extended or hyperextended position and results in partial or complete tears of the medial palmar fibrocartilage. In canine hyperextension injuries, antebrachiocarpal, middle carpal, carpometacarpal, and combined middle plus carpometacarpal instability is reported in 10%, 28%, 46%, and 16%, respectively. In order to select an adequate treatment, the identification of the involved joint levels is very important. Displacement into other directions (dorsal, medial, and lateral) is infrequent. If a single collateral ligament is torn, the

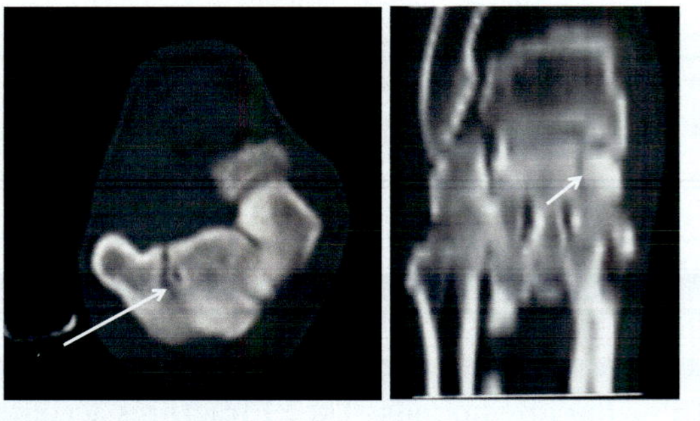

Figure 17.8 **Radial carpal bone fracture without known trauma in a young boxer. Although no fracture line was visible on the radiographs, on the transverse and reconstructed image, a clear fracture line can be detected (arrows).**

displacement usually results in a valgus or varus position (Figure 17.9).

Luxation and proximal displacement of the accessory carpal bone occur when the two ligaments attaching the bone distally to the base of the fourth and fifth metacarpal bones tear or undergo avulsion. Tears of the palmar carpal fibrocartilage and palmar transverse carpal ligament are often found simultaneously. Consequently, such a damage is seen in association with palmar luxation or subluxation of the antebrachiocarpal joint.

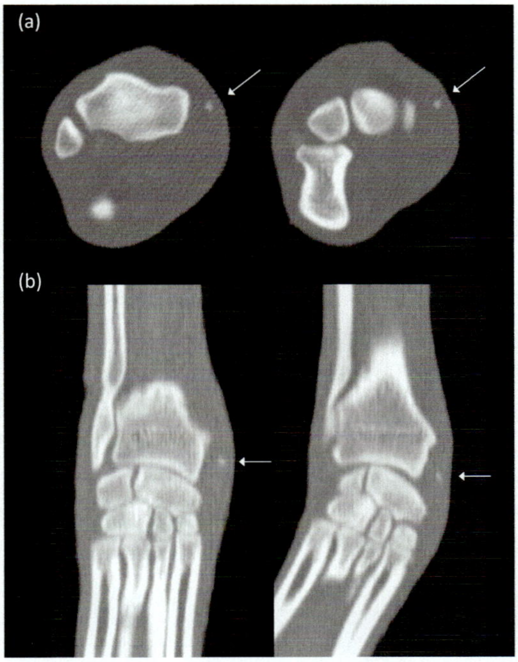

Figure 17.9 Carpus of a Whippet with ligamentous injury. In the transversal (a) and dorsal (b) images of the "neutrally" positioned limb, a small mineralized structure can be seen medial to the radius. Stress CT demonstrates the range of instability in the antebrachiocarpal joint due to lack of medial ligamentous support. The fragment (arrows) likely represents an avulsion fracture of the collateral ligament. No other fractures could be identified. Note the value of CT in the identification of small bone lesions. The fragment has a size of about 2 mm.

Individual bones of the carpus may luxate partially or completely. Frequently affected bones are the radial carpal bone and the fist or second carpal bone. Usually, they displace dorsally.

17.5.3 Degenerative disease

Due to the subtle nature of early changes and the limited value of radiography to detect and map related findings, CT is a valuable imaging modality to improve diagnostics.

17.5.3.1 Osteoarthrosis and degenerative joint disease (DJD)

Osteoarthritis is recognized in single or multiple joints. In the carpus, typical findings such joint effusion, thickening of the joint capsule, formation of osteophytes, subchondral bone sclerosis, perichondral cyst-like lesions, and signs of ankylosing arthropathy can be observed (Figure 17.10).

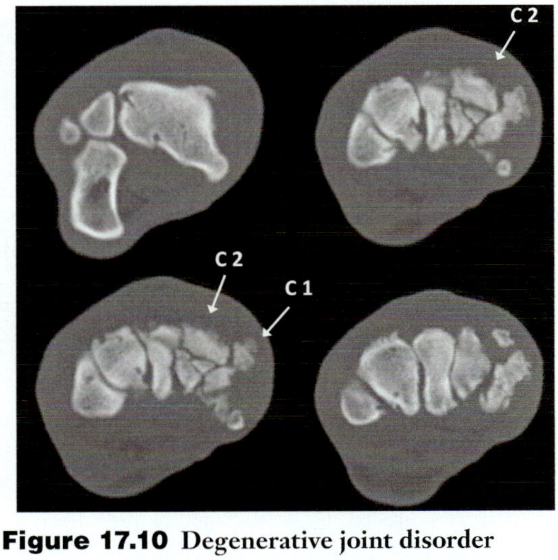

Figure 17.10 Degenerative joint disorder (DJD) in a 10-year-old Doberman pinscher. DJD developed due to non-treated fractures of the first and second carpal (arrows) bone dating back several months. CT images show new bone formation, multiple osteophytes, subchondral cyst-like lesions, sequestra, thickening of the joint capsule, and joint effusion.

17.5.3.2 Stenosing tenosynovitis of the abductor pollicis longus muscle

The abductor pollicis longus muscle originates on the lateral aspects of the proximal radius and ulna. It passes the extensor carpi radialis muscle at the level of the radial metaphysis, runs through a tendon sheath under the medial collateral ligament of the carpus, and finally attaches on the proximal aspect of the first metacarpal bone. The proliferation of connective tissue causes stenosis of the synovial sheath of the abductor pollicis longus muscle. As the primary reason for the underlying inflammation, chronic overstraining of the tendon due to overuse of the joint is discussed. Restricted movement of the tendon within the sheath causes lameness and pain predominantly while hyperflexing the carpus. There is soft-tissue swelling on the mediodorsal aspects at the level of the antebrachiocarpal joint. In advanced stage of disease, proliferative bony reactions are seen in the area of the radial styloid process (Figure 17.11).

17.5.3.3 Enthesiopathy of the short radial collateral ligament

In racing greyhounds, sprain of the short radial collateral ligament is described causing incomplete ligamentous tear and subsequent enthesiopathy at the site of the attachment of the ligament. Soft-tissue swelling over the medial distal radius, periosteal reaction, and dystrophic calcification are associated findings. Avulsion of parts of the origin of the ligament is also reported.

17.5.4 Neoplasia

Joint neoplasms can arise from any tissue of the joint or can originate from periarticular tissues and invade into the joint. Like in other joints, the most frequent primary joint neoplasm of the carpal joint is synovial cell sarcoma. It is more common in dogs than in cats. Synovial cell sarcomas arise from the synovium. The tumour is slowly growing and invades the bone. The soft-tissue swelling partly represents joint effusion. Bone invasion usually affects all neighbouring bones on either side of the joint. It is characterized by erosion of the cortical bone and multiple lytic areas with a cyst-like appearance in the adjacent cancellous bone. Metastases are seen in about 50% of the cases. It can be assumed that at the time of initial diagnosis in about 20% of the cases, distant metastases involving lymph nodes, lung, and spleen can be found; for staging, further examinations are required.

A number of neoplasms or tumour-like masses invading the carpus or located next to the joint and not being of synovial origin have been reported. In dogs, these include fibrosarcoma, giant cell carcinoma, hemangiopericytoma,

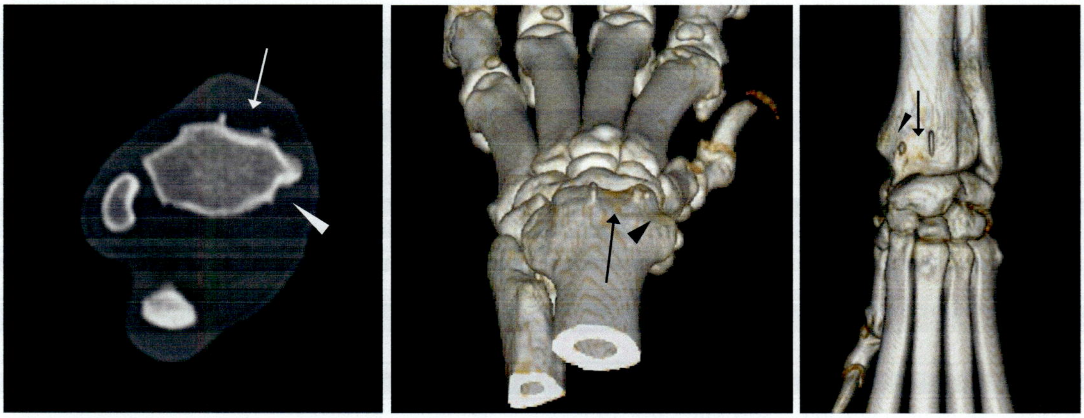

Figure 17.11 Stenosing tenosynovitis in a 10-year-old German Wirehaired Pointer dog. There are barnacle-like proliferative bone reactions forming rims at the surface of the radius for both the extensor carpi radialis muscle at the dorsal aspect (arrows) and the abductor pollicis longus muscle at the medial aspect (triangle).

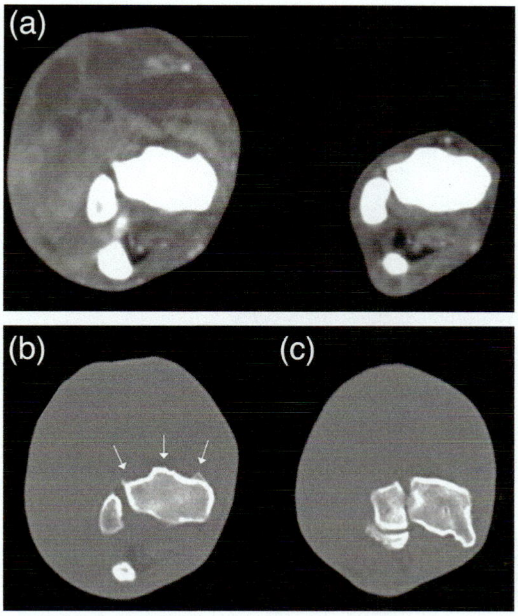

Figure 17.12 Soft-tissue tumour at the level of the left carpus in a 10-year-old Bernese Mountain dog. A mass of heterogeneous density surrounds the joint from the dorsolateral side. There is no evidence of joint effusion (a). There is mild periosteal reaction at the dorsal surface of the distal radius (b) (white arrows). The carpal bones are normal (c). Joint invasion is unlikely. It can be concluded that the mass originates from a tissue outside of the joint. Histopathology revealed a histiocytoma.

myxoma, lymphoma, and lipoma (Figure 17.12). In cats, fibrosarcoma and malignant fibrous histiocytoma have been described.

The features of the soft-tissue dense masses and the distribution of bony lesions are useful findings to detect the characteristics of the lesion. However, in many cases, a histologically oriented assessment cannot be derived from CT features. There is substantial overlap with findings seen in inflammatory and degenerative joint disease. CT proved to be a valuable tool to guide biopsies.

Furthermore, in the case of collapsing subchondral borders, primary and metastatic bone tumours can involve the joint. The distal radius and the ulna are predisposed sites of osteosarcomas (Figure 17.13).

17.5.5 Miscellaneous

17.5.5.1 Inflammatory joint disease

Inflammation of the synovial membrane with resultant synovial changes can be caused by immunologic factors or infectious agents. In immune-mediated polyarthropathy (rheumatoid arthritis), carpus and tarsus are sites of early manifestation. Immune-mediated joint disorders are classified as non-erosive or erosive. The morphological changes vary between those two groups. Whereas the non-erosive form exclusively presents signs of increased articular soft-tissue swelling, in the erosive arthropathy besides this, in the advanced stage of disease, bone erosion is evident. Bone destruction involves the subchondral and periarticular bone. Rheumatoid arthritis is rarely seen in dogs and cats. In dogs, bone destruction dominates the appearance. Joint laxity which results from lack of ligamentous support can be seen. Feline rheumatoid arthritis, occurring exclusively in young male cats, is characterized by reactive periosteal proliferation.

As in other joints, infectious arthritis usually is caused by bacteria that enter the carpus either by way of penetrating wounds (abrasions, lacerations, open fractures, and gunshots) or haematogenously. The severity of the injury depends on the type of bacteria and the duration of the infection. Infections with *Corynebacterium pyogenes*, *Clostridium species*, *Streptococcus*, and *Staphylococcus* result in distinctive bony destruction and in the case of arthritis in severe cartilage damage. Image findings depend on the stage and the characteristics of disease. In the majority of cases, severe lytic destruction of the involved bones dominates among other findings (Figure 17.14). CT can aid to guide biopsies. Further, the search for signs of a foreign body inoculation (detection of foreign material and evidence of a tract) could be an additional aim of a CT study.

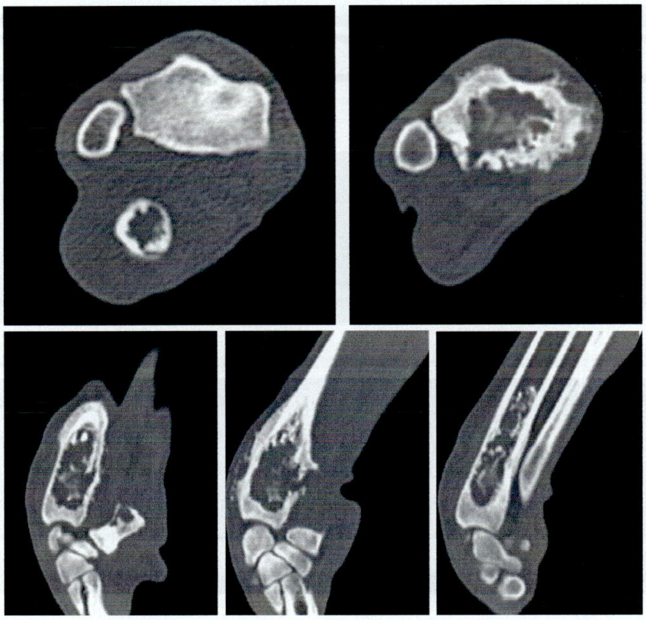

Figure 17.13 Site of an osteosarcoma 2 years after radiation therapy. The irradiated area shows severe lysis centred on the metaphysis of the radius, cortical bone destruction, neocortical bone formation, and amorphous periosteal reaction. Recently, the accessory carpal bone revealed bone lysis. In histopathology, it was found that it was radiation-induced necrosis.

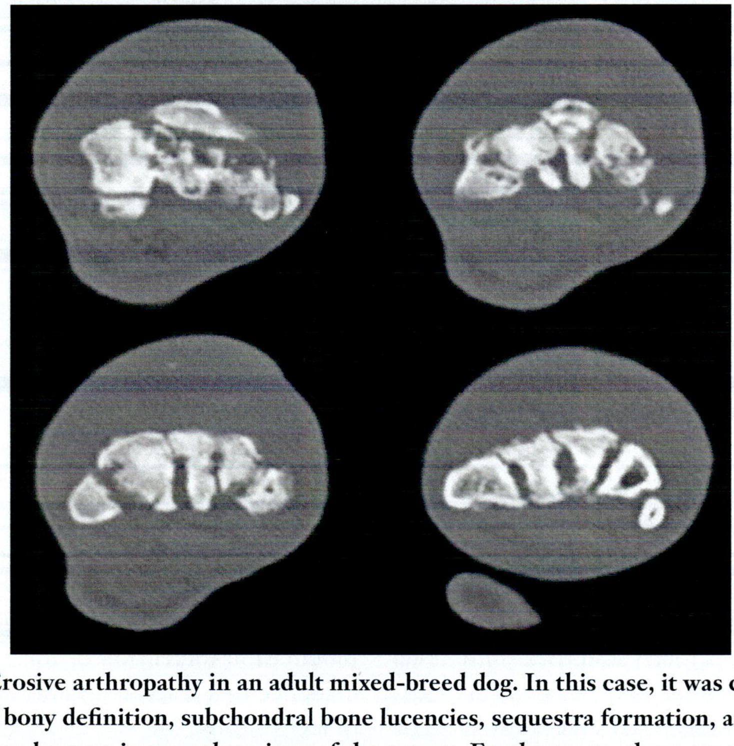

Figure 17.14 Erosive arthropathy in an adult mixed-breed dog. In this case, it was caused by septic arthritis. Loss of bony definition, subchondral bone lucencies, sequestra formation, and some new bone formation can be seen in many locations of the carpus. Furthermore, there are swellings of both the (intra-)articular and periarticular soft tissues.

The carpus is a joint where a number of disorders, seen in other joints, can manifest:

- Haemarthrosis
- Synovial osteochondromatosis
- Synovial cysts
- Villonodular synovitis.

FURTHER READING

Berg J, Gliatto JM, Wallace MK. 1990. Giant cell tumor of the accessory carpal bone in a dog. *Journal of the American Veterinary Medical Association* 197:883–885.

Brühschwein A. 2005. Anatomische Darstellung des caninen Karpalgelenkes mittels Magnetresonanztomographie und Computertomographie unter besonderer Berücksichtigung der Weichteilstrukturen. Inauguraldissertation (Doctoral thesis), Tierärztlichen Fakultät, Ludwig-Maximilians-Universität München.

Cetinkaya MA, Yardimci C, Sağlam M. 2007. Carpal laxity syndrome in forty-three puppies. *Veterinary and Comparative Orthopaedics and Traumatology* 20:126–130.

Chapman VA, Zeiner FN. 1961. The anatomy of polydactylism in cats with observations on genetic control. *Anatomic Record* 141:205–217.

Cornillie P, Van Lancker S, Simoens P. 2004. Two cases of brachymelia in cats. *Anatimia, Histologia, Embryologia* 33:115–118.

Craig LE, Julian ME, Ferracone JD. 2002. The diagnosis and prognosis of synovial tumors in dogs-35 cases. *Veterinary Pathology* 39:66–73.

Dismukes DI, Fox DB, Tomlinson JL, Essman SC. 2008. Use of radiographic measures and three-dimensional computed tomographic imaging in surgical correction of an antebrachial deformity in a dog. *Journal of the Armerican Veterinary Medical Association* 232:68–73.

Gaughan JM. 2007. What is your diagnosis? Lymphoma of the carpal joint. *Journal of the American Veterinary Medical Association* 231:203–204.

Gnudi G, Mortellaro CM, Bertoni G, Martini FM, Cantoni AM, Di Giancamillo M, Vignoli M. 2003. Radial carpal bone fracture in 13 dogs. *Veterinary and Comparative Orthopaedics and Traumatology* 16:178–190.

Grundmann S, Montavon PM. 2001. Stenosing tenosynovitis of the abductor pollicis longus muscle in dogs. *Veterinary and Comparative Orthopaedics and Traumatology* 14:95–100.

Guilliard MJ. 2001. Accessory carpal bone displacement in two dogs. *Journal of Small Animal Practice* 42:603–606.

Hanson JA. 1998. Radiographic diagnosis — canine carpal villonodular synovitis. *Veterinary Radiology & Ultrasound* 39:15–17.

Horvath I. 1983. Röntgenanatomische Untersuchungen zur postnatalen Entwicklung des Vordergliedmaßenskeletts der Hauskatze (Felis catus). Inauguraldissertation (Doctoral thesis), Tierärztlichen Fakultät, Ludwig-Maximilians-Universität München.

Johnson KA. 1987. Accessory carpal bone fractures in the racing greyhound. Classification and pathology. *Veterinary Surgery* 16:60–64.

Kim J, Blevins WE, Breur GJ. 2006. Morphological and functional evaluation of a dog with dimelia. *Veterinary and Comparative Orthopaedics and Traumatology* 19:255–258.

Langley-Hobbs SJ, Hamilton MH, Pratt JN. 2007. Radiographic and clinical features of carpal varus associated with chronic sprain the lateral collateral ligament complex in 10 dogs. *Veterinary and Comparative Orthopaedics and Traumatology* 20:324–330.

Lewis RE, van Sickle DC. 1970. Congenital hemimelia (agenesis) of the radius in a dog and a cat. *Journal of the American Veterinary Medical Association* 156:1892–1897.

Li A., Bennett D, Gibbs C,Carmichael S, Gibson N, Owen M, Butterworth SJ, Denny HR. 2000. Radial carpal bone fractures in 15 dogs. *Journal of Small Animal Practice* 41(2):74–79.

Lockwood A, Montgomery R, McEwen V. 2009. Bilateral radial hemimelia, polydactyly and cardiomegaly in two cats. *Veterinary and Comparative Orthopaedics and Traumatology* 22:511–513.

Macrì F, Lanteri G, Rapisarda G, Marino F. 2012. Unilateral forelimb partial aphalangia in a kitten. *Journal of Feline Medicine and Surgery* 14:272–275.

Mahoney PN, Lamb CR. 1996. Articular, periarticular and juxtaarticular calcified bodies in the dog and cat: A radiologic review. *Veterinary Radiology & Ultrasound* 37:3–19.

Mikić ZD, Ercegan G, Somer T. 1992. Detailed anatomy of the antebrachiocarpal joint in dogs. *Anatomic Record* 233:329–334.

Ober CP, Freeman LE. 2009. Computed tomographic, magnetic resonance imaging, and cross-sectional anatomic features of the manus in cadavers of dogs without forelimb disease. *American Journal of Veterinary Research* 70:1450–1458.

Pettitt R, Fox R, Comerford EJ, Newitt A. 2012. Bilateral angular carpal deformity in a dog with craniomandibular osteopathy. *Veterinary and Comparative Orthopaedics and Traumatology* 25:149–154.

Piermattei DL, Flo GL, Brinker WO, DeCamp CE. 2006. *Handbook of Small Animal Orthopedics and Fracture Repair*. St. Louis: Saunders Elsevier.

Schröder M. 1978. Beitrag zur Entwicklung des Skeletts der Vordergliedmaße beim Deutschen Schäferhund. Inauguraldissertation (Doctoral thesis), Tierärztlichen Fakultät, Ludwig-Maximilians-Universität München.

Shetye SS, Malhotra K, Ryan SD, Puttlitz CM. 2009. Determination of mechanical properties of canine carpal ligaments. *American Journal of Veterinary Research* 70:1026–1030.

Sis RF, Getty R. 1968. Polydactylism in cats. *Veterinary Clinics of North America: Small Animals Medicine, Small Animal Clinician* 63:948–951.

Szabo D, Ryan T, Scott HW. 2011. Carpal canal lipoma causing lameness in a dog. *Veterinary and Comparative Orthopaedics and Traumatology* 24:299–302.

Towle HA, Blevins WE, Tuer LR, Breur GJ. 2007. Syndactyly in a litter of cats. *Journal of Small Animal Practice* 48:292–296.

Turan E, Erden H. 2003. Computed tomography and morphometry of the carpal canal in the dog. *Annals of Anatomy* 185(2):173–178.

Vail DM, Powers BE, Getzy DM, Morrisin WB, McEntee MC, O'Keefe DA, Norris AM, Withrow SJ. 1994. Evaluation of prognostic factors for dogs with synovial cell sarcoma: 36 cases (1986—1991). *Journal of the American Veterinary Association* 205:1300–1307.

Yamate J, Aota M, Kuwamura M, Kotani T, Shimada T. 2006. Synovial sarcoma of the tendon and tendon sheath in a dog. *Journal of Veterinary Medical Science* 68:983–986.

CHAPTER 18

CT OF THE DIGITS

Ingmar Kiefer, Christian Niesterok, Eberhard Ludewig, Amalia Agut, and R. Latorre (Anatomy)

18.1 BRIEF INTRODUCTION TO DIGIT ANATOMY AND PATHOLOGY

The foot consists of the metacarpal/metatarsal bones and phalangeal bones. A number of sesamoid bones support muscle function. Flexor and extensor muscles are located on the dorsal and palmar/plantar aspects of the feet. In contrast to the bony structures, the visualisation of those soft tissue structures is challenging in CT. Contrast studies could help compensate for these shortcomings. The detection and differentiation of traumatic injuries, inflammatory condition, and neoplasms are major indications to perform a CT-study of the digits.

18.2 INDICATIONS FOR DIGIT CT

Radiography is the imaging tool of choice to evaluate the digits. In many cases, the information from radiography is sufficient to plan further treatment, e.g., to diagnose the presence and to characterise bony lesions. Whenever radiography fails, e.g., when soft tissue swellings must be characterised, ultrasonography is very helpful. Due to the advantages of these two modalities, being commonly available and therefore easily applicable, radiography and ultrasound are imaging modalities first used to perform an assessment of the digits. CT can add extra information, especially in cases

- with ongoing clinical symptoms,
- where the feature and/or the distribution of the lesion stay unclear, and
- where surgery requires cross-sectional images.

Using CT, subtle mineralised structures can be detected (e.g., alterations of the sesamoid bones), and soft tissue swellings can be delineated more precisely. It is critical to discuss the search for foreign bodies. Stones and metallic foreign bodies can be identified easily. However, the identification of wooden foreign bodies might be challenging. Some authors proved the feasibility of detecting wooden foreign bodies in a cadaver study. It could be shown that the visibility depends on the structure of the material, in particular on its water content. However, by the use of intravenous contrast injection, and fistulography fistular tracts can be highlighted more accurately or/and inflammatory conditions can be better delineated from normal soft tissue.

18.3 TECHNICAL ASPECTS

18.3.1 Positioning

The examination can be performed with the dog positioned either in ventral or in dorsal recumbency. In both cases, care should be taken to stretch out the limbs in a cranial or caudal manner. Furthermore, the limb must be in a physiologically extended position, to facilitate easier evaluation of the images. By making reconstructions, the limb orientation can be changed afterward. The investigation of both legs in a single scan is very helpful in order to distinguish anatomical conditions from pathology. Special care should be taken to position both extremities parallel and symmetrical to each. The authors usually examine the animals in a dorsal recumbency, with the feet fixed in a positioning that is orientated as symmetrical as

DOI: 10.1201/9781315118130-20

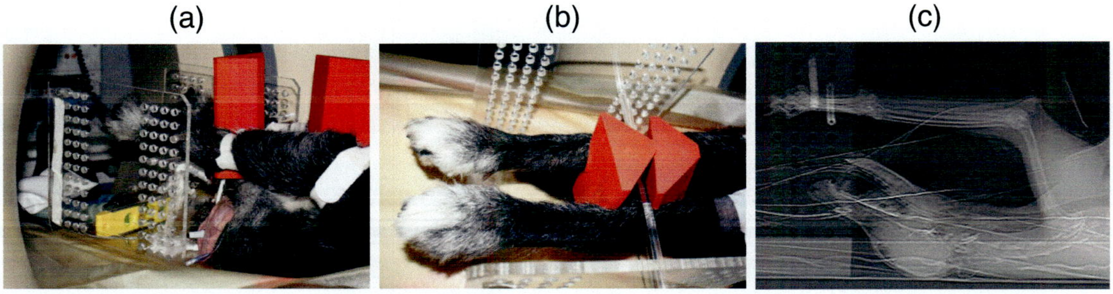

Figure 18.1 **Positioning device for the fixation of the feet (a,b). Scout image illustrating the orientation of the limbs (c).**

possible (Figure 18.1). Of course, examinations in lateral recumbency are possible as well. The advantage is that the used field of view (FOV) can be kept very small. Additional functional CT studies help demonstrate the degree of instability and identify potential causes.

18.3.2 Technical parameters

The settings applied are variable and case-dependent. If subtle bony changes are expected, a slice thickness of 1 mm should not be exceeded. In cases of soft tissue changes, a slice thickness up to 2 mm is adequate. The study is performed with a slice interval of 0.5 mm, leading to an adequate overlapping of data. The reconstruction matrix should not be less than 512 pixels. A matrix of 1,024 pixels achieves a higher spatial resolution, but it is doubtful whether it is really needed. Furthermore, it must be taken into account that many programmes are not able to work with a matrix exceeding 512 pixels. If soft tissue changes play an important role, a standard kernel is sufficient, whereas to evaluate bony changes, a high or ultrahigh kernel is required.

18.3.3 Contrast studies

Especially for soft tissue changes, contrast studies are necessary for identification and characterisation. Before performing a contrast study, a plane study has to be performed. This is important to differentiate between the contrast medium and mineralised structures. If possible, fistulas should not be explored prior to the contrast study in order to avoid artefacts or misinterpretation of the entrained air or iatrogenic damage.

For fistulography, water-soluble iodinated contrast media are used. To avoid increased beam hardening artefacts, the contrast medium should be diluted to a maximum of 100 mg iodine/mL. Mostly, even a lower iodine concentration is sufficient to visualise a canal and if necessary to perform a maximal intensity projection (MIP). For the representation of vascular anomalies, the concentration should not be less than 400 mg iodine/mL to ensure sufficient contrast. First, a sample injection is performed to reliably differentiate the vessels (artery and vein). The authors recommend the usage of bolus tracking systems after sample injection to register the appropriate time for the examination. To reliably detect the changes, a low injection rate (1 mL/s) is recommended. If the bolus tracking technique is not available, the sample injection should be used to determine the time the contrast medium needs to reach the region of interest. This time should later be utilised as delay time in the contrast study. It should be kept in mind that the scan time for a CT angiography must be adapted to follow the contrast flow in the vessels and to visualise the drain-off characteristics of the tissues. Therefore, the slice thickness needs to be adjusted if necessary. For a better differentiation of soft tissue from abscesses or other liquid-filled areas, a delay of 10 seconds after the application of contrast medium is usually sufficient. A cystic bony lesion can be caused by an aneurysm. To differentiate these from other conditions, dynamic examinations are recommended. In contrast to other conditions, an aneurysm is filled with contrast medium during the entire application.

18.4 NORMAL CT ANATOMY AND/OR ANATOMIC VARIATION(S)

Bones forming the paw are the metacarpal/metatarsal bones and the phalanges (Figures 18.2 and 18.3). The latter comprehend a proximal (first phalanx), a medial (second phalanx), and distal bone (third phalanx). The irregular structured third phalanx is conically shaped and carries the nail. The joints of the feet are diarthrodial joints characterised by a lining with joint cartilage and a joint capsule filled with synovial fluid. Medial and lateral ligaments support the joints. Several sesamoid bones support tendons. Proximal on the medial side of the carpus, a sesamoid bone is located in the tendon of abductor pollicis longus muscle. On the plantar aspect of the tarsometatarsal articulation in dogs, two sesamoids can be seen. One is the intraarticular tarsometatarsal bone, and the other is the lateral plantar tarsometatarsal bone. Adjacent to each metacarpophalangeal/metatarsophalangeal joint, there are three sesamoid bones. The single dorsal sesamoid is located within the extensor tendons, and there is a pair of palmar/wplantar sesamoids. These palmar/plantar sesamoids are numbered from medial to lateral beginning with the second forelimb/hindlimb digit. Flexor and extensor muscles are located on the dorsal and palmar/plantar aspects of the feet. The superficial digital flexor tendon and the deep digital flexor tendon insert at the second and third phalanx, respectively.

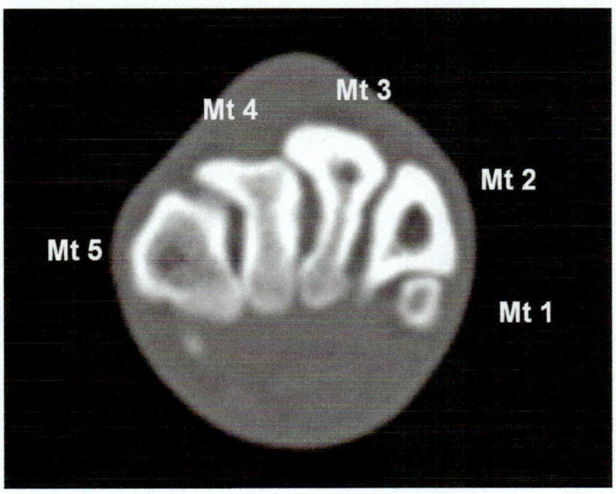

Figure 18.2 **Transverse CT slices showing the anatomy of the hindlimb of a dog on a level distal to the tarsus. Metatarsal bones 1–5.**

In contrast to the bones, it can be difficult to identify individual soft tissue structures (Figures 18.4 and 18.5). In large breed dogs, comparably good conditions exist to identify larger muscles at the level of the metacarpal/metatarsal bones. However, even in larger dogs, distal individual muscles and tendons normally cannot be clearly differentiated, except for pads (Figure 18.3). Due to the large amount of fat, they can be outlined from surrounding soft tissues. After intravenous application of contrast agents, vessels can be delineated.

Dog and cats exhibit a number of anatomic variances in the formation of the first forelimb/hindlimb digit. The first hindlimb digit in particular exhibits rudiments of varying degree (Figure 18.6).

18.5 PATHOLOGIC FINDINGS

18.5.1 Developmental

18.5.1.1 Supernumerary, missing, and deformed bones

Several developmental conditions of the digits can be seen in dogs and cats. Dewclaws are frequent conditions and occurs more often in certain breeds (e.g., Main Coon). In a number of dog breeds (e.g., Berger de Brie, Beauceron, and Kuvasz), even paired dewclaws are features of the breed standard. The appearance of dewclaws ranges from a loose skin-flat attachment up to the formation of a set of bones with a carpal/tarsal articulation. Unlike frontlimb dewclaws, hindlimb dewclaws tend to have little bone or muscle support.

Other malformations such as polydactyly, hypodactyly, dimelia, and ectodactyly are rare.

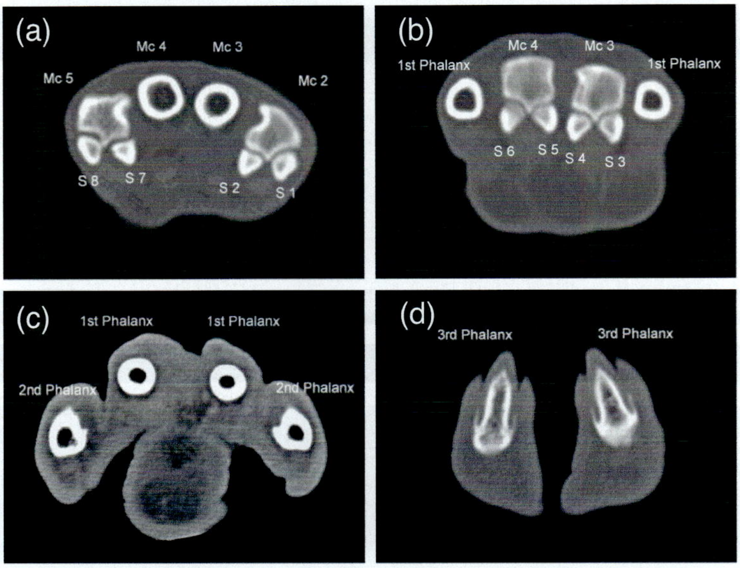

Figure 18.3 Transverse CT slices showing the anatomy of the frontlimb of a dog on certain levels. (a) Just proximal to the metacarpophalangeal joint of the second and fifth digits. (b) Just proximal to the metacarpophalangeal joint of the third and fourth digits. (c) At the level of the metacarpal pad. (d) Distal at the level of the third phalangeal bones of the third and fourth digits (Mc: metacarpal bone, s: sesamoid bone).

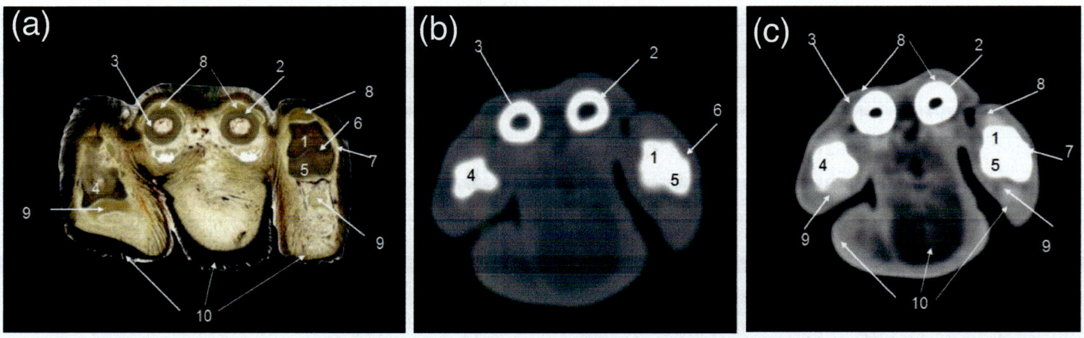

Figure 18.4 Transverse slices at the level of the proximal interphalangeal joint of left digits II and V. (a) anatomic section, (b) CT inbone window, (c) CT in soft tissue window. 1 = proximal phalanges (PP) II; 2 = proximal phalanges (PP) III; 3 = proximal phalanges (PP) IV; 4 = proximal phalanges (PP) V; 5 = middle phalanx (MP) II; 6 = proximal interphalangeal joint (PIJ); 7 = lateral collateral ligament (CL); 8 = common digital extensor tendon (CDE); 9 = deep digital flexor tendon (DDF); 10 = digital pad (DP).

18.5.1.2 Sesamoid disease

Multipartite (fragmented) sesamoid bones are of congenital origin or may result from osteonecrosis/osteitis (Figure 18.7). In large dogs (e.g., Rottweiler) the sesamoids 2 and 7 most commonly are affected. Doubts exist whether or not the condition is causing lameness in dogs. Local pain might demonstrate that the lesion represents an active process. Fractures of the sesamoid bones seem to be rare. They are seen in racing Greyhounds and other active large dogs.

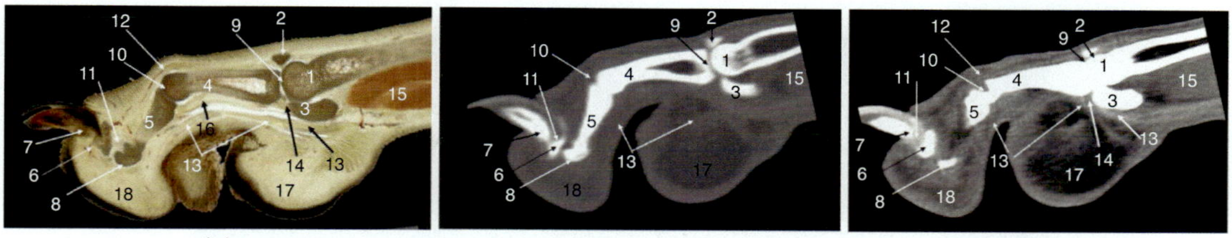

Figure 18.5 Sagittal slices of the left third digit. (a) Anatomic section; (b) CT in bone window; (c) CT in soft tissue window. 1 = metacarpal bone III; 2 = dorsal sesamoid bone (DSB); 3 = proximal palmar sesamoid bone (PPSB); 4 = proximal phalanx (PP) III; 5 = middle phalanx (MP) III; 6 = distal phalanx (DP) III; 7 = unguicular process (UP) III; 8 = flexor tubercle (FT); 9 = metacarpophalangeal joint (MPJ); 10 = proximal interphalangeal joint (PIJ); 11 = distal interphalangeal joint (DIJ); 12 = common digital extensor tendon (CDE); 13 = deep digital flexor tendon (DDF); 14 = sesamoidean ligament (SL); 15 = interosseous muscles (IM); 16 = straight sesamoidean ligament (SSL); 17 = metacarpal pad (MP). 18 = digital pad (DP)).

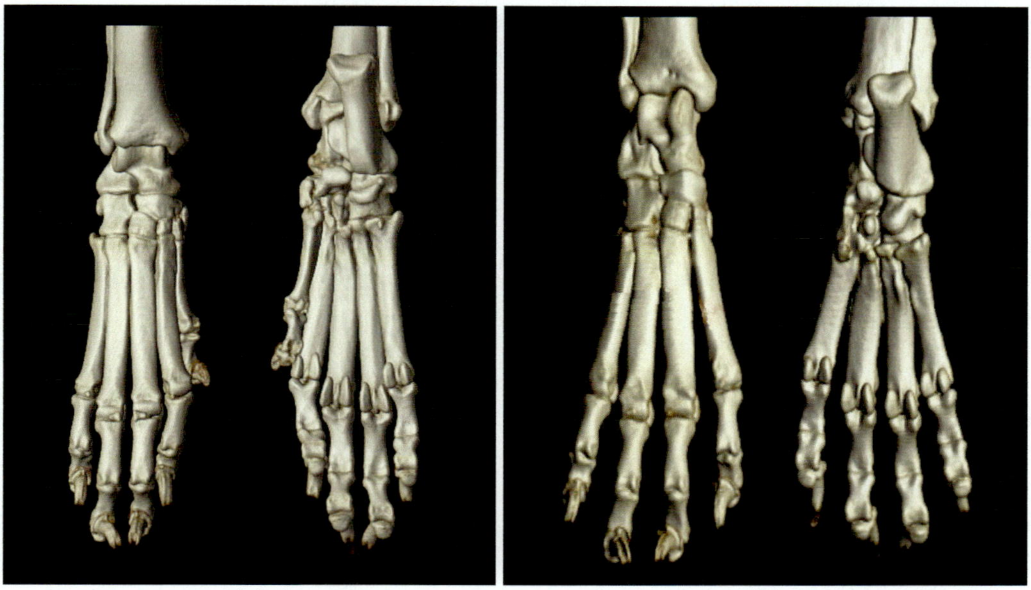

Figure 18.6 3D CT anatomy of the distal hindlimb in a Beagle dog (a) and in a French bulldog (b). While comparing the images, obvious differences in the conformation of the first digit can be seen.

A number of developmental disorders, attributed with changes in other regions, also manifest in digits:

- Mucopolysaccharidosis
- Scottish Fold cat osteochondrodysplasia
- Hypothyroidism
- Metaphyseal osteopathy (hypertrophic osteodystrophy)
- Epidermoid cyst

18.5.2 Traumatic

18.5.2.1 Fractures

Fractures include the metacarpal/metatarsal and the phalangeal bones and the sesamoids and are common. CT is seldom required to obtain a definitive diagnosis. In the majority of cases, radiography is able to detect and characterise injuries sufficiently. However, in complex trauma, with the involvement of a number of

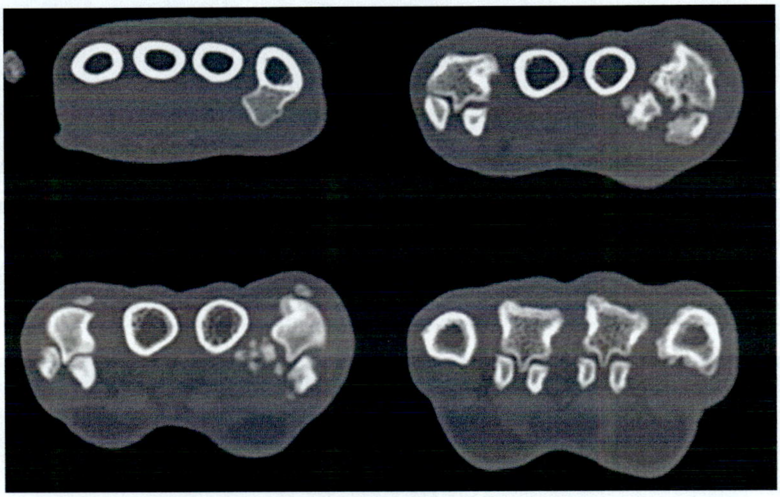

Figure 18.7 **Transverse CT images of the foot of a Great Dane. The 7th sesamoid bone cannot be outlined as a single bone. Instead, numerous small bony structures can be identified. There are irregular and well-organized periosteal reactions of both the metacarpal bone and the 1st phalanx. The joint seems not to be involved. Notice that there is mild soft tissue swelling. It should be considered that the process is active.**

bones, CT could deliver more detailed insights into the loss of structure and function of the traumatised digit. This might be important with respect to treatment planning and owner communication.

With the help of CT, even subtle fragments can detected. Furthermore, one can get a more detailed impression of the extent of soft tissue damage. CT angiography is a useful technique to visualise the vasculature of the affected limb.

These are arguments to consider CT, especially when multiple traumatic injuries are evident.

Despite the fact that CT is prone to artefacts caused by metallic implant, post-operative CT study controls can give reliable information, especially in cases with joint involvement, where this could be of special importance.

18.5.2.2 Luxations and deficits in ligamentous and tendons support

Traumatic luxations and injuries of ligaments and tendons are frequently seen in dogs and cats (Figure 18.8). In acute ligamentous or tendon injuries, initially soft tissue swelling due to, e.g., bleeding is evident. Over time, due to upcoming fibrosis, soft tissue swelling persists. At the sites of origin of damaged ligaments, enthesophytes can be seen. Mineralised scar tissue within torn fibrous tissue indicating chronicity. In the distal limb, these findings can be seen in damaged collateral ligaments of the metacarpo/metatarsophalangeal joint and the interphalangeal joints. In contrast, incomplete tear is hardly detectable with CT.

18.5.2.3 Osteitis in racing dogs

Racing dogs are prone to activity-related injuries of the extremities. Among those, the stress osteitis of metacarpal bones is common. Signs of stress osteitis are increased radioopacity of metacarpal diaphyses. It can result in comminuted fractures of the affected bone when not identified and treated.

The earliest and most serious findings can be seen on the frontlimb digits nearest to the centre of the racetrack. In dogs racing in an anticlockwise direction, Mc5 of the left and Mc2 of the right frontlimb are involved, whereas for those racing in a clockwise direction, Mc5 of the right and Mc2 of the left frontlimb are affected.

18.5.3 Degenerative disease

18.5.3.1 Osteoarthrosis

Degenerative joint disease can also affect the joints of the digits. It may affect individual joints of the digit or involve multiple joints. Osteoarthrosis can be idiopathic, age-related, or secondary to a traumatic injury. Common findings seen in CT are periarticular osteophytes, enthesophytes, or in a progressed stage of disease an ankylosing arthropathy. Generally, the subchondral bone has a normal structure. There is a controversial discussion about the clinical importance of osteoarthrosis of interphalangeal joints.

18.5.3.2 Non-erosive polyarthropathies

Non-erosive polyarthropathies are characterised by synovial thickening and joint effusion, causing articular soft tissue swelling. There are no signs indicating erosion of the subchondral bone. The disease is immune-mediated and can be found in a number of systemic disorders (e.g., lupus erythematosus).

18.5.3.3 Erosive polyarthropathies

Arthropathies with destructive changes of the subchondral bone occur in dogs (canine rheumatoid arthritis) and cats (feline rheumatoid arthritis, and feline chronic progressive polyarthropathy). At the beginning, patients show clinical indications of a non-erosive arthropathy. This is followed by the development of subchondral and periarticular erosive lesions. Typically, the changes can first be seen in the distal joints of the limbs. Finally, the disease results in ligamentous weakness with subsequent severe joint laxity.

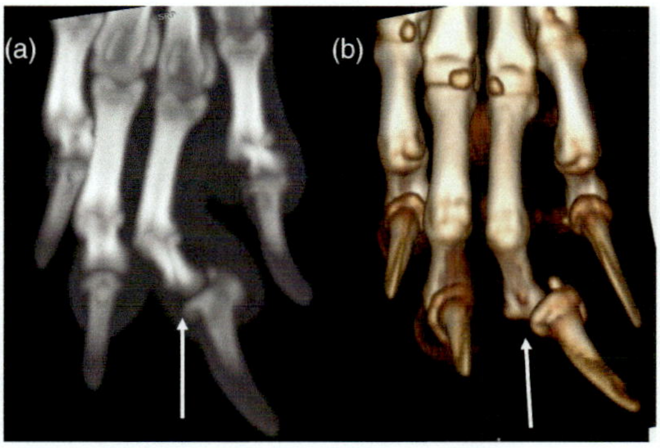

Figure 18.8 CT of a luxation of a toe. Dorsal reconstruction, MIP (a) and 3D (b) of a subluxation of the distal interphalangeal joint of the 3rd metatarsal (arrows). There is associated soft tissue swelling noticed.

18.5.4 Neoplasia

Occurrence of neoplasms of the digits is frequent, and the majority of tumours are malignant. In neoplasms of the digits, soft tissue swelling alone or in combination with bone lesions can be seen (Figure 18.9). A differentiation between a neoplastic or inflammatory origin or the intermediate tumour types

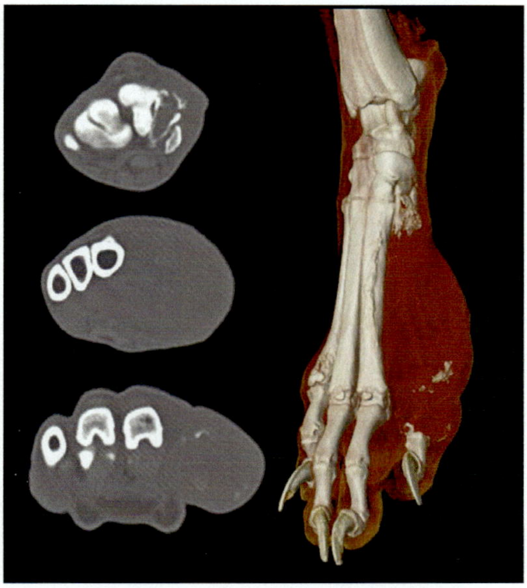

Figure 18.9 Transverse and 3D CT images of a tumour of the digital extremity, characterised by lytic destruction of the metatarsus and the first phalanx of the second digit. Within the largely swollen soft tissue, irregular mineralisation can be identified in the periphery. There is new bone formation at the dorsal aspect of the third metatarsal bone. Histopathology revealed that the lesion is an osteosarcoma.

solely based on imaging features is not possible. A definitive diagnosis requires FNA or biopsy. CT is able to identify the extent of tissue involvement, e.g., to define the lesion border.

18.5.4.1 Subungual squamous cell carcinoma

In dogs, there is a reported breed predisposition for Schnauzers and poodles. It is considered that subungual squamous cell carcinoma is rare in cats. In images, soft tissue swelling with or without bone involvement is evident (Figure 18.10). However, osteolysis of the phalanx may also result from paronychia or other neoplasms (e.g., bone-invasive soft-tissue tumours, primary bone tumours, and bone metastases). Therefore, also in dogs, regularly occurring melanomas, osteosarcomas and fibrosarcomas cannot be distinguished.

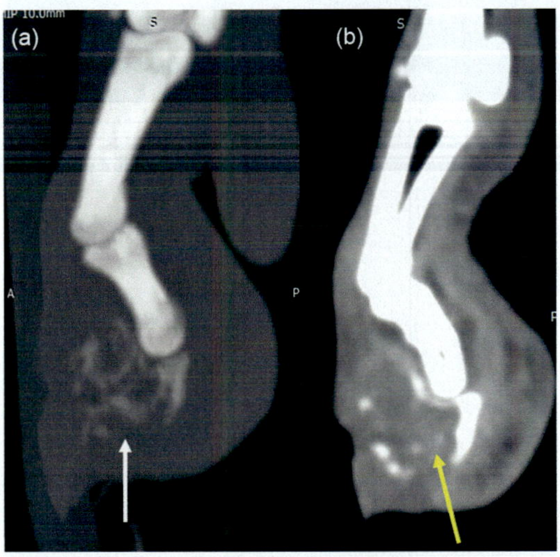

Figure 18.10 **Sagittal reconstructions in bone window (a) and soft tissue window after iv contrast administration (b) of a squamous cell carcinoma of a digit. There is metaplastic calcification and enhancement of the soft tissues (white arrow). The soft tissue swelling, and bony lysis of the 5th distal phalanx are obvious (arrows).**

18.5.4.2 Lung-digit syndrome

In cats, a primary lung tumour can metastasise to the digits. In such cases, one or more digits are swollen and osteolytic and/or proliferative bone lesions are evident.

18.5.4.3 Osteochondromas

Osteochondromas (cartilaginous exostosis) is a solitary benign lesion that can be located in different bones – among others, in the metacarpus and metatarsus. The metaphyses of long bones are most frequently affected. When multiple osteochondromas exist, the condition is termed multiple cartilagenous exostoses or hereditary multiple exostoses. Typically, osteochondromas are smoothly marginated bulbous bone lesions. In cats, the formation of osteochondromas may be related to the presence of feline leukaemia virus.

Mostly, osteochondromas are seen in growing dogs and cats. In most instances, they are incidental findings. They may undergo malignant transformation into chondrosarcoma and osteosarcoma.

Other benign tumours that can be found in the digits are histiocytoma, papilloma, and lipoma.

18.5.5 Miscellaneous

18.5.5.1 Pododermatitis, osteomyelitis, and septic arthritis

Most often, pododermatitis, osteomyelitis, and septic arthritis result from penetrating trauma. CT is very helpful in identifying inoculated foreign bodies being the carrier of microorganisms (Figures 18.11 and 18.12). In chronic cases with progressing symptoms of recurrent swelling and formation of fistula tracts, both CT and MRI could be used to delineate the region with the most severe tissue damage (Figure 18.13). Subsequent surgery aims to eliminate this hot spot. Whenever a foreign material is present

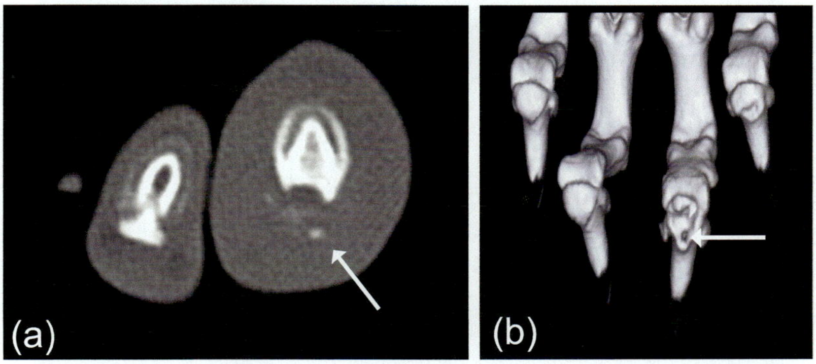

Figure 18.11 Transverse (a) and 3D (b) CT images of the digits. Foreign body in the fourth frontlimb digit in a young mixed-breed dog. There is severe soft tissue swelling encompassing the distal phalangeal bones. On the palmar aspect, a splitted small (2 mm) hyperdense structure can be identified (arrow). Another hyperdense structure can be seen within the axial soft tissues. Surgery confirmed the presence of a small stone. (Courtesy of Sibylle Kneissl, Veterinary University of Vienna.)

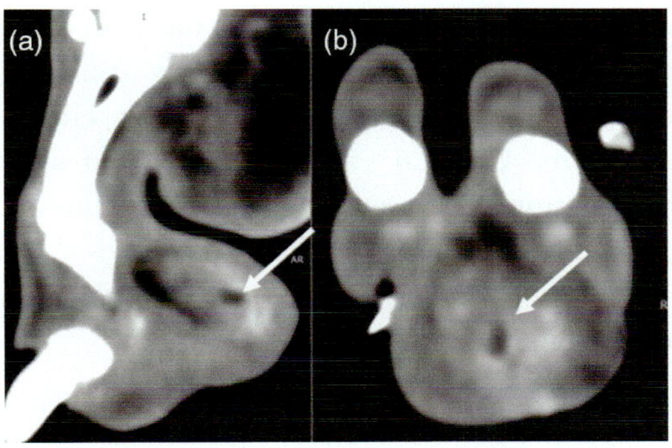

Figure 18.12 Sagittal reconstructed (a) and transverse (b) post IV contrast images in soft tissue window of a foreign body in the digit. There is soft tissue swelling between the 3rd and 4th digit, contrast enhancing and demonstrating a radiolucency, a filling defect, in the middle (arrows).

but undetected by CT (or MRI), it should be suspected that the foreign body is located within the area of severe soft tissue alteration. Haematogenous infection can be differentiated from infection due to penetrating wounds based on the history and physical examination.

18.5.5.2 Paronychia

The inflammation of the nail bed is caused by the increasing spread of microorganisms into the distal phalanx and the distal interphalangeal joint, resulting in loss of the nail. This again results in an ongoing loss of the nail barrier. Imaging findings are equivalent to findings we can see in neoplasms: soft tissue alterations in combination with bony reactions. Therefore, biopsy or FNA is required to obtain a definitive diagnosis.

18.5.5.3 Hypertrophic osteopathy (acropachy)

This is characterised by symmetric palisade-like periosteal proliferation on long bones. Early changes can be seen along the metacarpal/metatarsal bones and on the phalanges. When the disease progresses, it extends proximally, and in an advanced stage of disease, all bones of the limb are affected. Periosteal changes result from a vascular connective tissue proliferation of the periosteum. This in turn is mediated by a humeral or neurogen stimulation caused by a distant focus. Most often, intrathoracic mass lesions are focused upon, but extra-thoracic conditions (e.g., prostatic or ovary masses and pyometra) are identified as the origin. As far

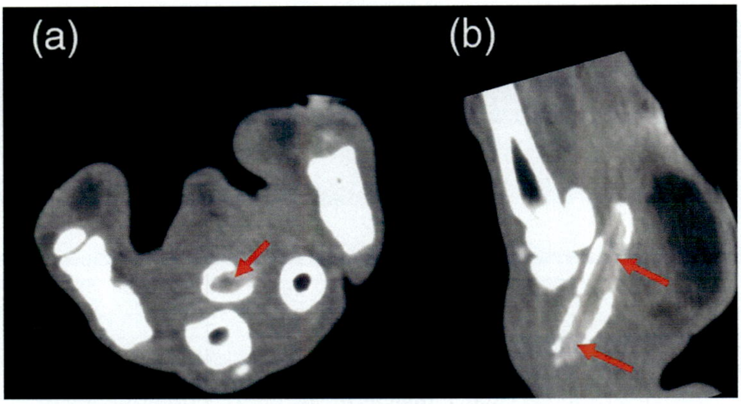

Figure 18.13 **Transverse (a) and sagittal reconstructed (b) images displayed in a soft tissue window with fistulography of the fistulous track of a linear foreign body in a digit. The linear foreign body is visible as a filing defect inside the track (arrows).**

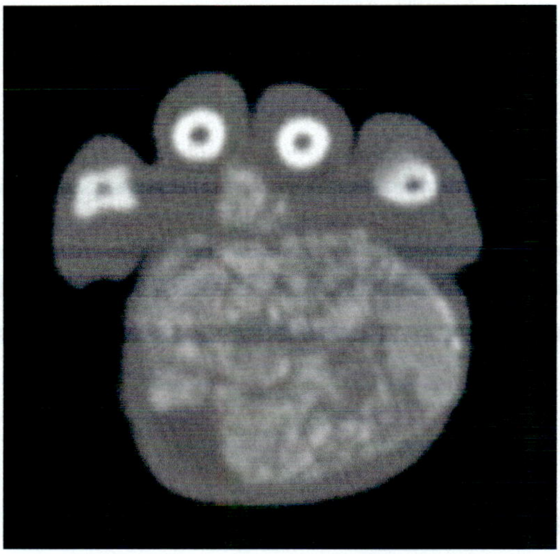

Figure 18.14 **Transverse CT image of calcinosis circumscripta in a 4-year-old dog. Note the cloddy mineralised material located in the metatarsal pad. There is only a mild swelling of the adjacent soft tissue.**

as the origin of the stimulus can be abolished, periosteal changes undergo regression.

18.5.5.4 Calcinosis circumscripta

Calcinosis circumscripta is characterised by focal deposition of stippled mineralized material within soft tissues. Typically, the lesion is solitary and well-marginated. In most cases, there is only a mild-to-moderate swelling of the surrounding soft tissue (Figure 18.14). Basically, calcinosis circumscripta can be seen in any soft tissue. However, the soft tissues adjacent to the metacarpal/metatarsal bones are often affected. The aetiology is unknown. In general, the lesions rarely cause clinical symptoms. Clinical symptoms are seen when the function of adjacent structures is influenced mechanically.

Other conditions with digital involvement are as follows:

- Hypervitaminosis A
- Hyperparathyroidism
- Panosteitis

FURTHER READING

Allan G, Nicoll R. Distal lims — carpus and tarsus. In: Barr FJ, Kirberger RM (eds) *BSAVA Manual of Canine and Feline Musculoskeletal Imaging*, Gloucester: BSAVA, 2006:187–214.

Belluco S, Brisebard E, Watrelot D, Pillet E, Marchal T, Ponce F. Digital squamous cell carcinoma in dogs: epidemiological, histological, and immunohistochemical study. *Vet Pathol* 2013;50:1078–1082.

Carpenter JL, Dayal Y, King NW, Jr, Moore FM. Distinctive unclassified mesenchymal tumor of the digit of dogs. *Vet Pathol* 1991;28:396–402.

Craig LE, Krimer PM, Cooley AJ. Canine synovial myxoma: 39 cases. *Vet Pathol* 2010;47:931–936.

de Rooster H, Risselada M, van Bree H. Excision arthroplasty of the interphalangeal joint as an alternative to digit amputation in two dogs. *J Small Anim Pract* 2007;48:169–173.

Esson D, Fahrer CS, Zarfoss MK, Dubielzig RR. Suspected uveal metastasis of a nail bed melanoma in a dog. *Vet Ophthal* 2007;10:262–266.

Furuya Y, Uchida K, Tateyama S. A case of glomus tumor in a dog. *J Vet Med Sc* 2006;68:1339–1341.

Guerin SR, Jones BR, Alley MR, Broome C. Multiple digital tumours in a Rottweiler. *J Small Anim Pract* 1998;39:200–202.

Guilliard MJ. Proximal interphalangeal joint instability in the dog. *J Small Anim Pract* 2003;44:399–403.

Kaufman KL, Mann FA. Short- and long-term outcomes after digit amputation in dogs: 33 cases (1999—2011). *J Am Vet Med Assoc* 2013;242(9):1249–1254. doi: 10.2460/javma.242.9.1249.

Kessler M. Tumoren der Zehen. In: Kessler M (ed) *Kleintieronkologie*. Stuttgart: Enke Verlag, 2013:428–431.

Kuroki K, Britt LG, Franklin SP, Henry CJ. Skeletal-extraskeletal angiomatosis in a dog. *Vet Pathol* 2010;47:944–947.

Liptak JM, Dernell WS, Rizzo SA, Withrow SJ. Partial foot amputation in 11 dogs. *J Am Anim Hosp Assoc* 2005;41:47–55.

Liu SK, Hohn RB. Squamous cell carcinoma of the digit of the dog. *J Am Anim Hosp Assoc* 1968;153:411–424.

Marino DJ, Matthiesen DT, Stefanacci JD, Moroff SD. Evaluation of dogs with digit masses: 117 cases (1981—1991). *J Am Vet Med Assoc* 1995;207:726–728.

Muir P, Norris JL. Metacarpal and metatarsal fractures in dogs. *J Small Anim Pract* 1997;38:344–348.

Muir P, Pead MJ. Chronic lameness after digit amputation in three dogs. *Vet Rec* 1998;143: 449–450.

Namitome K, Kano R, Sekiguchi M, Iwasaki T, Kaneshima T, Nishifuji K. Isolation of *Fusarium sp.* from a claw of a dog with onychomycosis. *J Vet Med Sc* 2011;73:965–969.

Ober CP, Freeman LE. Computed tomographic, magnetic resonance imaging, and cross-sectional anatomic features of the manus in cadavers of dogs without forelimb disease. *Am J Vet Res* 2009;70:1450–1458.

O'Brien MG, Berg J, Engler SJ. Treatment by digital amputation of subungual squamous cell carcinoma in dogs: 21 cases (1987—1988). *J Am Vet Med Assoc* 1992;201:759–761.

Papazoglou LG, Ellison GW, Farese JP, Bellah JR, Coomer AR, Lewis DD. Fusion podoplasty for the management of chronic pedal conditions in seven dogs and one cat. *J Am Anim Hosp Assoc* 2011;47:e199–e205.

Park K, Kang J, Subedi KP, Ha JH, Park C. Canine polydactyl mutations with heterogeneous origin in the conserved intronic sequence of LMBR1. *Genetics* 2011;179:2163–2172.

Patnaik AK. Histologic and immunohistochemical studies of granular cell tumors in seven dogs, three cats, one horse, and one bird. *Vet Pathol* 1993; 30:176–185.

Sanerkin NG, Capron TK. Mixed malignant tumour in a digit of a dog. *J Path Bacteriol* 1966;91:624–625.

Sonntag F, Mihaljevic M, Klumpp S, Kramer M. Gliedmaßen und Wirbelsäule. In: Mihaljevic M, Kramer M, Gomercic H (eds) *CT- und MRT-Atlas*. Stuttgart: MVS Medizinverlage, 2009:148–196.

Sprague W, Thrall MA. Recurrent skin mass from the digit of a dog. *Vet Clin Pathol* 2001;30: 189–192.

Sweeney PA. Digit amputation in dogs. *Vet Rec* 1998;143:540.

Towle H, Friedlander K, Ko R, Aper R, Breur G. Surgical treatment of simple syndactylism with secondary deep digital flexor tendon contracture in a Basset Hound. *Vet Comp Ortho Traumatol* 2007;20:219–223.

Whitley DB, Mansell JE. Multiple eccrine poromas in the paw of a dog. *Vet Dermatol* 2012;23:167–170.

Wrzosek M, Płonek M, Zeira O, Bieżyński J, Kinda W, Guziński M. Congenital bipartite atlas with hypodactyly in a dog: clinical, radiographic and CT findings. *J Small Anim Pract* 2014;55: 375–378.

CHAPTER 19

CT OF THE HIP, PELVIS AND THE SACROILIAC JOINT

Wolfgang Henninger, Amalia Agut and R. Latorre (Anatomy), and Ingrid Gielen

19.1 INTRODUCTION

The bony ring of the pelvis helps deliver forces from a single point of the caudal spine down in a two-sided motion and vice versa. This area of bones and muscles is heavily packed; the bones can be clearly seen on radiographs, but the pelvic cavity and the muscles around nerves and vessels cannot be seen.

CT can provide 3D information about the three pelvic bones (the iliac, pubic and sacral), the sacroiliac junction and the hip joints. In the pelvic cavity, the portions of the colon and rectum ending at the anus are seen; in males, the prostate and urethra can be seen ventrally, and in females, the cervix, urethra and vagina. Lymph nodes and vessels can also be documented. The outer pelvic musculature consists of gluteal and femoral muscles; the inner pelvis has iliopsoas and short pelvic muscles. Single-sided atrophy can be easily detected.

The sacroiliac joint can be subluxated or luxated by heavy trauma, such as high-rise syndrome or a car accident. Older dogs often suffer from degenerative joint disease with periosteal remodelling of the joint surfaces. Although traumatic insults of the hip joint (e.g., fracture and luxation) are usually diagnosed radiographically, CT can provide much additional information, such as incomplete central luxation with fracture of the socket that may be present or fissure lines along pelvic bones, combined with sacroiliac subluxation. Reconstructive surgery of the hip socket after a complex pelvic fracture is much easier when the surgeon has information about the position and size of the fragments seen on the CT images. Urethral rupture may also be present, but even a small leakage can be ruled out with urethrography using a positive contrast medium.

The dysplastic hip sometimes can be difficult to visualise in very young animals, but depth measurements in shallow sockets can support the decision of symphysiodesis. Early changes of epiphysiolysis in young, small-breed dogs can be seen quite clearly, as well as uni- or bilateral epiphysiorhexis of the femoral head in cats, although this should be clinically distinguished from femoral neck fracture.

Neoplasms that occur in the pelvic bones, such as early-stage osteosarcoma or multiple myeloma, and bony lysis of the femoral head can be clearly visualised with CT if they are masked on radiographs. Inflammatory conditions are rare and often the sequelae of traumatic conditions, such as osteomyelitis or synovitis.

19.2 INDICATIONS FOR USING CT

a. *Developmental*: sacroiliac asymmetry, epiphysiolysis and pubic symphysiodesis
b. *Traumatic*: sacroiliac (sub)luxation, hip socket fracture or (sub)luxation and epiphysiorhexis
c. *Degenerative*: DJD of sacroiliac joint and dysplastic hip joints
d. *Neoplastic*: osteosarcoma or lysis of the femoral head; sarcoma of any pelvic location (especially multiple myeloma).

DOI: 10.1201/9781315118130-21

19.3 TECHNICAL ASPECTS

19.3.1 Positioning

Dorsal recumbency in a special U-shaped positioning kit is the easiest method to get a stable and reproducible position of a small animal's body. The hind legs are placed on a pillow in extension and carefully fixed parallel to each other, similar to HD positioning, but less extended.

19.3.2 Technical parameters

Technical parameters are very much dependent on the type of the scanner; single-slice machines use a slice thickness of 3–5 mm in large-breed dogs and 2 mm in small-breed dogs and cats without tilting the gantry. Multidetector scanners usually scan all animals with 1 to 2 mm slice thickness, irrespective of the body size, to obtain thorough image resolution and also reformatted images, such as MPR or 3D-reconstruction.

19.3.3 Contrast medium

Intravenous (IV) contrast administration of a water-soluble iodine agent is important in sacroiliac junction and hip joint studies. A plain study should always be performed first, followed by the contrast scan, when 2 mL/kg body weight, 300–350 mg I/mL is usually administered during a venous phase. To achieve selective delineation of the hip joint(s), to evaluate the size of the socket's capsule, and to detect possible intra-articular bony or cartilage fragments, intra-articular administration of a diluted water-soluble iodine agent can be performed. Whichever product is used, the contrast medium should be diluted with sterile 0.9% physiologic saline solution (NaCl) to obtain a contrast medium concentration of approximately 80 mg of iodine per millilitre.

19.4 NORMAL ANATOMY

CT shows that a cat with normal anatomy has totally uniform hip joints, shallow sockets, and a round femoral head far too big for the acetabulum. The joint space is narrow and almost invisible. In dogs, the hip socket and the femoral head are much more diverse, with deep sockets that take in two-thirds of the femoral head (Figures 19.1 and 19.2). In dysplastic joints, the craniolateral rim is edged, sclerotic, flattened, or missing, depending on the degree of dysplasia, and the femoral head is displaced craniolaterally. Because the hip sockets and femoral heads have many different shapes, correct positioning of the hind legs is vital to avoid misinterpretation. Anatomic variations are similar, as radiographs have documented. The sacroiliac joint, with crescent-shaped surfaces on the ilium and the wings of the sacrum, is a combined synovial and fibrocartilaginous joint (Figure 19.3).

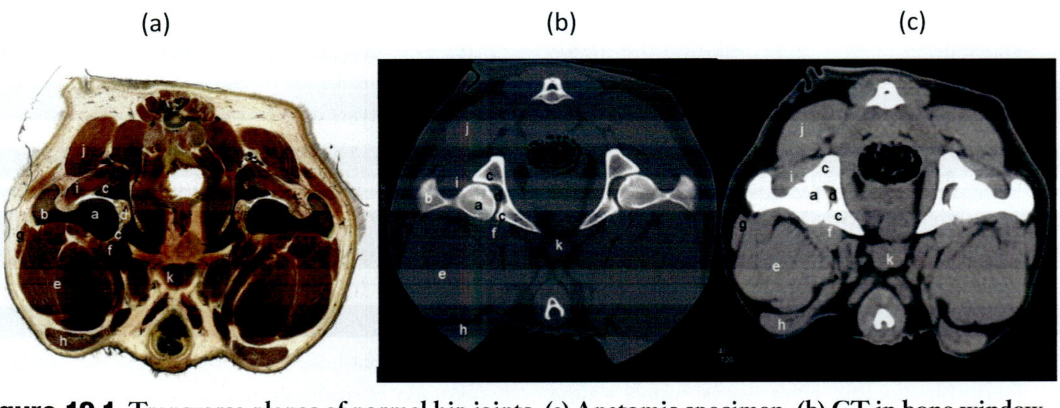

Figure 19.1 Transverse planes of normal hip joints. (a) Anatomic specimen, (b) CT in bone window, (c) CT in soft tissue window. a. head of the femur; b. greater trochanter; c. acetabulum; d. ligament of the femoral head; e. quadriceps femoris muscle; f. iliopsoas muscle; g. tensor fasciae latae muscle; h. sartorius muscle; i. gemelli muscle; j. middle gluteal muscle superficialis; k. rectus abdominis muscle.

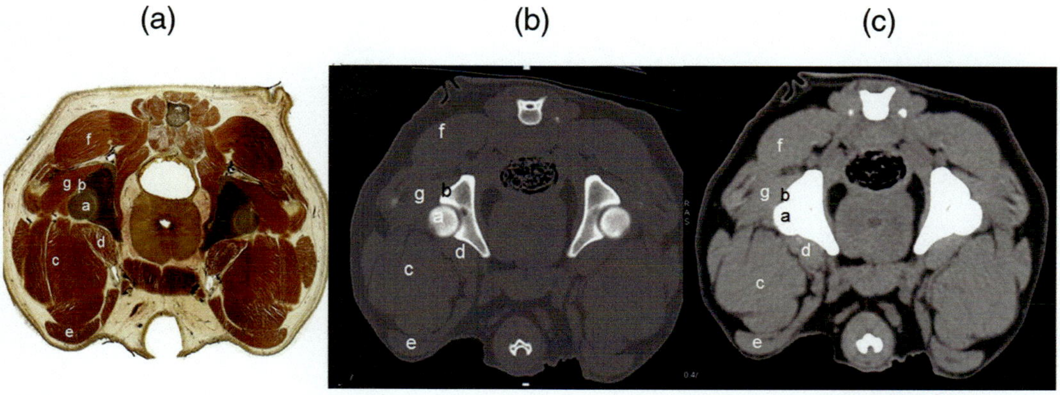

Figure 19.2 Transverse planes of normal hip joints. (a) Anatomic specimen, (b) CT in bone window, (c) CT in soft tissue window. a. head of the femur; b. acetabulum; c. quadriceps femoris muscle; d. iliopsoas muscle; e. sartorius muscle; f. middle gluteal muscle; g. deep gluteal muscle; h. rectus abdominis muscle.

19.5 PATHOLOGIC FINDINGS

19.5.1 Developmental disorders

19.5.1.1 Hip dysplasia and symphysiodesis

The development of the hip joint in 3–4 month-old dogs can be influenced by the premature closure of the pubic physis. Preoperative CT scans show the status of the shallow hip socket and the configuration of the craniolateral rim, in particular; if the femoral head is subluxated, a slight notch can be detected. The acetabular opening angle can be obtained to compare the effect of deepening the hip socket after symphysiodesis when the main physes have closed (usually at about 12–18 months) and the pelvis is fully developed. The symphysiodesis is considered successful when follow-up examination indicates an increase in the acetabular opening angle. CT evaluation is much more precise than the DAR projection of radiographs and also provides much more information about the entire joint and the femoral head (Figure 19.4).

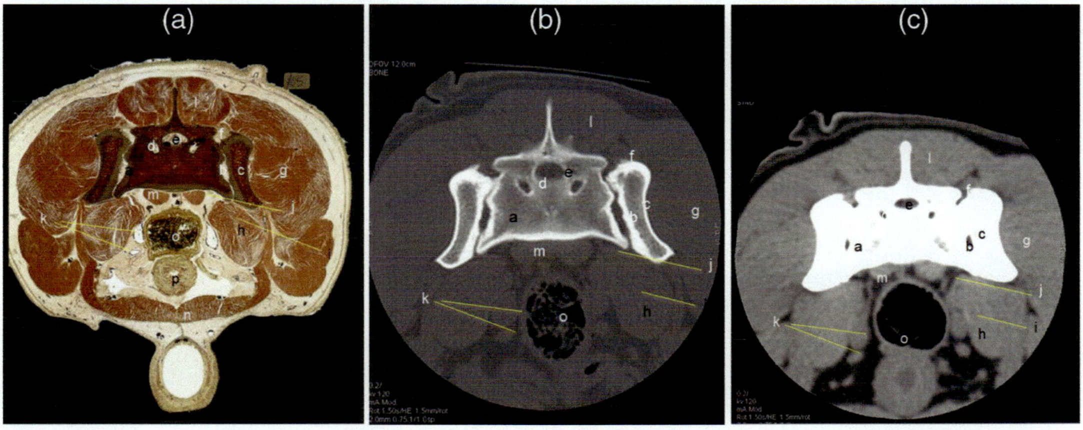

Figure 19.3 Transverse planes of the sacroiliac joint. (a) Anatomic specimen, (b) CT in bone window, (c) CT in soft-tissue window. a. wing of sacrum; b. sacroiliac joint; c. wing of ilium. d. dorsal sacral foramina; e. sacral canal; f. dorsal sacroiliac ligament; g. gluteus medius muscle; h. iliopsoas muscle; i. femoral nerve; j. lumbosacral trunk; k. external iliac vein and artery; l. dorsal sacrocaudal muscle; m. ventral sacrocaudal muscle; n. rectus abdominis trae; o. rectum. p. urinary bladder neck.

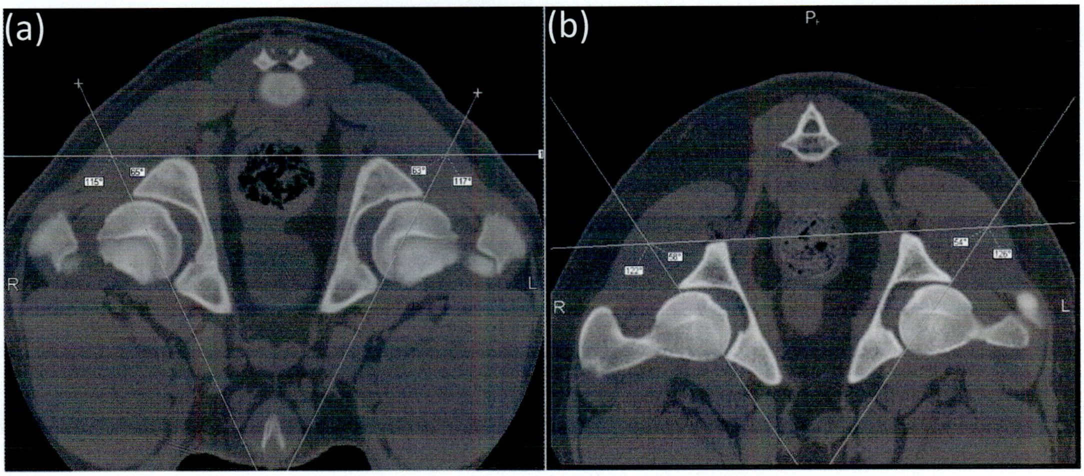

Figure 19.4 **Transverse CT images in bone window. Hip joints preoperatively for planned symphysiodesis of a 4-month-old male Beauceron (a) and 10-month follow-up postoperatively (b): the DAR angle is increased on both sides.**

19.5.1.2 Hip dysplasia and double or triple pelvic osteotomy (TPO)

In hip dysplasia patients, degenerative changes can be detected in an early stage and joint effusion can be evaluated. By using IV contrast associated synovial inflammation can be evaluated (Figure 19.5).

CT imaging can ensure successful preoperative planning in the mature dysplastic hip joint before pelvic osteotomy; similar to the procedure in young dogs, the acetabular opening angle is measured and the degree of possible rotation of the iliac bone virtually tested. Correct positioning is mandatory; both femurs

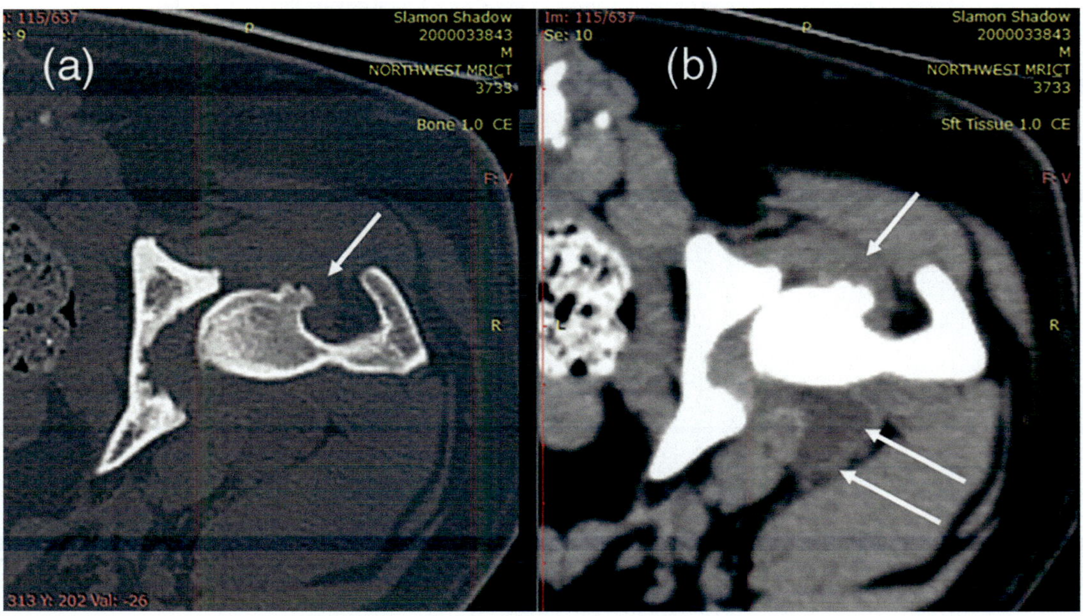

Figure 19.5 **Transverse images in bone-(a) and soft tissue algorithm (b) of a dysplastic dog displaying degenerative changes (arrows). On the soft tissue image (b) joint effusion can be appreciated (double arrow).**

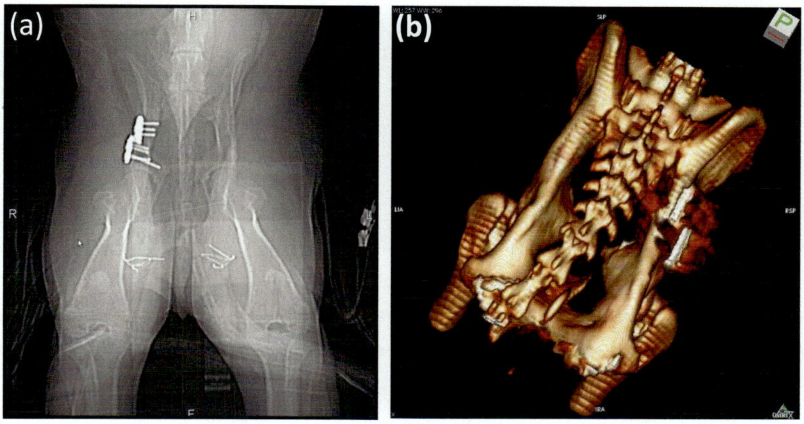

Figure 19.6 The scout view (a) of a 7.5-year-old male Labrador-Cross shows a metal implant having TPO; a surface-shaded 3D model of the pelvis (b) after bilateral TPO (implants of the left side had been removed) presents the implants of the right side in place, causing heavy streaking artefacts, but the hip socket is well covered by the acetabular rim.

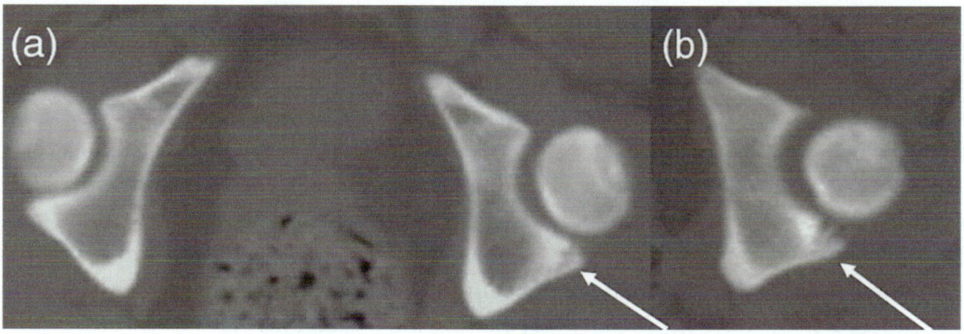

Figure 19.7 Transverse images in bone window (a & b) of a dysplastic dog showing damage of the left dorsal acetabular rim (arrows) making this dog unsuitable for TPO surgery.

must be positioned parallel in a slightly flexed position of about 60° from the horizontal plane with the dog in dorsal recumbency (Figure 19.6).

Computed tomography (CT) is not used routinely for the evaluation of canine hips. CT is mainly used to evaluate acetabular rim damage which is an important criterion and contraindication when triple pelvic osteotomy (TPO) is considered (Figure 19.7).

19.5.1.3 Epiphyseal dysplasia in the cat

Spontaneous femoral capital physeal fractures are reported in the cat. Heavy, early neutered males of 1.5–2.5 years. of age with delayed physeal closure are predisposed to the condition. CT show the often bilateral condition solely displaying a sharply bordered femoral head rotated in its socket and a sharply bordered femoral neck (Figure 19.8).

19.5.1.4 Necrosis of the femoral head/ Legg–Calve–Perthes disease

Aseptic or avascular femoral head and neck necrosis is commonly seen in miniature, toy and small-breed dogs at the age of 5–8 months. CT shows an early diagnosis of bone collapse and curvilinear zones of sclerosis. Irregularities of the epiphysis and joint effusion can be demonstrated (Figure 19.9).

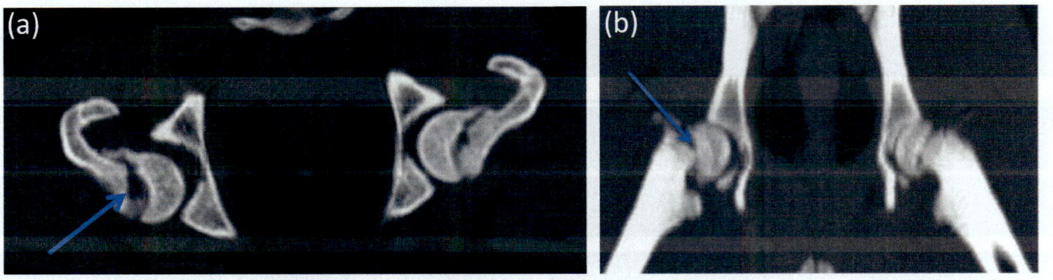

Figure 19.8 Transverse (a) and MIP (b) CT image of the hips of a Maine Coon, male, age 17 months with bilateral (right more severe) metaphyseal osteopathy. The femoral head remains in the acetabulum on both sides. A physeal fracture and step malalignment between femoral neck and head can be appreciated (arrows).

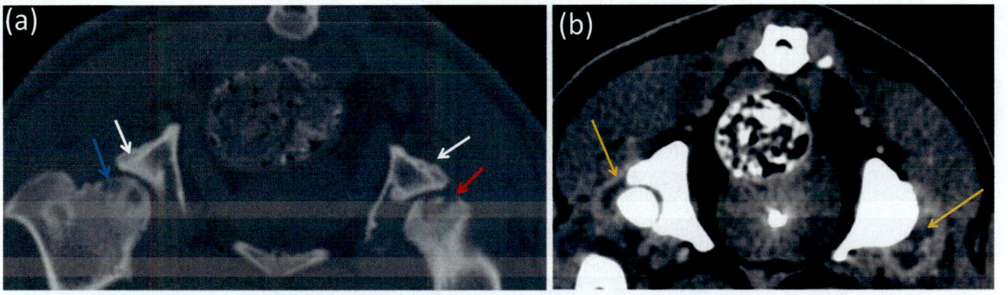

Figure 19.9 Transverse, bone window (a) and IV contrast, soft tissue window (b) at the level of the hip joints of a Jack Russel Terrier, male, 9 months old. Irregularity of the femoral head and subchondral hypodense areas are visible (blue arrow). The red arrow shows fragmentation of the femoral epiphysis. New bone formation at the dorsal acetabular ridge is noticed (white arrows). After IV administration, joint effusion and enhancement of the joint capsule can be appreciated (orange arrows). Histologically bilateral Legg–Calve–Perthes disease was confirmed.

19.5.2 Traumatic disorders

19.5.2.1 Hip joint fracture – central luxation

CT can perfectly show the insulted or fractured pelvis and simplify the orientation of fragments for reconstructive surgery of the hip joint, especially with 3D-techniques or multiplanar imaging.

19.5.2.2 Sacroiliac (sub)luxation: sacral wing fractures

Documentation of the degree of sacroiliac subluxation helps the surgeon reunite the plane joint surfaces at the correct point, the detection of hidden fractures or fissure lines, and documentation of the distance of dislocation in sciatic nerve disorder cases. In cases of chronic subluxation and/or fracture, the embedded sciatic nerve can be followed in the connective tissue with contrast-enhanced studies to provide possibilities in cases of pain with unknown origin and lameness (Figure 19.10).

19.5.2.3 Multiple pelvic fractures

Heavy trauma, such as a car accident, can cause single or multiple pelvic fractures; in these cases, the necessity for surgical stabilisation can only be established after CT examination and knowledge of the clinical hind-limb deficits. Again, multiplanar imaging, and 3D-reformation in particular, helps with an understanding of the orientation and reconstruction of the bony ring (Figure 19.11).

Urethrography and cystography should be used to show the course of the urethra in male dogs, which must be accurate for intact passage. Leakage of the contrast medium will be observed immediately, and soft-tissue necrosis, which is very painful, can be cured.

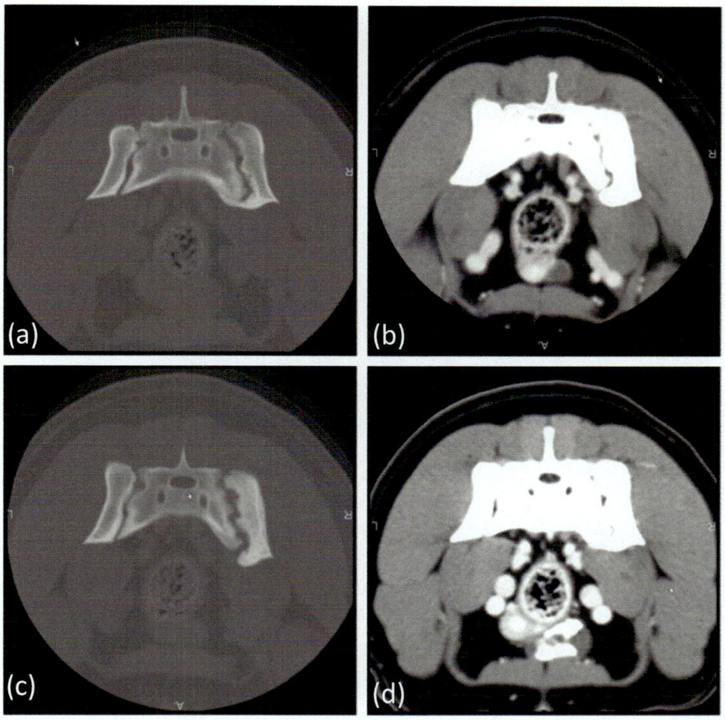

Figure 19.10 Fracture/subluxation of the right sacroiliac joint of a 9-month-old female Labrador retriever (a: bone window, b: soft tissue window after IV contrast administration) and 3-month follow-up after conservative treatment (c: bone window, d: soft tissue window after IV contrast administration): note the increased density of the iliac bone and the lateral periosteal reactions as well as the irregularly bordered joint space on the right side at the onset of pain; 3 months later, the bone returned to normal appearance, the irregular border of the tight joint space is outlined smoother and shows a bony bridge ventrally; in both sets, no significant soft tissue enhancement was noted.

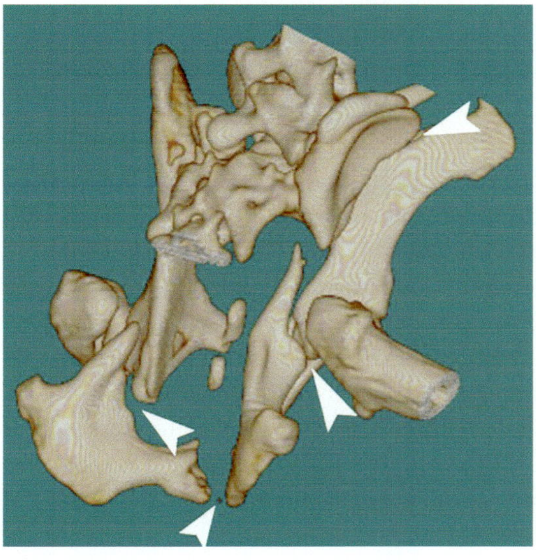

Figure 19.11 3D CT image of a pelvis showing multiple fractures. Arrow heads address the fractures and fragments.

19.5.2.4 (Postoperative) osteomyelitis

Changes of the iliac wing and body can occur either spontaneously or following triple pelvic osteotomy or surgical pinning. The periosteum of the bone becomes irregular and hyperproductive, and the bony medulla becomes sclerotic and loses its fatty content; in severe cases, gas bubbles can be detected in the altered soft tissue or the spongiosa. After contrast medium administration, hyperdense soft tissue demarcation may be present along the bones, with fluid filling around the bony structures that can spread to neighbouring joints or create a subcutaneous fistula. Severe muscular atrophy is always a sequela of misuse of the leg, which can be misinterpreted as neurogenic in origin. Sterile bone biopsy should be taken for

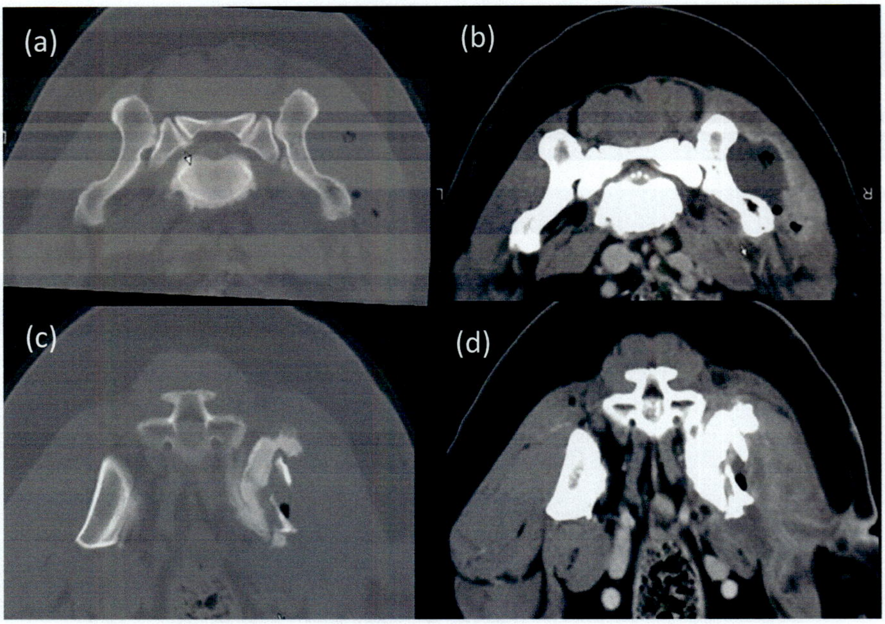

Figure 19.12 **Osteomyelitis of unknown origin at the iliac wing in a 3.5-year-old female Bullmastiff (a: bone window, b: soft tissue window after IV contrast administration) and 6-month follow-up (c: bone window, d: soft tissue window after IV contrast administration) under antibiotic treatment of abscessation caused by *S. marcescens*, *Cl. perfringens* and *E. coli* at the iliac bone. Note the gas within the altered gluteal muscle and the spongiosa of the iliac wing during onset; the follow-up shows resorption of the iliac bone with heavy periosteal formations after having created a deep fistula on the right side; gas content is still present in the inflamed tissue.**

bacteriologic examination and antibiogram (Figure 19.12).

19.5.2.5 Injury of the iliopsoas

The iliopsoas muscle is a double muscle (the iliacus and psoas major). These two muscles come together and form a common terminal tendon that attaches to the lesser trochanter (minor) of the femur. Iliopsoas strain (tear) results from excessive stretching of this muscle during highly athletic activities such as agility training.

Mineralization within the tendon or at the tendinous insertion may be observed in chronic cases, while avulsion of the lesser trochanter may be observed in acute traumatic events (Figure 19.13). Discrete lesions are detected earlier by CT, which can be used to confirm the diagnosis. Bilateral avulsions are possible. In active lesions, intravenous contrast administration may demonstrate enhancement within the insertion. Atrophy of the iliopsoas muscle is evident in chronic cases. (Myopathy of the iliopsoas muscle, see Chapter 23).

19.5.3 Degenerative disorders

19.5.3.1 Hip dysplasia and coxarthrosis

Remodelling of the dysplastic or traumatised hip joint results in various stages of coxarthrosis, an often painful and deforming condition that is usually seen on good-quality radiographs. In most cases, whether severe or not, CT can provide more information about the joint and its subchondral surfaces than radiographs. Lack of superimposition will allow detection of joint mice or other pathologies, such as mineralisation of bursae or enthesopathies (especially of the gluteus and iliopsoas muscle). To detect joint effusion, the IV iodine-contrast medium needs to be administered; an immediate scan is not

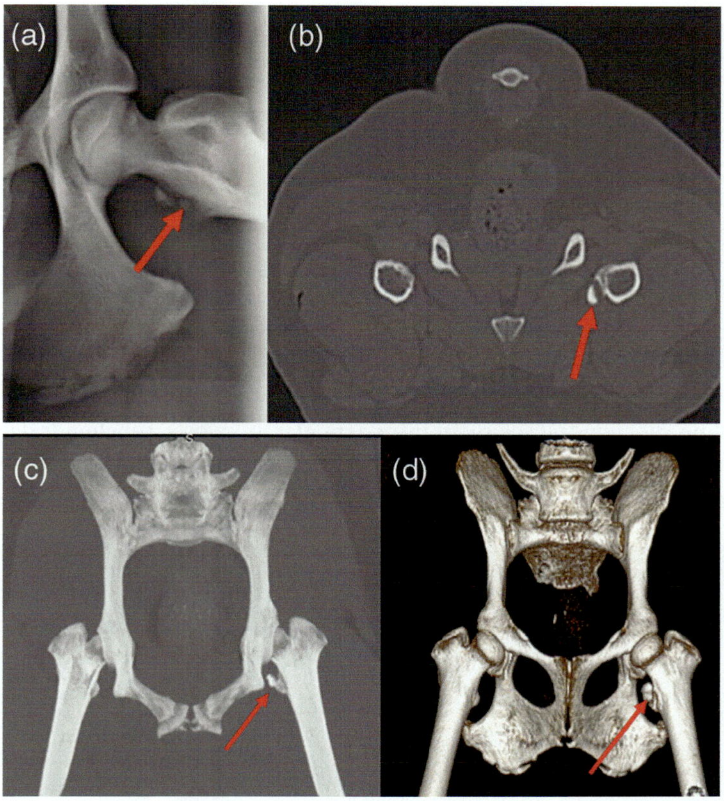

Figure 19.13 Ventrodorsal radiograph (a) of the left hip and a transverse CT image (b) in bone algorithm at the level of the trochanter minor. The red arrows show the mineralisations at the iliopsoas tendon at the trochanter minor. MIP CT (c) and 3D (d) CT images of the pelvis. Notice the avulsion fracture with mineralisations at the attachment of the iliopsoas muscle to the lesser trochanter (arrows).

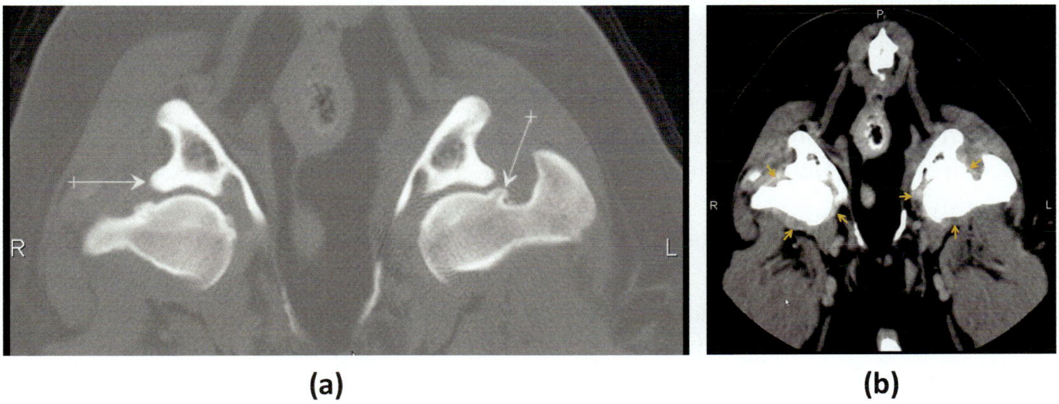

(a) (b)

Figure 19.14 Bilateral mild coxarthrosis in a 7.5-year-old male Labrador (a: bone window, b: soft tissue window after IV contrast administration): (a) the arrows point at the small subchondral arthritic ring around the left femoral neck and at the rounded and at the sclerotic craniolateral edge of the right acetabulum. (b) The bilateral increased joint filling is marked with arrowheads.

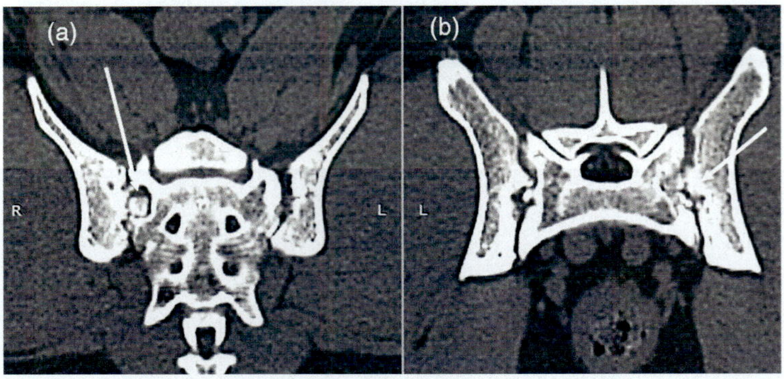

Figure 19.15 **Dorsal reconstructed (a) and transverse (b) images in bone window of Sacroiliac Joint Dysfunction including subchondral sclerosis, intra-articular ankylosis, subchondral cysts and, unique in the dog, subarticular clefts (arrows). In case of subluxation, periosteal reactions are usually present (for comparison, see Figure 19.10).**

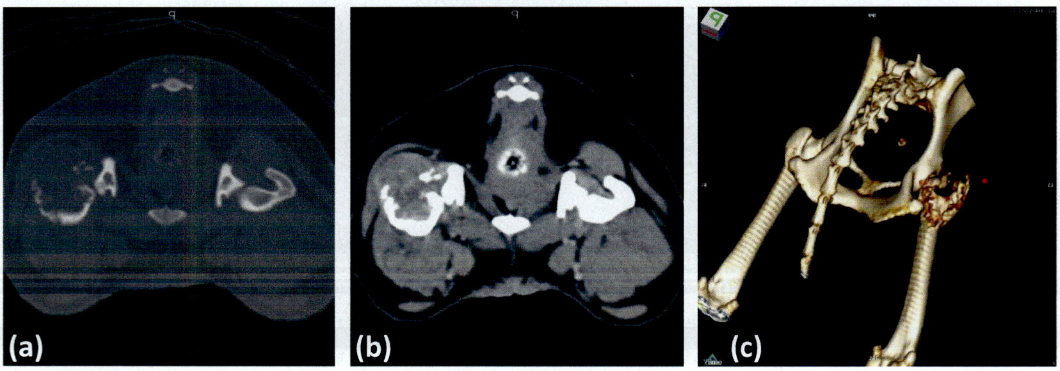

Figure 19.16 **Chondrosarcoma of the right femoral neck of a 1-year-old female German hunting terrier (a: bone window, b: IV contrast, soft tissue window): a large soft tissue swelling remodels the left femoral neck, the proximal diaphysis is widened (a), the femoral head resolved and the neoplastic tissue shows patchy contrast enhancement (b). A 3D-view (c) creates an overview of the bone where the femoral head is missing.**

necessary because the iodine molecules must be enhanced extracellularly in synovial tissue, which requires a few minutes (Figure 19.14).

19.5.3.2 DJD of the sacroiliac joints (pelvic ring anomaly)

The sacroiliac joint consists of ligamentous and synovial sections, which can both be affected by inflammatory changes. The ligamentous part is seen more commonly with an irregularly shaped joint space and bridging of the sacral wings to the iliac bone; the synovial part, however, shows sclerosis and sometimes exostoses. Degenerative Sacroiliac (SI) Joint dysfunction is a term used to describe pain in or around the region of a SI joint. Sometimes SI joint dysfunction can be the primary problem for a dog, and sometimes other injuries can contribute to SI joint dysfunction. CT findings include subchondral erosions, subchondral sclerosis, intra-articular ankylosis, subchondral cysts and, unique in the dog, subarticular clefts (Figure 19.15).

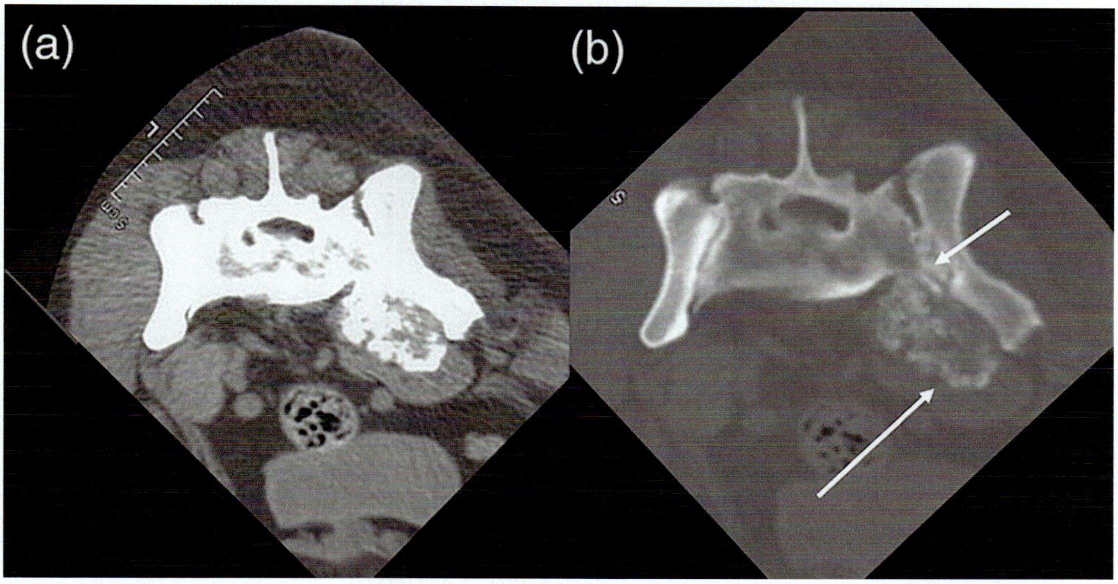

Figure 19.17 Transverse images in soft tissue-(a) and bone algorithm (b) of a neoplasia originating from the left sacroiliac joint (short arrow). This neoplastic process displays multiple granular opacities surround by soft tissue swelling (large arrow) and was confirmed as a chondrosarcoma.

19.5.4 Neoplastic disorders

19.5.4.1 Osteosarcoma

Tumours mostly arise from the femoral neck and are sometimes difficult to detect on radiographs, but with CT, lysis of the bone is more easily recognised because moth-eaten spongy tissue appears with periosteal reactions around the femoral neck, especially in older dogs. In very young dogs, chondrosarcoma may be present, resulting in a large soft-tissue mass as much of the femoral head is consumed. After contrast medium administration, all of the soft-tissue involvement of the mass and the bony remnant becomes visible (Figures 19.16 and 19.17).

19.5.4.2 Soft-tissue sarcoma

Soft-tissue sarcoma, a rare condition either at the hip joint or the sacroiliac joint, should be suspected when enhancing solid soft tissue fills the area of the joint capsule and/or pouches. In advanced cases, the cortex becomes infiltrated and resolved at different bone parts at the same time. CT-guided biopsy is required to confirm the malignancy.

19.5.5 Miscellaneous

Lipoma of the caudal femoral fossa is rare, but the pain and inactivity/misuse caused by the condition results in severe muscle atrophy of the affected hind leg. CT images will show hypodense fatty tissue (approximately –120 to –180 HU); the fat may be encapsulated by a thin lining in large formations, and the caudal femoral artery and vein that accompany the sciatic nerve distally are usually compressed. It is important to distinguish the homogeneity of the fatty content from infiltrative lipoma or liposarcoma; lipoma stays homogeneous and is low-attenuating before and after contrast administration, while infiltrative lipoma and liposarcoma have irregular boundaries and enhancing tissue that infiltrates within the fatty tissue itself or into its surroundings. The amount and type of muscle atrophy can be easily compared on each side. For surgical purposes, the dignity of the lesion, the entrance for its removal, and hot spots such as the course of the sciatic nerve and its vessels should all be documented in advance (Figure 19.18).

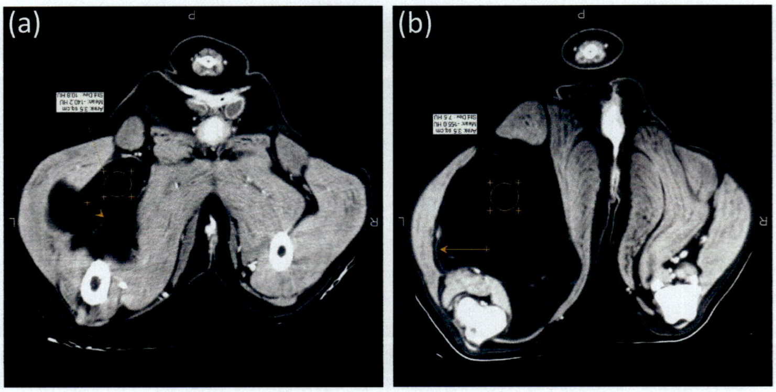

Figure 19.18 **Solid lipoma in the left thigh of an 8-year-old female Labrador retriever (transverse, IV contrast, soft tissue window images): arrow in figure (a) of the mid-thigh region points at a crossing vessel–nerve tract dividing the lipoma into a smaller proximal and a larger distal compartment; arrow in figure (b) shows the lateral displacement of the sciatic nerve at the level of the popliteal fossa. The measured density was 150 HU.**

FURTHER READING

Breit S, Knaus, I, Kunzel W. 2003. The gross and radiographic appearance of sacroiliac ankylosis capsularis ossea in the dog. *Research in Veterinary Science* 74(1): 85–92.

Conway WF, Totty WG, McEnery KW. 1996. CT and MR imaging of the hip. *Radiology* 198(2): 297–307.

Craig LE. 2001. Physeal dysplasia with slipped capital femoral epiphysis in 13 cats. *Veterinary Pathology* 38:92–97.

Farese JP, Todhunter RJ, Lust G, Williams AJ, Dykes NL. 1998. Dorsolateral subluxation of hip joints in dogs measured in a weight-bearing position with radiography and computed tomography. *Veterinary Surgery* 27(5):393–405.

Fujiki M, Kurima Y, Yamanokuchi K, Misumi K, Sakamoto H. 2007. Computed tomographic evaluation of growth-related changes in the hip joints of young dogs. *American Journal of Veterinary Research* 68(7):730–734.

Ginja M, Ferreira AJ, Jesus SS, Melo-Pinto P, Bulas-Cruz J, Orden M, Gonzolo-Orden J. 2009. Comparison of clinical, radiographic, computed tomographic, and magnetic resonance imaging methods for early prediction of canine hip laxity and dysplasia. *Veterinary Radiology & Ultrasound* 50(2):135–143.

Mahoney P. 2012. Musculoskeletal imaging in the cat what's normal? What's abnormal? *Journal of Feline Medicine and Surgery* 14(1):13–22.

Patricelli AJ, Dueland RT, Adams WM, Fialkowski JP, Linn KA, Nordheim EV. 2002. Juvenile pubic symphysiodesis in dysplastic puppies at 15 and 20 weeks of age. *Veterinary Surgery* 31(5): 435–444.

Thrall DE. 2007. *Textbook of Veterinary Diagnostic Radiology*. Elsevier Health Sciences, pp. 221–358.

Wise R, Jones J, Werre S, Aguirre M. 2022. The prevalence of sacroiliac joint CT and MRI lesions is high in large breed dogs. *Veterinary Radiology & Ultrasound* 63(6):739–748.

CHAPTER 20

COMPUTED TOMOGRAPHY OF THE STIFLE

Wolfgang Henninger, Amalia Agut and R. Latorre (Anatomy), and Ingrid Gielen

20.1 INTRODUCTION

The use of computed tomography (CT) of the stifle is controversial. In humans, magnetic resonance imaging (MRI) is currently the gold standard for joint examination (especially in cases of cranial or caudal cruciate ligament rupture (CCL), collateral ligament injury, or meniscal damage). However, many more joint conditions, such as angular limb deformity, trauma, arthritis, other inflammatory conditions, and bone and soft tissue neoplasms, are presented on a regular basis, and all can be diagnosed from CT images without much difficulty.

20.2 INDICATIONS FOR USING CT

a. *Developmental*: medial or lateral patellar subluxation or luxation (both resulting in angular limb deformity), OCD.
b. *Traumatic*: epiphysiolysis of the tibial plateau, femoral condylar fracture, fracture of the patella, avulsion fractures of intra-articular ligaments, collateral ligament lesion, patellar ligament lesion, gastrocnemius tendinopathy, and muscular haematoma.
c. *Degenerative*: DJD, secondary CCL rupture, enthesopathies.
d. *Neoplastic*: synovial cell sarcoma, osteosarcoma, soft tissue sarcoma, and lipoma.
e. *Inflammation/Infection*: juvenile gonitis, inflammatory fistula following tibial plateau levelling osteotomy (TPLO) or extracapsular ligament reconstruction and osteomyelitis.

20.3 TECHNICAL ASPECTS

20.3.1 Positioning

The patient is always "feet first" orientated to the CT gantry. Two ways of positioning for CT examination of the stifle joints are described. Some examiners prefer to position the animal in dorsal recumbency similar to the radiographic position for hip dysplasia examination. However, positioning the animal in sternal recumbency with the hind limbs extended caudally and parallel to each other is often more convenient (Figure 20.1). The latter has been proven to be a stable, reproducible method of achieving reliable results, especially in very large-breed dogs with weight exceeding 35 kg. To avoid painful extension of the lumbosacral transition area, the femurs shall include an angle approximately of 100°–120°, and the stifles and hind legs can rest horizontally on the table (Figure 20.2).

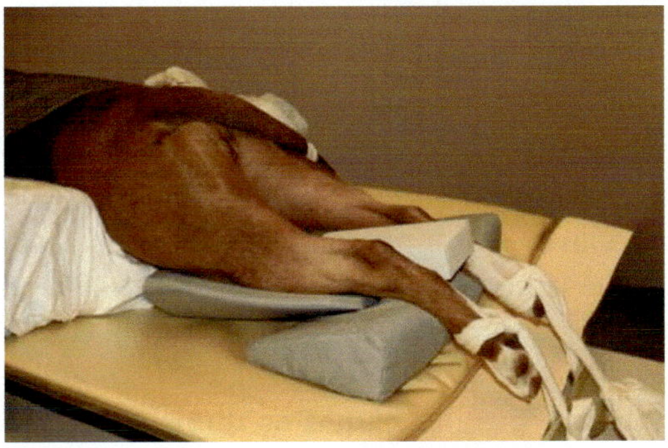

Figure 20.1 **Correct parallel positioning of the stifles with the dog/cat in dorsal recumbency; the tarsi rest on a pillow, and the extension of the stifle ligaments places a stress on them to allow better detection of injuries.**

DOI: 10.1201/9781315118130-22

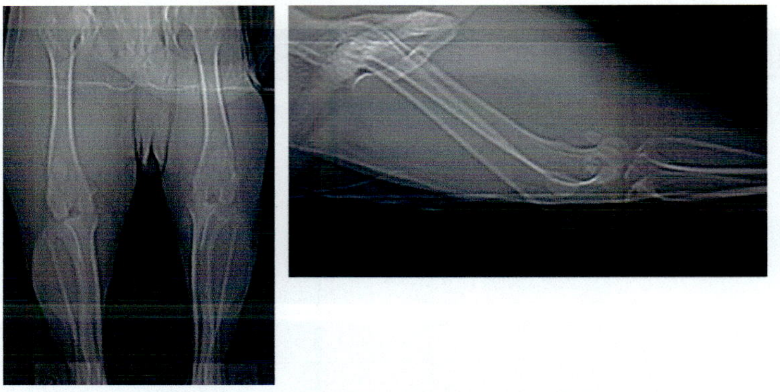

Figure 20.2 **Positioning and scout view of the animal in ventral recumbency, offering a more stable position, for CT of both stifle joints.**

When positioned in dorsal recumbency, the hind limbs are rested on a pillow at an approximate 45°–60° angle from the CT table. Both stifles must be positioned close to each other to minimize beam-hardening artefacts. The joint spaces should be at the same level, rotated slightly inwards similar to hip dysplasia positioning and fixed with a bandage to the pillow and the table, taking precaution with the pressure. In case of ligament injury or rupture, this leads to slight movement of the articular surfaces against each other, which can be easier picked up in the transverse images later.

20.3.2 Technical parameters

With single-slice CT, the x-ray beam is angled parallel to the surface of the tibial plateau; the scan range should cover the whole joint from the proximal pouches from the distal third of the femur to the proximal fifth of the length of the tibia. Multi-slice CT with isotropic resolution allows scanning of the stifle without tilting the gantry; angulation and orientation are performed during post-processing procedures. In multi-slice CT, the slice thickness should be as thin as possible, 1 mm or thinner and slice increment of up to 50%. Settings should be 100–120 kV and 100–200 mA, depending on the animal's weight and size. Reconstruction parameters are soft tissue and bone window algorithm.

20.3.3 Contrast medium

- Intravenous (IV) administration of 2 mL of iodine contrast (350-400 mg iodine/mL) medium per kilogram body weight should be used for any stifle problem; images with contrast enhancement provide better information about anatomy and pathology than plain images.
- Intra-articular (IA) administration (arthro-CT/CTA) of diluted non-ionic low-osmolar-type contrast medium such as iohexol has been reported to help with distinguishing the cruciate ligaments, evaluating the menisci and the surface of the cartilage. Although gross meniscal lesions and cruciate ligament damage can be evaluated, the problem with CTA is that in inflamed stifle joints, the injected contrast medium is very rapidly absorbed and diluted, making an accurate interpretation of possible lesions difficult. A mixture of 0.2 mL of epinephrine with conventional contrast agents in order to slow down resorption should be administered. For arthro-CT, the normal iodine concentration should be reduced till 80–100 mg I/mL to enhance visualization without contrast-obscuring structures. The original contrast medium can be diluted with saline, and for an average knee joint, a volume of 4 mL is recommended. During administration of IA

contrast, care should be taken to antisepsis and to avoid the fat pad. Gentle flexion and extension of the joint has been recommended to enhance contrast dispersion.

20.4 NORMAL ANATOMY

The distal end of the femur, including the femoral trochlea and condyles, the patella within the sulcus of the trochlea, the proximal tibial condyle, and the non-articulating head of the fibula can be easily visualized in a bone and soft-tissue window (Figures 20.3–20.5).

In dogs, the patella is thick and oval; in cats, it is flatter and more rounded (Figure 20.6a).

Multipartite patellae are developmental normal variants that occurs because of failure of the ossification centres to fuse (20.6b and c). This should be differentiated from patella fractures.

The fabellae (the lateral one is the largest), the medial sesamoids at the origin of the gastrocnemius muscles' heads, and the sesamoid bone of the popliteus muscle are recognized from their typical locations and positions. Bipartite and multipartite fabellae, often bilaterally, and bipartite popliteal sesamoid are considered congenital developing disorders because they are rounded on radiographic and CT images and are not linked to any clinical symptoms (Figure 20.7). The fat pad can be identified as a radiolucent structure caudally from the patellar ligament. Other soft-tissue structures, such as the cruciate ligaments, the menisci, and the collateral ligaments that form a soft-tissue cover around the subchondral bones,

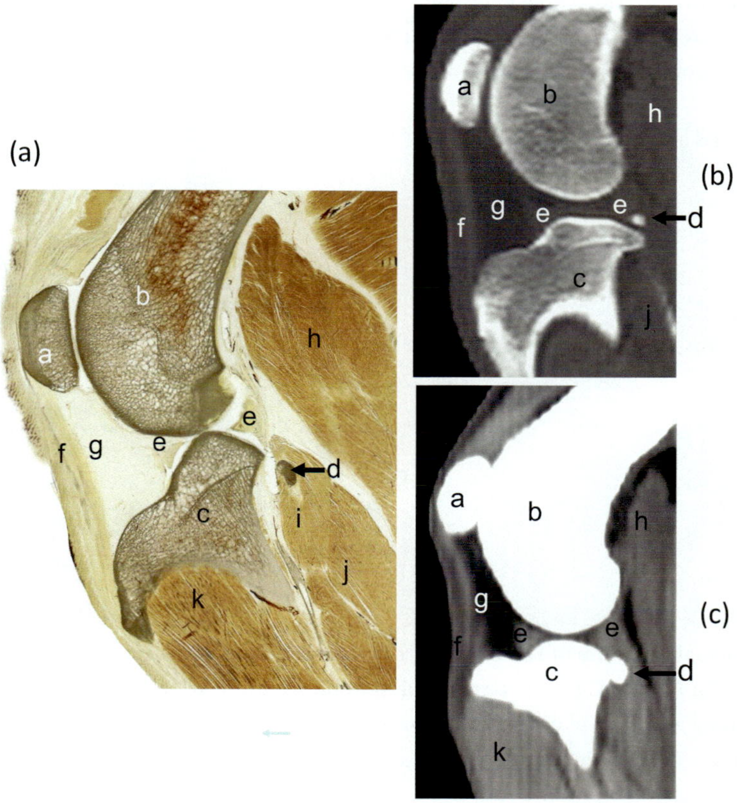

Figure 20.3 Normal anatomy of the canine stifle joint. (a) Anatomical sagittal slice; (b) sagittal CT reconstruction in bone window; (c) sagittal CT reconstruction in soft-tissue window. a. patella; b. femur; c. tibia; d. sesamoid bone popliteus muscle; e. lateral meniscus; f. patelar ligament; g. infrapatelar fat pad; h. gastrocnemius muscle lateral head; i. poplieus muscle; j. flexor digitorum lateralis; k. long digital extensor muscle.

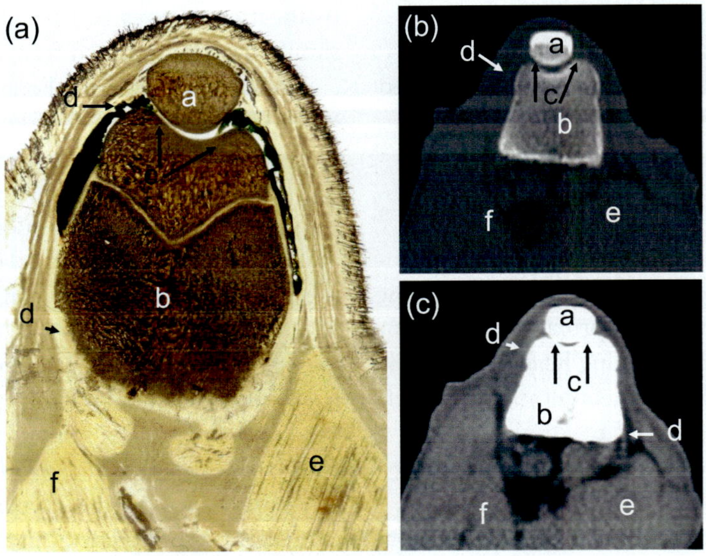

Figure 20.4 Normal anatomy of the canine stifle joint. (a) Anatomical sagittal slice with latex injected in the joint cavity, transverse CT images in bone window (b), and soft-tissue window (c) at the level of the proximal femoral trochlea. a. patella; b. femur; c. femoral trochlea; d. articular capsule; e. biceps femoris muscle; f. semimembranosus muscle.

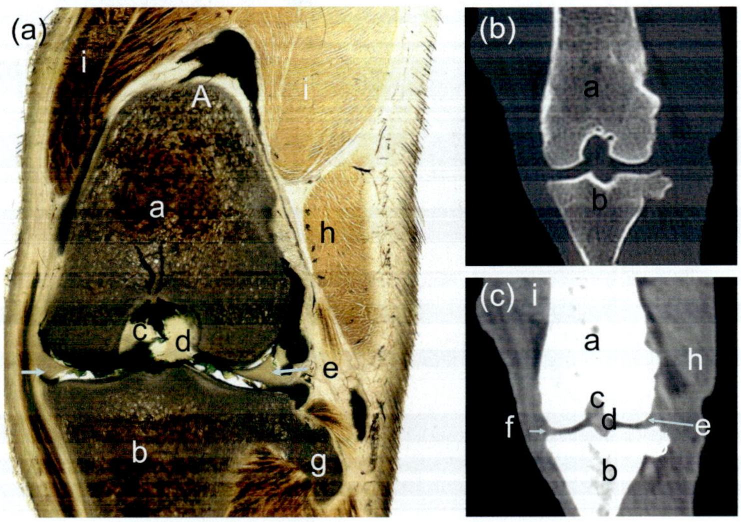

Figure 20.5 Normal anatomy of the canine stifle joint. (a) Dorsal anatomical slice with latex injected in the joint cavity, (b) Dorsal CT reconstruction in bone window, (c) Dorsal CT reconstruction in soft-tissue window at the level of the intercondylar notch. a. femur; b. tibia; c. caudal cruciate ligament; d. cranial cruciate ligament; e. lateral meniscus; f. medial meniscus; g. fibular head; h. biceps femoris muscle; i. quadriceps femoris muscle.

cannot be individually identified without applying IA contrast medium. The extension of the joint capsule and its proximal or distal pouches, including the proximal tendon sheath of the long extensor muscle, is visible only when filled with fluid or tissue for a variety of causes; however, it cannot be seen in the normal joint.

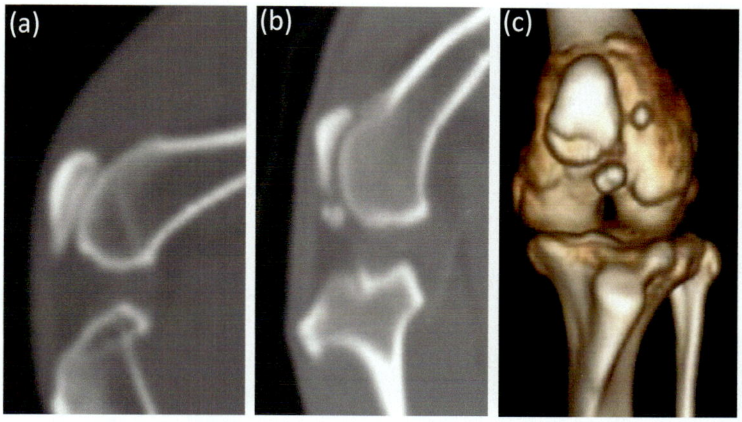

Figure 20.6 Normal anatomy of the feline stifle joint (a). Sagittal CT reconstruction in bone window. Notice the more flattened and more rounded shape of the patella. Sagittal reconstructions in bone window (b) and 3D CT images of multipartite patellae in a cat, these are considered as developmental normal variants.

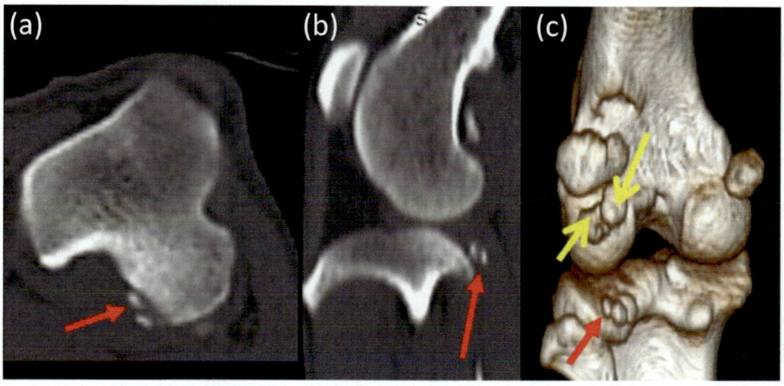

Figure 20.7 Transverse (a) at the level of the proximal tibia, sagittal reconstructed (b) CT images in bone algorithm and 3D CT image (c) of a dog's stifle joint. The popliteus sesamoid bone displays two separate ossification centres (red arrows).The same dog shows multipartite fabella which is a normal anatomical variant (yellow arrows). These findings were bilaterally noticed.

Calcification of the meniscal cartilage has been described in the cat (Figure 20.8), as well as related exotic species.

With arthrography, the cranial and caudal cruciate ligaments as well as the menisci can be visualized (Figure 20.9).

20.5 PATHOLOGIC FINDINGS

20.5.1 Developmental disorders

20.5.1.1 Medial patellar (sub)luxation

The ridges and the sulcus of the femoral trochlea are flat and even; the patella rides on the medial ridge or is medially displaced onto the side of the ridge (Figure 20.10); and the joint pouches are slightly enlarged and rarely filled with fluid. Surgical correction of this angular limb deformity faces a deepened sulcus where the wedge-shaped excision can be seen within the spongy bone even after a long period of time. A metal pin in the proximal tibia usually indicates lateral displacement of tuberositas tibiae. Following surgical intervention, or in cases of ongoing lameness, it is helpful to document the midline movement of the patella within the sulcus, with the knees fully extended and maximally flexed (Figure 20.11).

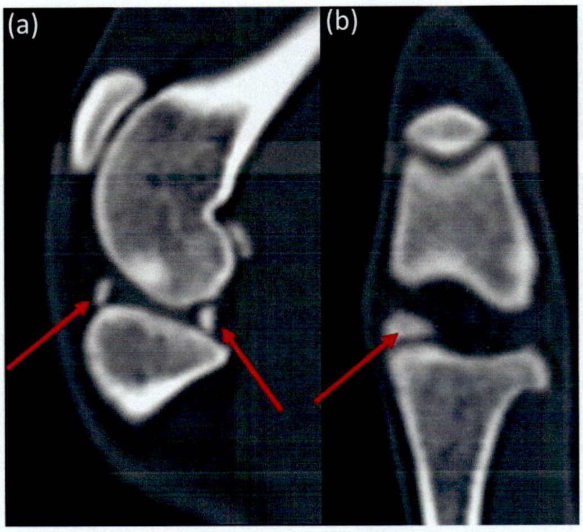

Figure 20.8 Sagittal (a) and dorsal (b) reconstructed CT images in bone algorithm of a cat's stifle showing meniscal mineralisation (arrows).

20.5.1.2 Lateral patellar (sub)luxation

Lateral displacement of the patella is more commonly seen in medium- and large-breed dogs. The patella can be seen either riding on top of the lateral ridge or completely displaced to the lateral side of the femoral trochlea, where the irregularly deformed long digit extensor muscle originates. Joint effusion and periosteal reactions of the joint surface may be present. In large-breed dogs, the angular limb deformity is severe and difficult to correct; therefore, CT examination of the entire limb, including the hip joints and tarsal bones, is recommended for 3D reformation. Following surgical intervention, a follow-up CT with the stifle joints extended and flexed can help document the patellar position. Follow-up examinations are always necessary to evaluate surgical results and to ensure a good prognosis.

20.5.1.3 OCD

In some animals, a delay of subchondral ossification of the lateral (or sometimes medial) femoral condyle can result in thickening of the cartilage, reduced trophy, and flap formation. The subchondral surface of the round femoral condyle may be flattened and in severe cases may be missing entirely (Figure 20.12). Joint effusion may be present, but is usually moderate. Superficial or deep injuries of the cartilage of the OC side are seen only after IA contrast medium administration (Figure 20.13). Joints affected with OCD eventually develop degenerative joint disease (DJD), which typically results in new bone formation along the trochlear ridges, at the patella and the tibial head, as well as thickening of the joint capsule combined with increased joint filling.

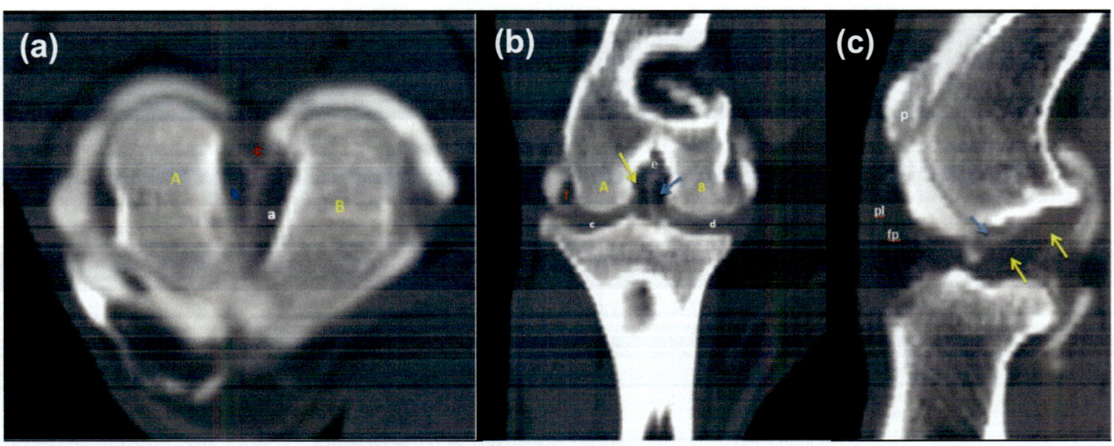

Figure 20.9 Arthro-CT images of a normal canine stifle joint. Transverse (a) dorsal reconstructed (b), B sagittal oblique (c), reconstructed images in bone algorithm. (A) lateral condyle femur; (B) medial condyle femur, (p) patella; (pl) patellar ligament; (fp) infrapatellar fat pad; (c) lateral meniscus; (d) medial meniscus; (e) meniscofemoral ligament; (t) popliteal tendon and sheath; (blue arrow) caudal cruciate ligament; (yellow arrows) cranial cruciate ligament.

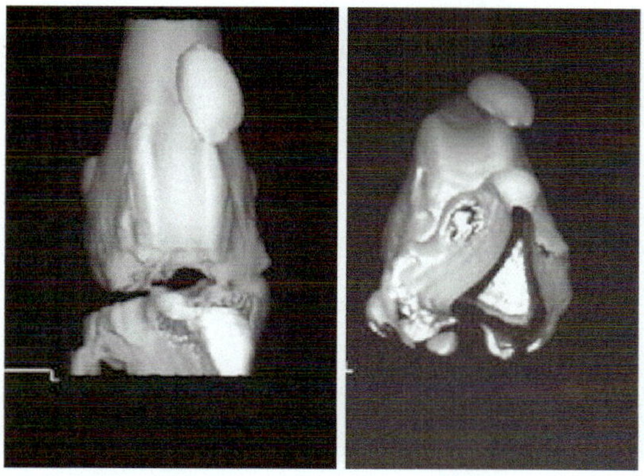

Figure 20.10 3D CT images of medial patellar luxation. Note the medial location of the patella.

20.5.2 Traumatic disorders

20.5.2.1 Epiphysiolysis and physeal fractures

Fractures can occur at the distal femur or at the tibial plateau due to very strong forces to the joint (hit by car or high-rise syndrome). In young (usually small-breed) dogs, the proximal tibial physis can rupture from moderate force, too (e.g., falling from the owner's arms). The diagnosis is often clear from radiographs, but malformations due to growth-plate damage may occur later and

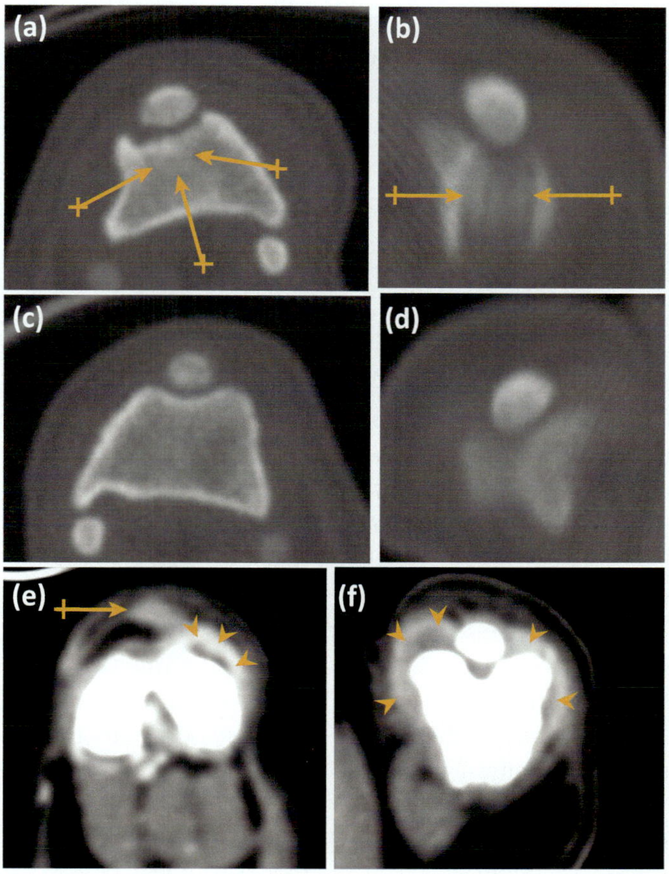

Figure 20.11 Effect of the sulcoplasty procedure and lateral transposition of the tibial tuberosity of the left stifle, resulting in normal patellar movement in a 7-year-old toy poodle with left hind limb lameness (a, b – left stifle extended and flexed, and Figure c, d – right stifle for comparison, both in a bone window, note the edge of the sulcoplasty marked with orange arrows). After IV administration of contrast (e and f), the thickened straight patellar ligament can be followed on image e, with the left stifle in extension (arrow heads). In flexion (image f), the increased joint effusion is labelled with arrowheads.

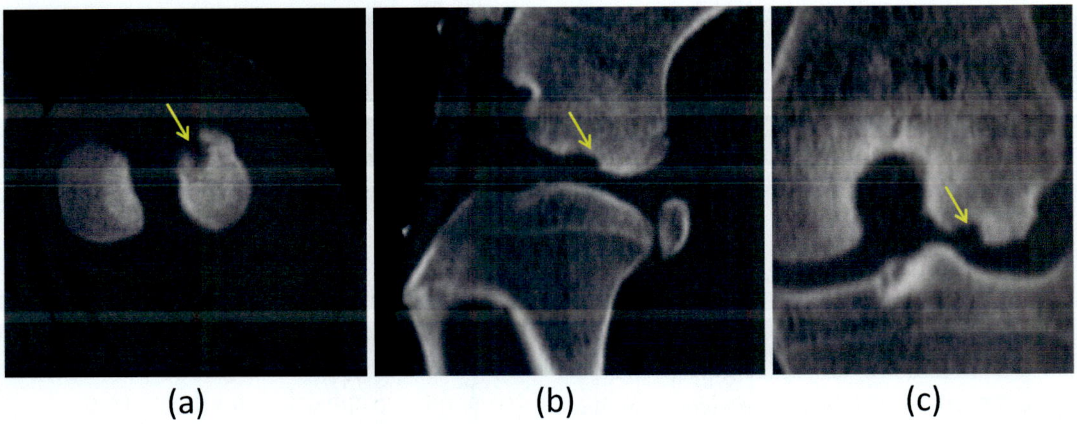

Figure 20.12 Transverse (a) and sagittal (b) and dorsal (c) reconstructed images in bone algorithm of an OCD lesion at the lateral femoral condyle of the left stifle of a 1-year-old Great Dane. The subchondral defect of the lateral femoral condyle surrounded by a sclerotic rim is clearly seen (yellow arrows).

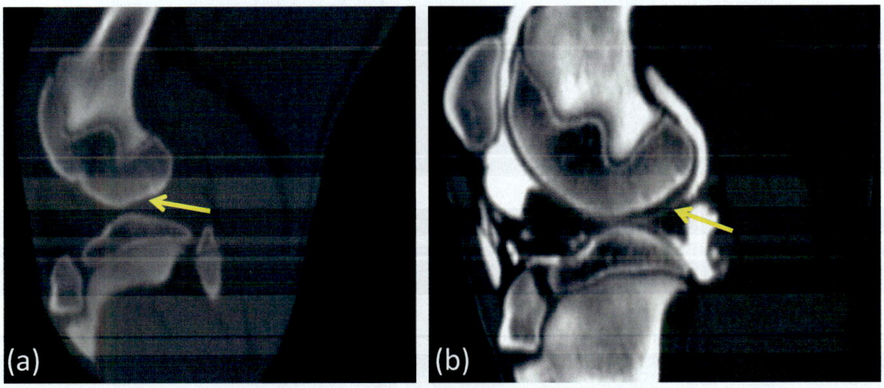

Figure 20.13 Sagittal reconstructions in bone window of a young dog with an OC lesion. (a) Notice the flattening of the condyle (arrow). (b) Arthro-CT of the same lesion showing thickened cartilage without flap formation, indicating that an OC lesion in present (arrow).

necessitate a CT series and correction with osteotomy (Figure 20.14).

20.5.2.2 Femoral condylar fracture

Severe trauma can lead to fracture of the femoral condyle or even unilateral condylar fracture. CT will show the fracture lines and the displacement of the fracture condyle, which are not always clearly visible with radiography.

20.5.2.3 Fracture of the patella or rupture of the patellar ligament

Before diagnosing a fractured patella, a dysplastic, multiple-section patella should be considered. A fractured patella will exhibit increased distance between the fragments and the fracture line (Figure 20.15, whereas a rupture of the patellar ligament (see 23.3, Chapter 23, Muscles) results in a patella alta because of the quadriceps muscle retraction.

20.5.2.4 Bone healing of IA fracture

The joint bone is well vascularized. During the healing procedure, resorption and apposition occur nearly simultaneously, resulting in a rather heterogeneous pattern of spongy structure, which gets homogeneous over time. The cortical bone is covered by periosteal

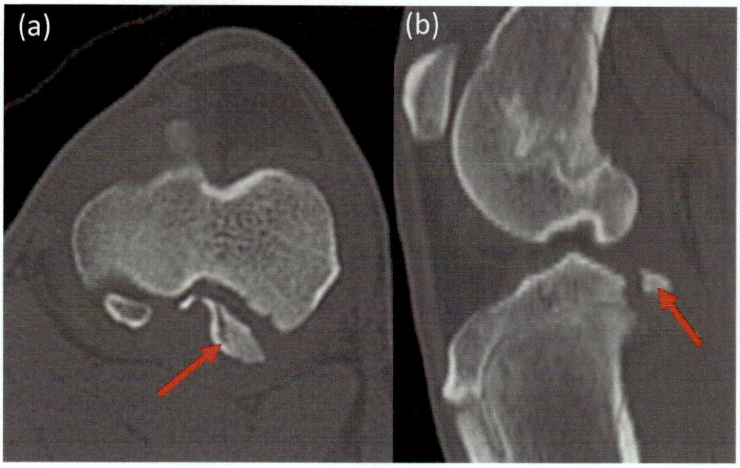

Figure 20.14 Transverse (a) and corresponding sagittal reconstruction (b) in bone window of a tear of the caudomedial tibial plateau in a 9-month-old female Great Schnauzer. The MPR is helpful to gain complete information about displaced bony fragments (arrows). (Courtesy Dr. Henning Schenk, Tierklinik Lüneburg, Germany.)

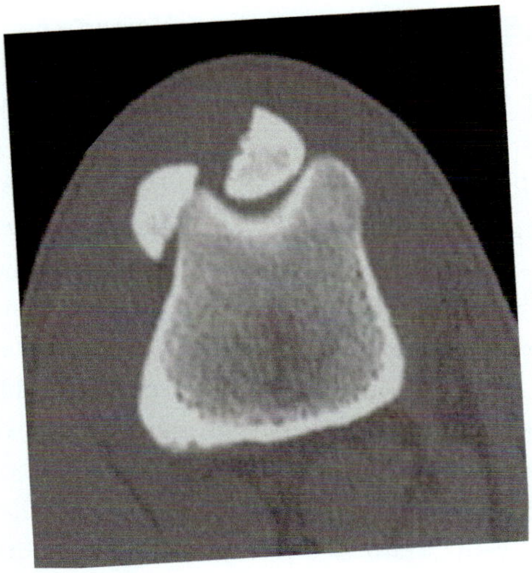

Figure 20.15 Transverse CT image in bone algorithm, of the patellar fracture with significant lateral fragment displacement.

reactions first and will be smoothed finally. Healing disturbances will be seen in insufficient bridging of fragments and bony lysis, as well as excessive periosteal reaction (additional information in 'bone infection and osteomyelitis').

20.5.2.5 Avulsions of IA tendons

Acute rupture of the cranial/caudal cruciate ligament: Avulsion fractures of the cranial cruciate ligament (CCL) (Figure 20.16).

Mild-to-moderate joint effusion is present, and the thickness of the joint capsule is normal and does not enhance during contrast medium administration; the joint surfaces are intact. The ossified sesamoid of the popliteus muscle is found displaced distally in a case of total cranial ligament (TCL) rupture (Figure 20.17). Subtotal ruptures of the ligaments may show effusion and can stay unremarkable. CT however is needed to rule out other differentials of lameness before surgery.

- Avulsion fractures of the caudal cruciate ligament in young dogs (Figure 20.18). Here are similar findings as described above.
- Avulsion of the popliteus tendon (Figure 20.19). A mostly bony fragment on the lateral part of the femoral condyle is present together with a focal soft tissue swelling.
- Avulsion of the long digitorum extensor muscle (Figure 20.20). It may occur as a single event, but the main changes result

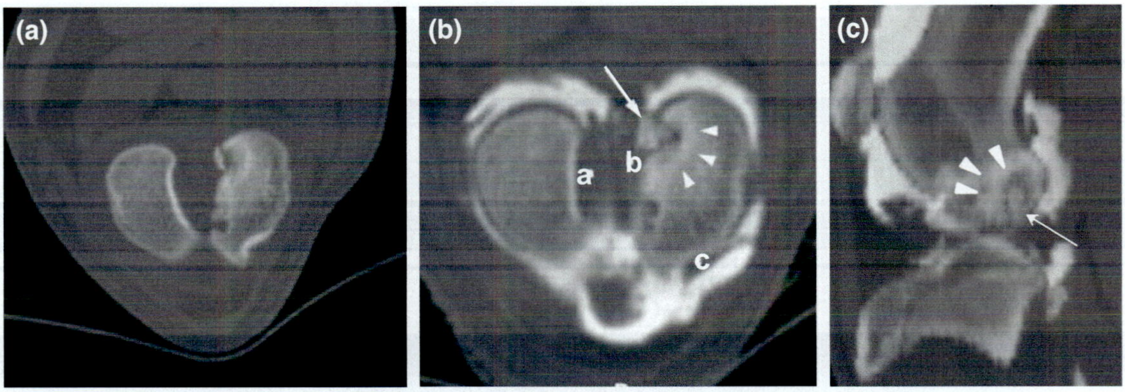

Figure 20.16 Avulsion of the cranial cruciate ligament in an 8-month-old Labrador Retriever. (a) Transverse, (b and c) arthro CT, transverse and sagittal reconstruction in bone window through the area of the femoral attachment (lateral femoral condyle) of the cranial cruciate ligament showing the avulsion of the attachment in bone window (white arrows). There is marked subchondral sclerosis (arrow heads) surrounding the defect, proving the chronicity of the lesion. There is no contrast surrounding the bony fragment, proving that the fragment is still partially attached within the subchondral bed. (a) Caudal cruciate ligament; (b) cranial cruciate ligament; (c) extensor digitorum longus muscle.

from the deformation in combination with the lateral subluxation of the patella (see description above).

- Avulsion of the collateral ligaments (Figure 20.21)
- Avulsion of the tuberositas tibiae (Figure 20.22)

20.5.2.6 Pseudoaneurysm and muscular hematoma

Orthopaedic surgery and placing implants around the distal femur or proximal tibia (like TPLO or TTA) may cause soft tissue damage and laceration postoperatively. This results in swelling in the caudal fossa and may cause haematoma. Also, severe trauma can result in haemorrhage or inflammation within any group of injured muscles. Plain CT will not always show changes of tissue density, but tissue damage will be indicated by inhomogeneous and patchy contrast uptake after IV contrast medium administration or the injured vessel can be identified. In these cases, higher Hounsfield units (plus 100 HU) are usually reached, compared with enhancement of soft tissue sarcoma, with less Hounsfield units (approximately plus 30–50 HU compared with plain scans).

20.5.3 Degenerative disorders

20.5.3.1 Degenerative joint disease (DJD)

The primary lesion may be congenital, traumatic, or neoplastic, or it may be the result of aging. CT images will show the reactions of the synovial membrane that result in effusion, thickening of the joint capsule, and bony spurring along the insertion area of the capsule to the cortical bone. The ability to detect DJD depends on technical factors and on the severity and also on the duration for which the joint has been affected. Changes occur unilaterally or bilaterally; findings can be detected more easily by comparing both sides (Figure 20.23).

20.5.3.2 Secondary cranial cruciate ligament (CCL) rupture

The preliminary diagnosis is based on clinical examination and radiographs, but CT can help confirm the diagnosis. In chronic CCL rupture cases, DJD changes are clearly visible. The affected joint will show effusion, an enhanced thickened joint capsule, and severe periosteal reactions along the subchondral bones of the trochlear ridge and the tibial plateau. Bony lyses at the origin of the cruciate ligament, or

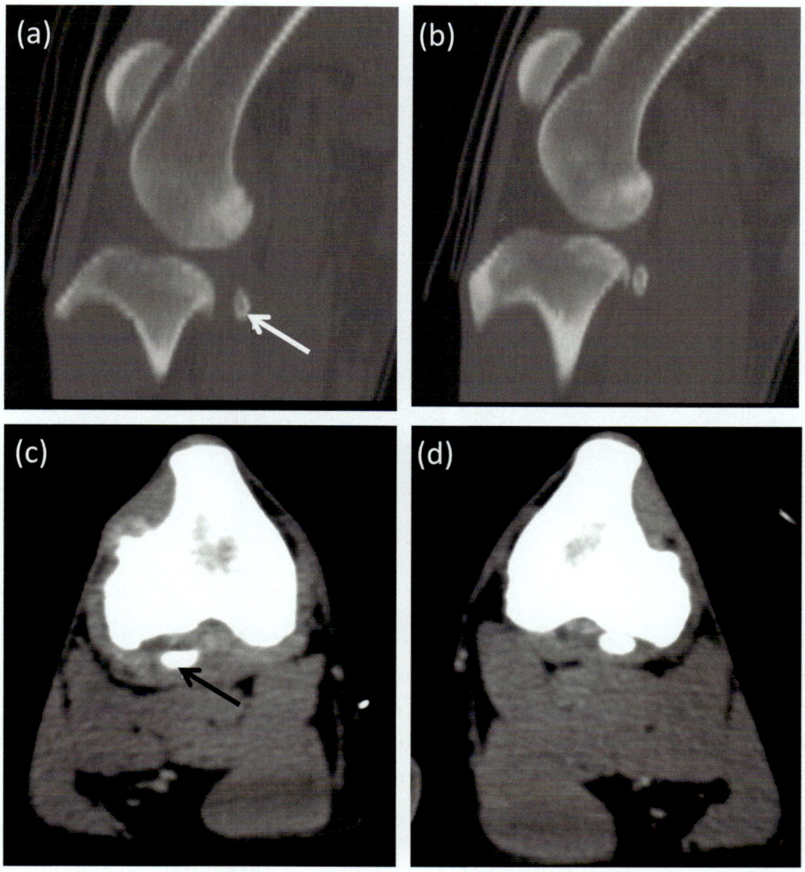

Figure 20.17 **CCL rupture of the right stifle in a 2-year-old male Labrador Retriever; the cranial displacement of the tibia results in a dislocation of the sesamoid of the popliteus muscle (white arrow). Sagittal reconstructions in bone window, (a) compared to (b), the contralateral stifle joint. The drop of the sesamoid bone is indicative for a complete rupture of the CCL. (c) and (d), transverse images in soft tissue algorithm after IV contrast administration. The displacement of the sesamoid can be picked up in the affected joint c (black arrow). Also, note the moderate muscle atrophy of the affected limb.**

the insertion area, are often seen. Displacement of the (ossified) popliteal sesamoid may not be so obvious. (see Figure 20.17).

Unfortunately, the menisci cannot be identified with plain CT, but diagnosis is not always a high priority if the rupture is stabilized with TPLO or TTA. Arthro-CT might be useful in demonstrating CCL pathology and meniscal damage (Figure 20.24).

20.5.3.3 Enthesopathies

Enthesopathies of the large muscles around the stifle joint are well-documented from radiographs. They can occur from the quadriceps muscle at the tuberosity of the tibia along the patellar ligament, and at the base and apex of the patella itself, which is commonly seen in combination with DJD in cats. CT helps identify these individual mineralizations at specific locations on muscular or ligamentous structures. An incomplete or complete tear of the lateral head of the gastrocnemius muscle in working dogs, especially in Border collies and related breeds, is a specific type of enthesopathy, which on radiographs may look like a simple periostitis that is commonly seen with arthritic changes or with a neoplastic condition; CT images, however, will show the entire tissue damage with

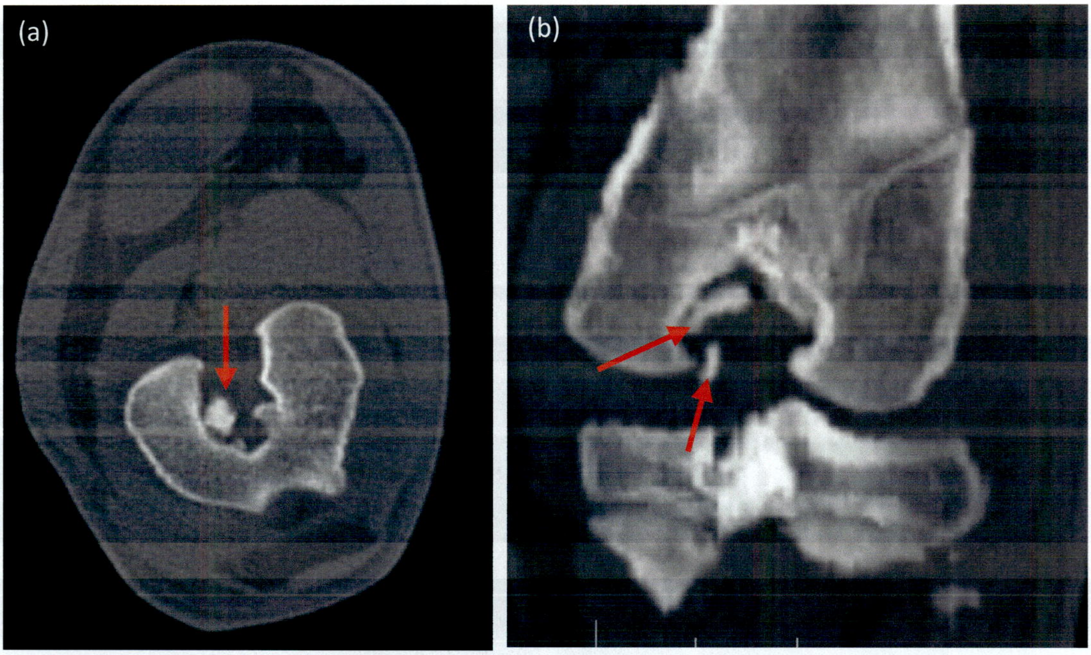

Figure 20.18 Avulsion of the caudal cruciate ligament. (a) Transverse CT image through the area of the femoral attachment (medial femoral condyle) of the caudal cruciate ligament showing the avulsion of the attachment in bone window. The dorsal reconstruction (b) shows two fragments as a result of the avulsion of the caudal cruciate ligament. The red arrows show the avulsed fragments.

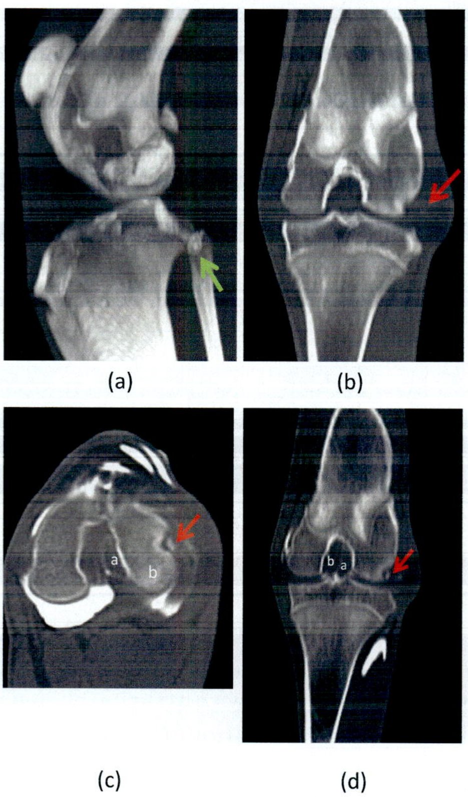

Figure 20.19 Avulsion of the popliteus tendon. Sagittal, MIP, (a) and dorsal (b) CT reconstructions, bone algorithm. The sesamoid bone of the popliteus muscle's origin tendon has been displaced caudodistally (green arrow). The subchondral defect at the attachment of the popliteus tendon to the femur is clearly visible. Multiple small opacities are present laterally in the soft tissues (red arrow). Transverse (c) and dorsal reconstructed (d) CT images after intra-articular administration of contrast medium. The fragment at the proximal attachment of the popliteus muscle is outlined by contrast medium (red arrows). Both crossed ligaments, the cranial (a) and caudal (b), outlined by contrast medium, are intact.

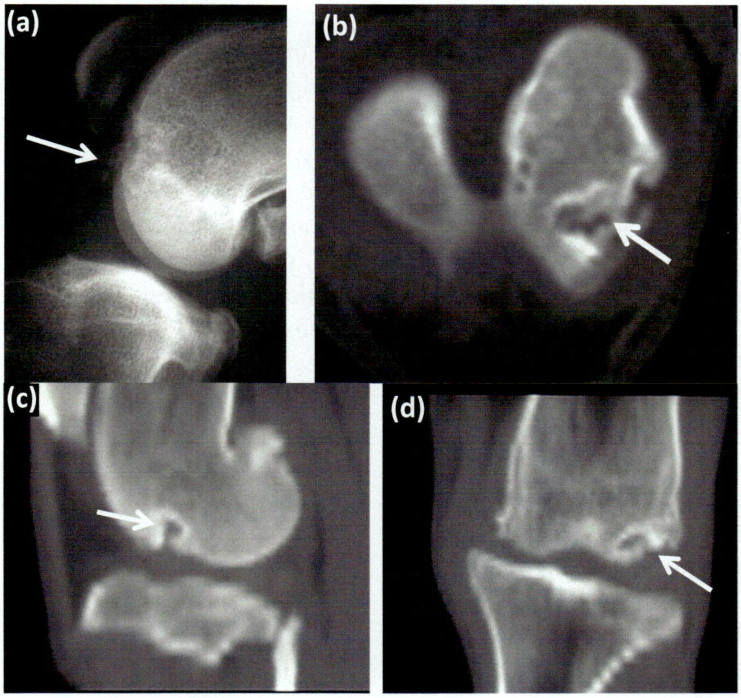

Figure 20.20 Avulsion of the extensor digitorum longus muscle. A medio-lateral radiograph (a) of the stifle of a Great Dane showing irregularities and calcifications in the area of the attachment of the extensor digitorum longus muscle (arrow). Transverse CT image (b) in bone window of the area of the attachment showing fragmentation and calcifications on the lateral femoral condyle (arrow). Sagittal (c) and dorsal (d) reconstructions showing the irregularity as a result of the avulsion of the extensor digitorum longus muscle.

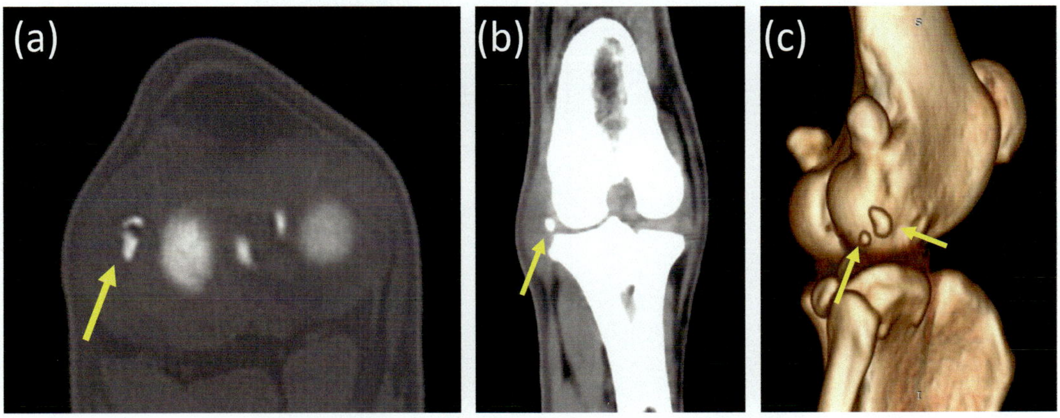

Figure 20.21 Transverse (a) and dorsal reconstructed CT images in soft tissue window (b), 3D CT (c) image of an avulsion fracture of the femoral attachment of the lateral collateral ligament (arrows). Notice the soft tissue swelling at the lateral side.

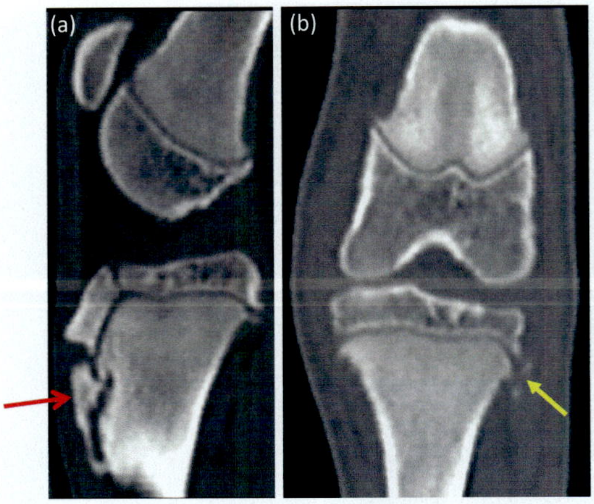

Figure 20.22 Sagittal (a) and dorsal (b) reconstructed images in bone window of the stifle of a puppy. The tuberositas tibiae is avulsed (red arrow), and the proximal epiphysis of the tibia shows fragments lateral to the growth plate (yellow arrow). This is most likely a Salter I fracture with slight displacement of the epiphysis towards the lateral side.

central necrosis and the tendon tissue repair in the proximal muscle area. This non-purulent condition is characterized by hypodense granulomatous tissue covered by enhancing layers of synovial and muscular tissue and only seen after contrast medium administration. The chronic inflammatory process leads to periosteal involvement at the origin of the gastrocnemius muscle and its lateral sesamoid bone (Figure 20.25).

20.5.4 Neoplastic disorders

20.5.4.1 Synovial cell sarcoma

Synovial cell sarcoma, the most common type of soft-tissue sarcoma at the stifle, is characterized by a thickened joint and moderate joint effusion. The various layers of the subcutaneous tissue, the fascia, and the joint capsule lose their borders and get homogeneously dense due to oedema, as observed in the plain

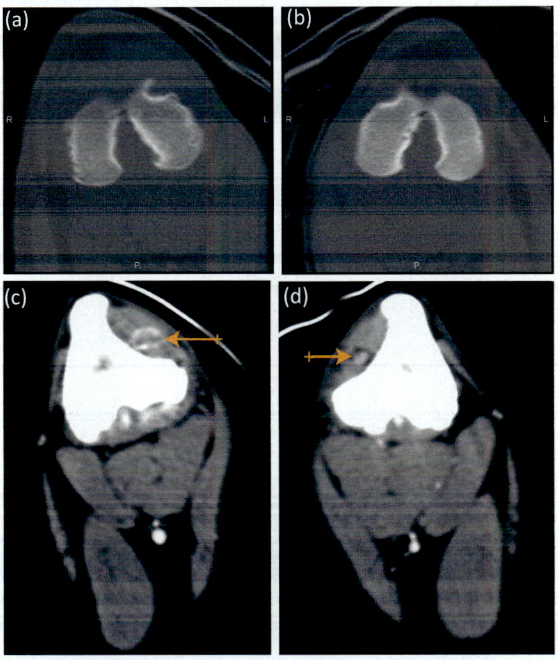

Figure 20.23 DJD of the left stifle (a, c) and the normal right stifle (b, d) of an 8-year-old female mixed-breed dog. (a and b) Transverse CT images at the level of the femoral condyles in bone window. New bone formation can be appreciated as irregular delineation of the bony structures in the affected left joint. (c and d) Transverse images in soft tissue algorithm after IV contrast administration. Contrast enhancement is present, and increased joint effusion enlarges the communicating tendon sheath of the long digital extensor muscle (left side in c, and normal appearance of the right side in d, both marked with arrows).

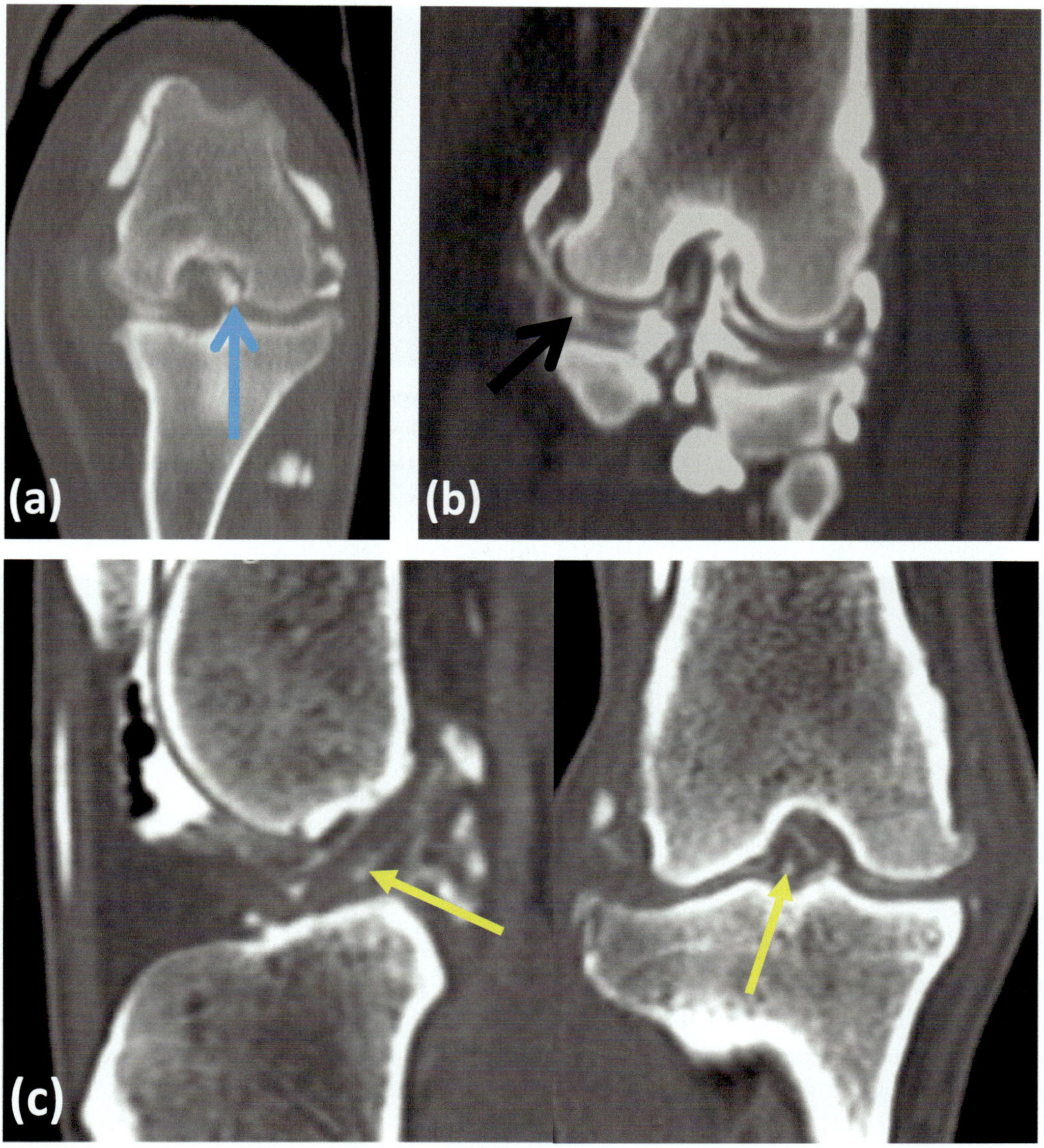

Figure 20.24 Arthro CT images of three different dogs. Dorsal reconstructed image (a) of a 3-year Rottweiler with a rupture of the cranial cruciate ligament. There is no delineation of the ligament seen, and a filling defect is present on the location of the ligament (arrow). Dorsal reconstructed image (b) of a medial meniscal tear in a canine stifle joint of a 5-year Rottweiler. The filling defect at the caudal horn of the medial meniscus is appreciated (arrow). Sagittal and dorsal oblique reconstructions (c) of a 6-year Labrador retriever with a small filling defect (arrows), at the cranial cruciate ligament. A partial rupture was confirmed.

imaging study. Post IV contrast, the heterogeneous enhancing soft-tissue mass usually fills all the proximal pouches and may also lead to displacement of the patella. The typical bony lysis of small to larger holes of the subchondral bone is seen in more advanced cases. The suspected diagnosis must be confirmed with FNA (Figure 20.26).

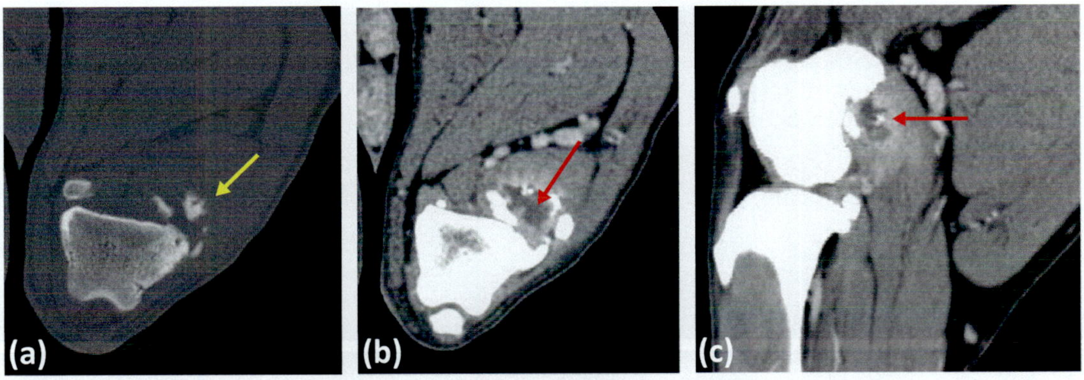

Figure 20.25 Avulsion of the gastrocnemius muscle. Transverse CT image (a), in bone window showing the avulsion of the lateral fabella of the left stifle (yellow arrow). Transverse- (b) and sagittal-reconstructed (c) CT images, post IV contrast in soft tissue window showing enhancement of the traumatized swollen, lateral left head of the gastrocnemius, and fluid pockets are inside the avulsed muscle (red arrows).

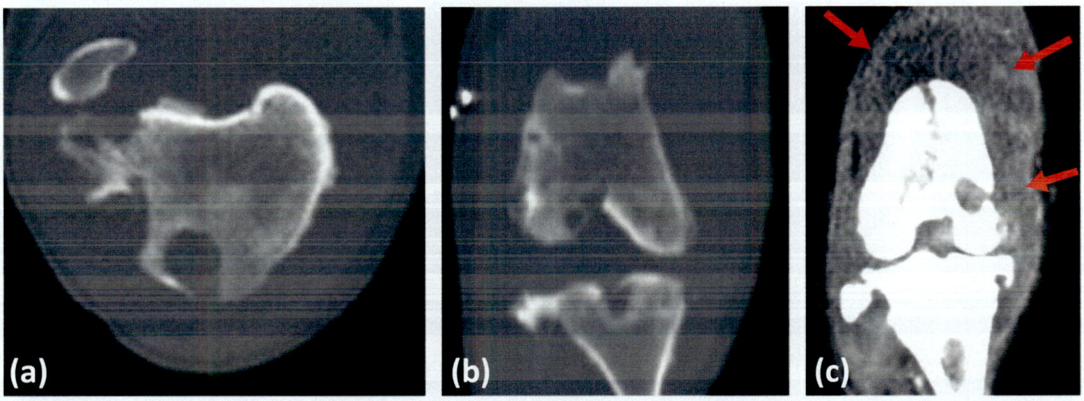

Figure 20.26 Synovial cell sarcoma in a feline stifle joint. Transverse (a) at the level of the proximal tibia and dorsal CT reconstruction (b) in bone window. Notice the multiple fingerprint-like osteolytic lesions in the femur and the tibia. In (c), transverse CT images post IV contrast in soft-tissue window show the enhancement of the thickened metaplastic synovial membrane (arrows).

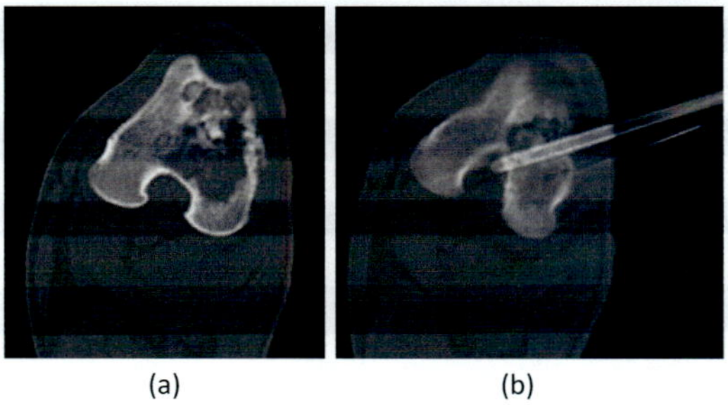

Figure 20.27 Osteosarcoma in a stifle joint. Transverse (a) CT image in bone window at the level of the distal femur showing the osteolytic appearance with marked thinning of the cortex. CT-guided biopsy (b) of this osteolytic lesion histologically confirmed an osteosarcoma. Before taking the biopsy, needles were placed for orientation.

20.5.4.2 Osteosarcoma

Osteosarcoma and chondrosarcoma may cause structural damage of the cortical or spongy bone of either the distal femur or the proximal tibia (Figure 20.27). Periosteal reactions of the compact bone may be absent in early disease cases, but with time, the malignant pathology leads to a "Codman's triangle" of the hyper-reactive periosteum and invasion of the soft tissue around the bone. Contrast medium administration can reveal the involved tissues around the bony destruction; they often will appear as a broad ring of the evading hyperdense soft tissue. The spread of the malignant tissue within the bony diaphysis can be seen as an enhancing soft tissue, which replaces the normal fatty spongy bone (Figure 20.28)

20.5.4.3 Synovial osteochondromatosis

In this very rarely reported condition, the enlarged joint space and pouches are filled with very small osseous and non-ossified chondromatous bodies, which cause worsening lameness over time(Figure 20.29). It shall be considered a benign differential to other sarcomas.

20.5.4.4 Lipoma

Abnormal proliferation of fatty tissue usually occurs at the hind limb in the caudal area of the thigh between the biceps femoris and the semimembranosus/semitendinosus muscle, often originating from the popliteal fossa. The enlarged cross-section of the thigh shows displaced muscles and vessels or nerves. Lipomas have a characteristic low attenuation on CT of –65 to –120 HU (Figure 20.30).

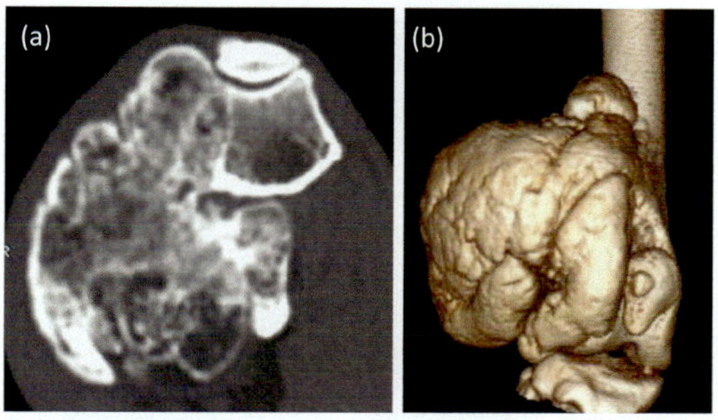

Figure 20.29 Transverse (a) and 3D (b) CT images of a feline stifle. At the lateral cortex of the distal femur, a massive new bone growth is visible that extends laterally and dorsally. This growth has varying bone densities alternating with hypodense zones but no osteolysis; only bone formation is visible. These images are compatible with cartilaginous exostosis—also known as osteochondromatosis. In cats, this is associated with feline leukaemia virus.

20.5.4.5 Other soft tissue sarcomas (mastocytoma, haemangiosarcoma, soft tissue sarcoma, etc)

Nodules of varying shapes and size may appear subcutaneously or between fascia, muscles, or within a certain muscle, which alter the anatomy and can cross tissue borders. Their contrast enhancement very much relies on their vascularization, and solid boundaries are advantageous for resection. Be aware of osseous involvement. Differentiating benign from malignant tissues is impossible, so biopsy is always needed.

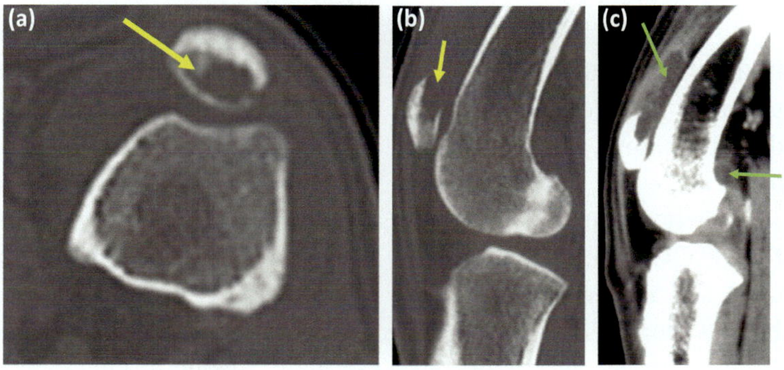

Figure 20.28 Transverse (a), sagittal reconstructed (b) in bone window, sagittal reconstructed, soft tissue window, after IV contrast administration (c). There is lysis within the patella. With destruction of cancellous bone and cortical disruption (yellow arrows). There is prominent joint effusion of the joint (green arrows). There is high suspicion of a primary bone tumor.

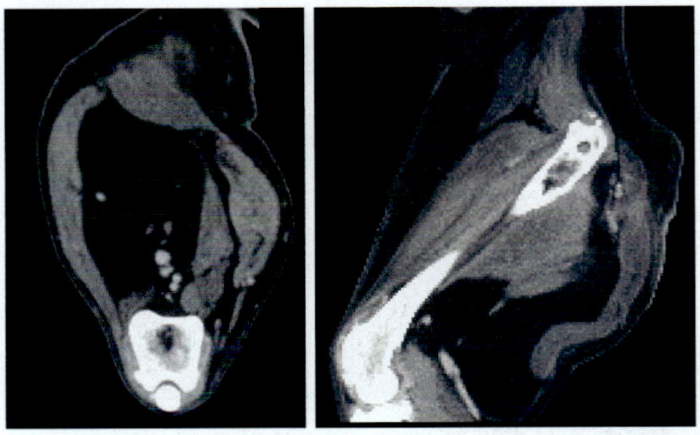

Figure 20.30 Lipoma in a canine hindlimb. Transverse (a) image in soft-tissue algorithm of the stifle after IV contrast administration. A hypodense well-delineated non-contrast-enhancing mass is seen. The sagittal reconstruction (b) shows clearly the extent of the lipoma.

20.5.4.6 Ganglion cyst

Hypoattenuating and fluid-filled cystic structures in the caudal fossa of varying size usually communicate with the joint space; IV contrast medium administration will only create a suspicion and IA contrast medium will reveal the origin; depending on size and location around the subchondral bone, periosteal reactions may appear. The lameness of the patient is caused by the decreased range of motion and the evolving periostitis.

20.5.5 Inflammation/Infection

20.5.5.1 Juvenile gonitis

The subchondral bone shows smooth surfaces and the physes are normal; the joint space can be significantly filled and the pouches swollen; the joint capsule enhances after the administration of the contrast medium. Diagnosis of the underlying infection (usually bacterial in origin) has to be proven by aspiration.

20.5.5.2 Inflammatory fistula/sinus tract

An inflammatory fistula can develop from a deep bite wound (cat's canine tooth), other perforating trauma, or even months after TPLO-surgery or extracapsular ligament reconstruction of the CCL. The affected joint is swollen and painful, and a drainage sinus may occur. The CT images clearly show the focal or generalized diffusely thickened tissues of the joint. Joint effusion may not be so obvious, but after IV contrast medium administration, the inflamed tissue is enhanced and shows the hypodense necrotic drain. With extracapsular ligament reconstruction, the lateral fabella may show periosteal proliferation (Figure 20.31).

20.5.5.3 Osteomyelitis

It may occur after IA fracture, deep bite wound, or TPLO surgery (Figure 20.32). While MRI imaging is impossible with metal implants, special CT reconstruction techniques allow reduction of the artefacts by metal pieces to get a reasonable image of the inflamed bony tissue. In CT images, the coarse trabecular pattern of inflamed bone is not the only sign of inflammation. Very often heavy periosteal reactions, soft tissue swelling, and even higher contrast uptake of the altered bone tissue are the main findings. A tissue sample is always needed to prove the type of infection and get a clue for effective antibiotic therapy.

20.5.5.4 Myositis – fibrotic gracilis myopathy

Cross-section imaging shows a unilateral thickened portion of the gracilis muscle (mainly the distal part); when compared to the other side, the muscle can be shortened on MPR images; the plain muscle structure is homogeneous and can be irregular post-IV contrast. Be aware that there might be contralateral muscular involvement of the semimembranosus/semitendinosus muscle. Mainly working and sports dogs (with German Shepherd Dogs highly ranked) are affected. In case of lack of this muscle, former onset of the disease, resulting in full or partial resection, shall be considered.

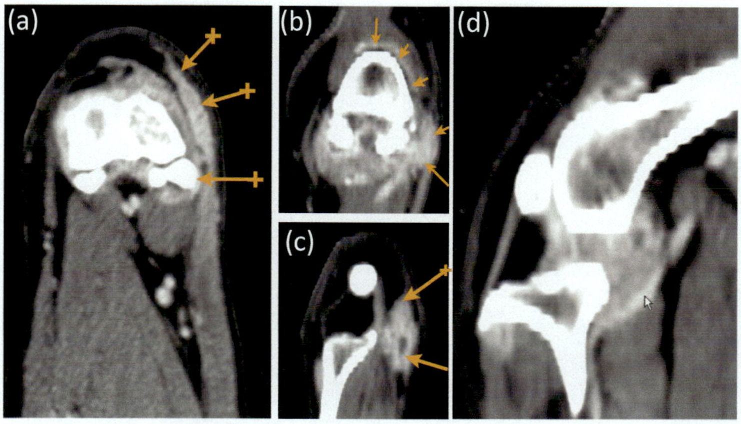

Figure 20.31 Sinus tract at the lateral left stifle following incompatibility with the suture material after extracapsular CCL repair in a 9-year-old female Entlebucher Dog. The entire sinus tract enhances after intravenous contrast medium administration on the soft-tissue images (arrows on the transverse image a). Severe hind limb muscle atrophy can be noticed. The dorsal reconstructions (b and c) at two different levels show the lateral and caudal involvement of the subcutaneous and ligamentous tissue along the fabellae (arrows pointing at the sinus tract, and arrowheads mark the joint effusion). On the sagittal reconstruction (d), cranial displacement of the tibia and severe joint effusion can be seen (arrowhead points at the caudal joint pouch).

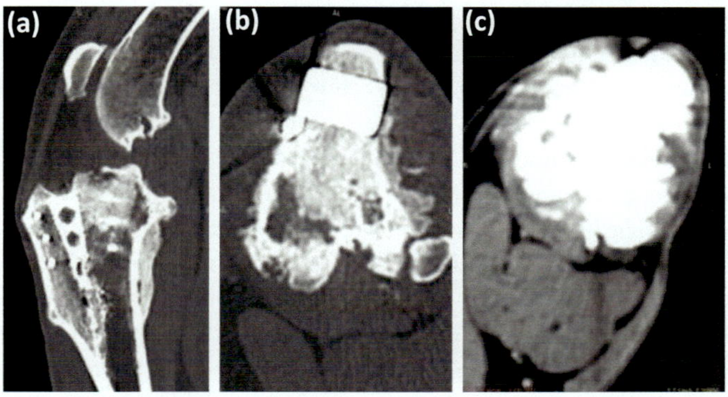

Figure 20.32 Sagittal reconstructed (a) and transverse (b) images in bone algorithm; transverse image (c) in soft tissue algorithm after IV contrast medium administration, post TPLO surgery. Alternating zones of severe osteolysis and spicula-like irregular periosteal new bone formation are present at the proximal level of the tibia (a, b). There is swelling of the soft tissues with contrast enhancement (c). These findings are suggestive of severe inflammatory/infectious reaction; however, the incidence of neoplasia could not be ruled out. Histological examination could not detect a tumour.

20.5.5.5 Tendinopathy of the patellar ligament

Patellar tendinitis, desmitis or tendinopathy may occur as a post-operative complication of TPLO.

Imaging reveals swelling and thickening of the patellar ligament (Figure 20.33).

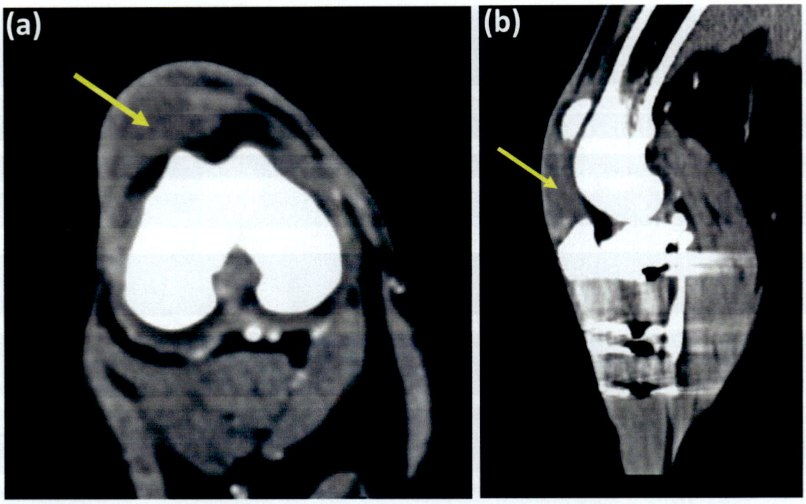

Figure 20.33 Sagittal reconstructed (a) and transverse (b) images in bone window, transverse image in soft tissue window after contrast administration of a stifle after TPLO surgery (Tibial Plateau Leveling Osteotomy). There is swelling and thickening of the patellar ligament (arrows).

FURTHER READING

Allberg CM, McEvoy FJ, Buelund LE, Miles JE. Stifle joint rotational variation in extended stifles in healthy dogs undergoing computed tomography examinations. *Vet Comp Orthop Traumatol.* 2020;33:137–141.

Brühschwein A, Burg-Personnaz J, Zöllner M, Reese S, Meyer-Lindenberg A. Comparison of CT-measured angles of pelvic limbs without patellar luxation of six canine breeds. *Front Vet Sci.* 2023;10:1194167.

Cieciora LC, Harms O, Freise F, Seifert H, Fehr M. Ex vivo evaluation of the cranial tibial artery and its compression through fragment rotation during tibia plateau levelling osteotomy: An angiographic three-dimensional reconstruction. *Vet Comp Orthop Traumatol.* 2022;35:220–229.

Coutin JV, Lewis DD, Kim SE, Reese DJ. Bifocal femoral deformity correction and lengthening using a circular fixator construct in a dog. *J Am Anim Hosp Assoc.* 2013;49:216–23.

De Rooster H, van Bree HV. Use of compression stress radiography for the detection of partial tears of the canine cranial cruciate ligament. *J Small Anim Pract.* 1999;40(12):573–576.

Dudley RM, Kowaleski MP, Drost WT, Dyce J. Radiographic and computed tomographic determination of femoral varus and torsion in the dog. *Vet Radiol Ultrasound.* 2006;47:546–52.

Fitch RB, Wilson ER, Hathcock JT, Montgomery RD. Radiographic, computed tomographic and magnetic resonance imaging evaluation of a chronic long digital extensor tendon avulsion in a dog. *Vet Radiol Ultrasound.* 1997;38(3):177–181.

Fox MG, Graham JA, Skelton BW, Blount KJ et al. Prospective evaluation of agreement and accuracy in the diagnosis of meniscal tears: MR arthrography a short time after injection versus CT arthrography after a moderate delay. *Am J Roentgenol.* 2016;207:142–149.

Gielen I, van Bree H. Computed tomography (CT) of the stifle. Part III, Chapter 20. In Muir P (Ed.) *Advances in the Canine Cranial Cruciate Ligament*, 2nd Edition. Wiley-Blackwell, pp. 141–155, 2018.

Gielen I, van Bree H, Van Ryssen B. Computed tomography (CT) in canine stifle disorders: A review of 30 cases. In *16th International Workshop for Small Animal Arthroscopy: Refresher Course: Stifle Joint*, p. 6, 2005.

Grand JG. Delayed life-threatening hemorrhage caused by cranial tibial artery pseudoaneurysm in two dogs. *J Am Anim Hosp Assoc.* 2024;60(2):74–80.

Harasen GLG, Simko E. Histiocytic sarcoma of the stifle in a dog with cranial cruciate ligament failure and TPLO treatment. *Vet Comp Orthop Traumatol.* 2008;21(4):375–377.

Hayes AM, Dennis R, Smith KC, Brearley MJ. Synovial myxoma: Magnetic resonance imaging in the assessment of an unusual canine soft tissue tumour. *J Small Anim Pract.* 1999;40(10): 489–494.

James DR, Webster N, White JD, Marchevsky AM, Cashmore RG, Havlicek M, Fearnside S, Black AP. Comparison of bone healing, as assessed by computed tomography, following tibial tuberosity advancement in dogs with and without autogenous cancellous bone grafts. *N Z Vet J.* 2017;65:270–276.

Kulendra E, Lee K, Schoeniger S, Moores AP. Osteochondritis dissecans-like lesion of the intercondylar fossa of the femur in a dog. *Vet Comp Orthop Traumatol.* 2008;21:152–155.

Lewis BA, Allen DA, Henrikson TD, Lehenbauer TW. Computed tomographic evaluation of the canine intercondylar notch in normal and cruciate deficient stifles. *Vet Comp Orthop Traumatol.* 2008;21:119–124.

Maley JR, Mertens WD, Bahr A. Osteomyelitis-related sequestrum formation in association with the combination tibial plateau levelling osteotomy and cranial closing wedge osteotomy procedure. *Vet Comp Orthop Traumatol.* 2010;23:141–147.

Marino DJ, Loughin CA. Diagnostic imaging of the canine stifle: A review. *Vet Surg.* 2010;39(3): 284–295.

Matchwick A, Bridges JP, Mielke B, Pead MJ, Phillips A, Meeson RL. Computed tomographic measurement of trochlear depth in three breeds of brachycephalic dog. *Vet Comp Orthop Traumatol.* 2021;34:124–129.

McEntee MC, Thrall DE. Computed tomographic imaging of infiltrative lipoma in 22 dogs. *Vet Radiol Ultrasound.* 2001;42(3):221–225.

Murata D, Sogawa T, Tokunaga S, Iwanaga T, Kawaguchi H, Miyoshi N, Momoi Y, Fujiki M, Miura N. Ganglion cysts arising from a canine stifle joint. *J Vet Med Sci.* 2014;76:457–459.

Petazzoni M, De Giacinto E, Troiano D, Denti F, Buiatti M. Computed tomographic trochlear depth measurement in normal dogs. *Vet Comp Orthop Traumatol.* 2018;31:431–437.

Pujol E, Van Bree H, Cauzinille L, Poncet C, Gielen I, Bouvy B. Anatomic study of the canine stifle using low-field magnetic resonance imaging (MRI) and MRI arthrography. *Vet Surg.* 2011;40:395–401.

Risselada M, Winter MD, Lewis DD, Griffith E, Pozzi A. Comparison of three imaging modalities used to evaluate bone healing after tibial tuberosity advancement in cranial cruciate ligament-deficient dogs and comparison of the effect of a gelatinous matrix and a demineralized bone matrix mix on bone healing — A pilot study. *BMC Vet Res.* 2018:22;164.

Samii VF, Dyce J. Computed tomographic arthrography of the normal canine stifle. *Vet Radiol Ultrasound.* 2004;45(5):402–406.

Samii VF, Dyce J, Pozzi A, Drost W, Mattoon JS, Green EM, Lehman AM. Computed tomographic arthrography of the stifle for detection of cranial and caudal cruciate ligament and meniscal tears in dogs. *Vet Radiol Ultrasound.* 2009;50(2):144–150.

Sasaki A, Hidaka Y, Mochizuki M, Honnami M. Computed tomographic measurements of the sulcus angle of the femoral trochlea in small-breed dogs with and without medial patellar luxation. *Vet Comp Orthop Traumatol.* 2022;35:314–320.

Selmic LE, Ryan SD, Boston SE, Liptak JM, Culp WT, Sartor AJ, Withrow SJ. Osteosarcoma following tibial plateau leveling osteotomy in dogs: 29 cases (1997—2011). *J Am Vet Med Assoc.* 2014;244(9):1053–1059.

Smith TJ, Baltzer WI, Löhr C, Stieger-Vanegas SM. Primary synovial osteochondromatosis of the stifle in an English Mastiff. *Vet Comp Orthop Traumatol.* 2012;25:160–166.

Soler M, Murciano J, Latorre R, Belda E, Rodríguez MJ, Agut A. Ultrasonographic, computed tomographic and magnetic resonance imaging anatomy of the normal canine stifle joint. *Vet J.* 2007;174:351–361.

Thrall DE. *Textbook of Veterinary Diagnostic Radiology.* Elsevier Health Sciences, pp. 221–358, 2007.

Tivers MS, Mahoney P, Corr SA. Canine stifle positive contrast computed tomography arthrography for assessment of caudal horn meniscal injury: a cadaver study. *Vet Surg.* 2008;37:269–277.

Tivers MS, Mahoney PN, Baines EA, Corr SA. Diagnostic accuracy of positive contrast computed tomography arthrography for the detection of injuries to the medial meniscus in dogs with naturally occurring cranial cruciate ligament insufficiency. *J Small Anim Pract*. 2009;50:324–332.

Togni A, Sievert C, Hurter K, Knell S. Chronic expanding haematoma in a cat. *J Feline Med Surg*. 2014;17(8):733–736.

Townsend S, Kim SE, Tinga S. Tibial plateau morphology in dogs with cranial cruciate ligament insufficiency. *Vet Surg*. 2018;47:1009–1015.

Triglia M, Horton H, Dobromylskyj M, Ferreira AJ, Robinson WP. Tibial plateau and stifle joint invasion with a subcutaneous mast cell tumor. *J Am Anim Hosp Assoc*. 2023;59:234–240.

CHAPTER 21

CT OF THE TARSUS

Ingrid Gielen and Henri van Bree

21.1 INTRODUCTION

Injuries to the tarsal joints are common in performance dogs. This joint acts as a shock absorber during weight bearing and is prone to injury due to its anatomic complexity and lack of muscular support. It is this complexity that creates a diagnostic problem for many veterinarians, and many tarsal injuries, particularly those that go undiagnosed or untreated, can result in an increased risk of osteoarthritis and potential long-term lameness. Tarsal injuries can result from either acute traumatic events or activities that cause sudden repetitive sprains to the joints' supportive structures.

Tarsal pain or instability causing hind limb lameness can result from a variety of aetiologies including: osteochondrosis (OC) and osteochondritis dissecans (OCD) of the talus, fractures, collateral ligament sprains or avulsion fractures, traumatic luxations, and septic and non-septic arthritis. Luxation, or dislocation, involves the disruption of multiple ligaments and the joint capsule. Common fractures associated with tarsal injuries in performance dogs include the attachment sites for the collateral ligaments (avulsion fractures) or bone fractures due to compression or shear forces. Fractures at these sites result in joint instability when they undergo stress on palpation or during weight-bearing. X-rays can be taken to evaluate the type and severity of injury. Although you cannot see ligamentous structures on x-rays, you can see bone fractures, luxations (dislocations or misalignments), and abnormal opening of a joint when they undergo stress (extension, flexion, varus, or valgus), which can help determine ligament integrity. Chronic or repetitive injuries may have bone spurs where ligaments are attached, and those will also be present in an X-ray. Tarsocrural osteochondrosis (TOC) is a less frequent cause of lameness in the hind limb, and lesions can be found on the medial and lateral trochlear ridge. The tibiotarsal joint in dogs is the third-most common joint affected by OCD, representing 9% of all OCD lesions found in dogs. Injuries to tarsal joints are common in performance dogs. These joints act as a shock absorber during weight-bearing and are prone to injury due to their anatomic complexity and lack of muscular support.

Although radiographic projections are often used in the diagnosis of tarsal lesions, computed CT allows a definitive diagnosis when the radiographic examination yields inconclusive findings. The use of CT avoids the superposition of surrounding (bony) structures and provides a more detailed view of the lesions. In OC/OCD, CT enables to determine the exact location, the extent of an injury, size and number of fragments, and information that's crucial for the surgeon.

The tarsal joint has a mainly hinge motion; its main motion is flexion-extension, together with a small gliding component at the proximal intertarsal joint. It is a very complex joint and contains several small bony structures overlying to each other. Diagnosing the tarsal pathology in dogs and cats remains difficult due to superimposition of the distal tibia, talus, fibula, and calcaneus. The complexity of the tarsocrural joint, especially in the young dog, is such that even with

DOI: 10.1201/9781315118130-23

an adequate number of appropriate radiographic projections, subtle changes involving both ridges of the trochlea tali can go undetected.

21.2 INDICATIONS FOR USING CT

It has been shown that radiographs, even if additional oblique and skyline views are taken, often are limited to assess a definitive diagnosis.

CT indications for the tarsal joint are as follows:

a. Diagnosing OC of the medial and lateral ridge of the talus and evaluation of the size, number, and location of the fragments.
b. Fragmentation of the medial malleolus of the tibia can be found radiographically in tarsi, which did not have evidence of osteochondrosis of the medial trochlear ridge but also in joints where both medial malleolar fragmentation and osteochondrosis of the medial trochlear ridge are present.
c. Evaluating the extent of fractures and fissures within the tarsus
d. To find the origin of radiographically detected fragments
e. Evaluating epiphyseal injuries
f. detailed description of dislocation and subluxation
g. Determination of degenerative changes.
h. Establishing neoplastic changes including synovial cell sarcoma.
i. Differentiation between synovial cell sarcoma and severe degenerative changes.
j. Miscellaneous bony- and soft-tissue lesions in the tarsal area.
k. Calcaneal tendon injuries

21.3 TECHNICAL ASPECTS

21.3.1 Positioning

It is recommended to anaesthetize the animals and position them in ventral recumbency on the CT-scanning table, with the tarsal joints in extension (Figure 21.1). This position allows a perfect symmetrical scanning of both joints, allowing the evaluation of the CT images of both tarsi at the same level. Eventually, a wedge can be placed in between the joints to maintain stability. Strings can be used to achieve extension of the limbs (Figure 21.1).

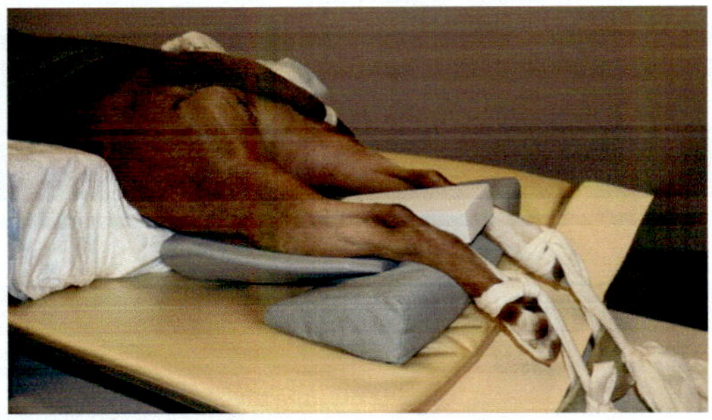

Figure 21.1 **Positioning of the dog in ventral recumbency during CT of both tarsal joints. Both tarsal joints are in extension and with perfect symmetry to allow comparison of both joints at the same level.**

21.3.2 Technical parameters

A lateral and caudo-cranial scout view are made to confirm correct positioning, and afterwards images in bone and soft tissue algorithm are processed. The joint is scanned from 2 cm proximally of the calcaneus till 2 cm distally of the intertarsal bones. A slice thickness of 0.8–1 mm in bone and of 1.5–2 mm in soft tissue algorithm is recommended. It is preferred to scan with an overlap of 50%, resulting in detailed multiplanar reconstructions in additional planes, which are frequently utilized in tarsal joints. Interleaved images reduce partial volume artefacts, increase the chance of missing tiny lesions at a slice's edge, and motion artefacts are diminished. The parameters used for this joint are 100–120 kVp and 140–160 mAs. The field of view (FOV) needs to be adjusted to increase in-plane resolution with a maximum of 15 cm. Individual images are reviewed using both a bone (WW=3,000 HU; WL=500 HU) and a soft-tissue window (WW=350 HU; WL=90

HU). Afterwards, CT reconstructions of the tarsocrural joint in sagittal and dorsal planes are made. The entire tarsocrural joint surface can be evaluated on these reconstructed images, without superimposition of any surrounding (bony) structure (Figure 21.2). Besides full anatomic detail of the bony structures, some soft tissues such as tendons and large blood vessels can also be evaluated, by adjusting the window and using the varying grey-scale formats. However, it is not possible to identify nerves and smaller blood vessels. In cases of trauma and intraarticular fractures, 3D images may be useful for treatment planning.

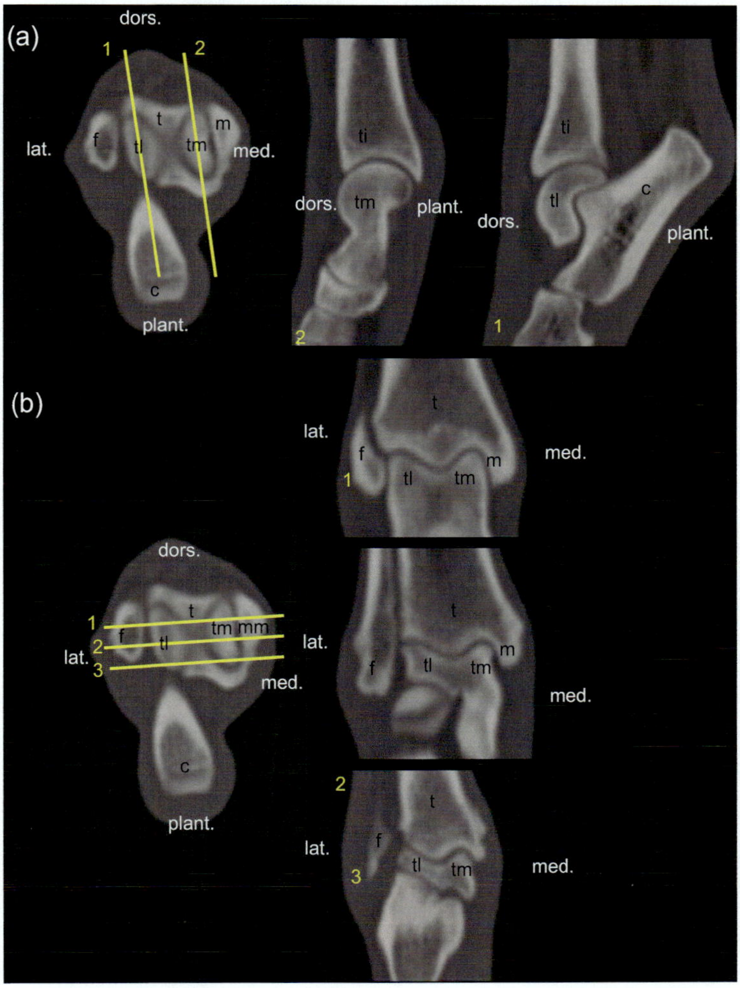

Figure 21.2 (a) Sagittal reconstructions, bone algorithm, from CT images of the tarsocrural joint of a sound large-breed dog. tl=Lateral ridge of the talus. tm=medial ridge of the talus. c=calcaneus. f=fibula. m=malleolus medialis, t=talus. ti=tibia. The sagittal images allowed a thorough evaluation of the lateral and medial part of the tarsocrural joint including the lateral and medial ridge. (b) Dorsal reconstructions, bone algorithm, from CT images of the tarsocrural joint of a sound large-breed dog. tl=lateral ridge of the talus. tm=medial ridge of the talus. c=calcaneus. f=fibula. m=malleolus medialis, t=talus. ti=tibia. On the dorsal reconstructions, the tarsocrural joint could be evaluated from the dorsal to plantar surface in detail.

21.3.3 Contrast

Administering intravenous contrast for a tarsal CT has a significant advantage in particular cases. The use of IV contrast significantly increases examination accuracy, allowing better tissue differentiation. It is mostly utilized to draw attention to variations between soft tissues that would otherwise appear to be identical, and it helps identify the presence of a neoplastic process or inflammation within the joint (Figure 21.3).

By arthro-CT (CTA), cartilage, subchondral lesions, ligaments, tendons, and bones are outlined, allowing their evaluation. The observation of separate fragments and the evaluation of fragment stability using CTA may be important in decision making in tarsal OCD (Figure 21.4). In addition to MR arthrography, CTA is mainly used in human medicine in joint imaging, especially in small joints. The tarsal joint in small animals is very small with very thin cartilage covering, requiring an imaging technique with good spatial resolution; hence,

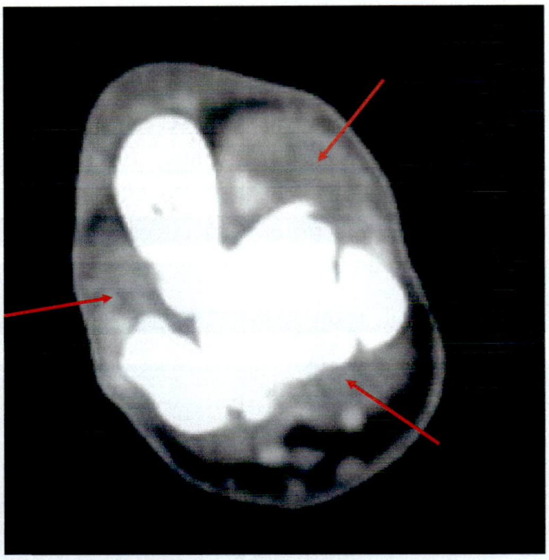

Figure 21.3 Transverse CT image in soft-tissue algorithm after IV administration contrast medium, allowing better differentiation of soft tissues. Note the distended joint capsule (red arrows) with inflammatory soft-tissue changes, synovitis.

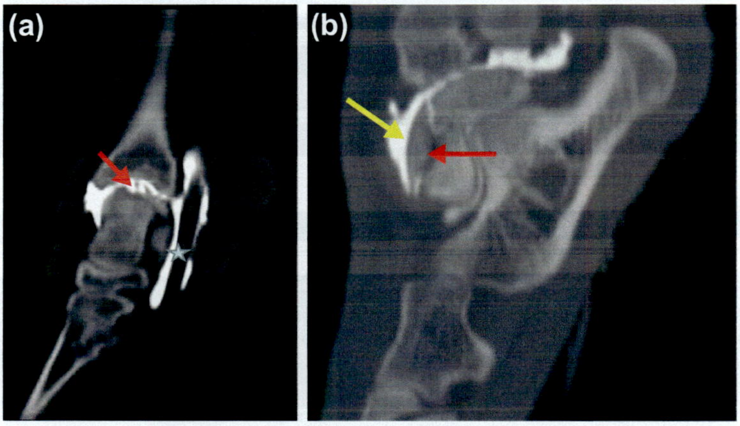

Figure 21.4 Computed tomographic arthrography of two different tarsal joints with an OCD lesion, sagittal reconstructed images, and bone algorithm. The proximal area of the medial talar ridge (a) shows loss of delineation of the cartilage, and the contrast medium is present in the subchondral defect and surrounding the detached fragment (red arrow). The delineation of the deep digital flexor muscle tendon (asterisk) surrounded by hyperdense contrast agent representing its tendon sheath, which is communicating with the joints space, is well visible. In the dorsal aspect of the lateral talar ridge (b), a large fragment is present. Contrast is visible in the proximal and distal part in between the fragment and the subchondral bed. There is no contrast at the middle of the fragment; it is still connected with fibrotic tissue (red arrow). The delineation of the thin cartilage is noticed (yellow arrow).

CT arthrography may be the study of choice in some circumstances, such as cartilage imaging and osteochondral lesion assessment given the tarsocrural thin cartilage covering.

21.4 CT-ANATOMY AND NORMAL VARIANTS

The joint is composed by seven bones, and they are arranged in two groups, a proximal and a distal group. There are three main joints: the talocrural joint (providing the greatest range of motion), the intertarsal joints with two subjoints (the proximal- and the distal intertarsal joints), and the tarsometatarsal joints. The talus is positioned most medially and the calcaneus most laterally in the proximal group. The central tarsal bone is positioned medially and distally in the proximal group. The tarsal bones I–III are relatively small in comparison with tarsal bone IV and are restricted to the medial half of the tarsus.

The joint capsule contains all the tarsal ligamentous structures, which holds the articulations together. The tarsal joint capsule extends like a sheath to the entire area of these joints. It has seven subdivisions, four medial and three lateral. The largest component surrounds the tarsocrural joint. The tarsal canal is located plantar to the talus and medial to the calcaneus. The medial and lateral plantar nerves, saphenous artery, and plantar branches, as well as the tendon sheath of the flexor hallucis longus muscle, are all located within this canal. The lateral (fibular) and medial (tibial) collateral ligaments, respectively, offer lateral and medial support. These collateral ligaments each contain three components. On the lateral side, the long lateral ligament, the calcaneofibular short ligament, and the talofibular short component are present. The other ligaments are the extensor retinacula, the flexor retinaculum, the plantar tarsal fibrocartilage, and the special ligaments, of which the long plantar ligament has most clinically significance. Strain of this ligament is mostly associated with lameness.

Tarsal joint muscles are responsible for movement in the tarsus. Because the flexor surface of the tarsal joint is positioned opposite to the flexor surface of the digits, the muscles that extend the tarsus will flex the digits. Most anatomical variants can be seen in German Shepherd Dogs and Rottweiler. A medial, bony spur in the shape of an incus emerging from the central tarsal bone may occur. A medially positioned triangular extra bone developed from an additional but fused apophysis of the central tarsal bone may be present. CT images processed at the most distal levels reveal the talus, calcaneus, and os tarsi centrale; the os tarsi centrale and os tarsale quartum; the os tarsale secundum, tertium, and quartum; the os tarsale primum, secundum, tertium, and quartum; or metatarsals II, III, IV, and V. CT images provide full anatomic details of the bony structures of the tarsal joint in dogs. Tendons and large blood vessels can also be evaluated; however, detailed soft tissue evaluation is limited. Studies are available describing the CT anatomy of the tarsal joint in comparison with the anatomic specimen (Figure 21.5).

21.5 PATHOLOGIC FINDINGS

21.5.1 Developmental

21.5.1.1 Osteochondritis dissecans (OCD)

In the dog, OCD of the tarsocrural joint is a well-known but uncommon cause of hind limb lameness. The OCD defect can involve the medial or the lateral trochlear ridge of the talus and results in instability, pain, lameness, and progressive degenerative joint disease (DJD).

In OC of the lateral ridge, the dorsal, dorsoproximal, or proximal part can be affected, whereas in OC of the medial ridge, the lesions occur on the most proximal part (Figure 21.6). If one wants to use a minimal invasive technique, such as arthroscopy to determine the best approach to access a lesion, its exact location is very important.

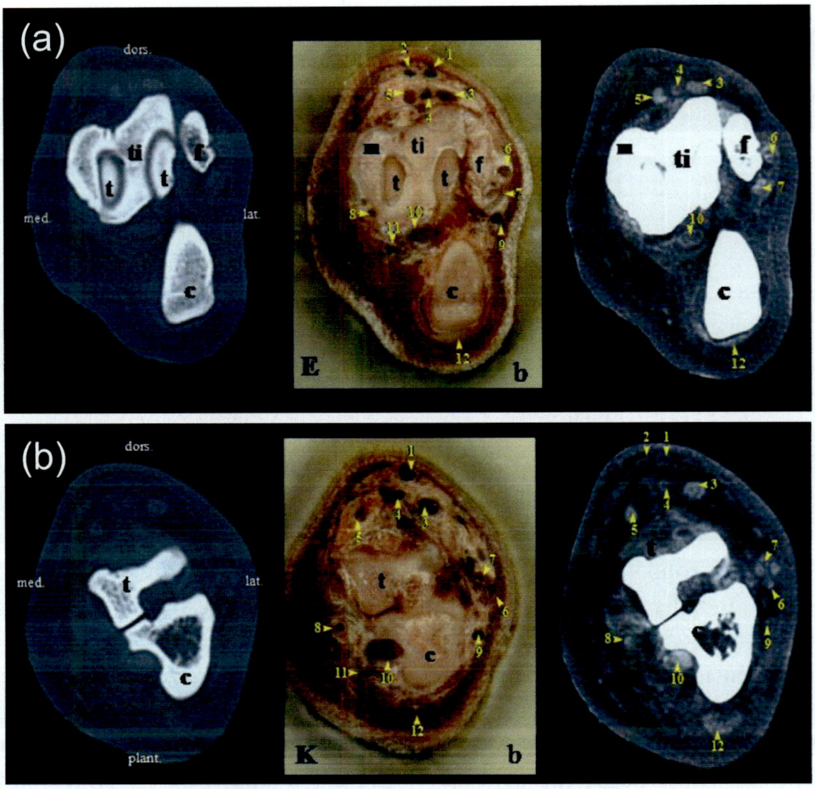

Figure 21.5 (a) Transverse CT images and photographs of the corresponding anatomic sections of the tarsal joint of a clinically normal dog at the level of the talar ridges. Images (left, bone image; middle, anatomic section; right, soft-tissue image) were obtained at the level of the tarsocrural joint. 1. plant=Plantar. Med=Medial. Lat=Lateral. Dors=Dorsal. 1=Ramus cranialis of the vena saphena lateralis. 2=Ramus cranialis of the vena saphena medialis. 3=Extensor digitorum longus muscle. 4=Tibialis cranialis artery. 5=Tibialis cranialis muscle. 6=Peroneus (fibularis) longus muscle. 7=Extensor digitorum lateralis muscle. 8=Flexor digitorum medialis muscle. 9=Ramus caudalis of vena saphena lateralis. 10=Flexor digitorum lateralis muscle. 11=Ramus caudalis of vena saphena medialis. 12=Flexor digito-rum superficialis muscle. 13=Extensor digitorum brevis muscle. 14=Talocalcaneal interosseous ligament (ligamentum talocalcaneum interosseum). c=Calcaneus. f=Fibula. m=Malleolus medialis. s=Sustentaculum tali. t=Talus. ti=Tibia. (b) Transverse CT images and photographs of the corresponding anatomic sections of the tarsal joint of a clinically normal dog at the distal level of the talus and calcaneus. Images (left, bone image; middle, anatomic section; right, soft-tissue image) were obtained at the level of the distal part of the talus and calcaneus. 1. plant=Plantar. Med=Medial. Lat=Lateral. Dors=Dorsal. 1=Ramus cranialis of the vena saphena lateralis. 2=Ramus cranialis of the vena saphena medialis. 3=Extensor digitorum longus muscle. 4=Tibialis cranialis artery. 5=Tibialis cranialis muscle. 6=Peroneus (fibularis) longus muscle. 7=Extensor digitorum lateralis muscle. 8=Flexor digitorum medialis muscle. 9=Ramus caudalis of vena saphena lateralis. 10=Flexor digitorum lateralis muscle. 11=Ramus caudalis of vena saphena medialis. 12=Flexor digitorum superficialis muscle. 13=Extensor digitorum brevis muscle. 14=Talocalcaneal interosseous ligament (ligamentum talocalcaneum interosseum). c=Calcaneus. f=Fibula. m=Malleolus medialis. s=Sustentaculum tali. t=Talus. ti=Tibia.

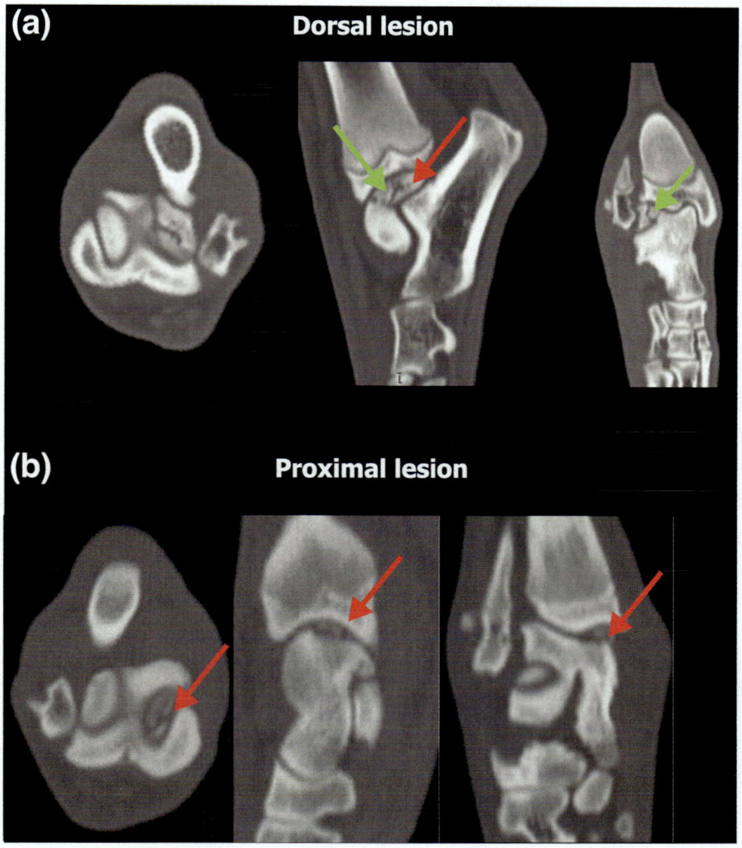

Figure 21.6 (a) Transverse-, sagittal-, and dorsal- reconstructed images, bone algorithm, of a tarsus of a 7-month-old Rottweiler with a lesion of osteochondrosis on the lateral trochlear ridge of the talus. Two major fragments are present on the lateral trochlear ridge. A large fragment is visible (red arrow), localized on the proximal part. The total lateral ridge of the talus is involved. A smaller, dorsally localized fragment is noticed (green arrows). (b) Transverse-, sagittal-, and dorsal- reconstructed images, bone algorithm, of an 8-month-old Labrador Retriever with a lesion of osteochondrosis on the medial trochlear ridge of the talus. The red arrows show the defect of the proximal part of the medial ridge. Multiple detached fragments are present.

It can often be difficult to diagnose TOC radiographically, and good technique and accurate positioning are essential for the radiographic visualization of the lesions. Most lesions of the medial trochlear ridge can be identified on the standard dorsoplantar radiograph, but the radiographic diagnosis of lateral OC is often difficult. The application of CT in canine TOC is extremely beneficial in the diagnosis of both lateral and medial TOC by giving a better general view of the lesion. The radiographic diagnosis of TOC can be difficult due to the complexity of the tarsocrural joint and the superimposition of different bony structures hiding its articular surface. When the lateral ridge is involved, there is supplementary superimposition of the calcaneus, probably the reason for later appearance of reports on lateral TOC than medial TOC. In 10% of the TOC cases, the radiographic diagnosis can be missed, even when an extended protocol including six views has been applied. When the same protocol has been used for diagnosing lateral TOC lesions, 27% of the lesions can be missed. In practice, the radiographic protocol is usually limited to two standard views (extended mediolateral and extended dorso-plantar) but 72% of these lesions can be missed. Treatment typically consists in removal of the loose cartilage flap, which should result in filling of the defect with fibrocartilage (Figure 21.7).

Significant differences exist between lateral and medial TOC lesions. The size of the lateral TOC lesions is significantly larger than these of the medial lesions.

Subsequent series have supported a traumatic aetiology for most osteochondral lesions of the talus, ranging from 75% to 100% for lateral lesions and 18%–80% for medial lesions. No proof exists for a traumatic aetiology for tarsocrural OCD in the dog, although traumatic fragmentation of the lateral talar ridge is reported as well.

It has been reported that some of the lateral lesions are thought to be avulsion fractures of the short part of lateral collateral ligaments associated with relatively large loose fragments situated proximally (Figure 21.8). Any possible relation between OCD and these avulsion fractures could not be evaluated. In the tarsocrural joint, the reported incidence of bilateral TOC is approximately 50%. Although clinical lameness in these cases is rarely symmetrical, 30% of the dogs presented with bilateral lesions. The lameness is present in the joint with the largest defects. The non-clinical TOC lesions are significantly smaller than the clinical ones.

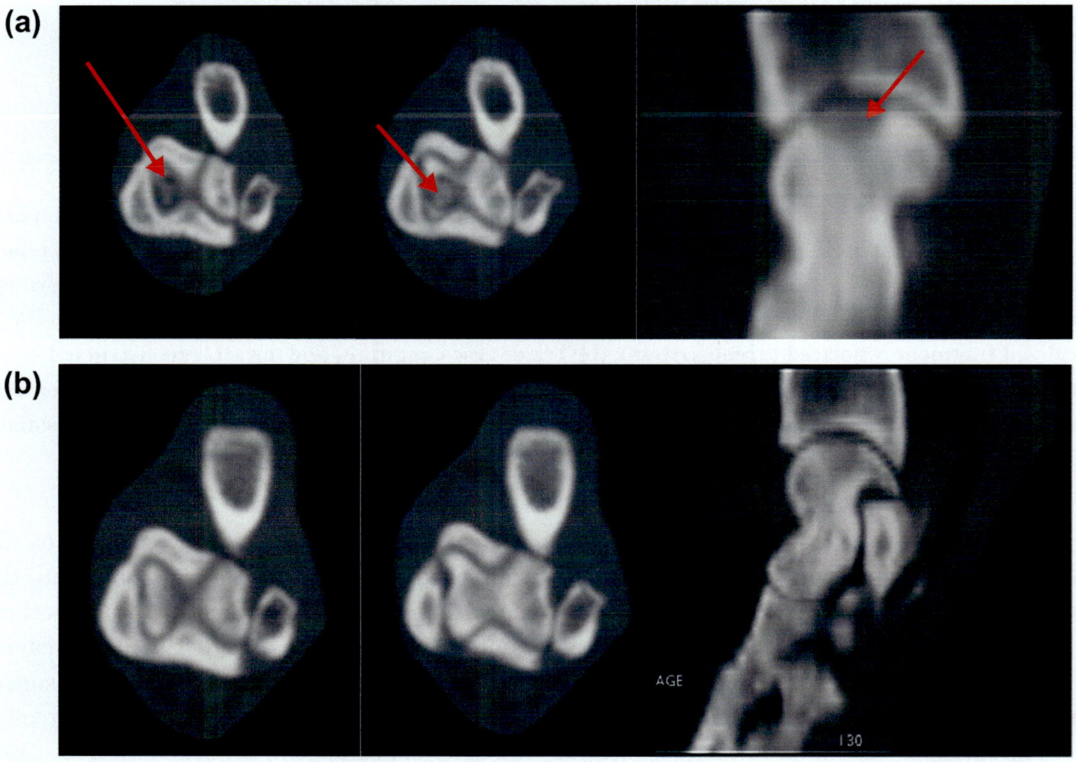

Figure 21.7 (a) Transverse- and sagittal- reconstructed CT images, bone algorithm, of a 14-month Golden Retriever with medial OCD of the talus. The separate loose fragment is clearly visible (red arrows). (b) Transverse- and sagittal- reconstructed CT images, bone algorithm, of the same Golden Retriever, 2 years and 8 months follow-up after removal of the loose fragment. Filling of the defect with fibrocartilage tissue is noticed.

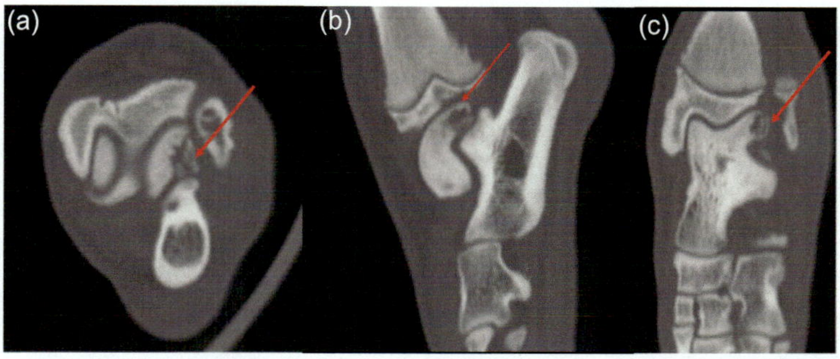

Figure 21.8 Transverse- (a), sagittal- (b), and dorsal- (c) reconstructed images, bone algorithm, of a tarsus of a 5.5-month-old Rottweiler with a lesion of osteochondrosis on the lateral trochlear ridge of the talus. There is a subchondral lesion associated with fragmentation in the latero-proximal area of the lateral talar ridge (red arrows). The medial ridge appears to be normal. There is associated soft tissue swelling of the joint. There is high suspicion of avulsion of the short part of the lateral collateral ligament.

Apart from a precise diagnosis of TOC, CT allows determination of the location, number, and size of the fragments. Establishing the exact location of the lesion, leading to the determination of the best approach to the lesion, is important for treatment planning (Figure 21.6). The choice between arthroscopy- and arthroscopically guided mini-arthrotomy can be determined in advance by using CT to measure the size of TOC defects. CT can also be used to determine the exact number of TOC fragments to be removed. More than 75% of TOC defects are associated with more than two fragments. The prognosis after surgical or arthroscopic removal depends on the size of the defect.

- OCD medial talar ridge
  On the transverse images, radiolucent defects surrounded with sclerosis are noticed. In most cases, several fragments associated with these defects can be observed. The defects are mostly located in the proximal and dorsal part of the ridge (Figures 21.6b and 21.9). This is important information for the surgeon; with the CT information, he knows exactly how many fragments need to be removed in which area. Dorsal and sagittal reconstructions allow complete inspection of the entire joint.
- OCD lateral talar ridge
  Most of the time, large fragments in the area of the lateral talar ridge are observed (Figure 21.6a and 21.10). Rarely fragments are seen in the distal part of the lateral ridge, which demands a different surgical approach.
- Free fragments may be present in the sheath of the deep digital flexor tendon, originating from tarsocrural OCD lesions (Figure 21.11).
- Fragmentation of the medial malleolus of the tibia can be found in dogs which do not have evidence of osteochondrosis of the medial trochlear ridge. In most of these cases, there is no clinical importance related to this finding. In other patients, fragmentation of the medial malleolus may be present in combination with TOC (Figure 21.12).

21.5.2 Trauma

21.5.2.1 Fracture

CT imaging could also be of use in cases of complex tarsal fractures, where the same problem of radiographic superimposition of the surrounding bony structures exists and could provide information contributing to adequate treatment. Sagittal and dorsal reconstructions and 3D images can be very useful in these cases (Figure 21.13).

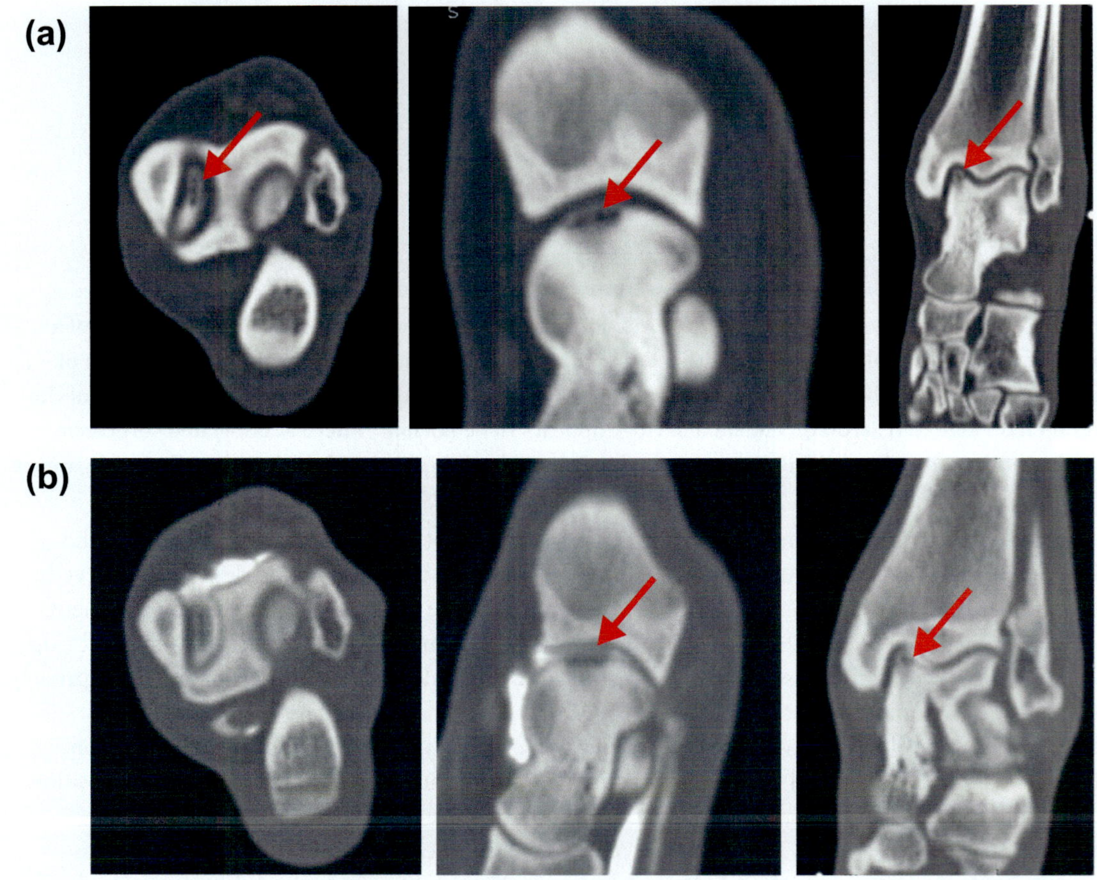

Figure 21.9 (a) Transverse-, sagittal-, and dorsal- reconstructed images, bone algorithm, of an 8-month-old Golden Retriever with a lesion of osteochondrosis at the medial trochlear ridge of the talus. There is a subchondral defect at the dorsoproximal area of the medial ridge. A large calcified flap is present in the subchondral bed (red arrows). (b) Transverse-, sagittal-, and dorsal- reconstructed images, bone algorithm, of the same dog after the administration of intra-articular contrast. The contrast medium does not appear under the fragment but passes over it (red arrows). There was high suspicion that the cartilage was not torn; OC lesion of the medial talar ridge. This finding was confirmed by arthroscopy.

The high sensitivity of fracture detection of CT is described, certainly in fractures of intertarsal bones and more specifically the central and the fourth tarsal bones. Computed tomography is superior to radiography in detecting small bone fragments and chip fractures. A fracture of the sustentaculum tali is usually difficult to diagnose on a radiograph. This type of fracture is clearly visible on CT (Figure 21.14). CT may be helpful in assessing complex metaphyseal or epiphyseal fractures (Figure 21.15).

Avulsion fractures and additional instability are usually radiographically diagnosed. Most of the time, the avulsed bone fragment is moved in the direction of the attached tendon, ligament, or joint capsule. For detection and further characterization, CT may be necessary (Figure 21.16). A detailed and definitive fracture diagnosis is helpful in surgical planning.

After surgery, CT provides accurate detail to evaluate fracture reduction and implants.

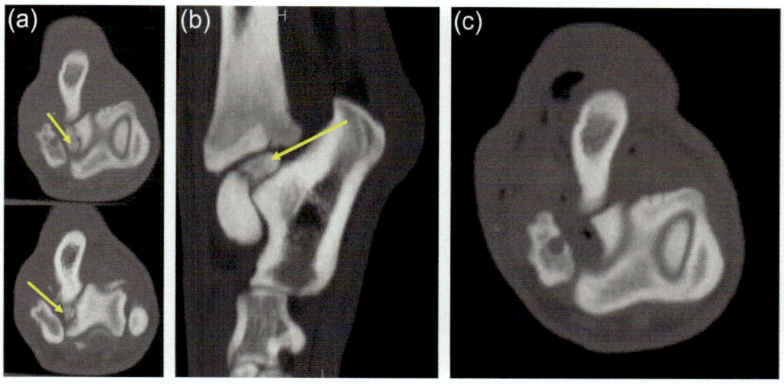

Figure 21.10 Preoperative transverse CT scan (a) at two levels, sagittal-reconstructed (b) and transverse image after removal of the fragments (c), bone algorithm, of the tarsus of a 10-month-old Rottweiler with an osteochondritis dissecans lesion of the proximal part of the lateral trochlear ridge of the talus. The yellow arrow on the transverse scan shows a large detached fragment. The sagittal reconstructed image shows the fragment on the proximal part of the lateral ridge (yellow arrow). The post-operative scan was made to check if all the fragments were removed.

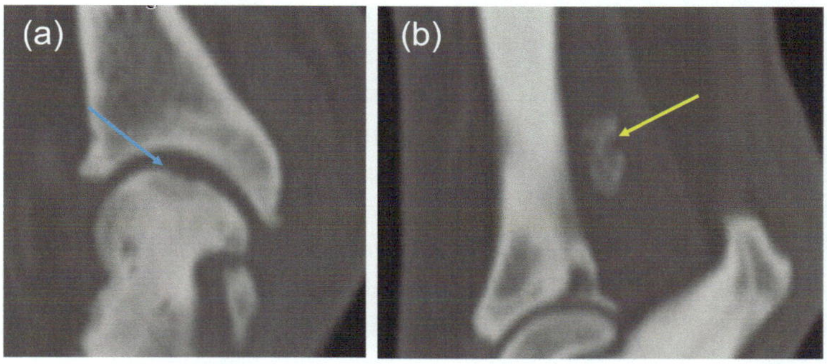

Figure 21.11 Sagittal (a)-reconstructed image at the level of the medial ridge of the talus and sagittal (b)-reconstructed image at the level of the lateral ridge of the talus, bone algorithm. There is an OCD lesion present at the medial ridge (blue arrow). A large opacity is seen in the sheath of the deep digital flexor tendon (yellow arrow).

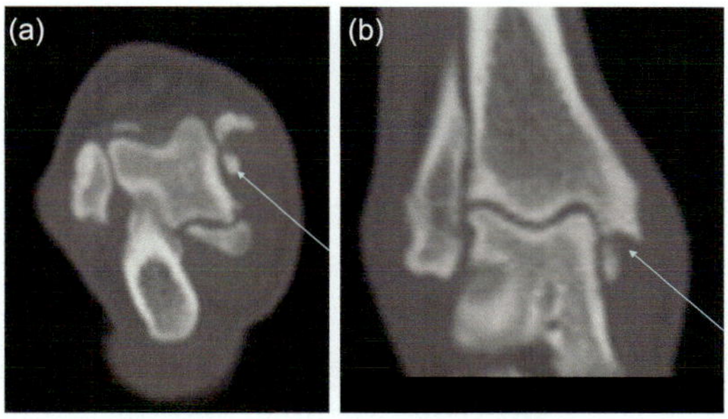

Figure 21.12 Transverse (a) at the level of the medial malleolus and dorsal reconstructed (b), bone algorithm of a 5-year-old Stafford. There is fragmentation of the medial malleolus present (arrows). Both talar ridges are intact. The joint is distended but does not show any instability.

21.5.2.2 Dislocation and subluxation

Several types of dislocation and subluxation may occur at the tarsal joint. These injuries are well established on a radiograph. However, subtle injuries may require CT. Dislocation of tarsal bones can easily be seen on the transverse CT images but can be even better evaluated on the dorsal and sagittal reconstructions (Figures 21.17–21.19).

Plantar ligament disruption and proximal intertarsal luxation are common injuries in the Shetland sheepdog and collie breeds, who seem to be predisposed to the degenerative from of the condition (Figure 21.20).

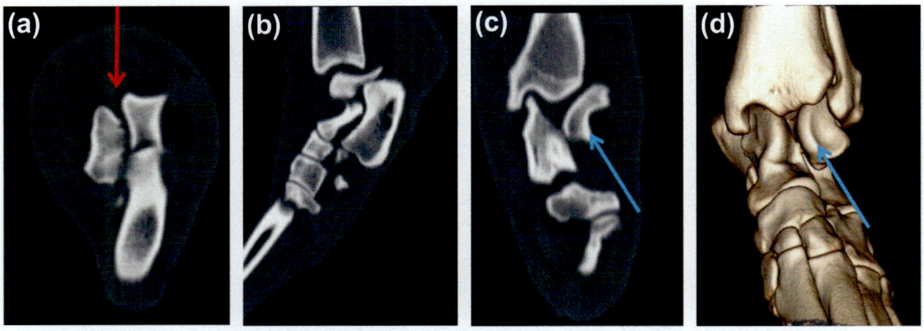

Figure 21.13 Transverse- (a), sagittal- (b), and dorsal-reconstructed (c), bone algorithm, 3D (d) images of an intra-articular fracture of the talus of a 1-y-old Belgian Shepherd. There is a fracture through the centre of the talus (red arrow), with luxation of and medial rotation of the lateral talar ridge (blue arrows). The tarsocrural joint is subluxated. Several smaller fragments are noticed. 3D CT provides useful information, especially in regard to the extent of complex fracture lines.

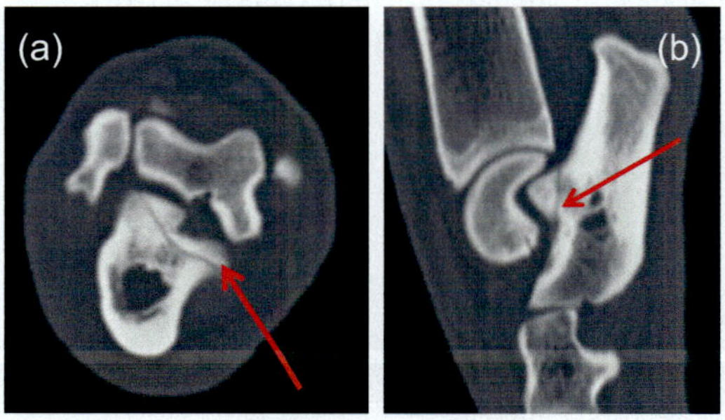

Figure 21.14 Transverse- (a) and sagittal- (b) reconstructed images, bone algorithm, of a sustentaculum tali fracture. A faint radiolucent line is noticed; there is minimal dislocation present (red arrows).

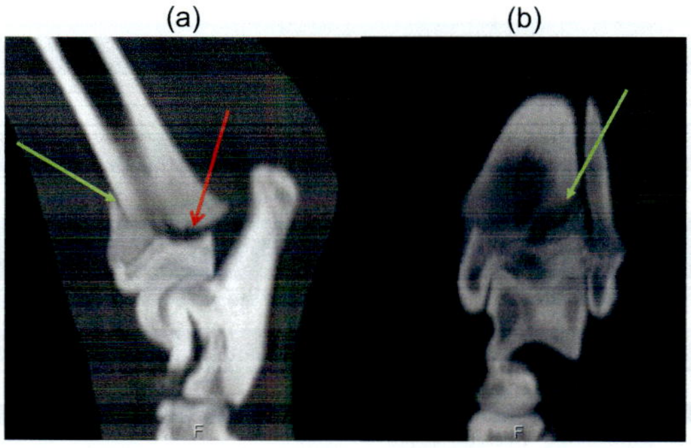

Figure 21.15 Sagittal (a) and dorsal (b) MIP images. Separation of the distal tibial epiphysis (red arrow) and oblique fracture of the distal tibia (green arrow) in a puppy. Salter–Harris fracture type 2.

21.5.3 Degenerative

21.5.3.1 Degenerative joint disease

Like in any other joint on CT, new bone formation can be detected in an earlier stage than on conventional radiographs. At the tarsus, particularly at the sulcus of the deep digital flexor tendon, new bone formation is observed, from the onset of TOC (Figure 21.21). Almost all joints show progressive degenerative joint disease, especially in TOC, regardless of the type of treatment. The clinical outcome is probably influenced by the degree of pre-surgical arthrosis, the increase in arthrosis over time, and especially the degree of post-surgical arthrosis. The progression of osteoarthrosis is most likely due to instability of the joint.

21.5.4 Neoplasia

21.5.4.1 Soft-tissue tumour: synovial cell sarcoma and histiocytic sarcoma

These are masses infiltrating in the soft tissues, muscles, cartilage, and adjacent bones. Bony lysis, soft tissue swelling, and thickening of the neoplastic synovial membrane are appreciated on CT. IV contrast causes enhancement of the neoplastic process and allows a more detailed demarcation and extent of the tumour (Figure 21.22).

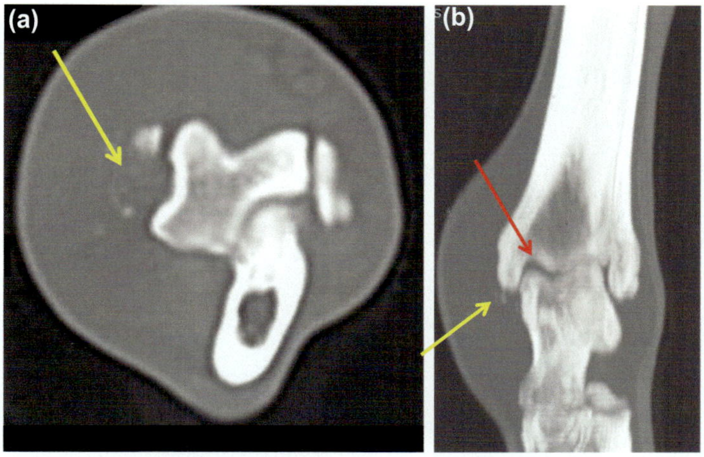

Figure 21.16 Transverse (a) and dorsal MIP (b) of a 5-year Border Collie, bone algorithm. There is prominent distension present on the medial side. The medial part of the tarsocrural joint space is enlarged, suggesting subluxation (red arrow). The medial malleolus shows a small defect associated with multiple calcified structures within the soft tissues (yellow arrows). These images are highly suggestive for an avulsion fracture of the short medial tarsal collateral ligament, probably of traumatic origin.

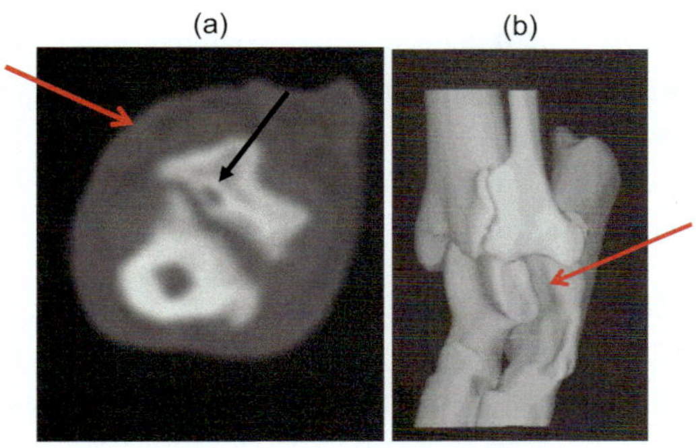

Figure 21.17 Transverse, bone algorithm, (a) and 3D-reconstructed image (b) of a tarsal joint of a cat. There is luxation between the talus and calcaneus (red arrows).

21.5.4.2 Primary bone tumour/metastasis

Primary bone tumours are rare in this area. On conventional radiographs, bone resorption of more than 30% can be appreciated. However, on CT, bone resorption can be seen in an earlier phase. Submillimetre-thickness CT slices can accurately define the extent of the tumour and surgical margins. By use of CT guidance, accurate bone biopsy can be performed, allowing an accurate diagnosis of the primary tumour and better treatment planning.

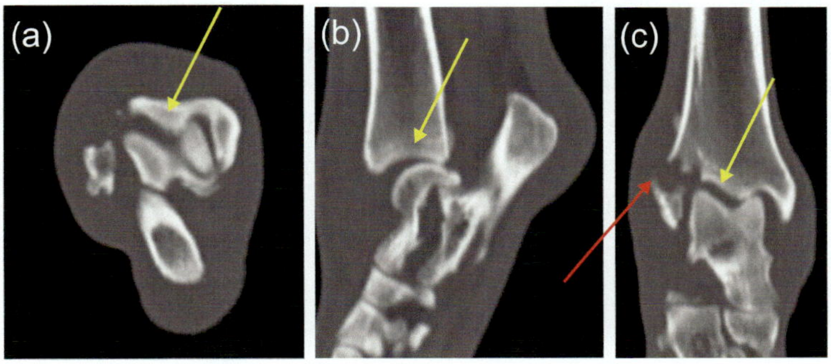

Figure 21.18 Transverse- (a), sagittal-, (b)- and dorsal- (c) reconstructed images, in bone algorithm, of the tarsus of a German pointer. There is a displaced fracture of the lateral malleolus (red arrow) with subluxation of the tarsocrural joint (yellow arrows). There is marked distension and slight arthrosis seen.

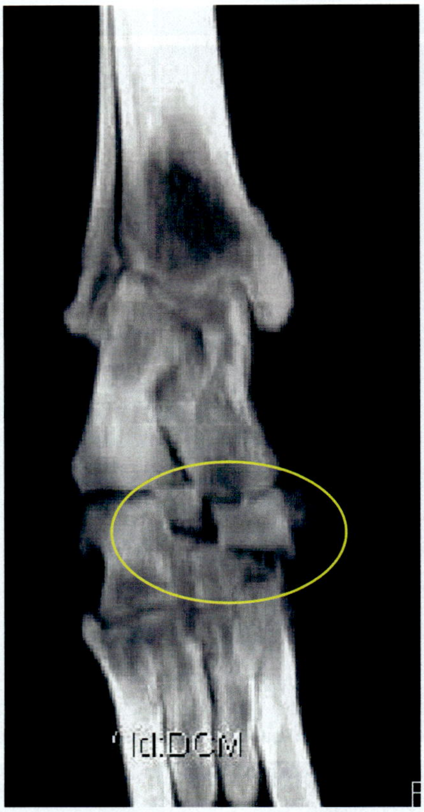

Figure 21.19 **MIP, dorsal image of the tarsus of a 9-year-old Belgian Shepherd with subluxation of the intertarsal bones. The joint space between the central bone and the fourth tarsal bone is enlarged. The third tarsal bone is displaced proximally. Multiple small fragments are present.**

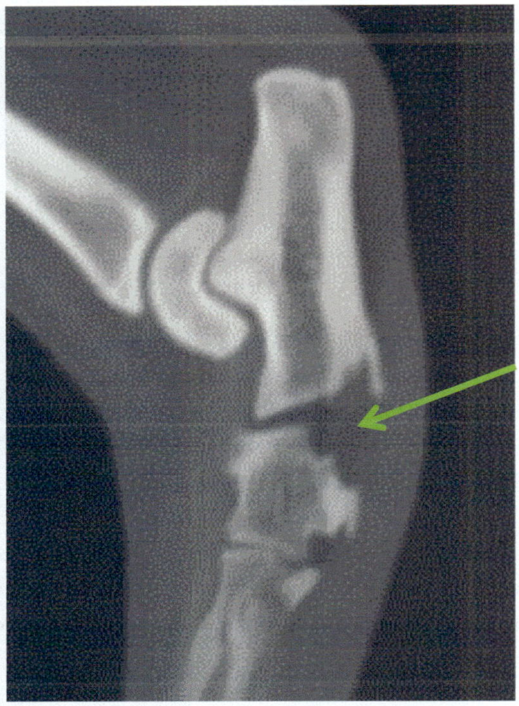

Figure 21.20 **Sagittal-reconstructed image in bone algorithm, of the tarsus of an 8-year-old Border Collie. There is subluxation present of the calcaneoquartal joint due to plantar–tarsal ligament rupture (green arrow).**

21.5.5 Pathologic findings – miscellaneous

21.5.5.1 Inflammatory joint disease

In cases of inflammatory joint disease, prominent subchondral bone loss is present, which may evaluate to extended bone destruction. Checking other joints for the eventual same appearance is recommended (Figure 21.23).

21.5.5.2 Calcaneal (Achilles) tendon injury

The common calcaneal tendon is a collective of several tendons that adhere to the tuber calcanei. The Achilles tendon is made up of the tendons of the gastrocnemius, the main component) and the tendon of the flexor digitorum superficialis muscles. Tendons of the biceps femoris, laterally, and the tendons of the semitendinous and gracilis are contributors. Injuries by impact may occur. CT enables the identification of enthesopathy, tendinopathy, and affections at the attachment on the tuber calcanei (Figures 21.24 and 21.25).

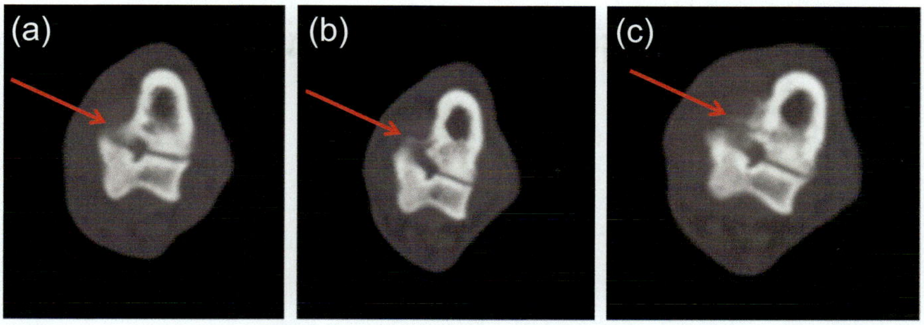

Figure 21.21 Transverse images i bone algorithm at the level of the sulcus of the deep digital flexor tendon showing the progression of osteoarthrosis of a dog suffering from TOC. At the moment of diagnosis (a) of TOC, slight osteoarthrosis was noticed at the sulcus of the deep digital flexor tendon. At the 6-month (b) and 2-year (c) follow-up, there is a drastic increase in osteoarthrosis at the sulcus (red arrows).

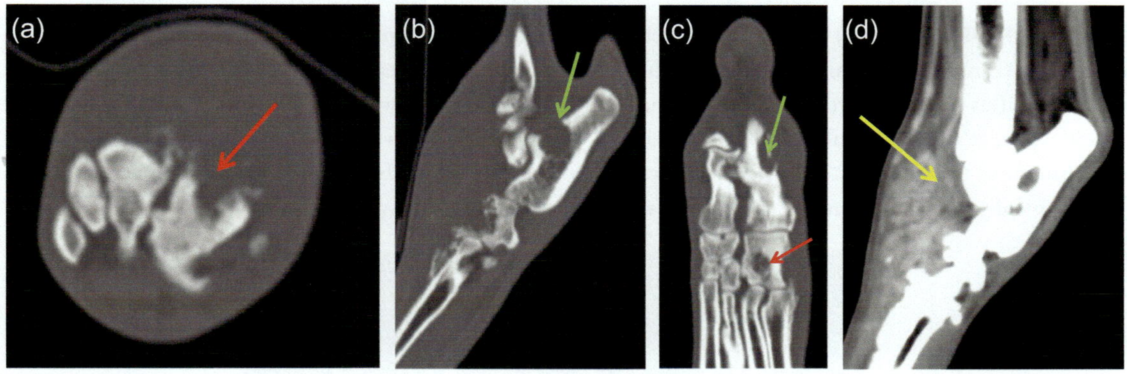

Figure 21.22 Transverse- (a), sagittal- (b), and dorsal-reconstructed (c) images in bone algorithm. Sagittal-reconstructed image (d) in soft tissue algorithm after the application of IV contrast medium, of a 5-year-old Labrador retriever with synovial cell sarcoma. Bone lysis is evident on the calcaneus (green arrows) and the fourth tarsal bone (red arrows) with periosteal reactions on the intertarsal bones. In the dorsal area, severe soft tissue swelling is present very inhomogeneously enhancing after IV contrast administration., with a highly vascularized component in the soft tissues (yellow arrow).

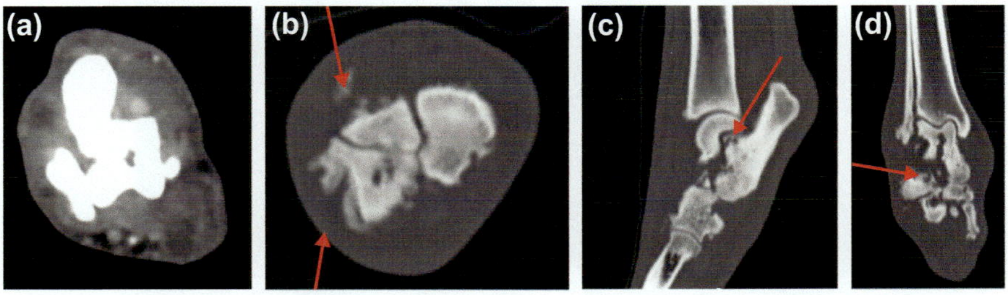

Figure 21.23 Transverse image after the administration of IV contrast, soft tissue algorithm (a) transverse (b), sagittal (c), and dorsal (d) reconstructions, in bone algorithm of a 8-year Belgian shepherd with polyarthritis. The tarsal joint shows severe distension of the joint capsule, and fluid pockets are present (blue arrows). Severe degenerative changes with subchondral lysis and periosteal reactions of the distal part of the talus and calcaneus, and the intertarsal bones are noticed (red arrows). The tarsocrural joint is normal. The same findings were seen in the other tarsal joint and in other joints. Inflammatory disease and polyarthritis were concluded.

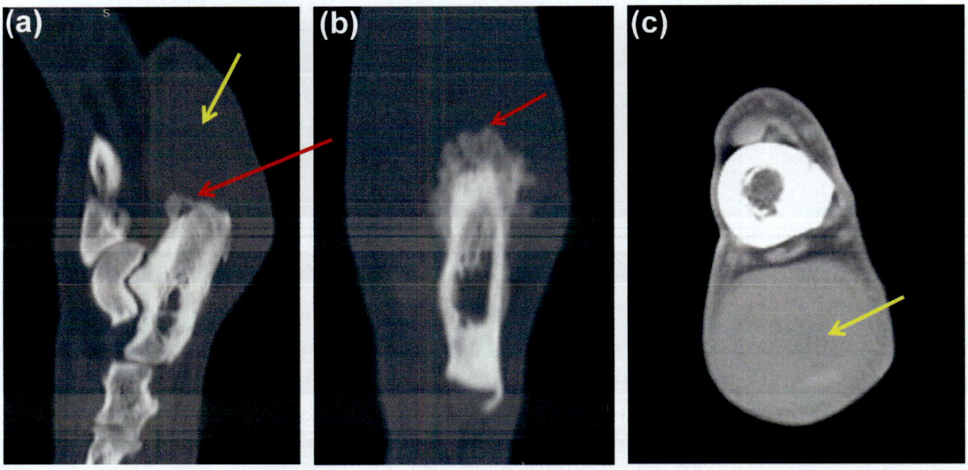

Figure 21.24 Sagittal (a) and dorsal (b) reconstruction, both in bone algorithm and (c) soft tissue algorithm transverse image at the level of the yellow arrow; of a 5-year Dobermann with avulsion of the Achilles tendon. Severe swelling of the Achilles tendon (yellow arrows) is present. There is calcification at the proximal part of the calcaneus with a fulminant irregular periosteal reaction at the attachment of the tendons. This periosteal reaction continues distally from the calcaneus (red arrows). This typically occurs frequently in the Dobermann, which show a typical stance with raised toes.

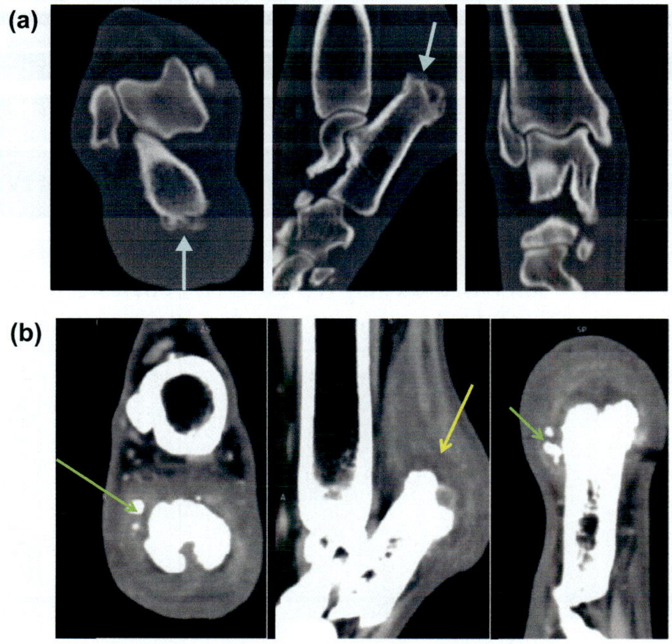

Figure 21.25 (a) Transverse- and sagittal-reconstructed images in bone algorithm of a 3-year-old mastiff with tendinitis of the Achilles tendon. The apex of the calcaneus shows irregular delineation (arrows). (b) Transverse-, sagittal-, and dorsal- reconstructed images in soft tissue algorithm after the application of IV contrast medium of the same dog. There are small calcifications present in the soft tissues (green arrows). There is staining of the thickened tendon insertion (yellow arrow). These images are compatible with tendinitis of the Achilles tendon. This may evaluate towards an avulsion of the insertion at the apex of the calcaneus.

FURTHER READING

Abako Piotr Holak, J.; Głodek, J.; Zhalniarovich, Y. Usefulness of imaging techniques in the diagnosis of selected injuries and lesions of the canine tarsus. A review. *Animals* 2021, 11(6), 1834; https://doi.org/10.3390/ani11061834

Armstrong, A.J.; Bruce, M.; Adams, R.; Kulendra, E.; Pease, T.; Perry, K.L. Injuries involving the central tarsal bone in nonracing dogs: Short-term outcomes and prognostic factors. *Vet. Surg.* 2019, 48, 524–536.

Ballegeer, E.A. Computed tomography of the musculoskeletal system. *Vet. Clin. N. Am. Small Anim. Pract.* 2016, 46, 373–420.

Beale, B.S.; Goring, R.L.; Herrington, J.; Dee, J.; Conrad, K. A prospective evaluation of four surgical approaches to the talus of the dog used in the treatment of osteochondritis dissecans. *J. Am. Anim. Hosp. Assoc.* 1991, 27, 221–229.

Bergh, M.S.; Piras, A.; Samii, V.F.; Weisbrode, S.E.; Johnson, K.A. Fractures in regions of adaptive modeling and remodeling of central tarsal bones in racing greyhounds. *Am. J. Vet. Res.* 2012, 73, 375–380.

Boudrieau, R.J.; Dee, J.F.; Dee, L.G. Central tarsal bone fractures in the racing Greyhound: A review of 114 cases. *J. Am. Vet. Med. Assoc.* 1984, 184, 1486–1491.

Breur, G.J.; Spaulding, K.A.; Braden, T.D. Osteochondritis dissecans of the medial trochlear ridge of the talus in the dog. *Vet. Comp. Orthop. Traumatol.* 1989, 2, 168–176.

Brunnberg, M.; Engelke, E.; Gielen, I.M.; van Bree, H.J.; Hoffmann, J.E.; Brunnberg, L.; Waibl, H.L.Cartilage thickness of the trochlea of the talus, with emphasis on sites predisposed to osteochondrosis dissecans, in clinically normal juvenile and adult dogs. *Am. J. Vet. Res.* 2011, 72, 10 Page1318–1324.

Buj, E.C.; Burton, N.; Mosley, J.R.; Meeson, R.L.; Major, A.; Whitelock, R.G.; Parsons, K., Langley-Hobbs, S.J. *Vet. Comp. Orthop. Traumatol.* 2021, 34, 68–73. Isolated Articular Fractures of the Canine Talus: Diagnosis and Signalment in Fourteen Dogs.

Butler, D.; Nemanic, S.; Warnock, J.J. Comparison of radiography and computed tomography to evaluate fractures of the canine tarsus. *Vet. Radiol. Ultrasound* 2018, 59, 43–53.

Carlisle, C.H.; Reynolds, K.M. Radiographic anatomy of the tarsocrural joint of the dog. *J. Small Anim. Pract.* 1990, 31, 273–279.

Carlisle, C.H.; Robins, G.M.; Reynolds, K.M. Radiographic signs of osteochondritis dissecans of the lateral ridge of the trochlea tali in the dog. *J. Small Anim. Pract.* 1990, 31, 280–286.

Corr, S.A.; Draffan, D.; Kulendra, E.; Carmichael, S.; Brodbelt, D. Retrospective study of Achilles mechanism disruption in 45 dogs. *Vet. Rec.* 2010, 167, 407–411.

Deruddere, K.J.; Milne, M.E.; Wilson, K.M.; Snelling, S.R. Magnetic resonance imaging, computed tomography, and gross anatomy of the canine tarsus. *Vet. Surg.* 2014, 43, 912–919.

Dingemanse, W.; Gielen, I.; Duchateau, L.; Van Bree, H. Comparison of morphological and clinical features between medial and lateral trochlear ridge talar osteochondritis dissecans in dogs. *Vet. Surg.* 2013, 42, 340–345.

Fitch, R.B.; Beale, B.S. Osteochondrosis of the canine tibiotarsal joint. *Vet. Clin. N. Am. Small Anim. Pract.* 1998, 28, 95–113.

Galateanu, G.; Aizenberg, I.; Hildebrandt, T.B.; Apelt, D. Short communications: Computed tomographic demonstration of central tarsal bone plantar process occult fracture in a dog. *Vet. Rec.* 2011, 169, 442.

Galateanu, G.; Apelt, D.; Aizenberg, I.; Saragusty, J.; Hildebrandt, T.B. Canine tarsal architecture as revealed by high-resolution computed tomography. *Vet. J.* 2013, 196, 374–380.

Gielen, I.M.; Van Bree, H.J.; Van Ryssen, B.; De Clercq, T.; De Rooster, H. Radiographic, computed tomographic and arthroscopic findings in 23 dogs with osteochondrosis of the tarsocrural joint. *Vet. Rec.* 2002, 150, 442–447.

Gielen, I.M.; De Rycke, L.M.; Van Bree, H.J.; Simoens, P.J. Computed tomography of the tarsal joint in clinically normal dogs. *Am. J. Vet. Res.* 2001, 62, 1911–1915.

Gielen, I. 2003. Computed tomography in the diagnosis and treatment of canine tarsocrural osteochondrosis. PhD thesis, Ghent University.

Gielen, I.; Van Ryssen, B.; Van Bree, H. Computerized tomography compared with radiography in the diagnosis of lateral trochlear ridge talar osteochondritis dissecans in dogs. *Vet. Comp. Orthop. Traumatol.* 2005, 18, 77–82.

Gielen, I.; Van Ryssen, B.; Coopman, F.; Van Bree, H. Comparison of subchondral lesion size between clinical and non-clinical medial trochlear ridge talar osteochondritis dissecans in dogs. *Vet. Comp. Orthop. Traumatol.* 2007, 20, 8–11.

Guilliard, M.J. Fractures of the central tarsal bone in eight racing greyhounds. *Vet. Rec.* 2000, 147, 512–515.

Guilliard, M.J. Central tarsal bone fracture in the border collie: Case report. *J. Small Anim. Pract.* 2007, 48, 414–417.

Guilliard, M.J. Third tarsal bone fractures in the greyhound. *J. Small Anim. Pract.* 2010, 51, 635–641.

Harasen, G.L.G. Fracture-luxation of the central tarsal bone in a dog. *Can. Vet. J.* 1999, 40, 195.

Harasen, G.L.G. Ruptures of the common calcaneal tendon. *Can. Vet. J.* 2006, 47, 1219–1220.

Hay, C.W.; Muir, P.; Johnson, K.A. Central tarsal bone fractures in two dalmatians. *Vet. Comp. Orthop. Traumatol.* 1995, 8, 222–225.

Hercock, C.A.; Innes, J.F.; McConnell, F.; Guilliard, M.J.; Ness, M.G.; Hodson, D.; Young, I.S. Observer variation in the evaluation and classification of severe central tarsal bone fractures in racing greyhounds. *Vet. Comp. Orthop. Traumatol.* 2011, 24, 215–222.

Hudson, C.C.; Pozzi, A. Minimally invasive repair of central tarsal bone luxation in a dog. *Vet. Comp. Orthop. Traumatol.* 2012, 25, 79–82.

Isaka, M.; Befu, M.; Matsubara, N.; Ishikawa, M.; Aono, H.; Namba, S. Type 1 Achilles tendon rupture caused by grooming trauma in a young dog. *Open Vet. J.* 2014, 4, 56–58.

Jeffcott, L.B. Osteochondrosis—An international problem for the horse industry. *J. Equine Vet. Sci.* 1996, 16, 32–37.

Johnson, K.A.; Howlett, C.B.; Pettit, G.D. Osteochondrosis in the hock joints in dogs. *J. Am. Anim. Hosp. Assoc.* 1980, 16, 103–113.

Kippenes, H.; Johnston, G. Diagnostic imaging of osteochondrosis. *Vet. Clin. N. Am. Small Anim. Pract.* 1998, 28, 137–160.

Lorinson, D.; Grosslinger, K. Central tarsal bone luxation in three non-racing dogs. *Vet. Comp. Orthop. Traumatol.* 2001, 14, 229–231.

Maley, J.R.; Dvorak, L.D.; Bahr, A. Diagnosis and management of a fracture of the lateral trochlear ridge of the talus in a dog. *Vet. Comp. Orthop. Traumatol.* 2010, 23, 284–288.

Mason, T.A.; Lavelle, R.B. Osteochondritis dissecans of the tibial tarsal bone in dogs. *J. Small Anim. Pract.* 1979, 20, 423–432.

Montgomery, R.D.; Hathcock, J.T.; Milton, J.L.; Fitch, R.B. Osteochondrosis dissecans of the canine tarsal joint. *Comp. Cont. Educ. Pract. Vet.* 1994, 16, 835–844.

Newell, S.M.; Mahaffey, M.B.; Aron, D.N. Fragmentation of the medial malleolus of dogs with and without tarsal osteochondrosis. *Vet. Radiol. Ultrasound* 1994, 35, 5–9.

Nielsen, C.; Pluhar, G.E. Outcome following surgical repair of achilles tendon rupture and comparison between post-operative tibiotarsal immobilization methods in dogs. 28 cases (1997—2004). *Vet. Comp. Orthop. Traumatol.* 2006, 19, 246–249.

Olsson, S.E. Lameness in the dog. A review of lesions causing osteoarthrosis of the shoulder, elbow, hip, stifle and hock joints. *Proc. Am. Anim. Hosp. Assoc.* 1975, 42, 363–370.

Post, C.; Guerrero, T.; Ohlerth, S.; Hässig, M.; Voss, K.; Montavon, P.M. Joint mice migration into the deep digital flexor tendon sheath in dogs. Clinical cases and anatomical study. *Vet. Comp. Orthop. Traumatol.* 2008, 21, 440–445.

Robins, G.M.; Read, R.A.; Carlisle, C.H.; Webb, S.M. Osteochondritis dissecans of the lateral ridge of the trochlea of the tibial tarsal bone in the dog. *J. Small Anim. Pract.* 1983, 24, 675–685.

Sjösström, L.; Håkanson N. Traumatic injuries associated with the short collateral ligaments of the talocrural joint of the dog. *J. Small Anim. Pract.* 1994, 35, 163–168.

Van der Peijl, G.J.W.; Schaeffer, I.G.F.; Theyse, L.F.H.; Dijkshoorn, N.A.; Schwencke, M.; Hazewinkel, H.A.W. Osteochondrosis dissecans of the tarsus in labrador retrievers: Clinical signs, radiological data and force plate gait evaluation after surgical treatment. *Vet. Comp. Orthop. Traumatol.* 2012, 25, 126–134.

Van Ryssen, B.; Van Bree, H.J. Arthroscopic evaluation of osteochondrosis lesions in the canine hock joint—A review of 2 cases. *J. Am. Anim. Hosp. Assoc.* 1992, 28, 295–299.

Wisner, E.R.; Berry, C.R.; Morgan, J.P.; Pool, R.R.; Wind, A.P.; Vasseur, P.B. Osteochondrosis of the lateral trochlear ridge of the talus in seven rottweiler dogs. *Vet. Surg.* 1990, 19, 435–439.

CHAPTER 22

CT OF LONG BONES

Gian Luca Rovesti (Intra-articular fractures), Michael Kowaleski (Angular deformities), Ingrid Gielen, and Henri van Bree

22.1 SHORT INTRODUCTION AND PATHOLOGY

Patients with bone defects and/or deformities must frequently undergo numerous investigations during their treatment. Radiography is the initial tool to investigate these cases. CT is the modality most often used, second to radiography. It is used for surgical planning, evaluating callus maturation, alignment measurement, length measurement, torsion measurement, and angiography.

22.2 INDICATIONS FOR USING CT

Orthopaedic surgeons rely on high-quality CT scans to evaluate acute and chronic injuries involving articular surfaces. Particularly, multiplanar reformation is helpful in revealing fracture patterns, displacement, and exact number of fragments. Fracture lines may be challenging to distinguish on transverse slices; they are simpler to understand when seen in the long axis. When implant failure is suspected, CT reconstructions are helpful for a detailed evaluation. In fracture healing, CT plays a role in the detection of callus prior to radiography. Additionally, union in long bone fractures can be confirmed when radiograph results are unreliable. Subtle changes in bone marrow, in spongy and cortical bone, and in the surrounding soft tissues are visible on CT. Three-dimensional volume rendering techniques are described for measuring torsion angles in long bones in cases of deformity (Figure 22.1).

22.3 TECHNICAL ASPECTS

22.3.1 Positioning

The extremities should be positioned symmetrically, and the opposite extremity should be scanned as well. The forelimbs should be

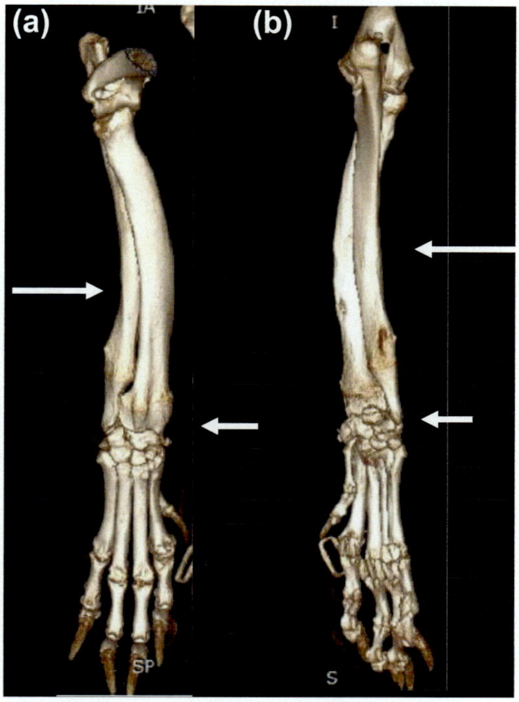

Figure 22.1 **3D CT images, craniocaudal (a) and caudocranial (b) view of the fore limb of a Leonberger of 9 months old, who had a distal radius/ulna fracture with slight dislocation (short arrow). After treatment with a cast the fracture healed. However the dog was still lame, showing exorotation of the foot and curving of the radius and ulna (long arrow).**

DOI: 10.1201/9781315118130-24

stretched gently forwards, with the patient in dorsal or sternal recumbency. For examining the hindlimbs, traction backwards with the patient in dorsal recumbency is recommended.

Beam-hardening artefacts occur by the passage of x-rays through dense material, such as implants. The end effect is a black artefact that radiates from the metal into the surrounding tissue, perhaps hiding pathology. By raising kVp, using metal artefact reduction reconstruction methods, decreasing the thickness of metal in the scan plane, carefully placing the patient, and/or tilting the scanner gantry, artefacts can be reduced/followed.

22.3.2 Technical parameters

Slices should be oriented perpendicular to the long axis of the involved bone.

The slice thickness depends on the size of the bone to be studied. When smaller lesions are investigated or in restricted anatomical areas, thin slices are necessary.

Both bone and soft tissue algorithms are utilized. Optimal 3D measurements require 0.6-mm slices with overlap.

22.3.3 IV Contrast

IV contrast is used to evaluate the enhancement of inflamed soft tissue.

It is helpful in the evaluation of tumour vascularity as well as assessment of tumour–vessel associations.

22.4 PATHOLOGIC FINDINGS

22.4.1 Diaphyseal fractures

Traditionally, long bone fractures are evaluated by conventional radiology. In some cases, additional CT is useful to demonstrate malunion, to locate bone fragments, show displacement of the affected bones, identify and outline the number of fractures, and characterize joint disorders (Figure 22.2). Several of these characteristics might be seen more clearly on CT than

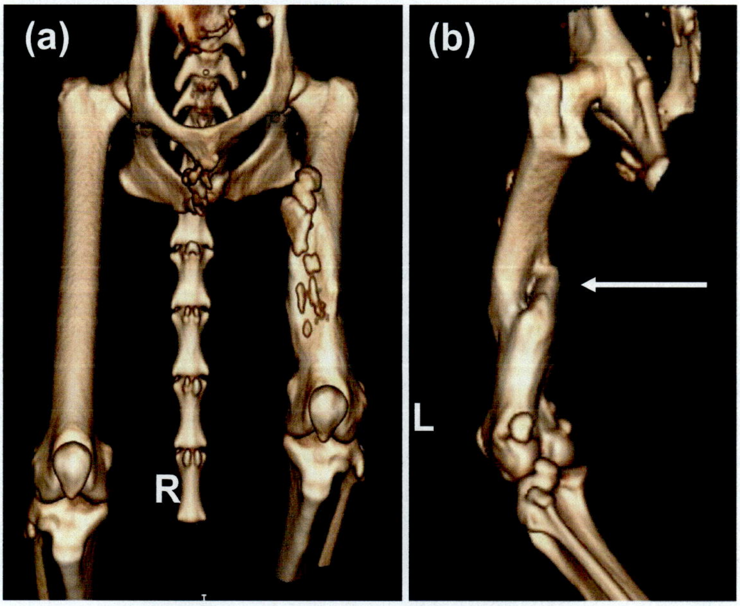

Figure 22.2 3D CT images, ventrodorsal (a) and dorsoventral (b) of the hind limb of a cat showing a mal-union of an old femoral diaphyseal fracture with overriding of the fragments (arrow) resulting in shortening of the left femur. Over the fracture line multiple calcified areas are present resulting from resorption of the original callus.

they could on a regular radiograph. Particularly, loose entities not visible on conventional radiography due to bone remodelling and sclerosis can be detected at CT. The extent of related soft tissue damage and fluid accumulation is also clearly seen by CT (Figure 22.3). Although not routinely performed, CT is accurate for identifying the location and integrity of the implant (Figure 22.4). IV contrast CT can be used in assessing the amount of vascularization in normal fracture healing and in non-union cases (Figure 22.5).

22.4.2 Intra-articular fractures

CT is not used usually to examine joints for fractures. Historically, radiographic assessment

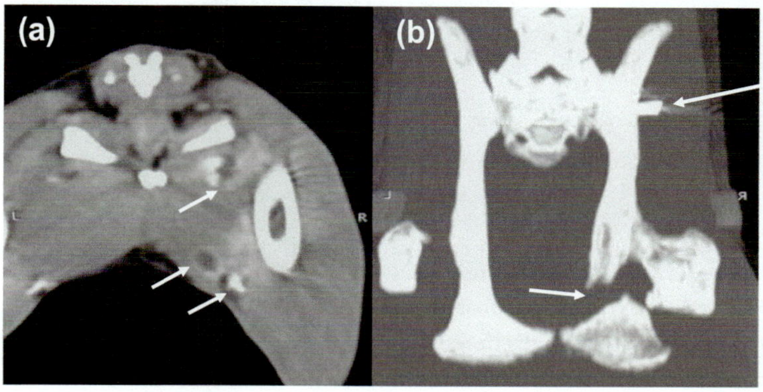

Figure 22.3 Transverse image with iv contrast administration (a) and dorsal MIP reconstruction (b) of the pelvis of a 4 year old cat who sustained a pelvic fracture and a sacroiliac luxation surgically treated 6 months ago (long arrow). On the contrast images (a) a diffuse enhancement of the soft tissues medial of the right femur can be observed. Several round areas showing ring enhancement associated with a hypodens centre are present (arrows). These images are suggestive for an infection with abscess formation. On the dorsal reconstruction (b) a non-union of the ischium with pelvic narrowing can be appreciated (short arrow).

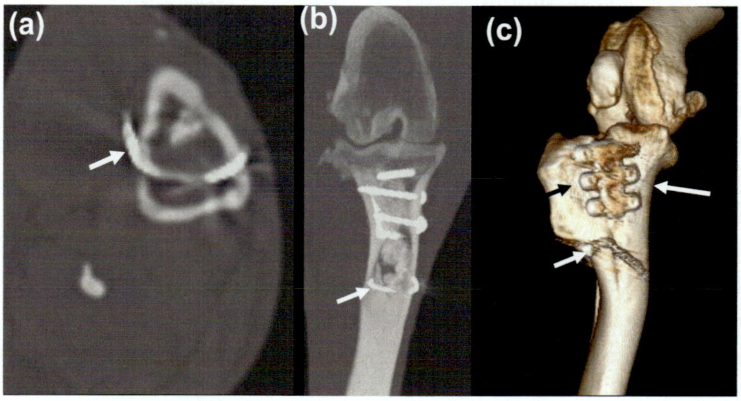

Figure 22.4 Transverse image (a) and dorsal (b) and 3D (c) reconstructions of a stifle joint with massive arthrosis surgically treated for cruciate ligament damage with TTA (black & white arrow) having a proximal tibial fracture as complication. This fracture was fixed with a cerclage wire (short white arrows). The cerclage wire was placed within the fracture line and the bad alignment of the fragments is causing a curvature.

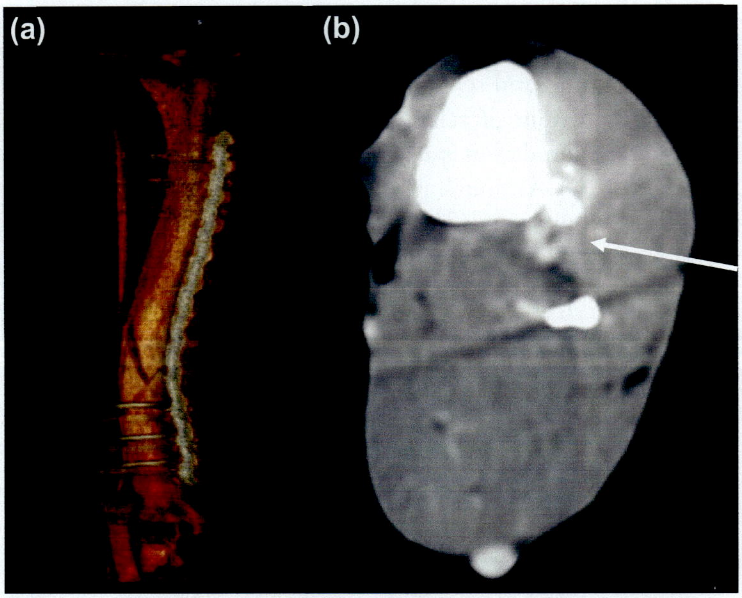

Figure 22.5 **3D CT (a) and transverse post contrast image (b) of a dog 6 weeks after surgery (plate and screws). (a) good fragment alignment can be noticed with a per primam bone healing. (b) vascularisation of the fibrous callus is visible (arrow).**

has been considered detailed enough for preoperative planning, which is true in most cases. Literature reports are quite scarce, and in most circumstances, the joints are examined for diseases other than fractures. Although the diagnosis of a joint fracture is considered straightforward, and mainly based on radiographic examination, a CT scan should be performed for these reasons:

- In complex treatable fractures, the 3D view of the fracture can greatly help the surgeon to correctly plan the surgical approach and fracture stabilization.
- In complex non-treatable fractures, evaluation of the technical aspects that suggest not approaching the joint beforehand is preferred, to avoid any useless surgical approach.

In addition to the standard capability of the technique to visualize joint structures without superimposition, 3D software reconstruction allows the surgeon to visualize the fracture fragments in a stereotactic way, thus greatly improving the correct understanding of their mutual positioning.

A joint-oriented approach that provides more details about the specific features of each joint is given as follows.

22.4.2.1 Shoulder

The most difficult area to evaluate by standard radiographic imaging is the glenoid cavity of the scapula. Though a fracture is usually easily identified on lateral projection, the most difficult area to appreciate is the medial wall of the scapular glenoid and neck. This area is usually fundamentally important for the choice of the surgical technique and should be carefully evaluated (Figure 22.6). The fractures on the sagittal and frontal planes of the glenoid and scapular neck are very difficult to stabilize, while those on the axial plane are technically less demanding.

22.4.2.2 Elbow

The most critical areas to evaluate are the epicondyles and the humeral condyle in bicondylar

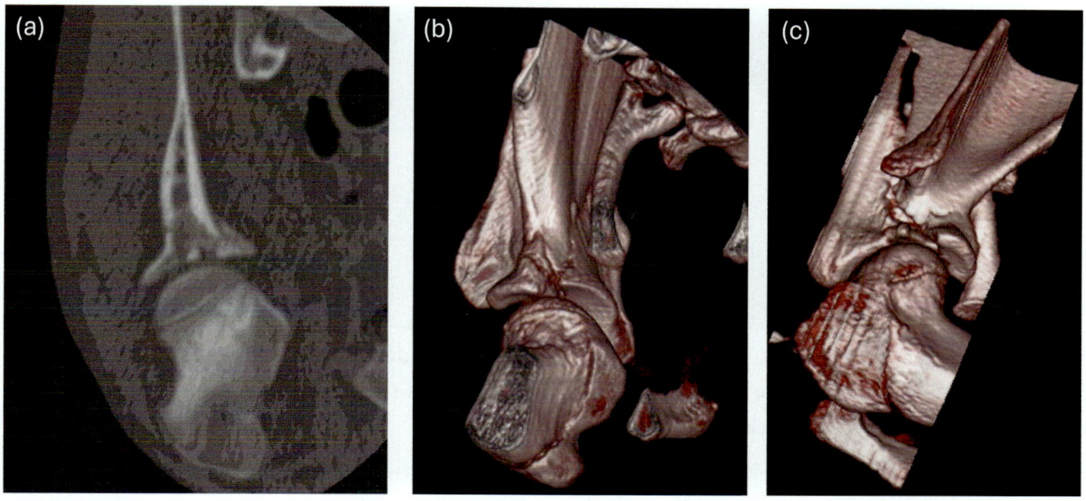

Figure 22.6 CT frontal scan showing a complex fracture of the lateral and medial borders of the glenoid cavity of the scapula (a). Three-D reconstructions show the fracture of the medial wall of the glenoid (b) and the comminuted intra-articular fracture of the scapula on the frontal plane (c).

fractures. The fracture is usually easily visualized on sagittal projection with the standard radiographic technique, but the most difficult area to appreciate is the median condylar surface. This area is usually fundamentally important for prognosis, especially in breeds predisposed to incomplete union of the humeral condyle (IOHC) (Figure 22.7). A pathological fracture due to IOHC usually carries a poor prognosis for fracture healing. Potential fractures of the medial coronoid process can be evaluated as well.

22.4.2.3 Carpus

CT scanning represents a specific indication for the diagnosis of slab fractures since they are difficult to detect on standard radiographic images. In some specific breeds, the fracture can be

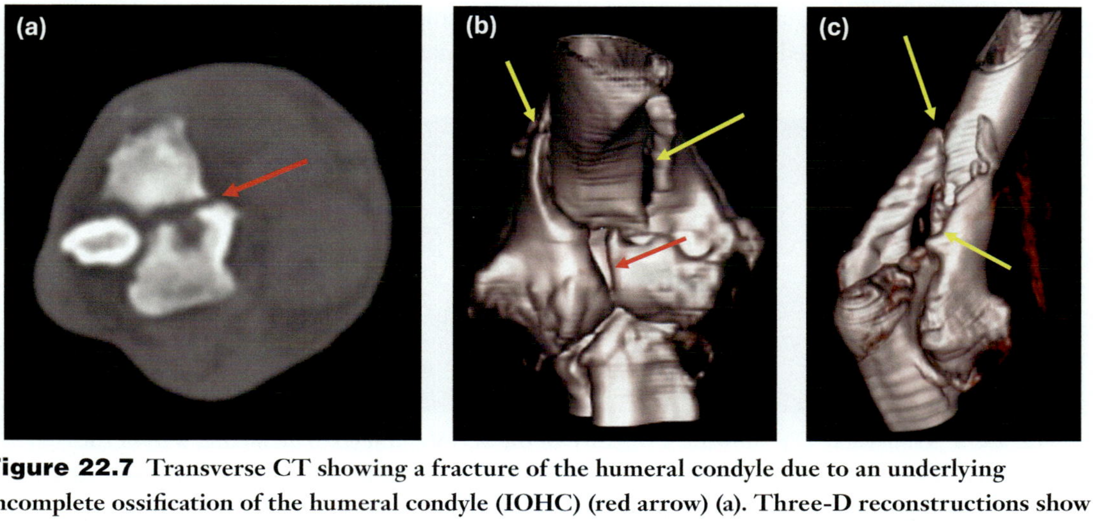

Figure 22.7 Transverse CT showing a fracture of the humeral condyle due to an underlying incomplete ossification of the humeral condyle (IOHC) (red arrow) (a). Three-D reconstructions show the fractures of the condyle (red arrow) (b) and both the epicondyles (yellow arrows) (c).

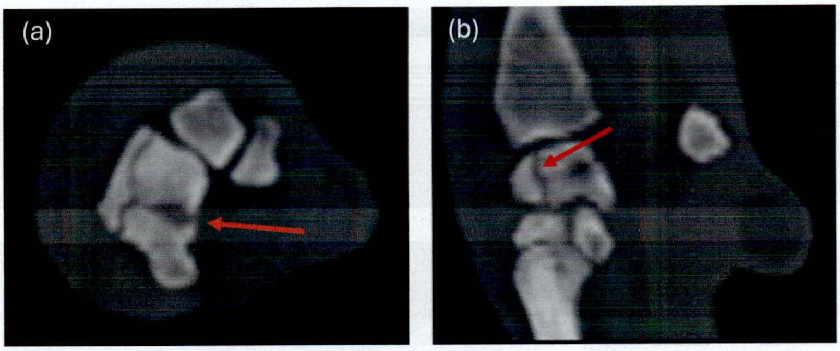

Figure 22.8 CT, transverse scan, showing a fracture of the radial carpal bone (a). The three-fragment fracture carries a poor prognosis. The fracture line on the transverse plane can be clearly seen on the sagittal reconstruction (b) (red arrows).

pathological, due to an incomplete ossification of the radial carpal bone. The prognosis for fracture healing is based upon the fracture features, which has a major impact on the return to function because most dogs affected by these fractures are active or working dogs. Usually, when more than one fracture line is present, the prognosis for return to working function is guarded to poor (Figure 22.8).

22.4.2.4 Hip

Acetabular fractures are challenging to evaluate by standard radiography, due to the superimposition of many structures. They are one of the most frequent indications for joint CT because of their high incidence compared with other joints and because of the quality of information that CT provides compared with the standard radiographic technique. The medial wall integrity is the most important structure for the evaluation of its surgical stabilization. When the medial wall is fractured in the axial and sagittal planes (Figure 22.9), the potential for osteosynthesis is greatly reduced because of the lack of bony purchase of the stabilizing elements on the medial wall. Usually, fractures on the frontal plane are technically less demanding.

22.4.2.5 Stifle

Major fractures of the articular components are usually clearly visualized by standard radiographic projections. Chip fractures are usually not caused by direct trauma and are more challenging to diagnose. Ligaments represent one of the most important joint components and account for the anatomical complexity of the stifle joint. When

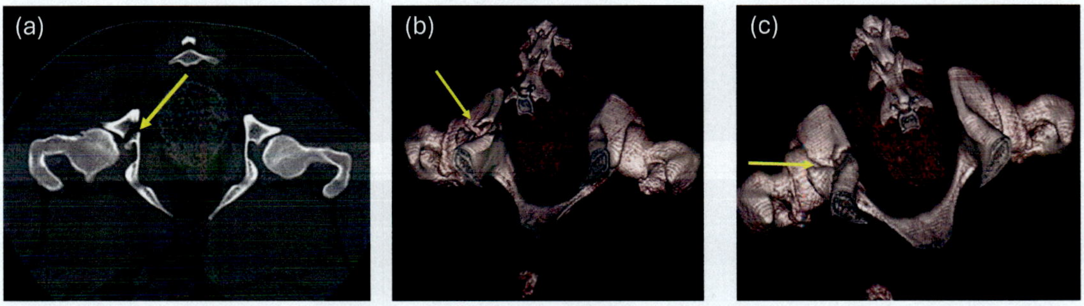

Figure 22.9 Transverse CT showing a fracture of the acetabulum (a). The medial wall of the acetabulum is fractured longitudinally (b), and a third fragment on the caudal portion of the dorsal acetabular rim is present (yellow arrows) (c). The osteosynthesis of this fracture is challenging, and the prognosis for full recovery is guarded.

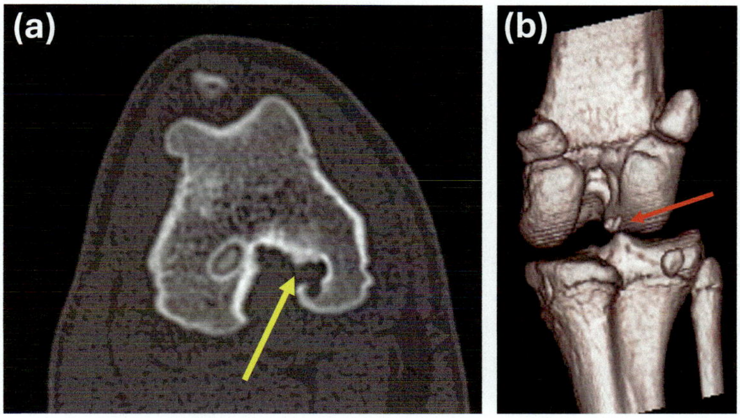

Figure 22.10 Transverse CT of an avulsion fracture of the proximal insertion of the cranial cruciate ligament in a young dog (yellow arrow) (a). Three-D reconstruction shows the avulsion fracture chips (red arrow) and the area on the medial side of the lateral femoral condyle where the chips originated (b). Repositioning of chips of this kind is extremely challenging.

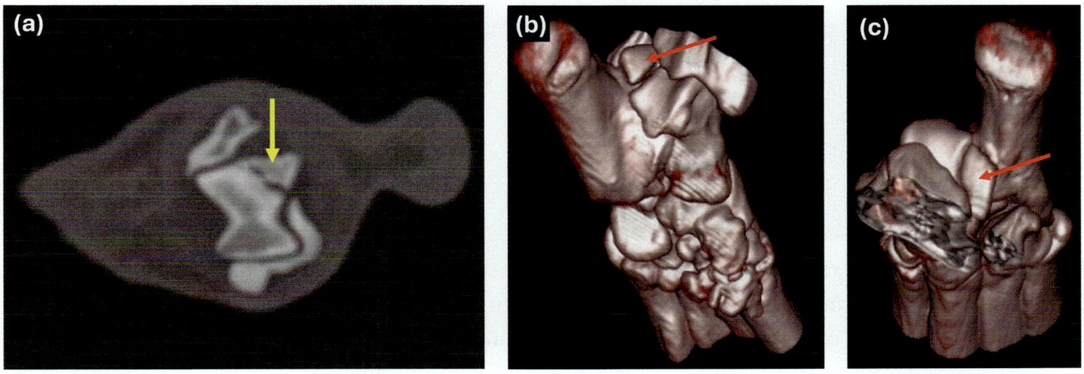

Figure 22.11 Transverse CT of a fracture of the caudo-lateral portion of the lateral talar ridge of the tarsocrural joint (yellow arrow) (a). Three-D reconstructions, caudo-medial (b) and cranial views (c) allow a precise visualization of the fracture (red arrows), which helps with preoperative planning.

the joint is overloaded and the ligaments overdistended, avulsion chip fractures at the insertional sites can be produced. These fractures and their site of origin are not easily detectable on plain radiographs. The caudal portion of the intercondylar notch, the origin of cruciate ligaments, is the most critical area to examine. Apart from the visualization of small chips, the area of origin of the avulsion can be determined (Figure 22.10).

22.4.2.6 Tarsus

Due to the anatomical complexity of the relationship between the tibial cochlea and talar ridges, small fractures of the talus are often difficult to visualize on standard radiographic projections. The caudo-lateral part of the talo-crural joint is the most critical area to examine because the calcaneus superimposition obscures it. Usually, the fragments are moderately displaced (Figure 22.11) and clinical signs may be quite subtle. Because the joint is difficult to approach, a thorough evaluation of the fracture location and its relationships with adjacent structures is mandatory during preoperative planning.

22.4.3 Infection/Neoplasia

In small animals, the differentiation of malignant bone tumours and acute osteomyelitis can be difficult. Both can produce a wide spectrum of destructive and productive bony changes. Many of their radiographic features can be common to both disease processes, especially in the early stages. This means that a diagnosis based solely on the radiographic appearance will often be erroneous. Therefore, additional laboratory data (biopsy and histological examinations and thoracic radiographs) are mandatory. A few guidelines include primary bone tumours are usually monostotic; metastatic neoplasms are usually polyostotic. Furthermore, osteomyelitis can also affect multiple bones. Primary bone tumours are usually located in the metaphyseal region of long bones (proximal humerus, distal radius and ulna, distal femur, proximal/distal tibia, and proximal ulna). A soft tissue mass and aggressive bony changes with periosteal new bone formation are present, and there is usually no involvement of the adjacent joint. Characteristics of a bone lesion that indicate

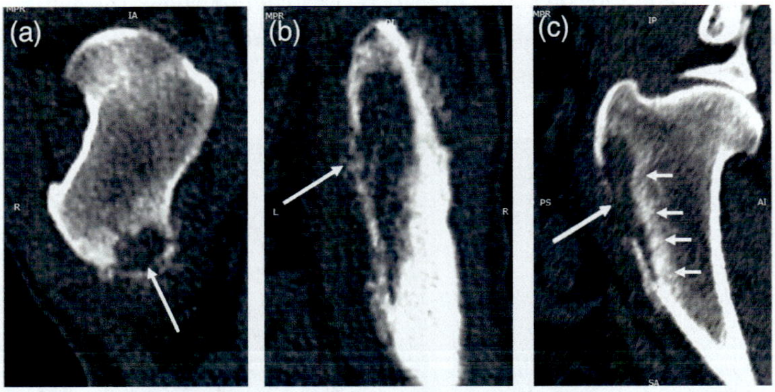

Figure 22.12 **Transverse image (a) and dorsal (b) and sagittal (c) reconstructions of a primary bone tumour of the proximal humerus. Predominantly lytic tumour breaking through the cortex (arrows) and with a gradual zone of transition between normal and abnormal bone (small arrows).**

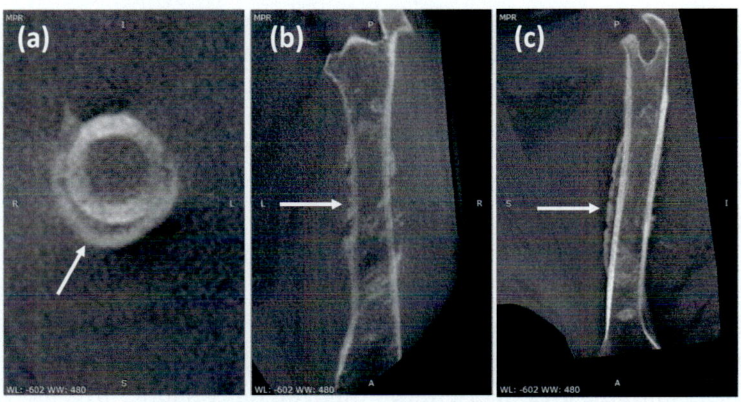

Figure 22.13 **Transverse image (a) and dorsal (b) and sagittal (c) reconstructions of a case of an acute osteomyelitis caused by mycosis. The femoral diaphysis shows a palisade type of periosteal reaction, which seems to be separated from the underlying cortex (arrows). There is also lysis from the subchondral bone present. The femur shows a heterogenous mottled aspect mostly in the medulla of the distal part. In this case, several bones were involved.**

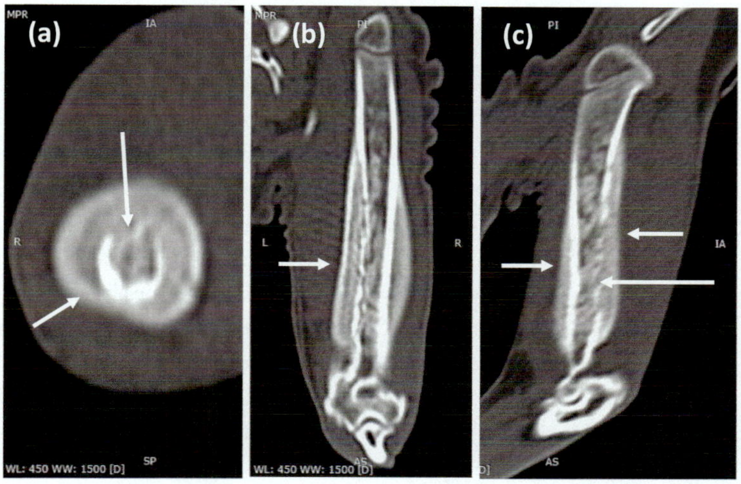

Figure 22.14 Transverse image (a) and dorsal (b) and sagittal (c) reconstructions of a case of a chronic osteomyelitis in an 8-month-old cat. There is soft tissue swelling. There is a thickened cortex due to a chronic periosteal reaction (short arrows). The medulla displays a mottled aspect, and there is cortical destruction (long arrows).

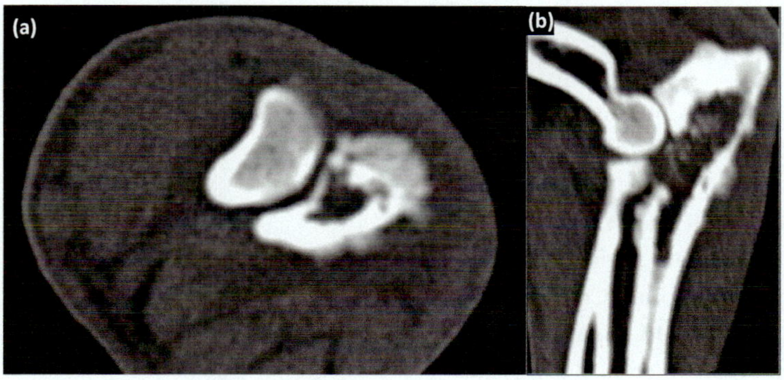

Figure 22.15 Transverse (a) and sagittal reconstructed (b) images of a bone metastasis of a 13-year-old dog, suffering from a confirmed bronchial carcinoma. There is lysis and periosteal new bone formation of the proximal third of the ulna. The bone is weakened and prone to a pathologic fracture.

that it is active and aggressive include the following: prominent soft tissue swelling; focal or diffuse, moth-eaten or permeative bone lysis with poorly defined margins, bone production with ill-defined margins, pathologic fracture, poorly demarcated lesion, gradual zone of transition between normal and abnormal bone, and rapid rate of change on follow-up radiographs (Figures 22.12–22.14). The CT features of primary bone tumours and osteomyelitis are the same as the radiographic features; however, it enables definition of soft tissue and bone changes that are not detected by conventional radiography, and subtle new bone formations and bone lysis are better identified on CT images when compared to conventional radiography.

The most common primary bone tumour in dogs is osteosarcoma. Other bone tumours include chondrosarcoma, fibrosarcoma, and haemangiosarcoma. CT imaging with- and without IV contrast has been used to assess the length of the neoplastic process when limb

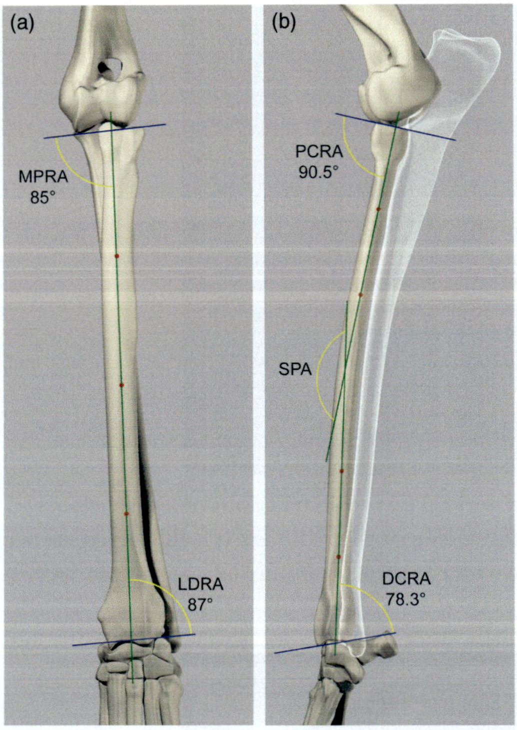

Figure 22.16 Frontal (a) and sagittal (b) plane alignment of the canine radius determined by the CORA methodology. In the frontal plane (a), the Medial Proximal Radial Angle (MPRA) and the Lateral Distal Radial angle (LDRA) have been determined. The Proximal Cranial Radial angle (PCRA) and the Distal Caudal Radial Angle (DCRA) have been determined in the sagittal plane (b).

sparing techniques have been envisaged and to be sure that all margins had been removed. However, from different reports, it was concluded that there is a wide variation between patients, and relying on CT alone would have resulted in incomplete excision of the tumour. Skeletal metastases are frequently attributable to carcinoma. Most metastases are detected by staging of the patient (Figure 22.15).

22.4.4 Angular deformities

Limb deformity correction and limb alignment to correct skeletal malalignment are commonly employed in small animal orthopaedic surgery. Until recently, a unified system that was applicable to all deformities in all long bones was lacking in veterinary surgery. Recently, the Center of Rotation of Angulation (CORA) methodology has been described for use in deformity planning and correction in people, and several authors have adapted this system for use in dogs. The CORA methodology utilizes an axis drawn along the diaphysis of the long bone and joint reference lines drawn across the articular surfaces at specific anatomic landmarks to develop reference angles for the proximal and distal joints of each long bone (Figure 22.16). Two axes exist in each bone; the anatomic axis is drawn along the centre of the diaphysis of the bone, while the mechanical axis is drawn from the centre of the proximal joint surface to the

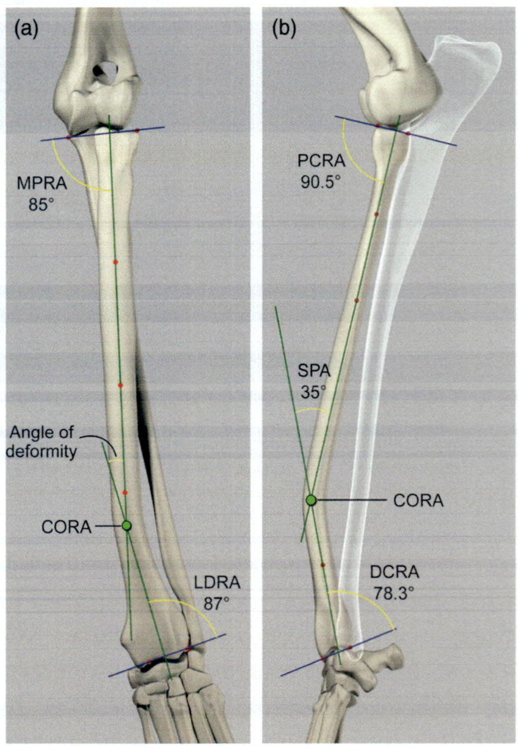

Figure 22.17 A uni-apical deformity has been quantified in the frontal (a) and sagittal (b) planes using the reference angles from Figure 22.16. The location of the CORA in the frontal and sagittal planes has been identified.

centre of the distal joint surface. In bones such as the canine femur, the mechanical and anatomic axes are different, while in other bones, such as the canine tibia, the axes are identical. The intersection between the joint reference line and the bone axis determines the joint reference angle.

Breed-specific joint reference angle ranges have been developed from images of normal dogs (Figure 22.16), and these can be used to aid in the planning of deformity correction. Utilizing breed-specific normal values, the location and magnitude of uni-planar and multi-planar deformities can be identified and quantified in the frontal, sagittal, and transverse planes. In the event that breed-specific values are not available, the opposite normal limb can be utilized to obtain joint reference angles for the individual patient. If both limbs are affected, the mean joint reference angles from the literature can be used. The joint reference angles are utilized to set the anatomic or mechanical axis for proximal and distal bone segments. The Center Of Rotation of Angulation (CORA) is located at the intersection of these anatomic axes, and its magnitude can be measured at this intersection (Figure 22.17). In many deformities, the CORA is uni-apical, occurring at a single location; however bi-apical multi-apical deformities do exist. In some cases, the CORAs occur in the same plane, and in other cases, they are in different planes. Correction is typically undertaken at the location of the deformity in the frontal plane.

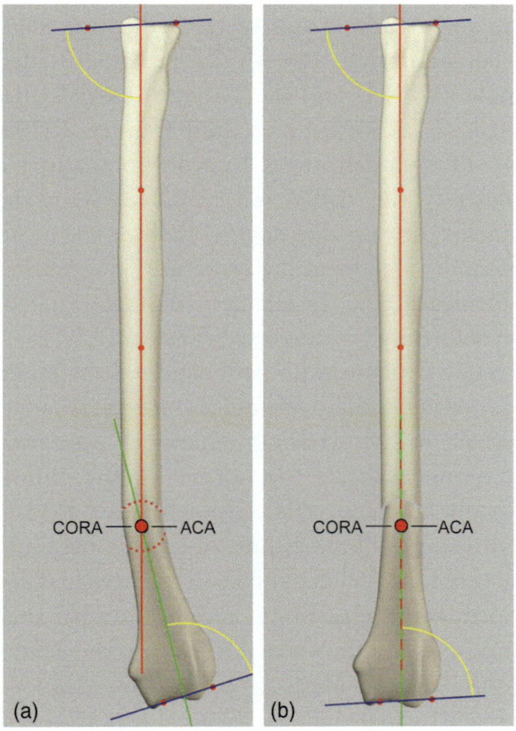

Figure 22.18 The CORA and Angular Correction Axis (ACA) are situated at the same location in this example. The red dotted line represents the potential radial osteotomy positions proximal and distal to the CORA and ACA (a). Following radial osteotomy proximal to the CORA and ACA and angular correction, the distal radial anatomic axis (dashed green line) is aligned with the proximal radial anatomic axis (red line), indicating that anatomic alignment has been restored (b). Alternatively, the osteotomy could have been performed on the red dotted line distal to the CORA and ACA (see Figure 22.18a).

A closing wedge, opening wedge, or radial osteotomy can be performed. Each has advantages and disadvantages; for instance, the opening wedge increases limb length at the expense of stability, while the closing wedge provides a more stable construct with shortening of the bone. A radial osteotomy can be utilized close to joints as the osteotomy location and Angular Correction Axis (ACA) are at different locations (Figure 22.18); however, torsional deformity can be less readily correct with a radial osteotomy as compared to a linear osteotomy since the bone surfaces are flat in the later method. Stabilization can be performed utilizing bone plates and screws, linear external fixation, or circular external fixation.

22.4.4.1 Methods of deformity assessment

Deformity can exist in any plane or any combination of planes. In reality, the deformity typically exists in a single plane that is oblique to the traditional frontal, sagittal, and transverse planes. However, imaging is rarely performed in the actual plane of the deformity. The types of deformities that commonly occur include angulation, translation, torsion, and limb length discrepancy. The imaging modality and method best suited to quantify the deformity vary depending on the deformity type, location, and complexity, i.e., the combination of co-deformities that exist. The opacity of the three-dimensional model that most easily allows identification of anatomic landmarks depends on the bone or bones of interest. In the case of antebrachial growth deformity and tibial deformity, a model that simulates a radiographic view is used, while in the case of femoral deformity, an opaque model is best.

22.4.4.2 Antebrachial growth deformity

The common constellation of deformities present in cases of antebrachial growth deformity includes distal radial valgus, excessive radial procurvatum, external radial torsion, ulnar shortening, retraction of the ulnar styloid process, and distal humero-ulnar subluxation. Application of the CORA methodology to quantify the deformity is hindered by the presence of excessive procurvatum and torsion since the proximal and distal radial articular surfaces commonly cannot be accurately identified simultaneously in a single image, such as a frontal (coronal) plane view of a three-dimensional reconstruction, or a cranio-caudal radiographic view (Figure 22.19a). Creating a three-dimensional volumetric reconstruction and performing the measurements *in silico* can overcome this difficulty.

The patient is placed in dorsal or ventral recumbency, with the forelimbs extended. Transverse, 1-mm slices are obtained from the distal humerus to the carpus. A three-dimensional volumetric reconstruction of the transverse images is created, and the adjacent humerus and carpal bones are deleted from the volume using crop and/or scissor tools, creating an unobstructed reconstruction of the radius and ulna. The radius and ulna are positioned in a frontal plane view. The distal end of the volume is rotated cranially ("out of the screen") until the distal radial articular surface can be identified. The articular surfaces of the radius are best seen if the contrast and brightness of the volume are adjusted to simulate a radiographic view. The distal radial joint reference line is drawn. The volume is rotated back into the frontal plane view, and the proximal joint reference line and anatomic axis of the radius are identified and drawn. The Medial Proximal Radial Angle (MPRA) and the Lateral Distal Radial Angle (LDRA) are measured (Figure 22.19b) to confirm a deformity. In order to quantify a deformity, the MPRA and LDRA are set to the value of the opposite normal limb or the mean reported value from the literature. The CORA is identified, and the location and magnitude of the deformity are measured (Figure 22.19c).

Sagittal plane alignment can be determined by rotating the model into a position simulating a mediolateral radiographic view. Since procurvatum exists in the normal canine radius, a single anatomic axis cannot be used for the entire length of the radius. In the normal limb, the proximal and distal joint reference lines are drawn, and the proximal and distal radial anatomic axes are drawn by identifying the centre of the radial diaphysis at two to three locations in the proximal and distal segments of the bone (Figure 22.16b). The Proximal Cranial Radial Angle (PCRA) is identified at the intersection of the proximal joint reference line and the proximal radial anatomic axis. The Distal Caudal Radial Angle (DCRA) is identified at the intersection of the distal joint reference line and the distal radial anatomic axis. The Sagittal Plane Alignment (SPA) is calculated at the intersection of the proximal and distal anatomic axes.

In the case of a sagittal plane deformity, the proximal and distal joint reference lines are drawn. If significant valgus or varus deformity obscures the identification of the articular surfaces, the three-dimensional model can

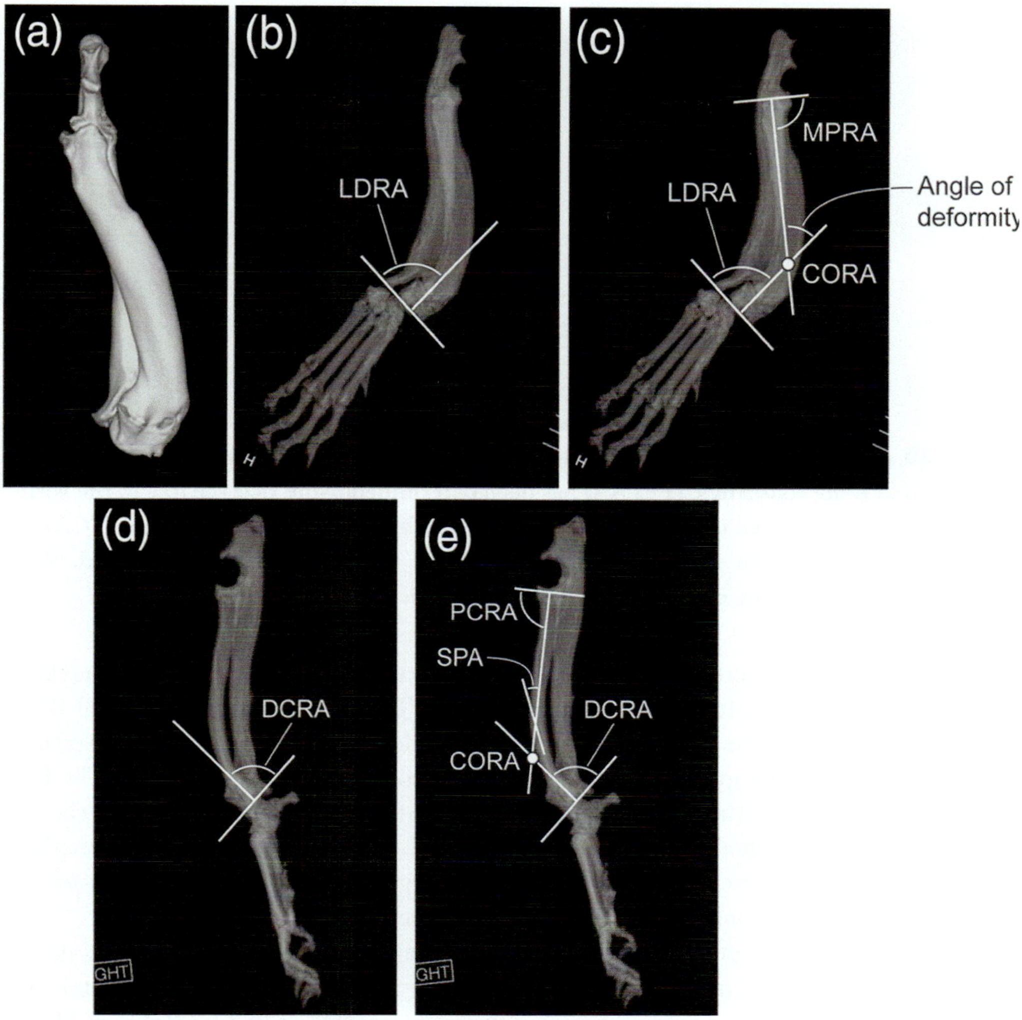

Figure 22.19 (a) A three-dimensional reconstruction of the radius and ulna positioned in the cranio-caudal view. Note that neither the proximal or distal articular surfaces of the radius can be distinguished, due to the radial valgus, torsion, and excessive procurvatum. (b) Frontal plane view of the radius and ulna; note that the contrast and brightness have been adjusted to create a "radiograph"-like appearance. The distal radial joint reference line has been drawn with the image positioned to optimize the observation of the distal articular surface. The Lateral Distal Radial Angle (LDRA) has been set to the value of the opposite normal limb, the distal radial anatomic axis has been drawn, and the model has been repositioned so that the proximal articular surface can be observed. (c) The proximal joint reference line has been drawn, the proximal anatomic axis has been drawn using the Medial Proximal Radial Angle (MPRA) from the opposite normal leg, and the CORA and angle of deformity have been identified. (d) Sagittal plane view of the radius and ulna. The distal radial joint reference line has been drawn with the image positioned to optimize identified of the distal articular surface. The Distal Caudal Radial Angle (DCRA) has been set to the value of the opposite normal limb, the distal radial anatomic axis has been drawn, and the model has been repositioned so that the proximal articular surface can be observed. (e) The proximal joint reference line has been drawn, and the proximal anatomic axis has been drawn using the Proximal Cranial Radial Angle (PCRA) from the opposite normal leg. The sagittal plane alignment is set to the value of the opposite normal limb, and the CORA has been identified. The magnitude of the deformity can be measured at the CORA.

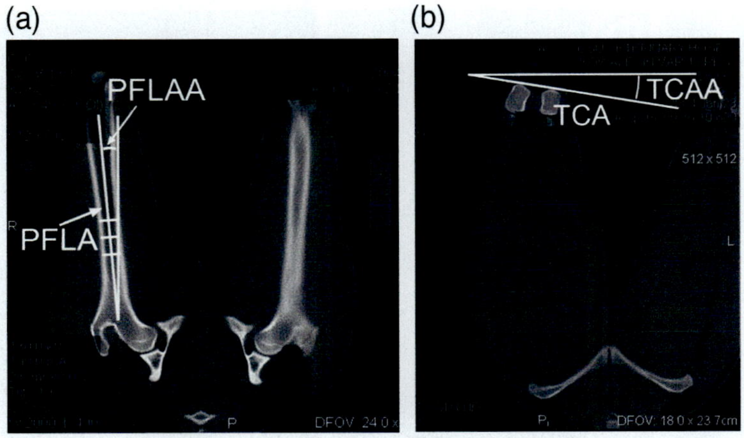

Figure 22.20 **(a) The Proximal Femoral Long Axis (PFLA) is identified on this CT slice, and the Proximal Femoral Long Axis Angle (PFLAA) has been determined. Note that the line in the vertical plane is the reference line used to determine the axis angle. (b) The Transcondylar Axis (TCA) is identified on this CT slice, and the Transcondylar Axis Angle (TCAA) has been determined. Note that the line in the horizontal plane is the reference line used to determine the axis angle.**

be rotated in or out of the screen as needed to allow the axes to be drawn. It is important to avoid rotating the model in the cranial or caudal direction during these manipulations, as this will alter the orientation of the axes once they have been drawn. The proximal and distal radial anatomic axes identified by setting the PCRA and DCRA to the normal value, and the SPA is determined at the intersection of the proximal and distal anatomic axes (Figure 22.19d and e).

22.4.4.3 Femoral deformity

Angular deformities can be quantified using individual CT slices of the limb, as described by Dudley. This method utilizes frontal plane slices of the femur to quantify the deformity. The proximal femoral anatomic axis is identified in the proximal femur, and a vertical reference line is used to account for angulation of the femur from true vertical (Figure 22.20). The Proximal Femoral Long Axis Angle (PFLAA) is recorded; an abaxial deviation of the proximal femur is considered a positive value, and an axial deviation is considered a negative value. The Transcondylar Axis (TCA) is identified in the distal limb, and a horizontal reference line is used to account for angulation of the condyle from true vertical. The Transcondylar Axis Angle (TCAA) is recorded; an axial deviation is considered a positive value, and abaxial deviation is considered a negative value. The Femoral Varus Angle (FVA) is calculated as the sum of the PFLAA and the TCAA. This method has been largely replaced by three-dimensional volumetric reconstruction.

Quantification of femoral varus deformity using a three-dimensional volumetric reconstruction was described by Robb and has been shown to be more accurate and precise than the radiographic method described by Tomlinson, although both remain clinically acceptable. The benefit of three-dimensional volumetric reconstruction is that the image (volume) can be rapidly manipulated into the appropriate position to simulate the analogous radiographic views.

To obtain a smooth, three-dimensional model, 1-mm-thick slices are preferred. The patient is positioned in dorsal recumbency with the hind limbs extended. In typical cases, the imaging is initiated just cranial to the femoral head (distal ilium) and continued to the tarsus if the deformity is confined to the femur and tibia, or phalanges, if hock or metatarsal deformity is present. Transverse CT slices are reconstructed

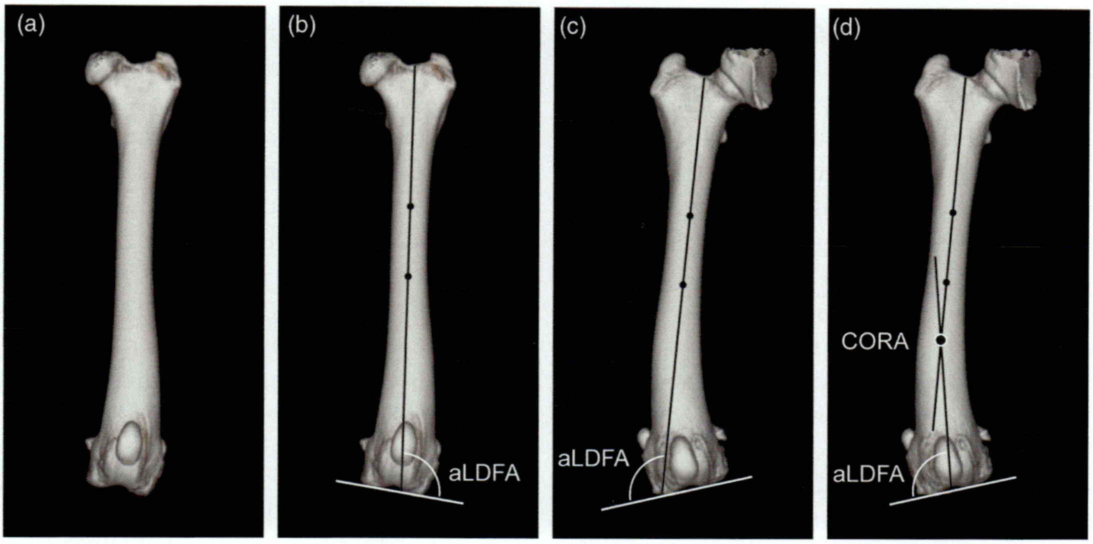

Figure 22.21 (a) The opaque model of the femur used for femoral deformity assessment. (b) The Femoral Anatomic Axis (FAA), distal joint reference line, and anatomic Lateral Distal Femoral (aLDFA) angle are determined in a normal femur; note that the FAA traverses along the lateral edge of the intercondylar notch adjacent to the medial edge of the lateral femoral condyle. (c) The distal joint reference line and anatomic Lateral Distal Femoral (aLDFA) angle are determined in an abnormal femur; note that the aLDFA is greater than normal. (d) The aLDFA has been set to the value of the opposite normal limb (b), the distal femoral anatomic axis has been determined, the proximal femoral anatomic axis has been identified, and the CORA has been determined.
The magnitude of the deformity can be measured at the CORA.

into an opaque three-dimensional model. It is not critical that the slices are precise transverse images; in fact, any imaging plane can be used. The adjacent acetabulum and tibia are cropped from the model, leaving only the femur, patella, and fabella in the volume. The femoral volume is positioned to simulate a well-positioned cranio-caudal radiographic view, with the fabella bisected by the adjacent femoral cortex, the lesser trochanter partially visible (about 50% visible), and the patella centred on the femur if it is not luxated. Since the axis utilized in locating

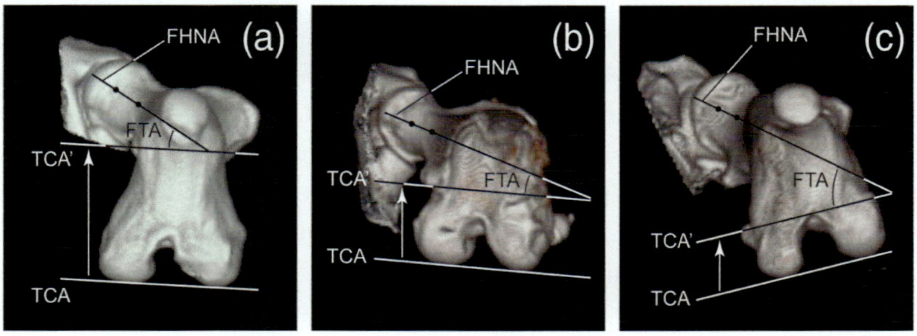

Figure 22.22 (a) Three-dimensional model of a normal femur positioned in the axial view. The Femoral Head And Neck axis (FHNA) and TCA have been identified. The Femoral Torsion Angle (FTA) can be measured at the intersection of the FHNA and TCA' (the TCA has been moved upward within the image to create the TCA'; this have been done so that the intersection of the lines is located within the image box); the angle is approximately 27°. Three-dimensional models of femora with external (b; FTA less than 27°) and internal (c; FTA greater than 27°) femoral torsion.

and quantifying the deformity in the femur can be identified from the bone surface, and an opaque model is typically used (Figure 22.21a).

The femoral anatomic axis (FAA) is determined by identifying the centre of the femoral diaphysis at 33% and 50% of the femoral length, measuring from proximal to distal. The FAA connects these two points. Owing to the slight femoral varus in normal dogs, the FAA typically traverses along the lateral edge of the intercondylar notch, adjacent to the medial edge of the lateral femoral condyle. The magnitude of femoral varus (or valgus) is determined by measuring the anatomic Lateral Distal Femoral Angle (aLDFA) at the intersection of the FAA and the distal joint reference line of the femur, similar to the radiographic method described by Tomlinson (Figure 22.21b). In a femur with pathologic femoral varus or valgus, the aLDFA is not within the range of normal values (Figure 22.21c). To quantify the deformity, the proximal FAA is determined as before, but it is not drawn to the distal extent of the bone, and the aLDFA is set to the breed-specific normal value or the value of the opposite normal limb to identify the distal FAA (Tomlinson et al., 2007). The femoral deformity is determined at the Centre Of Rotation of Angulation (CORA) located at the intersection of the proximal and distal anatomical axes of the femur (Figure 22.21d).

Femoral torsion can be quantified from the axial view of the femur. To obtain the axial view, the distal end of the volume is rotated cranially ("out of the screen") until the caudal articular surface of the femoral condyles is visible; the femoral head and neck should remain visible with the femur in this position (Figure 22.22a–c). The femoral torsion angle (anteversion angle) is determined by the intersection of the TCA and an axis drawn through the centre of the femoral head and neck, the Femoral Head and Neck Axis (FHNA). It should be noted that this method quantifies femoral head and neck anteversion and femoral diaphyseal torsion simultaneously. The reported range for the normal anteversion angle is quite broad and varies between studies; no breed-specific normal values are available at this time. The mean anteversion angle reported in one study was 27° (range 12°–40°), and this is the normal value used by the author (Figure 22.22a). The decision to treat must be made on a case-by-case basis; for example, in the case of medial patellar luxation, if the patient's femoral torsion angle is less than 27°, torsional correction during corrective osteotomy is considered (Figure 22.22b). If the patient's femoral torsion angle is greater than 27°, correction is not performed as this may exacerbate medial patellar luxation; however, it

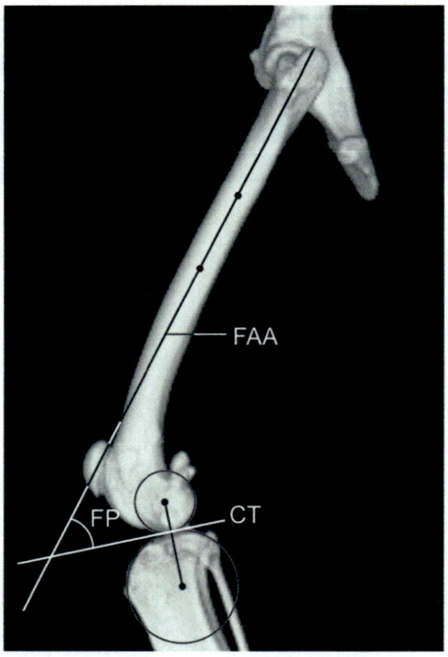

Figure 22.23 **Quantification of femoral procurvatum using the common tangent between the femoral and tibial articular surfaces at their contact point. The centres of circles corresponding to the femoral and tibial articular surfaces have been drawn, the centres of the circles have been connected with a line, and a perpendicular representing the common tangent has been drawn. The angle between the proximal FAA and the common tangent (CT) is measured to determine Femoral Procurvatum (FP).**

is corrected in cases of lateral patellar luxation (Figure 22.22c).

Femoral procurvatum is difficult to measure since no reliable landmarks exist in the distal femur, particularly in the presence of femoral deformity. The common tangent between the femoral and tibial surfaces at their contact point, as reported for determination of the patellar tendon angle in tibial tuberosity advancement (TTA) surgical planning, offers a reliable distal femoral axis that is relatively unaffected by slight changes in the femoro-tibial joint angle. Determining the centre of the femoral diaphysis at two to three points in the proximal femur, and connecting these points with a line identifies the proximal femoral anatomic axis. Drawing circles that correspond to the articular surfaces of the femoral condyles and the tibial plateau can be used to determine the common tangent of the femoro-tibial contact point. The centre of the circles is connected with a line, and a perpendicular to this line is drawn at the femoro-tibial contact point; this is the common tangent. The orientation of the proximal FAA, and the Common Tangent (CT) can be used to quantify Femoral Procurvatum (FP, Figure 22.23).

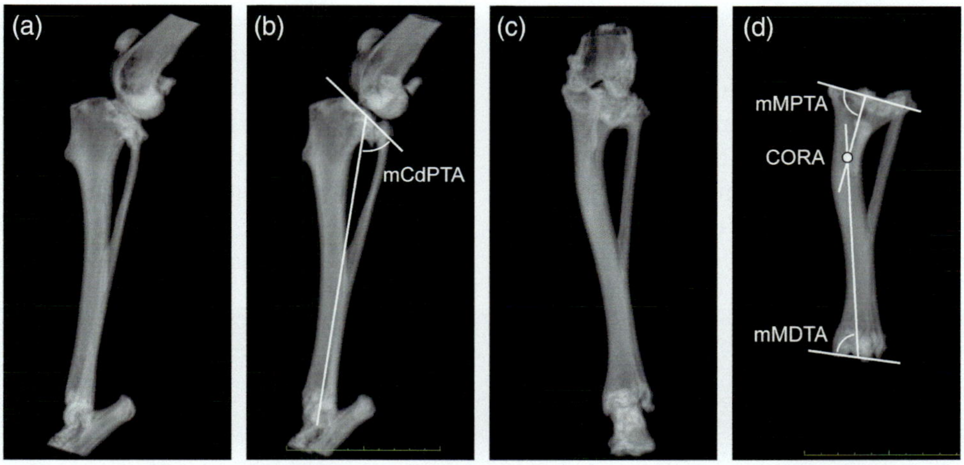

Figure 22.24 **Sagittal view of the three-dimensional model of the tibia of a patient with proximal tibial valgus and excessive tibial plateau angle; note that the oblique nature of the proximal tibial articular surface obscures the anatomic details of the region (a). (b) The model has been repositioned on the screen by rotating the proximal tibia medially ("out of the screen"); note that the proximal tibial articular surface is now clearly visible. The proximal joint reference line and tibial mechanical axis have been drawn, and the mCdPTA has been identified. Note that the mCdPTA is high (excessive tibial plateau angle). (c) Frontal plane view of the tibial model; note the proximal tibial valgus and that the proximal tibial articular surface is obscured due to the excessive tibial plateau angle. (d) The model has been repositioned on the screen by rotating the proximal tibia cranially ("out of the screen"); note that the proximal tibial articular surface is now clearly visible. The proximal joint reference line has been drawn, the mechanical Medial Proximal Tibial Angle (mMPTA) has been set to the value of the opposite normal limb, and the proximal mechanical axis has been determined. The distal joint reference line has been drawn, the mechanical Medial Distal Tibial Angle (mMDTA) has been set to the value of the opposite normal limb, and the distal mechanical axis has been determined. The CORA is located at the intersection of the proximal and distal mechanical axes; its magnitude can be determined at this location.**

22.4.4.4 Tibial deformity

Excessive tibial plateau angle, varus, valgus, and torsion are the most common tibial deformities. The tibial plateau angle can be readily quantified from radiographic images; however, when angular or torsional deformity is present, CT is the preferred imaging modality. The patient is positioned in dorsal recumbency, and transverse, 1-mm slices are obtained from the distal femur to the distal tarsus. If deformity of the femur or tarsus is suspected, imaging is initiated cranial to the femoral head (distal ilium) and continued until the phalanges.

A three-dimensional model is created from the transverse images, and the femur and tarsal bones are removed using crop and scissor tools. The distal articular surface of the tibia must be preserved to accurately locate the tibial mechanical axis. Since intra-articular anatomic landmarks are used to perform measurements of the tibia, a three-dimensional model with opacity similar to that of a radiographic image is best.

The tibial plateau angle is the primary angle of interest in the sagittal plane. The three-dimensional model is positioned in a sagittal view, and the proximal joint reference line (tibial plateau axis) and tibial mechanical axis (tibial long axis) are drawn. The mechanical Caudal Proximal Tibial Angle (mCdPTA) or the tibial plateau angle are measured and compared to the mean reported angles. In the case of significant proximal tibial varus or valgus, the proximal end of the model can be rotated laterally or medially (in or out of the screen) until a skyline view of the tibial plateau is obtained (Figure 22.24a and b). The mechanical Cranial Distal Tibial Angle (mCrDTA) can also be measured, but this is of limited clinical value.

The three-dimensional model is positioned in a frontal plane view, simulating a cranio-caudal radiographic projection. The proximal and distal joint reference lines are drawn, and the tibial mechanical axis is drawn connecting the centre of the proximal and distal articular surfaces. Since the proximal tibial articular surface is angled from proximal to distal, the greater the tibial plateau angle, the higher obscurity of the articular surface in a true frontal plane view (Figure 22.24c). Rotating the proximal end of

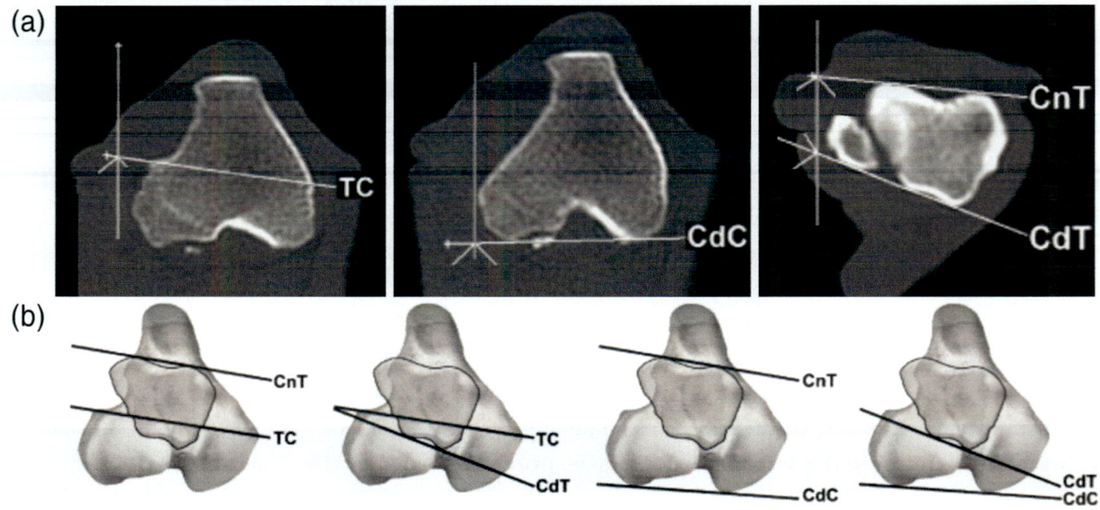

Figure 22.25 Proximal and distal tibial axes identified on representative computed tomographic images. Transcondylar axis (TC), caudal condylar axis (CdC), distal cranial axis (CnT), and distal caudal tibial (CdT) axis. Note that the lines in a sagittal plane are the reference lines used to determine the axis angle (a). (From Aper et al., 2005.) (b) Graphic representation of each axis pair; note that the distal articular surface is outlined in black and is superimposed over the proximal articular surface. TC, transcondylar axis; CdC, caudal condylar axis; CnT, cranial tibial axis, and CdT distal caudal tibial axis. (From Aper et al., 2005.)

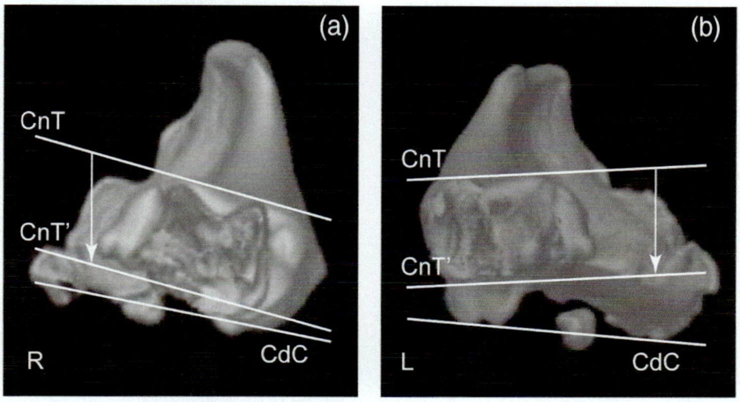

Figure 22.26 **Disto-proximal view of the three-dimensional model of the right (a) and left (b) tibia of the same patient, as in Figure 22.24. The caudal condylar axis (CdC) and cranial tibial axis (CnT) have been identified, and tibial torsion can be quantified by measuring the angle between CnT' and CdC. Note that there is no clinically significant difference in tibial torsion between the right (normal) and left (affected) limbs.**

the model cranially ("out of the screen") will allow a more tangential view of the articular surface, increasing the accuracy of the varus or valgus measurement (Figure 22.24d). Obtaining a tangential view of the tibial plateau is likely important to accurately place the proximal joint reference line in all cases but is particularly important in cases of an excessive tibial plateau angle (Figure 22.24). The mechanical Medial Proximal Tibial Angle (mMPTA) is measured at the intersection of the proximal joint reference line and the mechanical axis, and the mechanical Medial Distal Tibial Angle (mMDTA) is measured at the intersection of the distal joint reference line and the mechanical axis. These values are compared to published normal values or the values of the unaffected opposite limb to determine if significant varus or valgus is present. In the case of tibial deformity, the proximal and distal joint reference lines are drawn, and the proximal and distal mechanical axes are determined by setting the mMPTA and the mMDTA to the breed-specific value or that of the normal opposite limb (Figure 22.24d). The CORA is located at the intersection of the proximal and distal mechanical axes, and the magnitude of the CORA can be measured at this location.

Tibial torsion can be quantified using individual proximal and distal transverse slices, as described by Aper. In this method, the relative orientation of a proximal and distal tibial axis is used to determine the magnitude of tibial torsion. The proximal axes that have been described are the transcondylar (TC) and caudal condylar (CdC) axes; the distal axes are the cranial tibial (CnT) and the caudal tibial (CdT) axes (Figure 22.25). This method of tibial torsion determination has been shown to be more accurate and less subject to positioning artefact than the method described by Slocum. The same axes and methodology can be applied to a three-dimensional model rather than individual slices (Figure 22.26).

22.4.4.5 Tarsal deformity

Metatarsal rotation or rotation of the pes is a relatively uncommon disorder of large- and giant-breed dogs. Although the deformity appears to affect the pes, the level of the deformity is typically the proximal intertarsal joint. Quantification of the torsional deformity can be performed using an axis drawn along the cranial surface of the distal tibia and metatarsal bones (Figure 22.27). The deformity is commonly bilateral, and no reference ranges for normal angles are currently available. The correction entails de-rotation at the level of the proximal intertarsal joint, and external skeletal fixation or internal fixation is utilized to stabilize the proximal intertarsal arthrodesis.

CT imaging of deformity correction is more rapid and accurate than standard radiography owing to the infinite ability to modify the position of the three-dimensional model. Although the normal joint reference angles should not differ considerably from those reported for radiographic imaging, breed-specific normal ranges based on three-dimensional CT reconstruction would be ideal to optimize the accuracy of deformity planning.

22.4.5 Miscellaneous

Panosteitis is a disease of the medullary bone affecting rapidly growing large- and giant-breed dogs, typically between 5 and 12 months of age, although it has been reported in 5-year-old dogs. Some studies have found that German shepherd dogs and male dogs are overrepresented. It may occur in more than one bone at a time or may spread, causing lameness to shift from one bone or leg to another. Radiographic signs of panosteitis may not appear for days to weeks behind clinical signs. The earliest radiographic sign of panosteitis is a decrease in opacity around the nutrient foramen. Later signs include an increased density of long bones within the medullary canal and loss of the normal trabecular bone pattern (Figure 22.28). Periosteal and endosteal new bone may also be seen in more severe cases (Figure 22.29). Radiographs of the affected limb may be compared with those of the contralateral limb to assist in diagnosis, and lesions may be found in several long bones within the same animal.

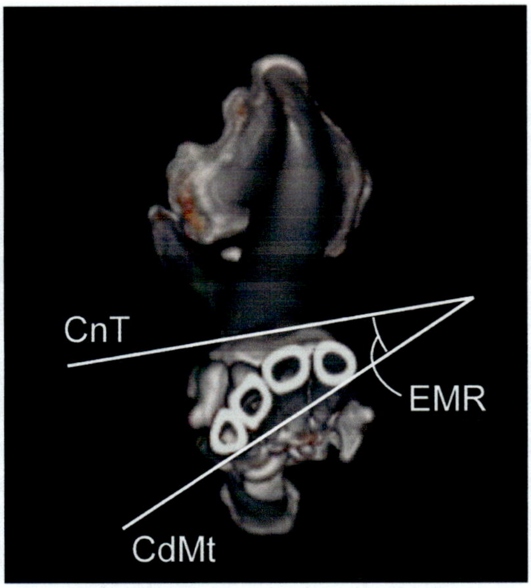

Figure 22.27 **Disto-proximal view of the three-dimensional model of a patient with metatarsal rotation. The distal cranial tibial axis (CnT) and a Caudal Metatarsal axis (CdMt) have been drawn; there is approximately 35° of External Metatarsal Rotation (EMR).**

Hypertrophic osteodystrophy (HOD), although uncommon, typically affects young growing large- and giant-breed dogs. The disease is also known as canine metaphyseal osteopathy and has an unknown multifactorial

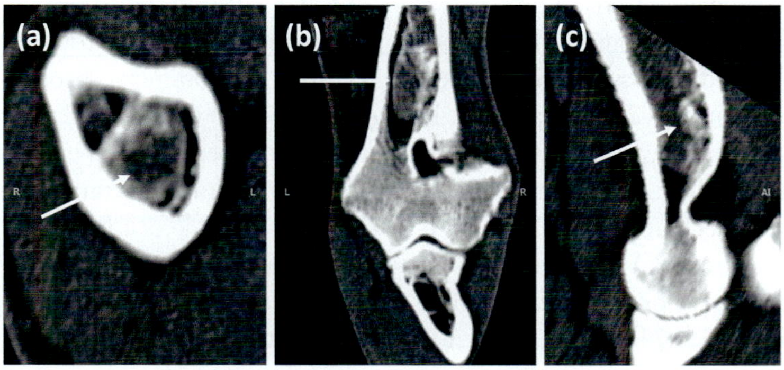

Figure 22.28 **Transverse image (a) and dorsal (b) and sagittal (c) reconstructions of a case of panosteitis in the distal part of the humerus. Notice the increase in density within the medullary canal and loss of the normal trabecular bone pattern (arrows).**

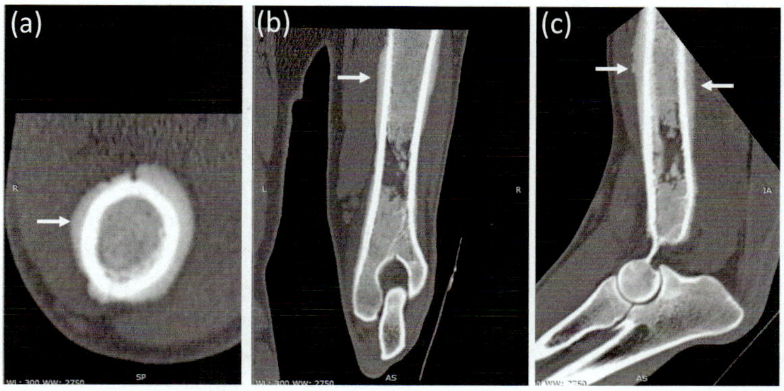

Figure 22.29 Transverse image (a) and dorsal (b) and sagittal (c) reconstructions of a fulminant case of panosteitis in the humerus. Notice the increase in density within the medullary canal associated with periosteal and endosteal new bone formation (arrows).

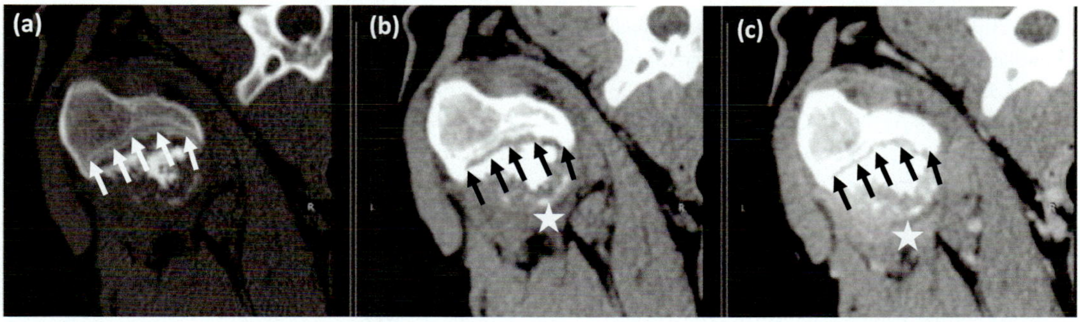

Figure 22.30 Transverse image in bone (a), soft tissue (b), and after IV contrast administration (c) of a case of Hypertrophic Osteo Dystrophy (HOD). Notice the moth-eaten lysis (arrows) along the proximal humeral physis interspaced with areas of hyperattenuating bone. There is swelling of the soft tissues (b) enhancing after IV contrast administration (asterisk). There were several bones involved.

pathogenesis. Many underlying aetiologies have been suggested, including vaccine associations, nutritional deficiencies, microorganisms, and genetic predispositions. Radiographic appearance of lesions is typically of a periosteal hyperplasia and calcification, creating a "moth-eaten" appearance. Lesions are usually symmetric and may occur on any long bone metaphysis; however, other locations have been reported, including the mandible, cranium, and ribs. A line of reduced opacity adjacent to the physis can be seen as the predominant finding; this has been described as a double physeal sign. The physis itself appears normal, and mild metaphyseal flaring may be present. Swelling of the soft tissues is not uncommon (Figure 22.30).

FURTHER READING

Apelt D, Kowaleski MP, Dyce J. 2005. Comparison of computed tomographic and standard radiographic determination of tibial torsion in the dog. *Veterinary Surgery* 34(5):457–462.

Aper R, Kowaleski MP, Apelt D et al. 2005. Computed tomographic determination of tibial torsion in the dog. *Veterinary Radiology and Ultrasound* 46 (3):187–191.

Avery DM, Matullo KS. 2014. Distal radial traction radiographs: interobserver and intraobserver reliability compared with computed tomography. *Journal of Bone Joint Surgery* 96:582–588.

Bergh MS. 2015. Panosteitis. *Orthopedics* 13:10–11. Accessed 28 Apr 2020 from https://www.cliniciansbrief.com/article/panosteitis

Borrelli J Jr, Goldfarb C, Catalano L, Evanoff BA. 2002. Assessment of articular fragment displacement in acetabular fractures: a comparison of computerized tomography and plain radiographs. *Journal of Orthopaedic Trauma* 16(7):449–456; discussion 456–457.

Bruinsma WE, Guitton TG, Warner JJ, Ring D. 2013. Interobserver reliability of classification and characterization of proximal humeral fractures: a comparison of two and three-dimensional CT. *Journal of Bone Joint Surgery* 95:1600–1604.

Butler D, Nemanic S, Warnock JJ. 2018. Comparison of radiography and computed tomography to evaluate fractures of the canine tarsus. Vet Radiol Ultrasound. 59:43–53.

Chun R. 2005. Common malignant musculoskeletal neoplasms of dogs and cats. *Veterinary Clinics Small Animal Practice* 35:1155–1167.

Crosse KR, Worth AJ. 2010. Computer-assisted surgical correction of an antebrachial deformity in a dog. *Veterinary and Comparative Orthopaedics and Traumatology* 23(5):354–361.

Cunningham B, Jackson K, Ortega G. 2014. Intraoperative CT in the assessment of posterior wall acetabular fracture stability. *Orthopedics* 37:328–331.

Dismukes DI, Tomlinson JL et al. 2007. Radiographic measurement of the proximal and distal mechanical joint angles in canine tibia. *Veterinary Surgery* 36(7):699–704.

Dismukes DI, Tomlinson JL et al. 2008. Radiographic measurement of canine tibial angles in the sagittal plane. *Veterinary Surgery* 37(3):300–305.

Dudley RM, Kowaleski MP, Drost WT et al. 2006. Radiographic and computed tomographic determination of femoral varus and torsion in the dog. *Veterinary Radiology and Ultrasound* 47(6):546–552.

Fox DB, Tomlinson JL et al. 2006. Principles of uniapical and biapical radial deformity correction using dome osteotomies and the center of rotation of angulation methodology in dogs. *Veterinary Surgery* 35(1):67–77.

Hill M, Scudder CJ, Glanemann B, Drees R. 2015. Hypertrophic osteodystrophy in a dog imaged with CT. *Veterinary Record Case Reports* 3:1–4.

Hoffman DE, Kowaleski MP, Johnson KA et al. 2011. Ex vivo biomechanical evaluation of the canine CrCL deficient stifle with varying angles of stifle joint flexion and axial loads after TTA. *Veterinary Surgery* 40(3):311–320.

Katsura M, Sato J, Akahane M, Kunimatsu A, Abe O. 2018. Current and novel techniques for metal artifact reduction at CT: practical guide for radiologists. *RadioGraphics* 38(2):450–461.

Karnik KS, Samii VF, Weisbrode SE, London CA, Green EM. 2012. Accuracy of computed tomography in determining lesion size in canine appendicular osteosarcoma. *Veterinary Radiology and Ultrasound* 53(3):273–279.

Kroner K, Cooley K, Hoey S, Hetzel SJ, Bleedorn JA. 2017. Assessment of radial torsion using computed tomography in dogs with and without antebrachial limb deformity. *Veterinary Surgery* 46(1):24–31.

Lambert R, Wendelburg K. 2010. Determination of the mechanical medial proximal tibial angle using a tangential radiographic technique. *Veterinary Surgery* 39(2):181–186.

Lusetti F, Bonardi A, Eid C, Brandstetter de Belesini A, Martini FM. 2017. Pelvic limb alignment measured by computed tomography in purebred English Bulldogs with medial patellar luxation. *Veterinary and Comparative Orthopaedics and Traumatology* 30(3):200–208.

Mateusz C, Florkow MC, Koen Willemsen K, Mascarenhas V, Oei E, van Stralen M, Seevinck PR. 2022. Magnetic resonance imaging versus computed tomography for three-dimensional bone imaging of musculoskeletal pathologies: a review. *Journal of Magnetic Resonance Imaging* 56(1):11–34.

Meola SD, Wheeler JL, Rist CL. 2008. Validation of a technique to assess radial torsion in the presence of procurvatum and valgus deformity using computed tomography: a cadaveric study. *Veterinary Surgery* 37(6):525–529.

Nakanishi Y, Omokawa S, Shimizu T, Nakano K, Kira T, Tanaka Y. 2013. Intra-articular distal radius fractures involving the distal radioulnar joint (DRUJ): three dimensional computed tomography-based classification. *Journal of Orthopaedic Science* 18:788–792.

Newman M, Voss K. 2017. Computed tomographic evaluation of femoral and tibial conformation in English Staffordshire Bull Terriers with and without congenital medial patellar luxation. *Veterinary and Comparative Orthopaedics and Traumatology* 30(3):191–199.

Oblak ML, Boston SE, Woods JP, Nykamp S. 2015. Comparison of concurrent imaging modalities for staging of dogs with appendicular primary bone tumours. *Veterinary and Comparative Oncology* 13(1):28–39.

Paley D and Herzenberg JE. 2002. (ed) *Principles of Deformity Correction*. Berlin, Springer-Verlag.

Pineda C, Rolando Espinosa R, Pena A. 2009. Radiographic imaging in osteomyelitis: the role of plain radiography, computed tomography, ultrasonography, magnetic resonance imaging, and scintigraphy. *Seminars in Plastic Surgery* 23(2):80–88.

Scheyerer MJ, Pietsch C, Zimmermann SM, Osterhoff G, Simmen HP, Werner CM. 2014. SPECT/CT for imaging of the spine and pelvis in clinical routine: a physician's perspective of the adoption of SPECT/CT in a clinical setting with a focus on trauma surgery. *European Journal of Nuclear Medicine and Molecular Imaging* 41(Suppl 1):59–66.

Shinji M, Takanobu O. 2017. Practical use of imaging technique for management of bone and soft tissue tumors. *Journal of Orthopaedic Science* 22(3):391–400.

Siddiqui NA, Galizia MS, Almusa E, Omar IM. 2014. Evaluation of the tarsometatarsal joint using conventional radiography, CT, and MR imaging. *Radiographics* 34:514–531.

Slocum B, Slocum TD. 1993. Tibial plateau leveling osteotomy for repair of cranial cruciate ligament rupture in the canine. *Veterinary Clinics of North America: Small Animal Practice* 23:777.

Suzuki D, Ono H, Furuta K, Katayama T, Akahane M, Omokawa S, Tanaka Y. 2014. Comparison of scapholunate distance measurements on plain radiography and computed tomography for the diagnosis of scapholunate instability associated with distal radius fracture. *Journal of Orthopaedic Science* 19:465–470.

Talbott JL, Boston SE, Milner RJ, Lejeune A, de M. Souza CH, Kow K, Bacon NJ, Hernandez JA. 2017. Retrospective evaluation of whole body computed tomography for tumor staging in dogs with primary appendicular osteosarcoma. *Veterinary Surgery* 46(1):75–80.

Tomlinson JL, Fox DB et al. 2007. Measurement of femoral angles in four dog breeds. *Veterinary Surgery* 36(6):593–596.

Willems A, Içli C, Waarsing JH, Bierma-Zeinstra SMA, Meuffels DE. 2022. Bone union assessment with computed tomography and statistical associations with mechanical or histological testing: a systematic review of animal studies. *Calcified Tissue International* 110(2):147–161.

Yasukawa S, Edamura K, Tanegashima K, Seki M, Teshima K, Asano K, Nakayama T, Hayashi K. 2016. Evaluation of bone deformities of the femur, tibia, and patella in Toy Poodles with medial patellar luxation using computed tomography. *Veterinary and Comparative Orthopaedics and Traumatology* 29(1):29–38.

Zlotorowicz M, Czubak J, Caban A, Kozinski P, Boguslawska-Walecka R. 2013. The blood supply to the femoral head after posterior fracture/dislocation of the hip, assessed by CT angiography. *The Bone and Joint Journal* 95: 1453–1457.

CHAPTER 23

CT OF THE MUSCLES

Stephan Kaiser

23.1 INTRODUCTION

The skeletal musculature makes up 50% of the body mass of dogs. Myopathy in dogs may be underestimated because of the failure to establish a definitive diagnosis, poor recognition of muscular changes when accompanied by concurrent, more severe trauma, inflammation or neoplasia and due to problems with the classification of myopathy.

23.2 INDICATIONS FOR USING CT

Magnetic resonance imaging (MRI) is the diagnostic method of choice when examining the musculature. Sonography is also used for the extremities. Both methods offer good soft part contrast. CT, however, also provides very good soft part differentiation when used in combination with an intravenously (IV) administered contrast medium. CT especially offers very good visualization of mineralized lesions, adipose tissue tumours, the blood supply and contrast medium kinetics with regard to changes in the muscles as well as very good differentiation from the surrounding areas and bone structures. Using the multi-slice technique, larger sections of the body can be examined with regard to muscular involvement in a short examination time, as, for example, in poly-traumatized patients or during tumour screening. In this way, CT provides orthopaedists and surgeons with valuable information on the extent of the changes in the musculature and adjacent body regions.

23.3 TECHNICAL ASPECTS

23.3.1 Positioning

The positioning is specific for each body part. The animal should be positioned symmetrically. The limbs should be positioned parallel and with equal angulation of the joints on both sides, if possible. A comparison with the contralateral limb allows for easier identification of lesions.

23.3.2 Technical parameters

Muscles and tendons have a narrow inherent object contrast and require viewing with a narrow window setting, which enhances noise. Therefore, a thicker slice width of 3 to 5 mm should be selected.

23.3.3 Contrast

Contrast-enhanced CT images are obtained routinely in patients with clinically documented musculoskeletal disease. The contrast medium is administered as a bolus injection of 600 mg of iodine/kg body weight with an injection rate of 2 mL/s immediately prior to the scanning sequence. For standard CT image acquisition to detect musculoskeletal disease, the contrast medium can be administered by hand or a power injector. Regardless of whether the contrast medium is administered, the use of saline flush must be considered.

DOI: 10.1201/9781315118130-25

23.4 NORMAL CT ANATOMY AND/OR ANATOMIC VARIATIONS

23.4.1 Musculature

Muscles have a homogenous soft part density of 35–55 HU. The density may change according to the varying composition of the musculature, fat, connective tissue and water. Normal muscles accumulate only little contrast medium. CT visualization of muscles changes through the various sectional planes. Due to interfascial adipose tissue, some muscle groups can be well distinguished from one another on the CT scan. Most of the blood vessels and nerves run through the interfascial adipose tissue. Lymph nodes are also located in the adipose tissue between the muscles. After administration of contrast medium, the vessels can be well distinguished from the nerves due to their contrast flow. The nerves and high contrast of blood vessels in the adipose tissue offer a very good CT visualization of the anatomical course of those tubular structures.

23.4.2 Tendon

In the sagittal plane, the tendon has a homogenous soft tissue texture. In a transverse image, the tendon appears round to ovoid. Tendon sheaths and synovial bursae appear as hypoattenuating sheaths. Especially with inflammatory processes, there is a homogenous contrast enhancement in the periphery of the enlarged hypodense soft-tissue tendon sheath (Figure 23.1).

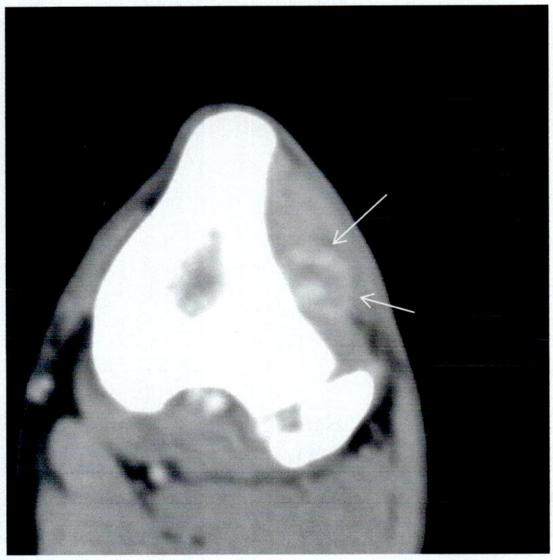

Figure 23.1 Transverse post-contrast CT image at the level of the left tibial tuberosity of a dog with a tenosynovitis of the long digital extensor tendon. There is a hypodense ring between the hyperdense tendon and the enhancement of the tendon sheath (white arrows).

23.5 PATHOLOGICAL FINDINGS

23.5.1 Trauma

23.5.1.1 Muscle

Traumatic injury lesions include contusions, laceration, strains, ischaemia, and complete ruptures of the skeletal muscle. It is estimated that a total of 30% of injuries seen in typical human sports medicine practice are acute, stretch-induced muscle injuries. Furthermore, they have been reported to be the most common injury seen in general medical practices. According to veterinary literature on small animals, muscle injuries account for only 5% of reported musculoskeletal disease. Muscle injuries in dogs may be underestimated because of the failure to establish a definitive diagnosis and poor recognition of muscular damage. The CT features of enlargement of the injured muscles, variable hypoattenuating lesions within the muscle parenchyma and nonuniform contrast enhancement of the affected muscles are similar to those CT abnormalities found in human athletes with acute muscle strains and likely correspond to intramuscular oedema. Inflammation and oedema are the primary pathologic lesions present in the acutely strained muscle, which may also be accompanied by variable degrees of haemorrhage depending on the severity of the strain (Figure 23.2). Muscular inflammation and oedema appear as hypoattenuating lesions on CT because of the different linear attenuation coefficients of water and muscle. Some forms of mild, acute muscular injuries, such as strains,

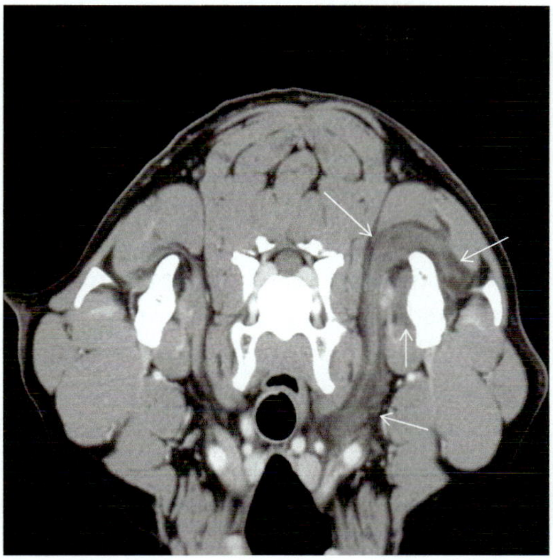

Figure 23.2 Transverse post-contrast CT image at the C6 level of a dog with an acute muscular trauma. Asymmetrical enlargement of the left pectoralis superficialis, subscapularis and supraspinatus muscle visible (arrows). The traumatic injured musculature is hypoattenuated with mild non-uniform enhancement. The hypoattenuating musculature is considered to indicate an increased fluid content, consistent with oedema and inflammation.

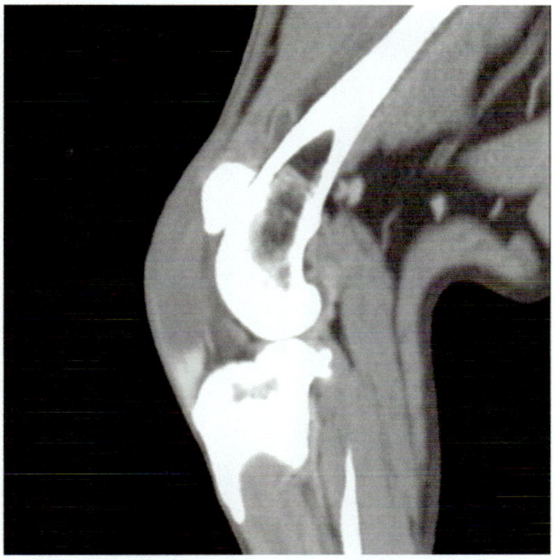

Figure 23.3 Sagittal reconstructed post-contrast CT image of the left stifle of a dog with a complete rupture of the left patella ligament. There is a hypodense area between the patella and the distal end of the tendon with homogeneous contrast medium enhancement. The hypodense area is due to the haematoma between the tendon stumps. A patella alta is also visible.

may result in submacroscopic changes within the muscle that are not readily identifiable with ultrasound or CT. A mechanical lameness may remain as a result of fibrous contraction of an injured muscle. Acute trauma or chronic repetitive trauma may be the cause for fibrotic myopathy. Muscular contracture or fibrotic myopathy is a chronic, progressive disorder of fibrosis and severe muscle contracture. A strain injury may result in the appearance of dystrophic calcification at the injury site, which can be seen on a CT scan.

23.5.1.2 Tendon

On CT, it is usually possible to differentiate a complete from a partial tendon rupture. The ruptured area is characterized by an interruption of the normal tendon. The haematoma between the tendon stumps is usually homogenous hypodense. The ends of the tendon are distended and accumulate contrast medium (Figure 23.3). If a rupture is located directly at the origin or insertion of the tendon, bone fragments may be seen as mineralized structures. In a partial rupture, most of the tendon structure is still recognizable, while other parts of the tendon have lost their typical CT appearance. Often, there is a linear enhancement with small hypoattenuating lesions in the affected enlarged tendon and the muscle-tendon junction (Figure 23.4). If the tendon has a tendon sheath, in the transversal image, a hypodense ring of fluid between the tendon and the peritendineum suggests a partial tear. In chronic diseases, the hypodense area is transformed into more unorganized, inhomogeneous, hyper- to hypodense tissue with a much larger tendon diameter compared to its original size and non-uniform enhancement pattern (Figure 23.5a and b).

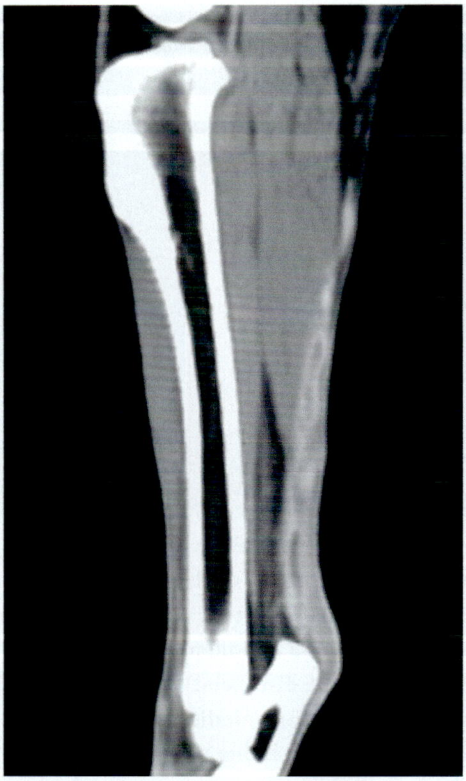

Figure 23.4 Sagittal-reconstructed post-contrast CT image of a dog with traumatic musculotendinopathy of the Achilles tendon. There is a linear enhancement with small hypoattenuating lesions in the affected enlarged Achilles tendon and muscle–tendon junction. A disruption of the normal architecture within the myotendinous unit is visible.

23.5.2 Specific traumatic muscle injuries

23.5.2.1 Myopathy of the iliopsoas muscle

The iliopsoas muscle is the strongest of the inner hip muscles. In dogs, it arises ventrally from the transverse process of the second and third lumbar vertebrae and attaches at the lesser trochanter of the femur. The nerve roots of the fourth and fifth lumbar nerves unite in the iliopsoas muscle to form the femoral nerve. The femoral nerve runs through the iliopsoas muscle, and both leave the abdominal cavity. Clinically, varying degrees of pelvic limb lameness and monoparesis may occur in dogs with iliopsoas muscle injuries. Signs of pain are usually elicited during internal rotation and extension of the affected coxofemoral joint. In dogs and cats, the psoas muscles are easily visualized on abdominal and pelvic CT scans. In non-contrast images, there is asymmetric enlargement of the affected iliopsoas muscle when compared with the unaffected side. Often, variably sized hypoattenuating lesions are also visible within the parenchyma of the affected iliopsoas muscle. There is a non-uniform contrast enhancement of the enlarged muscle (Figure 23.6a and b). Ultimately, histopathologic confirmation would be required to appropriately characterize the nature of the muscular pathology.

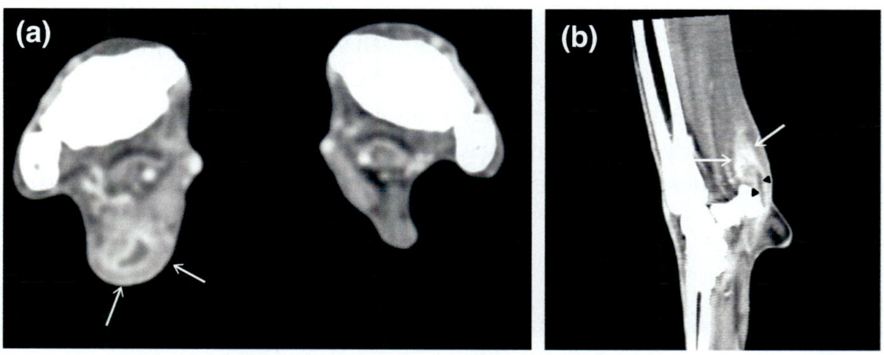

Figure 23.5 (a) Transverse post-contrast CT image at the level of the carpus of a dog with flexor carpi ulnaris tendinopathy at the level of the carpus. Note the enlargement of the right flexor ulnaris (arrow). There is also a small hypoattenuating lesion with visible ring enhancement. (b) Sagittal reconstructed post-contrast CT image of a dog with flexor carpi ulnaris tendinopathy. Note the non-uniform enhancement of the enlarged flexor ulnaris tendon at the muscle–tendon junction (arrows). There is also a linear hypoattenuating lesion (arrowhead) palmar to the accessory carpal bone visible.

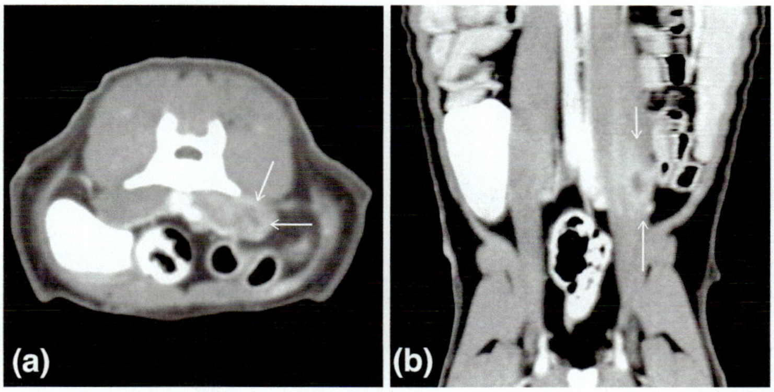

Figure 23.6 (a) Transverse post-contrast CT image of the caudal lumbar spine at L6 of a cat with traumatic injury. Note the non-uniform enhancement of the left iliopsoas musculature (arrows). Compare with the contralateral muscle. (b) Reconstructed dorsal post-contrast CT image of the sublumbar musculature. The left iliopsoas muscle is asymmetrically enlarged, and an area of hypoattenuation is seen in the muscular belly (arrows), which suggests the presence of oedema or inflammation within the muscle parenchyma.

23.5.2.2 Musculotendinopathy of the gastrocnemius muscle

The gastrocnemius muscle is divided into a lateral and a medial head, which are covered by strong tendinous leaves and infiltrated by tendinous strands. Both heads arise out of a large tendon from the lateral and medial supracondylar tuberosity of the femur and contain a large sesamoid bone that articulates with the corresponding condyle. The medial and lateral heads fuse distally, with the tendon attaching at the tuber calcanei. Total or partial avulsion of the lateral head without known prior trauma are among the injuries of gastrocnemius muscle origin in dogs. Affected dogs show high grade lameness, sometimes with a plantigrade stance. Distal placement of the corresponding sesamoid bone of the gastrocnemius head is usually apparent. Especially herding dogs such as Border Collies and Australian Shepherds are diagnosed with myopathy of the iliopsoas muscle. In non-contrast images, asymmetric enlargement of the lateral head of the gastrocnemius muscle in comparison with the medial head can be observed. Often, variably sized hypoattenuating lesions are also visible within the parenchyma of the affected head of the gastrocnemius muscle. There is a non-uniform contrast enhancement of the enlarged muscle. In some cases, a ring-enhancing lesion with a hypodense centre appears (Figure 23.7a and b). Mild changes in the clinically normal contralateral limb can also be observed. Typically, no signs of joint effusion or instability of the stifle are present. The secondary mineralizations at the sesamoid bone and in surrounding tissue (Figure 23.8) indicate a chronic component.

23.5.3 Neoplastic

23.5.3.1 Primary muscle and adipose tissue tumors

Rhabdomyomas and rhabdomyosarcomas are very rare (Figure 23.9a and b). Tumours originating from the intermuscular adipose tissue are more common. Depending on their size and extension, lipomas, liposarcomas and infiltrative lipomas may cause lameness. Due to the fact that the reported recurrence rate following surgical excision of infiltrative lipomas ranges from 36% to 50%, it is important to be able to differentiate between lipomas and infiltrative lipomas. CT provides an adequate assessment of the extent of disease. It should be considered prior to surgery or irradiation of lipomatous mass that cannot be determined from physical examination to be non-infiltrative. Furthermore,

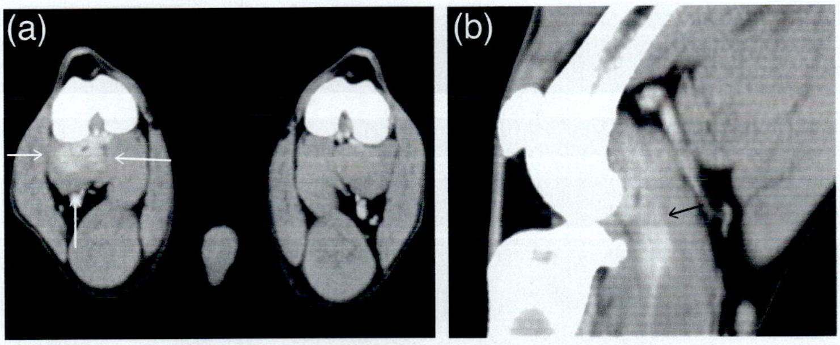

Figure 23.7 (a) Transverse post-contrast CT image at the level of the femoral condyles of a dog with a gastrocnemius musculotendinopathy. Note the increased volume of the lateral head of the gastrocnemius muscle in the right stifle (arrows). Compare with the contralateral muscle. There is a non-uniform enhancement of the lateral head of the visible gastrocnemius muscle. There is also a small hypoattenuating lesion with a ring enhancement in the affected head of the gastrocnemius muscle. (b) Sagittal reconstructed post-contrast CT image of a dog with a gastrocnemius musculotendinopathy. There is a non-uniform enhancement and also a small hypoattenuating lesion with a ring enhancement (arrow) in the affected head of the gastrocnemius muscle. Typically, there are no signs of joint effusion.

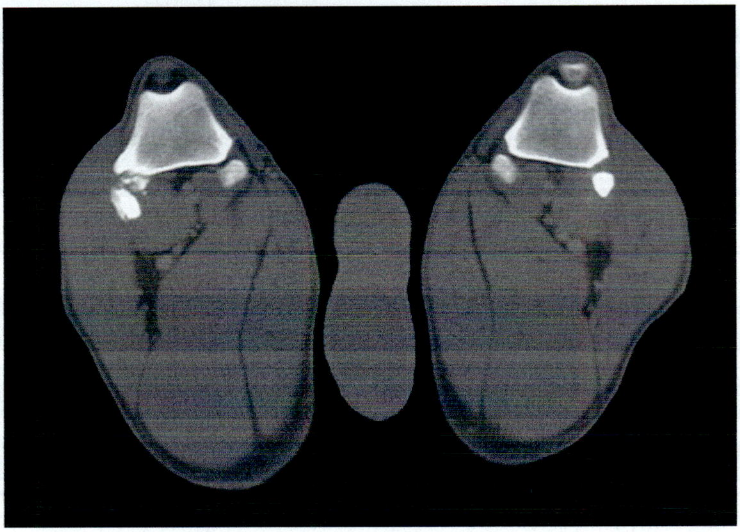

Figure 23.8 Transverse CT image at the level of the femoral trochlea of a dog with a gastrocnemius musculotendinopathy. A mineralization at the lateral sesamoid bone and in the surrounding tissue caudal to the right femur can be identified.

imaging prior to treatment will hopefully avoid euthanasia at the time of surgery due to inability to resect the tumour. Infiltrative lipomas are locally aggressive with infiltration of the surrounding normal muscle and fibrous tissue, but do not metastasize. A definitive diagnosis is made based on the histologic identification of normal-appearing adipose tissue with infiltration of the adjacent tissue. Typical locations of infiltrative lipomas are the thoracic wall, muscles of the head and neck, paraspinal musculature and the extremities. Although the value varies by the specific body location, on CT scans, lipomas appear as homogeneous hypoattenuated masses with a CT number ranging from –65 to –120 HU. Thus, it is often helpful to have a direct comparison with the attenuation of surrounding normal fat. On CT imaging, the

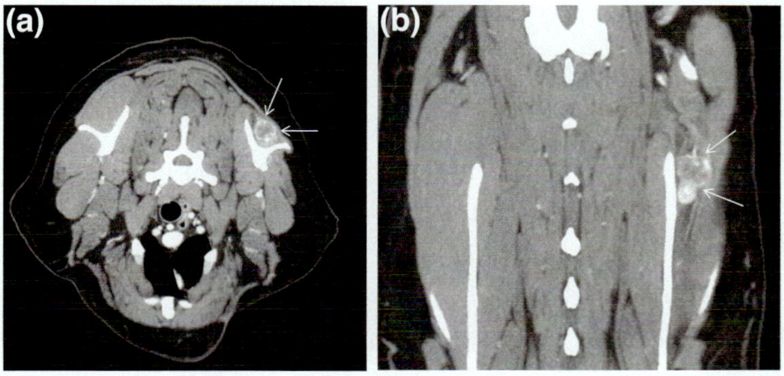

Figure 23.9 (a) Transverse post-contrast CT image of a dog with a rhabdomyosarcoma of the scapula. A small contrast-enhancing mass with ring enhancement (arrows) is identifiable in the left supraspinatus muscle. Marked atrophy of the supraspinatus and infraspinatus muscle is visible. (b) Dorsal reconstructed post-contrast CT image shows the mass (arrows) extending in in the left supraspinatus muscle.

majority of infiltrative lipomas are well-defined or moderately well-defined. Fat opacity masses infiltrating surrounding tissues can be identified using CT (Figure 23.10a and b). It is only problematic to delineate the infiltrative lipoma in portions of the CT images where the infiltrative lipoma interdigitated with normal body fat. Typically, infiltrative lipomas do not show contrast enhancement, except within the fibrous capsule. Dogs that have previously undergone surgery could benefit from contrast administration as contrast enhancement may help delineate the surgical margin for inclusion in the radiotherapy field.

23.5.3.2 Metastasis

In domestic animals, muscular metastases have been reported to be rare, with sporadically described occurrences in animals with carcinomas (pulmonary, mammary, and prostatic), lymphoma (Figure 23.11a and b) and mast cell

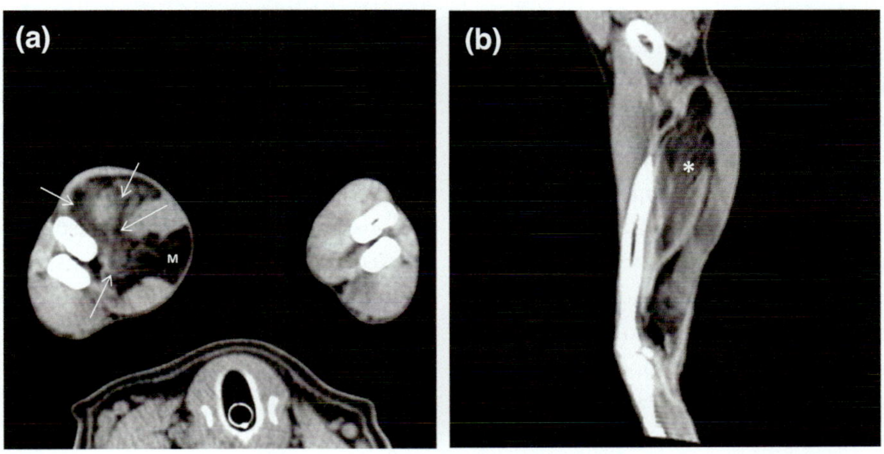

Figure 23.10 (a) Transverse post-contrast CT image of a dog with an infiltrative lipoma of the right forelimb. There is a large fat opacity mass (M) that infiltrates the surrounding muscle tissue (arrows). The mass is well-defined and moderately well-defined on the CT image. (b) Sagittal-reconstructed post-contrast CT image of a dog with a right forelimb infiltrative lipoma. The musculature of the right forelimb has fine linear striations due to infiltration of the muscle by the tumour (asterisk).

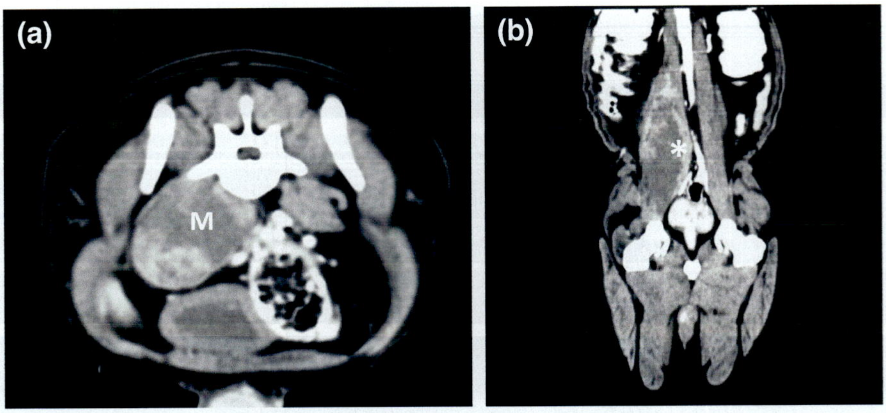

Figure 23.11 (a) Transverse post-contrast CT image of the caudal lumbar spine at L7 of a dog with an undifferentiated round cell tumour of the right iliopsoas muscle. Asymmetrical enlargement of the right iliopsoas musculature is visible (M). Note the non-uniform enhancement with an area of hypoattenuation. Compare with the contralateral muscle. (b) Reconstructed dorsal post-contrast CT image through the sublumbar musculature. The right iliopsoas muscle is enlarged, and an area of hypoattenuation (asterisk) is seen in the muscular belly, which suggests the presence of tumour necrosis.

tumours. The low metastatic rate to muscles may be explained by the cellular environment of the skeletal muscle being less hospitable than that of other organs. In a previous report, 2.2% of 1,201 oncologic patients who underwent total body CT were found to have muscular metastases. Oncologic patients with both lameness and reluctance to move support the use of whole-body CT for tumour staging. Whole-body CT enables visualization of muscular metastatic lesions and guidance of biopsy or cytology sample collection. Muscular metastases exhibited varying patterns within the same patient (Figure 23.12), possibly attributed to the temporal stage of metastatic development. Muscular metastases were mainly ring-enhancing lesions with a histologically confirmed hypodense necrotic centre (Figure 23.13a and b). Following in terms of frequency, lesions also appeared heterogeneous, had hyperdense nodular patterns and contained areas of multifocal intramuscular mineralization. In one study, 24 out of 66 dogs had cardiac metastasis, mainly secondary to carcinoma. The frequency of cardiac metastasis in dogs may be affected by the high incidence of haemangiosarcoma. In CT image, there is often a nodule present within the heart causing a filling defect (Figure 23.14b).

23.5.4 Miscellaneous

23.5.4.1 Infection

CT and MR imaging are the most sensitive techniques for early diagnosis of muscle infection in humans. Advantages of CT are the ability to identify bone destruction and small accumulations of air or calcification. CT imaging permits the severity and extent of changes associated with muscle infection to be characterized and allows the location, number and any communication of sinus tracts to be documented. On CT image, there is typically an enlargement of the muscle and a hypoattenuating areal with marked ring enhancement (Figure 23.15). These lesions are consistent with muscle abscessation or necrosis. A definite diagnosis, however, can only be made by puncture or biopsy. CT imaging appears to be a valuable guide for biopsy of the lesion and surgical planning.

23.5.4.2 Dystrophic

Calcinosis circumscripta is an uncommon syndrome of dystrophic or ectopic idiopathic mineralization characterized by deposition of calcium salts in the soft tissue. Typically large young dogs such as the German Shepherd are affected, and the disease occurs most commonly

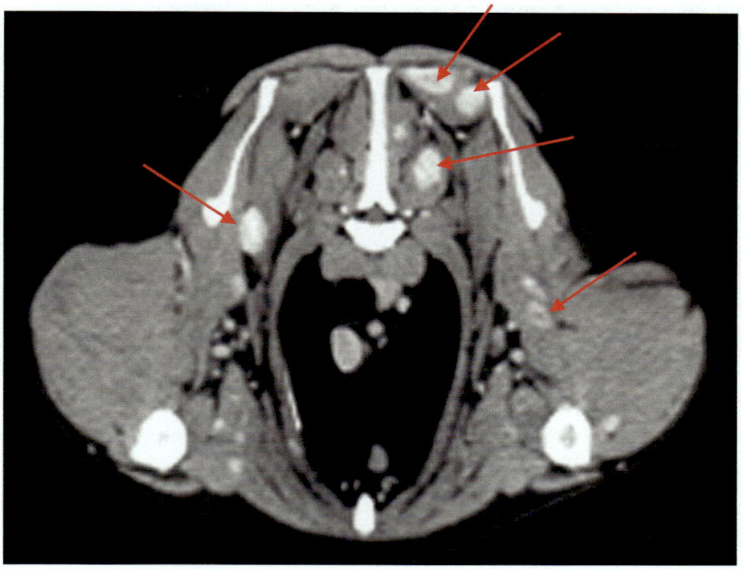

Figure 23.12 **Transverse post-contrast CT image of a dog with muscle metastasis due to mammary carcinoma. Variable patterns of CT characteristics are present in the musculature of the neck and thoracic vertebral wall. The contrast study shows oval lesions with strong contrast enhancement and a hypoattenuating centre and round nodules with ring enhancement (red arrows).**

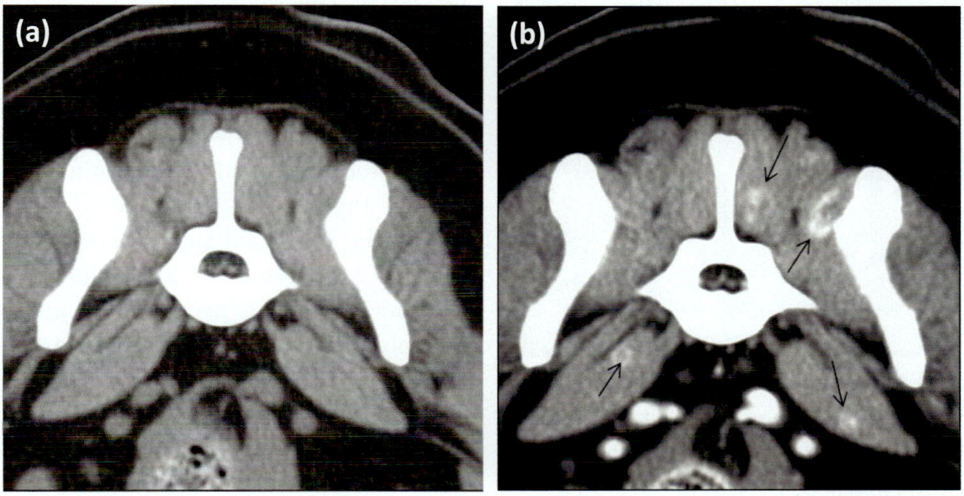

Figure 23.13 **Transverse post-contrast CT image at L7 of a dog with muscle metastasis due to splenic haemangiosarcoma. Precontrast (a) and post-contrast study (b). Note that in the precontrast CT, the lesions are isodense with normal musculature tissue. The contrast study shows ring-enhancing lesions in the iliopsoas and paraspinal muscles (arrows).**

on the feet and tongue. Complete excision of the calcified tissue is the treatment of choice. Prior to surgical excision, the extent and margins of the calcified tissue can be precisely determined with CT. The CT image shows solitary and multiple diffuse calcified areas in the muscles or soft part tissue (Figure 23.16a and b). Myositis ossificans circumscripta or ossifying myopathy is the heterotopic formation of bone in muscle due to metaplastic evolution of a fibrotic contracture or myopathy. The condition generally affects the caudal muscles of the thighs,

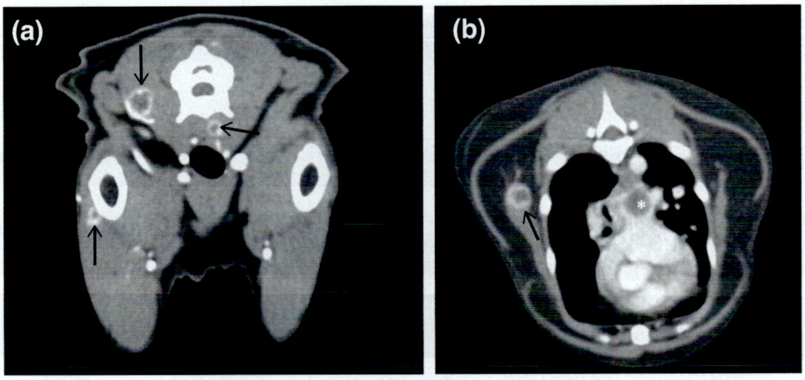

Figure 23.14 **(a) Transverse post-contrast CT image of a cat with muscle metastasis due to squamous cell carcinoma. Variably sized nodules with strong ring enhancement (arrows) are present in the musculature of the neck and right forelimb. (b) A nodule is present within the left atrium causing filling defect (asterisk). In this cat, skeletal muscular metastases were also present in the musculature of the thoracic wall (arrow).**

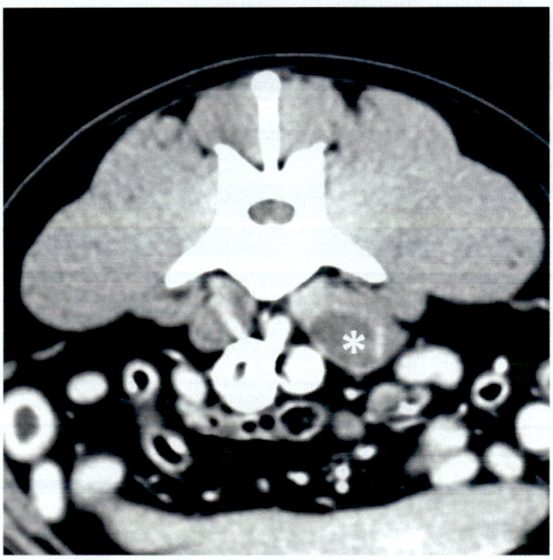

Figure 23.15 **Transverse post-contrast CT image of the caudal lumbar spine at L4 of a dog. The left iliopsoas muscle is asymmetrically enlarged, and a hypodense area in the muscle with ring enhancement is seen (asterisk). Compare with the contralateral muscle. The diagnosis confirmed by biopsy was a pyogranulomatous inflammation.**

especially the biceps femoris, semitendinosus, semimembranosus, gracilis, sartorius and gluteal muscles in the region of coxofemoral joint. The CT shows in the muscle a shell of mature, well-marginated bone with a peripheral ring of calcific density around a less dense, central area.

23.5.4.3 Inflammatory

Inflammatory muscular lesions have variable contrast enhancement, which is often related to the severity, extent, aetiology and temporal duration of the inflammation. CT may be a useful adjunct in the diagnosis of masticatory myositis in dogs, including selection of sites for diagnostic muscle biopsy and ruling out other conditions with similar clinical signs. Masticatory myositis is an inflammatory myopathy of the muscles of mastication (masseter, temporalis, pterygoid and digastric) in the dog. Patients can present with masticatory muscle swelling or atrophy, painful facial muscles and common extreme difficulty in opening the jaw. On CT images, changes in size (atrophy, swelling or both) were common in all masticatory muscles, except the digastric muscle. The dogs have typically marked inhomogeneous contrast enhancement in the temporalis, masseter and pterygoid muscles (Figure 23.17). Enlargement and inhomogeneous enhancement of a muscle is interpreted to be consistent with oedema and active inflammation. Also, rim enhancement with a non-enhancing core can be visible (Figure 23.18), suggesting necrosis. Often, enlargement of the head and neck lymph nodes is visible.

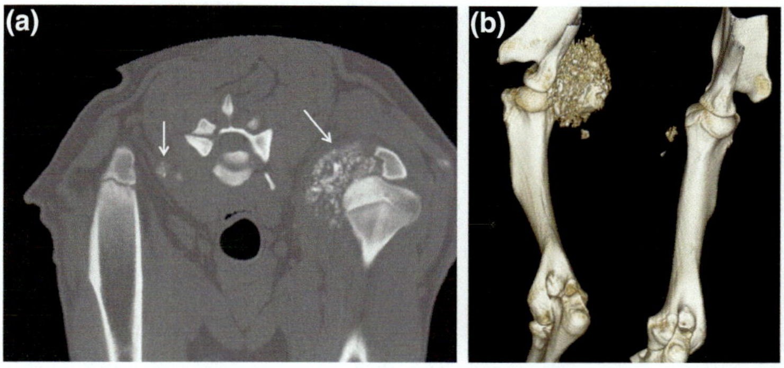

Figure 23.16 (a) Transverse post-contrast CT image of a young German Shepherd with a calcinosis circumscripta within the tissue of the shoulders. Note the multiple diffuse mineralizations in the musculature caudomedial to the shoulders (arrows). Compare with the contralateral muscle. (b) Volume-rendered image of shoulder joints shows the calcified lesions.

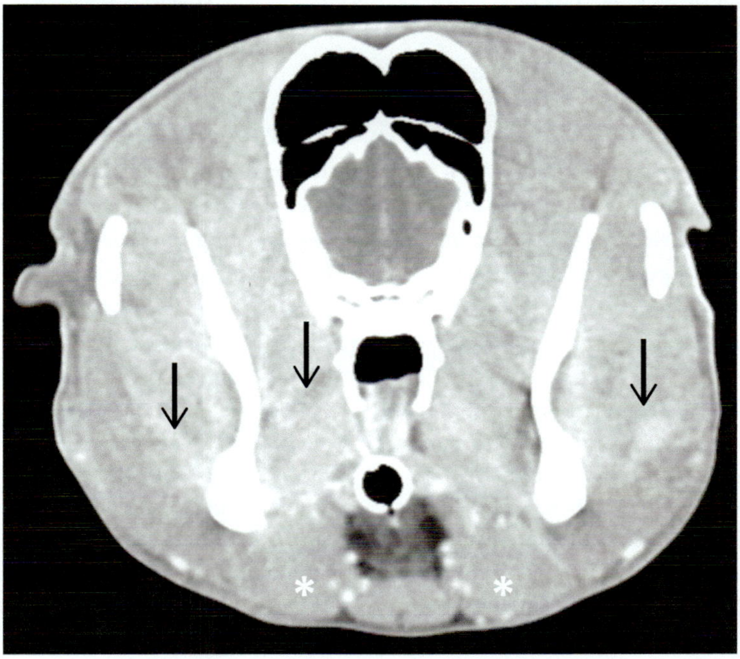

Figure 23.17 Transverse post-contrast CT image of the head of a 1.5-year-old dog with a masticatory myositis. There is a marked inhomogeneous contrast enhancement of the masseter, the temporalis and the pterygoid muscle (arrows). Note that the digastric muscle is hypoattenuated compared with the other masticatory muscles (asterisk).

FURTHER READING

Adrega Da Silva C, Bernard F, Bardet JF, Theau V, Krimer PM. 2009. Fibrotic myopathy of the iliopsoas muscle in a dog. *Veterinary and Comparative Orthopaedics and Traumatology*. 22(3):238–242.

Agut A, Anson A, Navarro A, Morciano J, Soler M, Belda E., Pallares FJ, Laredo FG. 2013. Imaging diagnosis-infiltrative Lipoma causing spinal cord and lumbar nerve root compression in a dog. *Veterinary Radiology & Ultrasound*. *54*(4):381–383.

Aupperle H, März I, Ellenberger C, Buschatz S, Reischauer A, Schoon HA. 2007. Primary and secondary heart tumours in dogs and cats. *Journal of Comparative Pathology*. 136(1):18–26.

Bergman PJ, Withrow SJ, Straw RC, Powers BE. 1994. Infiltrative lipoma in dogs: 16 cases

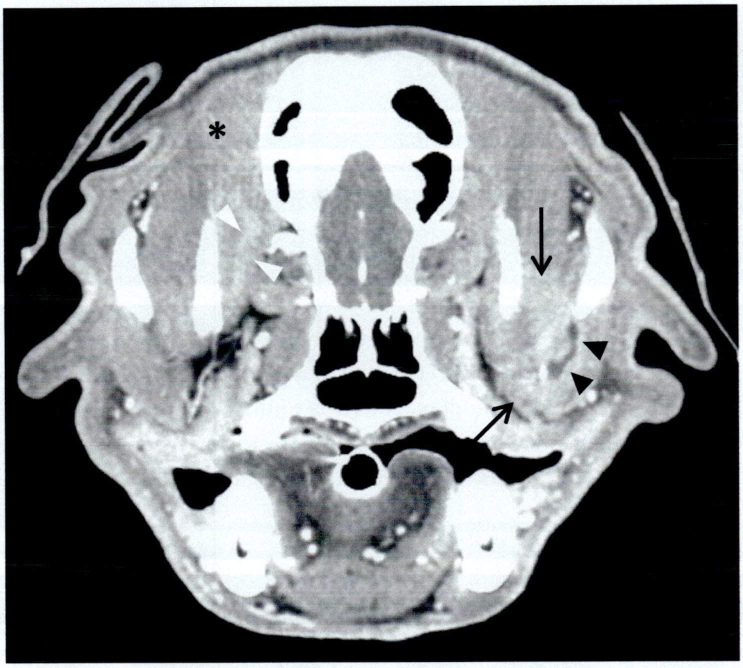

Figure 23.18 **Transverse post-contrast CT image of the head of a 2-year-old dog with masticatory myositis. There is a marked swelling and inhomogeneous contrast enhancement in the left masseter muscle (arrows). Notice the rim enhancement with a non-enhancing linear lesion in the left masseter muscle (arrowhead), suggestive of an area of necrosis. Mild left temporalis atrophy is evident (asterisk).**

(1981—1992). *Journal of the American Veterinary Medical Association*. 205:322–324.

Breur G, Blevins W. 1997. Traumatic injury of the canine iliopsoas muscle. *Journal of the American Veterinary Medical Association*. 210:1631–1634.

Bullen LE, Evola MG, Griffith EH, Seiler GS, Saker KE. 2017. Validation of ultrasonographic muscle thickness measurements as compared to the gold standard of computed tomography in dogs. *PeerJ*. 25;5:e2926.

Fitch RB, Jaffe MH, Montgomery RD. 1997. Muscle injuries in dogs. *Veterinary and Comparative Orthopaedics and Traumatology*. 19:947–957.

Garrett WE. 1996. Muscle strain injuries. *American Journal of Sports Medicine*. 24:2–8.

Gettinger E, Smeak D, Marolf A. 2017. Paramedian submandibular approach for removal of foreign bodies in the pterygoid muscle in two dogs. *Journal of the American Animal Hospital Association*. 53(4):221–226.

Hermanson JW. 1993. The muscular system: muscles of the pelvic limb. In: Evans HE (ed) *Miller`s Anatomy of the Dog*, W.B. Saunders Company, Philadelphia, PA, pp. 369–376.

Holloway A, Dennis R, McConnell F, Herrtage M. 2009. Magnetic resonance imaging features of paraspinal infection in the dog and cat. *Veterinary Radiology & Ultrasound*. 50(3):285–291.

Johnson JA, Austin C, Breuer GJ. 1994. Incidence of canine appendicular musculoskeletal disorders in 16 veterinary teaching hospitals from 1980 through 1989. *Veterinary and Comparative Orthopaedics and Traumatology*. 7:56–69.

Kirkendall DT, Garrett WE. 2002. Clinical perspectives regarding eccentric muscle injury. *Clinical Orthopaedics and Related Research*. 403:81–89.

Maddox TW, May C, Keeley BJ, McConnell JF. 2013. Comparison between shoulder computed tomography and clinical findings in 89 dogs presented for thoracic limb lameness. *Vetrinary Radiology and Ultrasound*. 54(4):358–364.

McEnttee MC, Thrall DE. 2001. Computed tomographic imagine of infiltrative lipoma in 22 dogs. *Veterinary Radiology & Ultrasound*. 42(3):221–225.

Negus S, Sihdu PS. 2000. MRI of retroperitoneal collections: a comparison with CT. *British Journal of Radiology*. 73:907–912.

Nielsen C, Pluhar GE. 2005. Diagnosis and treatment of hind limb muscle strain injuries in 22 dogs. *Veterinary and Comparative Orthopaedics and Traumatology*. 18:247–253.

Reiter AM, Schwarz T. 2007. Computed tomographic appearance of masticatory myositis in dogs: 7 cases (1999—2006). *Journal of the American Veterinary Medical Association*. 231(6):924–930.

Rossmeisl JH, Rohleder JJ, Hancock R, Lanz OI. 2004. Computed tomographic features of suspected traumatic injury to the iliopsoas and pelvic limb musculature of a dog. *Veterinary Radiology & Ultrasound*. 45(5):388–392.

Seely S. 1980. Possible reasons for the high resistance of muscle to cancer. *Medical Hypotheses*. 6:133–137.

Speer KP, Lohnes J, Garrett WE. 1993. Radiographic imaging of muscle strain injury. *American Journal of Sports Medicine*. 21:89–96.

Stahl C, Wacker C, Weber U, Forterre F, Hecht P, Lang J, Gorgas D. 2010. MRI features of gastrocnemius musculotendinopathy in herding dogs. *Veterinary Radiology & Ultrasound*. 51(4):380–385.

Steiss J. 2002. Muscle disorders and rehabilitation in canine athletes. *Veterinary Clinics of North America Small Animal Practice*. 32:267–285.

Taylor J, Tangner CH. 2007. Acquired muscle contractures in the dog and cat. A review of the literature and case report. *Veterinary and Comparative Orthopaedics and Traumatology*. 20:79–85.

Valentine BA. 2011 Skeletal muscle in pathologic basis of veterinary disease. In: McGavin D (ed) *Pathologic Basis of Veterinary Disease*, Mosby Press, St. Louis, pp. 973–1040.

Vansteenkiste DP, Lee KC, Lamb CR. 2014. Computed tomographic findings in 44 dogs and 10 cats with grass seed foreign bodies. *Journal of Small Animal Practice*. 55(11):579–584.

Vignoli M, Terragni R, Rossi F, Frühauf L, Bacci B, Ressel L, Capitani O, Marconato L. 2013. Whole body computed tomographic characteristics of skeletal and cardiac muscular metastasic neoplasia in dogs and cats. *Veterinary Radiology & Ultrasund*. 54(3):223–230.

Wess G. 2013. Kardiale und mesenteriale Tumoren. In: Kessler M (ed) *Keintieronkologie*, Enke Verlag, Stuttgart, ch 24, p. 388.

CHAPTER 24

CT OF THE SPINE

Wolfgang Henninger, Nuria Corzo-Menéndez and Peter Smith (Spinal subluxation/fractures and Atlantoaxial subluxation/instability, Arachnoid Diverticula), S De Decker (Cervical spondylomyelopathy, Spinalmalformations), C. Falzone (Lumbo-sacral diseases), Amalia Agut and R. Latorre (Anatomy), and Ingrid Gielen

24.1 INTRODUCTION

The spine is one of the most common areas where computed tomography (CT) is used to obtain a strong diagnosis on neurological disease and decide between surgical and non-surgical treatment choices. Some consider CT simpler than magnetic resonance imaging (MRI), which may be true to some point, but it ultimately benefits the clinical workup. CT machines are more readily available and easier to use; multi-slice technology has greatly improved image resolution so that high-quality images can now be obtained quickly, which is especially useful in emergency cases. The CT image favourably compares with the radiograph's black, white and grey shades and is much easier to interpret than other cross-sectional modalities. However, the incredible amount of detail shown in a series of CT images can overwhelm a less experienced reader and confirm a well-known radiologist's saying, "You only see what you know," which is also very true with spinal cases.

24.2 INDICATIONS FOR USING CT

These include the following:

a. All cases of painful acute or chronic disc extrusion.
b. Suspected space-occupying masses that arise from tissues other than a disc (dorsal ligaments, articular facets, chondromas in growing animals or osteomas).
c. Traumatized patients with a suspected vertebral fracture, luxation or haemorrhage.
d. Traumatized patients with suspected vertebral fracture or luxation due to former decompressing surgery and/or stabilizing treatment.
e. Any case of suspected discospondylitis (e.g., due to foreign body or of unknown origin).
f. Thickening of the nerve roots (brachial or lumbar/sciatic plexus).
g. Swelling of the spinal cord of unknown origin.
h. Congenital malformation of the vertebra or pediculi (so-called breed-associated anomalies like wedge-shaped vertebra, hemivertebra, facet joint asymmetry and spina bifida).
i. Narrowing of the vertebral foraminae, particularly in cases of cauda equine syndrome.
j. Spondylitis and spondylosis (also DISH), combined with degenerative joint disorder (DJD) of the articular facets.
k. Suspected subarachnoid cyst (Myelo-CT) or a herniating cyst from an intervertebral joint.
l. All sorts of suspected osteosarcoma, chondrosarcoma, multiple myeloma and haemangiosarcoma
m. Follow-up examinations to the indications given above.

24.3 TECHNICAL ASPECTS

24.3.1 Positioning

The spine is best examined with the animal lying on its back in a U-shaped radiolucent foam profile. The dorsal (supine) recumbency

DOI: 10.1201/9781315118130-26

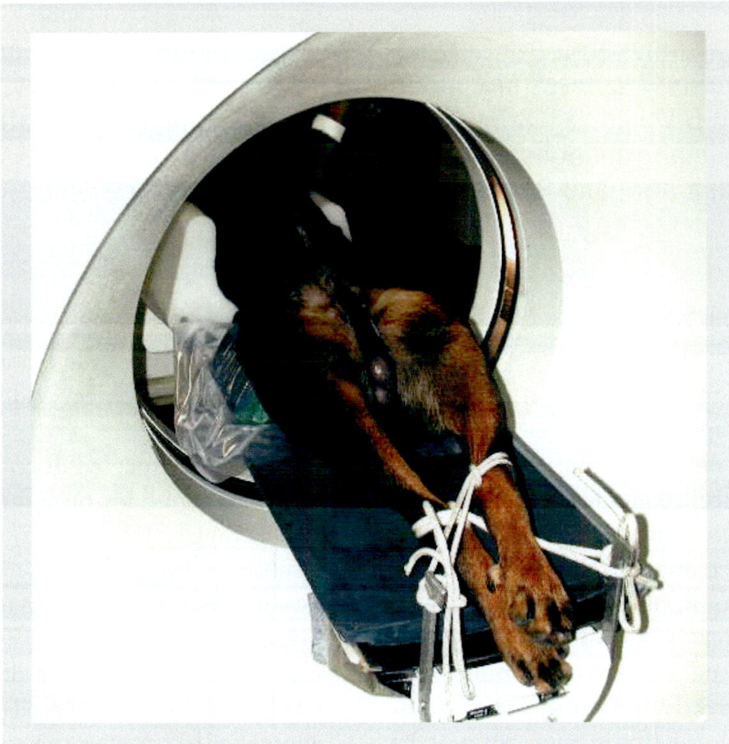

Figure 24.1 **A dog positioned in dorsal recumbency for CT of the lumbar and lumbosacral regions. The animal is lying in a U-shaped radiolucent foam profile with the hind legs extended.**

with the animal's front legs extended overhead, and the hind legs extended (Figure 24.1) caudally provides a stable position to let the animal breathe normally, while the spine is motionless. This allows to acquire reproduceable good-quality images. Scanning while the animal holds its breath also minimizes shading or dishing effects within the vertebral canal. Most of the companion animals are of moderate size; therefore, the patient can be positioned "feet first" on to the table, which means that the hind legs are put into the gantry first and the front part of the animal is outside. For the anaesthetist, this sort of positioning also allows easier access to the animal to follow vital parameters and to act more quickly in an emergency situation.

The front legs can rest on a radiolucent pillow on the animal's lower jaw. The hind legs should not only be pushed down onto the CT table but also rested on a pillow and the femur with the knees carefully extended not to make cauda equina or hip problems worse. While it is out of question that this position of the extended hind legs allows examination of the lumbar and sacral area best, the position of the front legs in cases of scanning cervical spinal disease is a subject of much debate. The authors strongly prefer to examine the cervical spine with the front legs to be extended over the animal's head to form a cylindrical shape of the body, whose diameter is the same from the head down to the shoulders and thoracic inlet. Many examiners prefer the front legs pulled caudally (humanoid position) to gain artefact-free images of the cervical spine, and they may be right for the cranial part, but not for the more important caudal part of C5-6-7-Th1. Positioning the front legs caudally leads to a sudden increase in body thickness due to the laterally placed shoulders at the cervico-thoracic transition, which in turn causes poor image quality and increased radiation dose, mainly mAs.

Positioning the legs caudally bends the brachial plexus, its peripheral nerves and their accompanying vessels around the first rib, which may cause partial volume artefacts and mimic recognition of early nerve thickening. Therefore, the authors strongly advise to place the front legs cranially for examination of the plexus brachialis region. Image interpretation and the perception of different diameters can be improved by stretching the nerves and vessels cranially to compare both sides. As a transverse scanning procedure, CT picks up thickened and roundish structures more easily than if they are elongated. For the anaesthetist, on the other hand, in very large dogs, caudal positioning of the front legs may provide an easier access to the cephalic vein to administer narcotic drugs and intravenous (IV) contrast medium when they have to be positioned "head first" into the gantry.

In animals with a less moveable shoulder part, positioning the front limbs caudally alongside the body may lift up the cervical spine from the bed, and the position of the vertebrae will differ from one segment to the next, which can create steps; a similar effect can be produced by placing a pillow under the animal's head. In single-slice scanners, tilting the gantry may help, and in multi-slice scanners, the images have to be adjusted at the workstation afterwards or during image interpretation. To sum up, there are several reasons to save time beforehand during proper positioning and to minimize uprising artefacts.

24.3.2 Technical parameters

The list of suspected differentials and the technical features of the scanner significantly influence the settings.

For the planning of a scanning procedure on a topogram/scoutview, single-slice scanners are orientated along the vertebral endplate perpendicular to the vertebral canal to get the full height of the disc and the vertebral canal. For thoracic and lumbar spine exams, the perpendicular orientation is easy; however, for cervical scanning, this can be tricky – for planning, it is recommended to use the lateral overview (topogram) and transfer the perpendicular angulation from the thoracic spine cranially over the cervical spine, which will perfectly fit the appearance of the intervertebral disc and vertebral height.

Slice thickness in cats and small dogs is about 1 mm and 2–3 mm for larger dogs, always overlapping slices. For an overview of any spinal condition, 2 mm slice thickness is used for cats and small dogs and 3–5 mm slice thickness for large dogs; very thin slices can be taken later of the region of interest for more accurate resolution. In cats and small dogs, 100 kVp and about 120 mA are used, while 120 kVp and 140 mA (based on the body volume) are required for larger dogs. Obese animals can severely prolong scanning intervals, and the high tube heat must be considered before rescanning any region following IV contrast administration.

Multi-slice scanners allow perfect orientation of the scanned body volume during various post-processing procedures, so tilting the gantry is not so necessary. This does not mean that a radiographer should not be careful while positioning and orientating the patient. Slice thickness also depends on resolution and patient size; it is 0.5 or 1 in smaller patients and up to 2 mm with little increase in slice thickness for overlapping reconstruction in larger ones. The tube rotation is very fast (possibly less than 1 second) and is usually combined with settings of around 120 kVp and automatic mA, based on the body volume. Plain series are reconstructed in bone and soft tissue algorithm, and post-contrast series need to be reconstructed only in soft tissue algorithm.

24.3.3 Contrast medium

IV contrast medium administration in CT is always essential in diagnostic studies; otherwise, the study outcome may be limited.

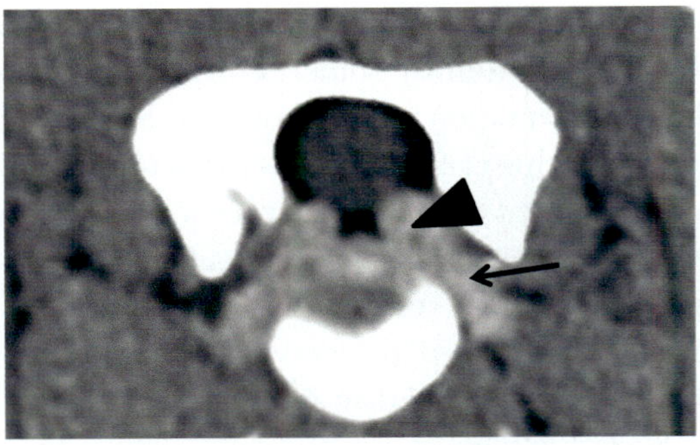

Figure 24.2 **Transverse CT image, post-contrast, in soft-tissue window. In the cervical area, the internal vertebral venous plexus can be prominent (arrow head). Also, the intervertebral vein can be noticed (arrow).**

According to recent literature reports, contrast medium shall be helpful only in neoplastic cases, but the authors do not limit it to these patients. For CT, a bolus of about 2 mL (400 mg Iodine/mL) of the iodinated contrast medium per kilogram of body weight should be administered. With multi-slice scanners, an early (arterial)-phase study is easily performed for angiographic purposes, but a second, late-phase venous study is recommended. This because contrast uptake in soft tissue and musculoskeletal tissues often takes longer to be enhanced compared to well-perfused organs. The inflamed tissue also gets more bright over time, so repeating the series after 5–10 minutes is recommended. On the other hand, single-slice scanners need a longer duration of examination and often only get a late-phase enhancement. The goal for all purposes however is to finish contrast administration quickly enough so that the angiographic enhancement is distinguishable. During the angiographic phase, the ventral internal venous plexus does undergo marked enhancement at the cervical spine but also along some thoracic part, less along the lumbar spine and finally better seen at the lumbosacral transition. These vessels are located ventrolateral to the spinal cord section on both sides and are pronounced at the disc level at the cervical spine. They form bright, hyperdense spots all along the spinal cord, sometimes bilateral and sometimes unilateral (Figure 24.2). During plain scans, this may at first be mistaken for nerve root section because their shape and soft tissue attenuation are identical.

Enhancement of the spinal cord or nerve roots may be a sign of an inflammatory or neoplastic condition; when accompanied by other findings, a more benign or more malignant differential may be considered. High enhancement of the para- and perivertebral soft tissue and musculature may suggest inflammation or necrosis, which is very typical with fistulae or discospondylitis; however, low enhancement of the soft tissue can be normal due to normal vascularization but also be present in suspected neoplastic conditions. Soft tissue biopsy, therefore, is always recommended.

24.4 NORMAL ANATOMY

In the occipito-atlanto-axial joint, the dens of the axis must be positioned in the ventral half of the atlas body; the articular facets of the axis seem to loosely contact the caudal part of the atlas and can easily look incongruent because the spine is not correctly positioned when the long axis is tilted to one side.

The oval-shaped intervertebral disc gives a shallow impression on the dorsal aspect according to the shape of the neighbouring bony endplates, which is consistent through the whole spine, from C3 to the last lumbar vertebra. The horizontal appearance of the dorsal aspect is considered normal at the C2/3 and L7/S1 levels. The intervertebral space consistently has the same thickness as the nucleus pulposus and its fluid content (Figure 24.3).

The longitudinal crest, which is present on the dorsal surface of the vertebral body, can be pronounced in the lumbar region (Figure 24.4a). One or more vascular channels pass through the vertebral body and contain the basivertebral veins, which connect the internal and external ventral vertebral venous plexuses with each other (Figure 24.4b).

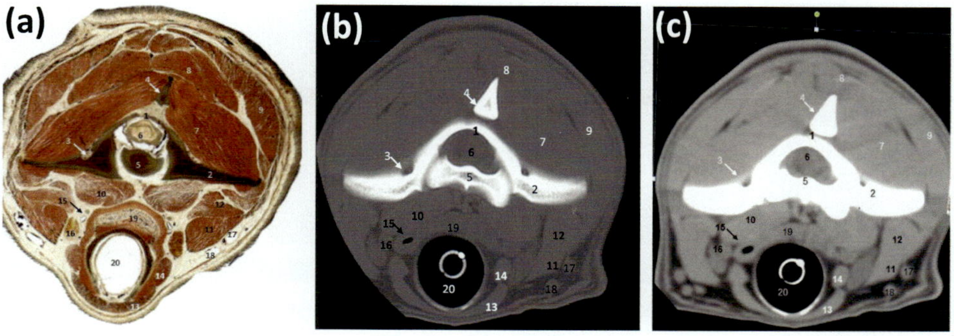

Figure 24.3.1 Transverse images at the level of C1 and C2. (a) Anatomic section, (b) Bone window, (c) Soft tissue window: 1. Dorsal arch of atlas; 2. Wings of atlas (transverse process); 3. Transverse foramen; 4. Spinous process of axis; 5. Body of axis; 6. Myelencephalon; 7. Obliquus capitis caudalis muscle; 8. Semispinalis capitis M muscle; 9. Splenius capitis muscle; 10. Longus capitis muscle; 11. Sternocephalicus muscle; 12. Cleidomastoideus muscle; 13. Esternohyoideus muscle; 14. Esternothyroideus muscle; 15. Common carotid artery, vagosympathetic trunk; 16. Medial retropharyngeal lymph node; 17. Maxillary vein; 18. Linguofacial vein; 19. Laryngopharynx; 20. Larynx.

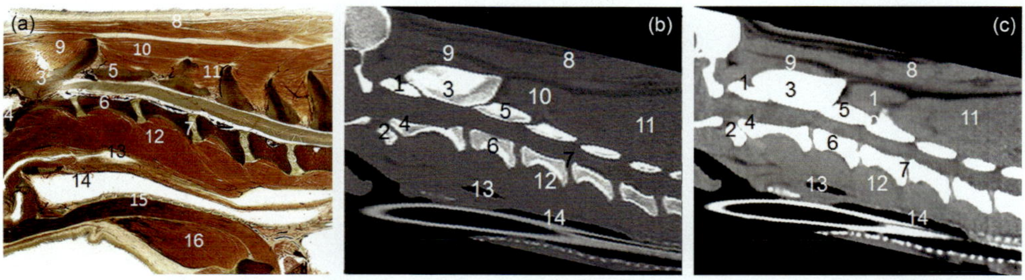

Figure 24.3.2 Sagittal images of the cervical spine: 1. Dorsal arch of atlas; 2. Ventral arch of atlas; 3. Spinous process of axis; 4. Dens of axis; 5. Vertebral arch of C3; 6. Body of C3; 7. Intervertebral disc; 8. Biventer cervicis muscle; 9. Obliquus capitis caudalis muscle; 10. Spinalis cervicis muscle; 11. Multifidus muscle; 12. Longus colli muscle; 13. Esophagus; 14. Trachea; 15. Esternohyoideus and esternothyroideus muscles; 16. Esternocephalicus muscle.

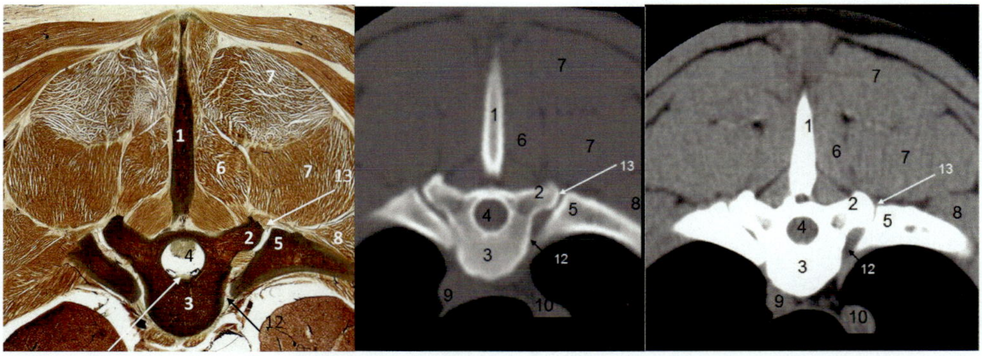

Figure 24.3.3 Transverse images at the level of the fourth thoracic vertebra: 1. Spinous process; 2. Transverse process; 3. Vertebrae body; 4. Vertebral foramen; 5. Rib; 6. Spinalis and semispinalis thoracis muscle; 7. Longissimus thoracis muscle; 8. Iliocostalis muscle; 9. Azygos vein; 10. Aorta; 11. Dorsal longitudinal ligament; 12. Capitis costae joint; 13. Costotransverse joint.

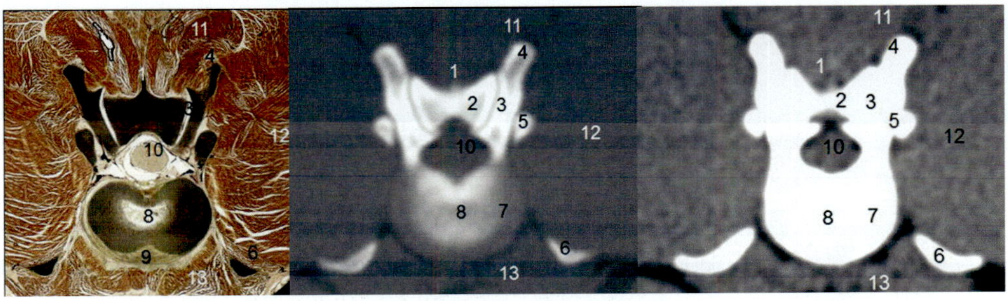

Figure 24.3.4 Transverse images of the intervertebral disc space between two lumbar vertebrae: 1. Spinous process; 2. Craneal articular process; 3. Caudal articular process; 4. Mammilary process; 5. Accesory process; 6. Transverse process; 7. Vertebrae body; 8. Intervertebral disc: nucleus pulposus; 9. Intervertebral disc: anulus fibrosus; 10. Vertebral formanen, spinal cord; 11. Multifidus muscle; 12. Longissimus lumborum muscle; 13. Sublumbar muscles (psoas major).

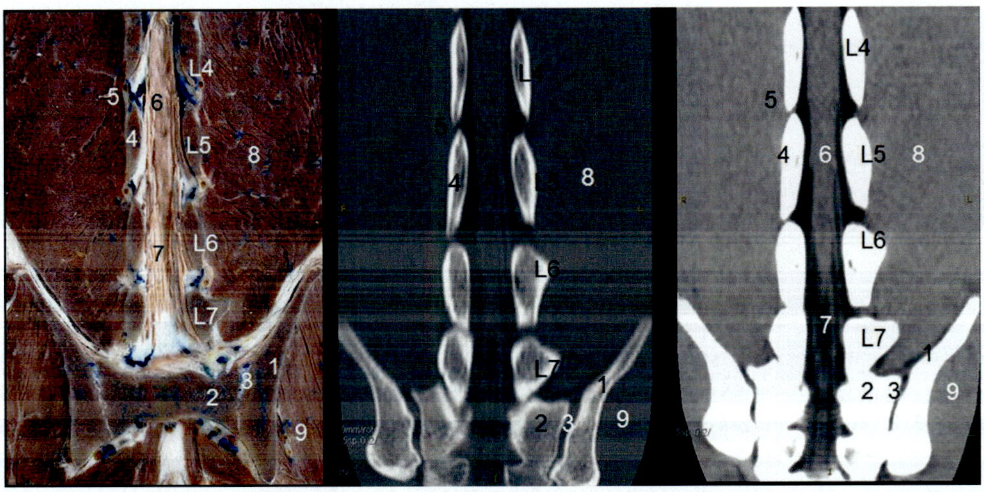

Figure 24.3.5 Dorsal images from L4 to the sacrum: 1. Wings of ilium; 2. Wings of sacrum; 3. Sacroiliac joint articular surface; 4. Vertebral arch; 5. Intervertebral space; 6. Spinal cord; 7. Nerve roots of the cauda equina; 8. Longissimus and iliocostalis muscles; 9. Gluteus muscle.

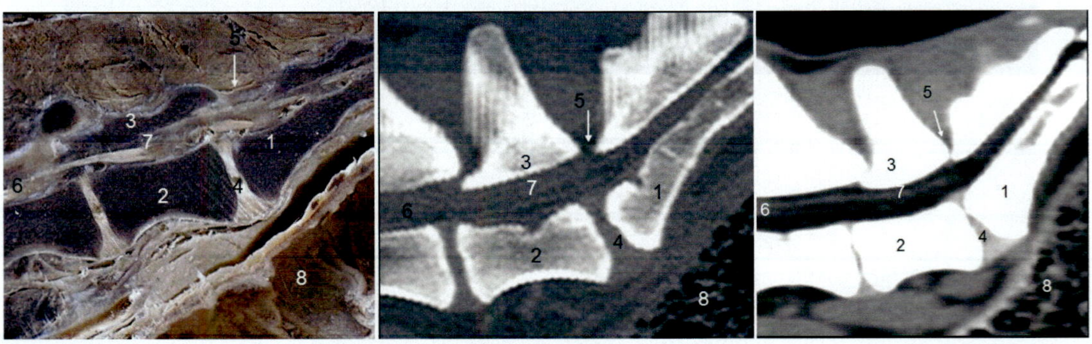

Figure 24.3.6 Sagittal images at the level of the lumbosacral area: 1. Sacrum; 2. Body of vertebra L7; 3. Lamina of vertebra L7; 4. Intervertebral disc of lumbosacral joint; 5. Lumbosacral interarcuate space; 6. Spinal cord; 7. Nerve roots of the cauda equina; 8. Rectum.

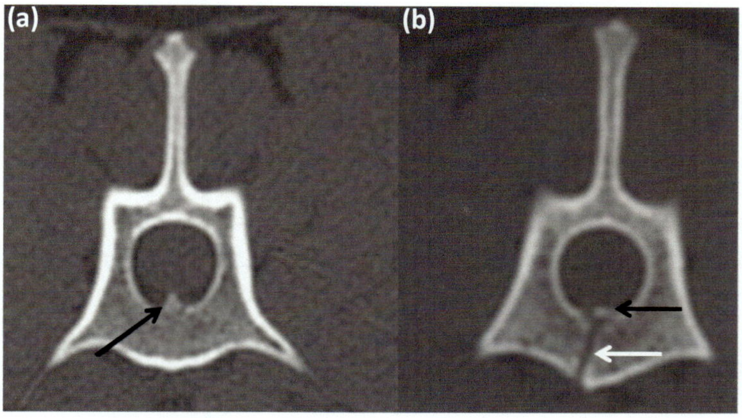

Figure 24.4 (a) The longitudinal crest on the dorsal surface of the vertebral body can be pronounced in the lumbar region (black arrows). (b) Through the vertebral body, vascular channels pass. They contain the basivertebral veins that connect the internal and external ventral vertebral venous plexuses with each other (white arrow).

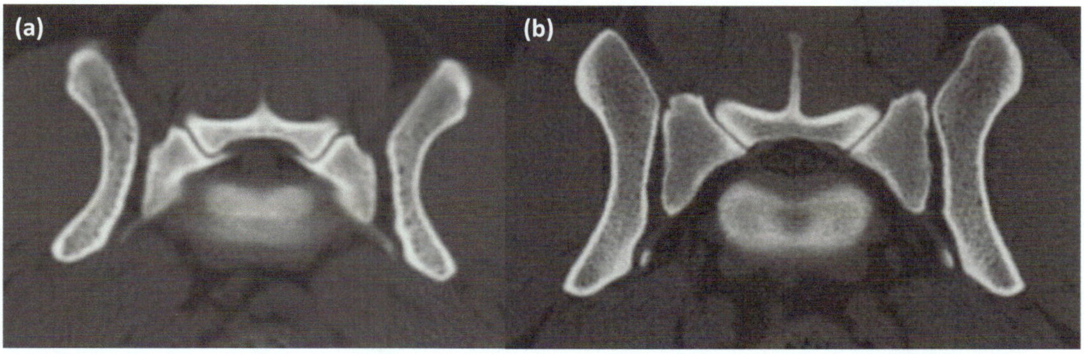

Figure 24.5 Transverse images in bone window at the level of the lumbosacral vertebrae. The articular facet joints of the English Cocker (a) have a more curved shape. In the German Shepherd (b), straight articular facet joints are observed.

The articular facets of the caudal and cranial articular processes of each vertebra are smoothly outlined and homogeneously compact and dense. They form a thin inter-articular joint space, either straight or slightly bent, with a notch at the lamina base where symmetrical spots of soft tissue that form articular pouches can be seen, mainly along the lumbar vertebra. When looking for lumbosacral instability, special attention should be paid to the more curved shape of the articular facets at the lumbosacral junction and joint width in some dog breeds (Figure 24.5).

Half, bloc or wedge-shaped vertebral bodies are abnormal formations in most breeds but considered normal for Boston terriers, Mops, French bulldogs, English bulldogs and their crossbreeds. Also, these conditions are often combined with synostotic spinal processes and asymmetric offspring of the ribs. Asymmetric offspring of the last ribs is an important issue to correctly orientate the surgeon in case of disc disease.

24.5 PATHOLOGIC FINDINGS

24.5.1 Acute

24.5.1.1 Spinal subluxation/fractures

Vertebral body trauma is relatively common in companion animals, with the majority caused by traffic accidents. Damage to the nervous system, including the spinal cord and nerve roots, is the most common problem following vertebral trauma. The nature of the injury dictates which structures are most likely to be damaged: for example, trauma to a limb is more likely to cause nerve root damage, potentially causing avulsion of the nerve roots from the spinal cord, whereas trauma directly to the vertebrae is more likely to cause spinal cord injury. When any section of the spinal cord is injured, both the neuronal cell bodies in the grey matter and axons in the white matter are damaged.

Most injuries will have two components—contusion and compression.

- Contusion refers to the bruising at the time of injury, which leads to local cell death and a focal area of necrosis. With spinal cord injury, in particular, the region of necrosis enlarges following the initial insult, caused by complex physiological changes in the surviving cells around the specific damaged area. Clinical assessment and diagnostic imaging are, therefore, essential to thoroughly evaluate the damage to areas other than the vertebral column.
- Compression is commonly seen secondary to trauma, either through traumatic extrusion of the intervertebral disc material into the spinal canal or distortion of the spinal canal caused by vertebral fracture and/or subluxation. Determining the severity of the compression and removing its source are critical to improve the outcome and can readily be assessed using plain myelography, CT myelography and/or magnetic resonance imaging (MRI). The latter two modalities provide cross-sectional analysis of the compression's severity, which gives them an advantage over plain myelography. Modern high-field MRI scanners offer good details of bony structures; however, they cannot yet provide the details of CT images.

24.5.1.1.1 Aim of CT

Diagnostic imaging is performed to provide a thorough evaluation of the primary injury site and to determine whether other injuries that may impact the outcome exist. Once concurrent injuries have been stabilized, determining the necessity for spinal surgery is key. Therefore, diagnostic investigations following trauma should address these questions:

- Are additional injuries (e.g., pneumothorax, lung contusion, thoracic or abdominal haemorrhage, rib fractures and long bone fractures) likely to impact recovery?
- Does static compression of the vertebral column need to be resolved?
- Will instability of the vertebral column possibly lead to further compression and/or pain?

MRI will provide more details about the severity and extent of any spinal cord changes following vertebral trauma, but CT is carried out much faster and will provide answers to these key questions.

24.5.1.1.2 CT features

The entire CT study should be systematically examined, including images in the bone window and in soft-tissue window to maximize the study's potential. After checking the transverse series, it is important to look at the images in all three planes using multiplanar reconstructions with both the bone and soft-tissue window settings. The application of contrast media (e.g., myelography/intravenous) is usually not required.

24.5.1.1.2.1 Key points in the search for lesions

Bone window

- Abnormal lucent lines with or without bone displacement affecting any of the bony structures (in particular, the spinous processes, articular processes, laminae, pedicles, vertebral body, end plates and physes in young patients). Minimally displaced facet joint fractures can be difficult to diagnose in patients with skeletal immaturity (Figures 24.6–24.8).

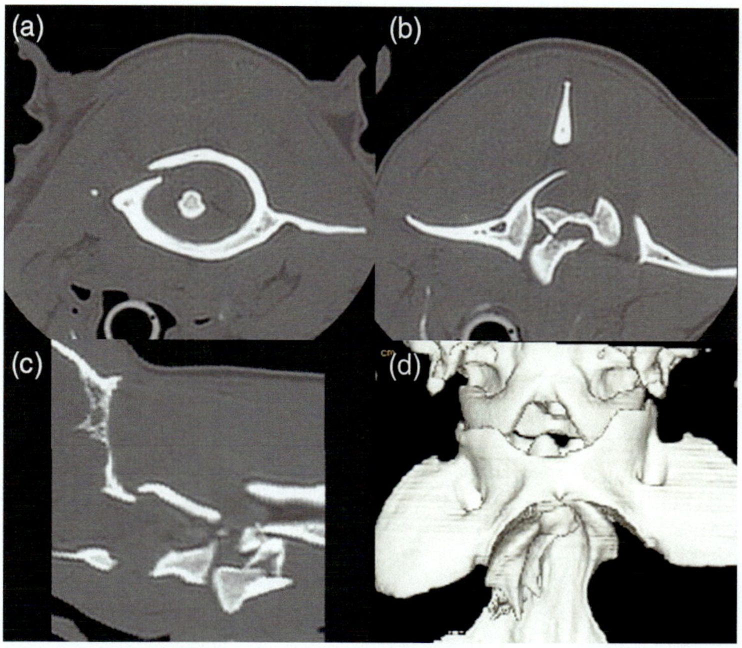

Figure 24.6 Images from a 5-year-old Lurcher with acute-onset tetraparesis and neck pain after a suspected traumatic event. There is a comminuted fracture in the right condyle of C2, with craniodorsal displacement of C2 causing narrowing of the spinal canal, and C1–C2 subluxation. Note the fractured condyle is displaced ventrally (C). This dog required surgical stabilization and achieved a good outcome. (a) Transverse image displayed in bone window though the dens. (b) Transverse image displayed in bone window though the atlantoaxial joint. (c) Right-sided parasagittal multiplanar reconstruction (MPR). (d) 3D reconstruction.

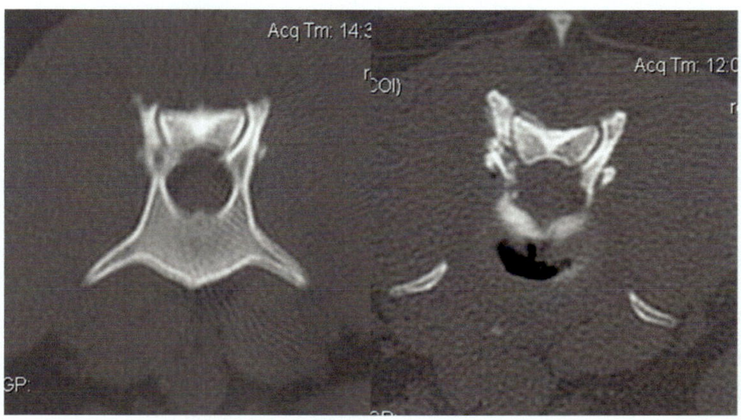

Figure 24.7 Transverse images displayed in a bone window from a dog with a transverse fracture of the facet joint on the right side at the thoracolumbar junction. Two of the three compartments of the vertebral column are intact and, on this basis of this one image, the fracture is not intrinsically unstable. The image on the left is taken at the time of the injury; the image on the right 9 weeks after conservative treatment showing signs of bone healing.

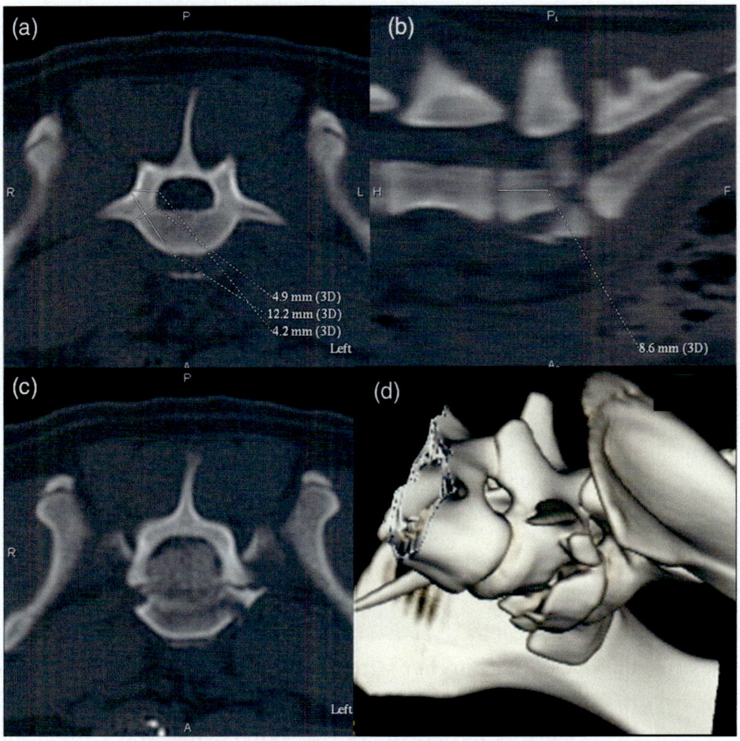

Figure 24.8 Displaced comminuted fracture of the vertebral body of L7 with consequent lumbosacral luxation; note one of the large fragments is dorsally displaced into the vertebral canal, causing nearly complete occlusion (b and c). According to the three-compartment model, this fracture is unstable, and the patient required surgical stabilization. Figures (a) and (b) show how the CT software allows measurements to be made that aid in surgical planning and implant selection and placement. (a) Transverse image in a bone window at the level of a normal segment—caudal L6. (b) Midline parasagittal image in a bone window showing intact L6 and sacrum and fractured L7 vertebra. (c) Transverse image in a bone window at the level of caudal L7, showing the vertebral canal obliterated by a fragment of the L7 vertebral body, displaced dorsally. (d) The oblique 3D reconstruction adds relatively little new information but assists in surgical planning.

- Vertebral body integrity and alignment, including the degree and direction of any displacement; the spinal canal should be carefully examined for any evidence of narrowing. (Figures 24.6 and 24.8)
- Widening, narrowing and disruption of the intervertebral disc space and the presence of extruded disc material in the spinal canal, which is more visible if the material is mineralized.
- The presence of hyperattenuating/hyperdense material in the spinal canal, which can represent disc material, bone fragments or haematoma (Figure 24.8).
- A generalized degree of bone mineralization because dogs with poorly mineralized bone are more susceptible to vertebral trauma and pathological fracture (Figure 24.9).

Soft-tissue window

- The integrity of the surrounding musculature/soft tissues—intravenous contrast can be beneficial.
- The integrity of the intervertebral disc (annulus and nucleus) and associated

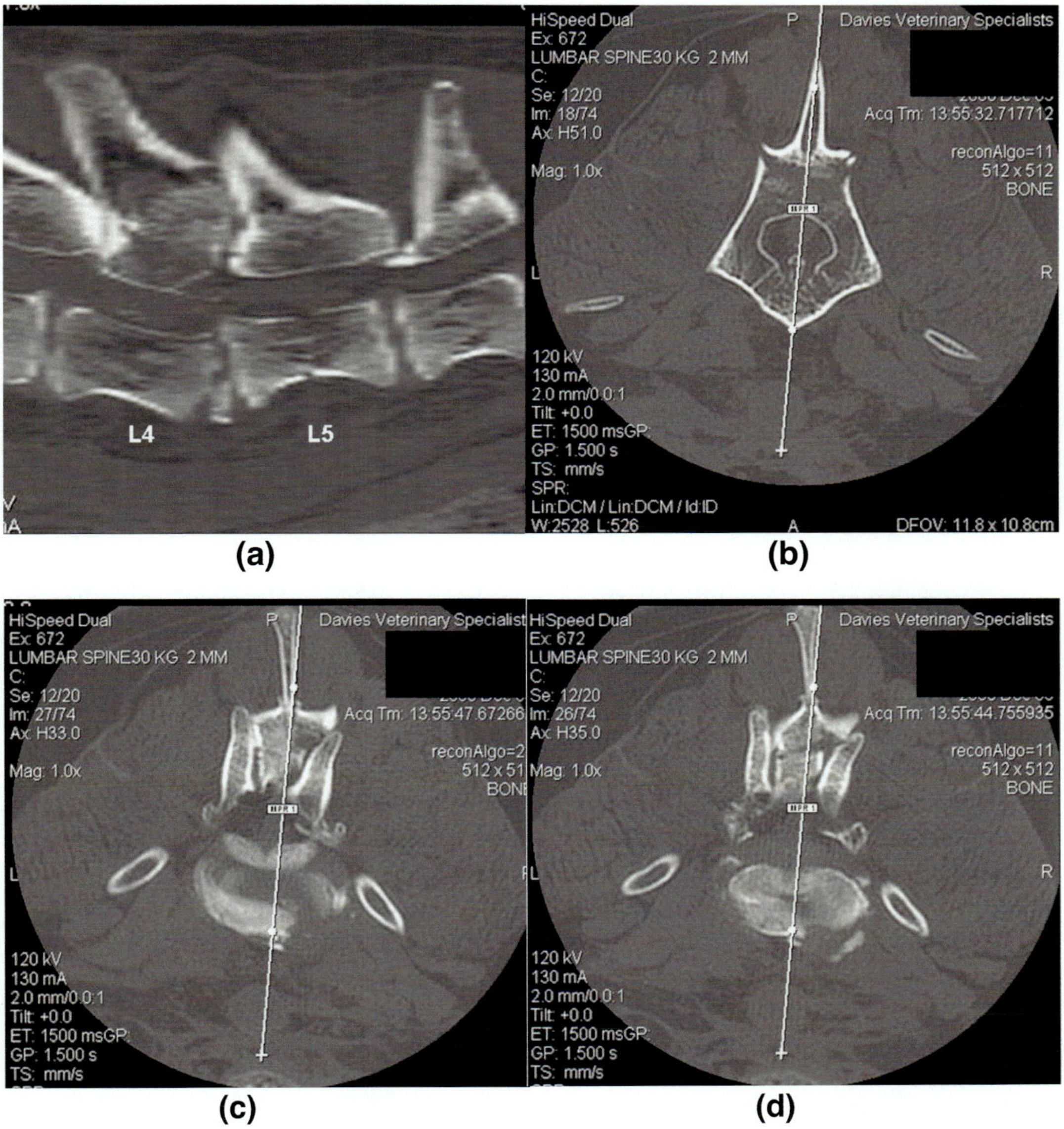

Figure 24.9 CT images showing fracture of L4 and luxation at the junction between L4 and L5 in a 10-year-old crossbreed dog with back pain. The fractures have an unusual folding/green-stick conformation suggestive of osteopenia and pathological fracture. Note the very thin cortices and reduced mineral density of all the bones shown in all images, typical of osteopenia. (a) Mid-sagittal MPR of the lumbar vertebral column; bone window; (b–d) Transverse images; bone window.

ligaments (dorsal and ventral longitudinal ligaments), which may be difficult to assess accurately on CT.

- Spinal cord compression—the spinal cord should be examined for displacement and the lack of visualization of epidural fat. Compression can be caused by blood/haematoma (hyperdense), disc material (hyperdense to hypodense, depending on the degree of mineralization/degeneration) and bone fragments. Myelography increases the sensitivity for detecting spinal cord compression.

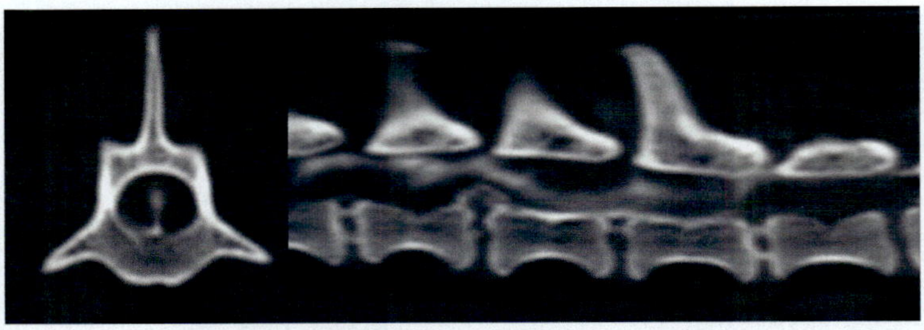

Figure 24.10 **Transverse and midsagittal reconstruction of the lumbar spine in the bone window after an unsuccessful myelogram (injection of 0.4 mL of iohexol) at L4–L5. Contrast and gas are present in the epidural space (L7–S1) and some contrast in the central canal from L6 until T13. There is a very large amount of contrast medium at the level of L3–L4 in the middle of the spinal cord. The entire spine is subjectively swollen, and there is a decrease of the normal fat surrounding the spinal cord. A moderate amount of disc material is protruding in the vertebral canal, partially mineralized, at the level of L3–L4 creating a moderate compression on the spinal cord. There is a high suspicion of myelomalacia of the spinal cord with leakage of contrast medium at that level.**

- Spinal cord changes and intramedullary swelling, which can be hyperdense due to contusion/haemorrhage or hypodense due to oedema, and which can be difficult to assess on CT. Intravenous contrast can be helpful. Myelography is also beneficial to indirectly show spinal cord swelling, with attenuation of the contrast columns. Myelomalacia, which can develop when the spinal cord is severely injured, can be identified on CT myelography by contrast material leakage into the spinal cord parenchyma.
- Poor delineation of the nerve roots/spinal nerves may indicate avulsion (brachial or lumbar plexus). Myelography can again help evaluate the nerve roots. (This will be covered in another section.)

24.5.1.1.3 Vertebral column stability assessment

The damaged vertebra should be thoroughly evaluated to assess whether the lesion is intrinsically unstable. Many structural elements contribute to the vertebral column's stability, and damage to too many elements will cause instability and necessitate external splinting or internal fixation. Stability is commonly assessed radiographically by dividing the vertebral body into three compartments (Figure 24.11) and evaluating each compartment's integrity; damage to two or more will cause instability, raising the possibility of further spinal cord contusion and/or compression. Figure 24.11 shows the three compartments:

- *Dorsal*: spinous processes, the supraspinous, interspinous and flaval ligaments, the articular facets and the dorsal part of the pedicles;
- *Middle*: the ventral part of the pedicles, the dorsal longitudinal ligament, the dorsal portion of the annulus and the dorsal part of the vertebral body;
- *Ventral*: the lateral and ventral portions of the annulus, the nucleus pulposus, the ventral longitudinal ligament and the ventral part of the vertebral body.

24.5.1.2 Atlantoaxial subluxation/instability

Atlantoaxial subluxation is a relatively common cause of clinical signs in dogs that presents as both neck pain and a cervical myelopathy. The condition can follow a traumatic incident that disrupts the various connective tissue elements joining the atlas and axis, causing laxity between the two bones and permitting damage to the spinal cord and/or the nerve roots at the

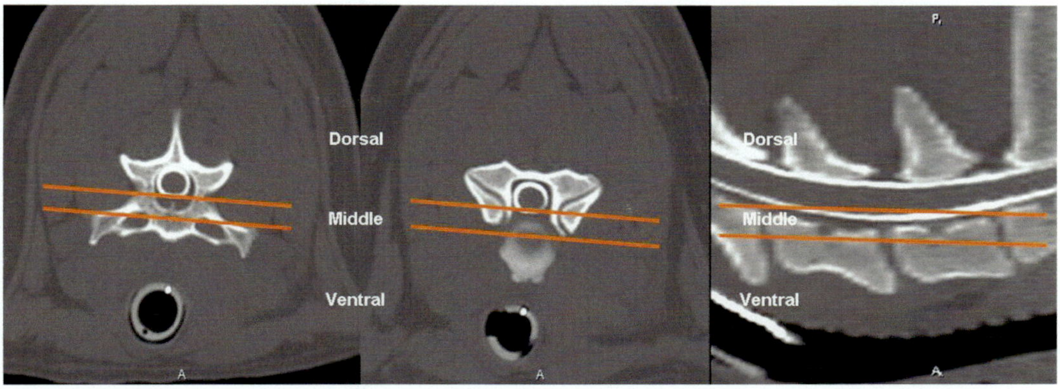

Figure 24.11 Two transverse images and a midsagittal reconstruction of the cervical spine in a normal dog showing the limits of the three compartments.

trauma site. The more common cause, however, is an intrinsic bony anatomical defect that leads to chronic instability, compression and an increased risk of acute subluxation. These cases may present as acute or chronic problems with progressive clinical signs.

24.5.1.2.1 Anatomy of the atlantoaxial joint

The atlantoaxial joint is synovial. The normal dog has several ligamentous structures connecting the two bones:

- The dorsal atlantoaxial ligament runs from the spinous process of the axis to the dorsal lamina of the atlas—cutting this ligament markedly increases the rotational instability between the two bones.
- The apical ligament of the dens, that extends from the cranial extremity of the dens to the ventral foramen magnum (Figures 24.12 and 24.13).
- Two alar ligaments arise from a similar site on the apex of the dens, diverge to run obliquely and cranially and attach to the occipital bones medial to the condyle (Figure 24.13).
- The transverse ligament of the dens (Figures 24.12 and 24.13)—a robust structure overlying the dens and running

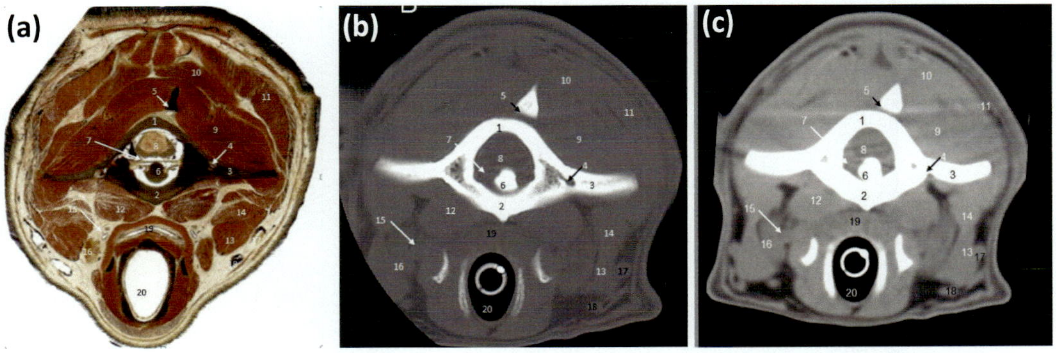

Figure 24.12 (a–c) Transverse images at the level of the atlantoaxial joint: 1. Dorsal arch of atlas; 2. ventral arch of atlas; 3. wings of atlas (transverse process); 4. transverse foramen; 5. spinous process of axis; 6. Dens (odontoid process); 7. transverse ligament of atlas; 8. myelencephalon; 9. obliquus capitis caudalis muscle; 10. semispinalis capitis muscle; 11. splenius capitis muscle; 12. longus capitis muscle; 13. sternocephalicus muscle; 14. cleidomastoideus muscle; 15. common carotid artery, vagosympathetic trunk; 16. medial retropharyngeal lymph node; 17. maxillary vein; 18. linguofacial vein; 19. laryngopharynx; 20. larynx.

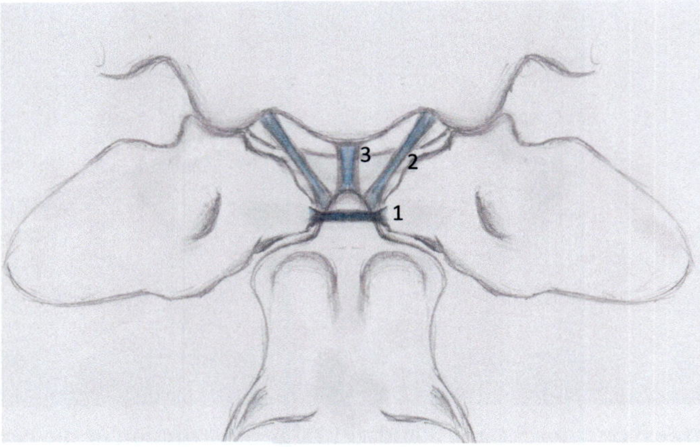

Figure 24.13 Ventrodorsal image of the canine the atlantoaxial joint with presentation of the ligaments. 1. Transverse ligament of the atlas, 2. Alar ligament, 3. Apical ligament of the dens.

in the transverse plane—has the important function of preventing the dens from moving dorsally into the spinal canal. A bursa is present between the dens and the transverse ligament.

24.5.1.2.2 Anatomical abnormalities

Toy breed dogs are predisposed to atlantoaxial subluxation. They commonly exhibit abnormalities of the dens, with hypoplasia; sometimes they have no discernible dens, creating instability in the dorsoventral plane. Regions of incomplete ossification in the atlas have been described recently, frequently in the neural arch of the atlas in the dorsal midline (Figure 24.14). Such abnormalities are over-represented in hunting dog breeds and have been associated with an increased risk of atlantoaxial subluxation.

24.5.1.2.3 Aim of CT

Diagnostic imaging is performed to confirm the diagnosis, reveal any underlying anatomical or pathological factors that may have triggered the problem and help determine whether the problem is likely to recur and whether surgery will be necessary. The following questions should be addressed:

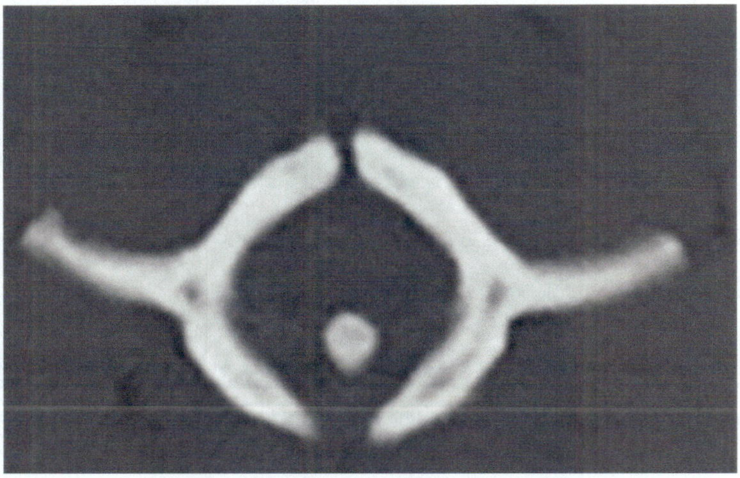

Figure 24.14 Transverse slice of the atlas showing the wide dorsal suture with blunted ends of the neural arch. Also, the intercentrum is presenting a gap in the middle part. There was no history of trauma and no evidence of neurologic symptoms in this dog.

- Are the bones of the region anatomically normal?
- Is a bone fractured?
- Is the spinal canal narrowed?
- Does the spinal cord show any evidence of compression?
- Is any instability indicated?

These assessments, in the context of the clinical evaluation and history, should help in deciding whether surgery will be necessary and establishing a reliable prognosis.

24.5.1.2.4 CT features

The entire CT study should be systematically examined, including scrutiny of the images in the bone window for the bone algorithm and the soft-tissue window for the soft-tissue algorithm to maximize the study's potential. After obtaining the transverse series, it is important to look at the images in all three planes using multiplanar reconstructions with both the bone and soft-tissue window settings. Three-dimensional reconstruction of the images will further help with the understanding of any anomaly. The contrast medium is not required; indeed, cisternal puncture is contraindicated because of the risk of the dens further compressing the spinal cord.

24.5.1.2.5 Key points in the search for lesions

Bone window

- The space between the dorsal margins of C1 and C2 and/or the odontoid process and intercentrum of the atlas is wider with atlantoaxial subluxation (Figure 24.15).
- Odontoid process changes (dens) may be absent (Figure 24.16), hypoplastic (Figure 24.17), dysplastic or fractured, or show abnormal angulation (dorsal) or dorsal protrusion (if supporting ligaments fail) (Figure 24.15).
- Other bone anomalies affecting the occipital bone or atlas, such as incomplete ossification, can predispose to the disease and are best evaluated on transverse CT images (Figure 24.14).
- The anatomical relationship between all the bones involved and bone opacity (i.e., is the dens in a position where spinal cord compression is likely and/or does the association between the two bones suggest instability (Figure 24.15).

Soft-tissue window

- The integrity of the ligaments (apical and alar ligaments and transverse ligament of the dens) if visible.
- Spinal cord compression—look for gross displacement and evidence of epidural fat signal disruption.

Intramedullary spinal cord changes (if visible); changes in signal may suggest oedema, haemorrhage and contusion.

24.5.1.3 Disc herniation

24.5.1.3.1 Key facts

Acute disc herniations may provide little change to anatomical detail; therefore, watch out for subtle abnormalities

Narrowing of intervertebral disc space, best seen in sagittal reformation, in combination with a tiny gas bubble within the intervertebral disc space

Crescent-like hyperattenuating/hyperdense dura mater, most unilateral and fading cranial or caudal to the lesion due to haemorrhage (Figure 24.17).

Spinal cord swelling, where the cord section increases and consumes the surrounding epidural fat is commonly found.

Use myelo-CT for further examination when plain and IV contrast did not correspond to the presumed neurologic site (a damaged subarachnoid space may be marked by an intramedullary contrast spot). Especially in toy breed dogs with few changes in extradural pathologic tissue, myelo-CT can be extremely helpful.

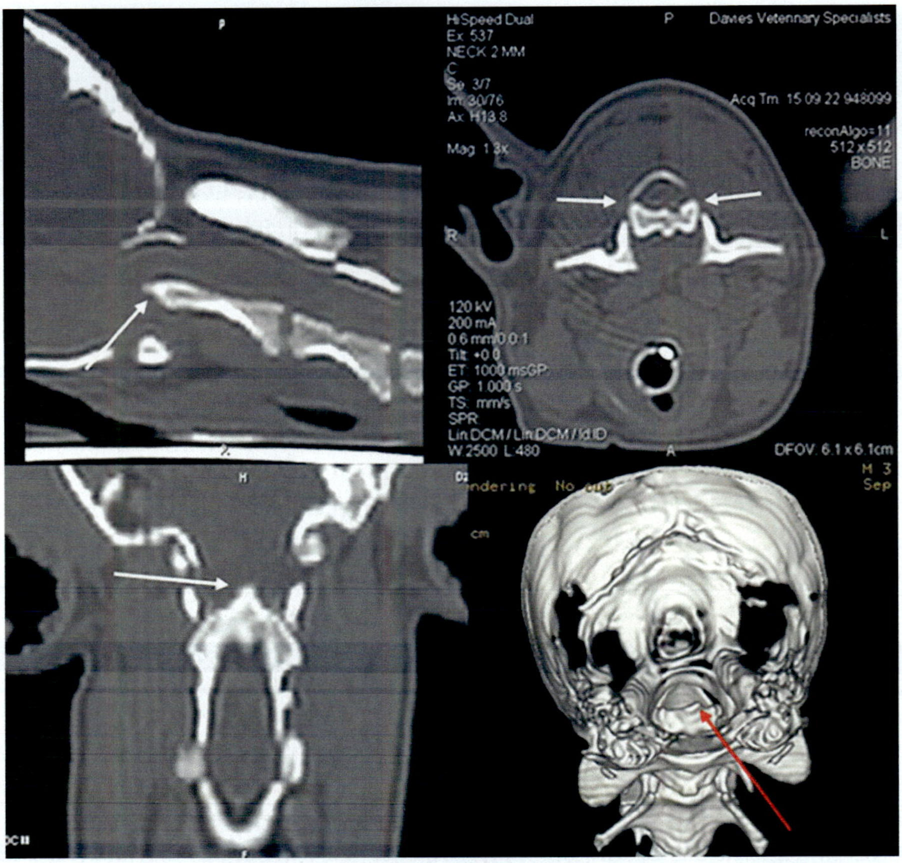

Figure 24.15 CT images of a 3-year-old Yorkshire terrier tetraparesis due to atlantoaxial subluxation. There is cranial and dorsal (arrows, b) displacement of C2 relative to C1, with protrusion of the odontoid process dorsally into the spinal canal (arrow, a). Note the small and asymmetrical shape of the odontoid process, representing hypoplasia (arrow, a, c and d); the spinal canal is markedly narrowed at this level. (a–c) Midline parasagittal, transverse and dorsal MPR images at the level of the C1–C2 joint; bone window. (d) 3D reconstruction oriented in a rostrocaudal direction to demonstrate the abnormal odontoid process protruding into the spinal canal; note the similarity in shape to the odontoid process, as illustrated in figures (a) and (c) (arrow, d).

Due to the quick scanning modalities, the whole spine can be easily examined and should be done in any case of mixed upper or lower motoneuron deficits.

Disc material always appears hyperdense (about 80–100 HU) to the spinal cord section due to chondroid attenuation of the nucleus and shows no contrast medium enhancement. Unfortunately, in acute settings, this is often mixed with haemorrhage and therefore harder to differentiate. The annulus ruptures during disc extrusion, resulting in pronounced swelling and compression and possible displacement of the spinal cord section. Displacement of the spinal cord is seen to be easier and more circumscribed when the protruded disc material contains mineral deposits than the more liquid chondroid nucleus alone. The differentiation of each individual tissue involved (disc, spinal cord, blood, oedema or fat) may be difficult to distinguish then. Measuring Hounsfield Unit (HU) can be helpful (Figure 24.18). A clear delineation between the disc and spinal cord section

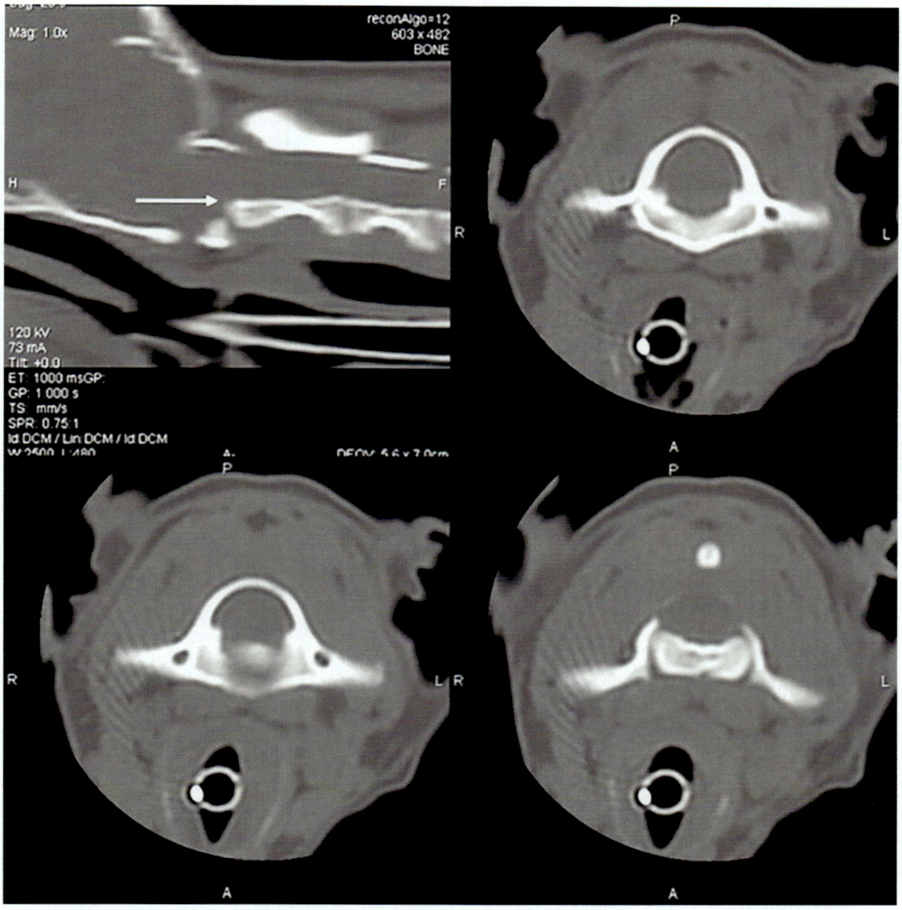

Figure 24.16 **CT images of a 7-month-old chihuahua with tetraparesis and low head carriage over the preceding 4 months, showing atlantoaxial subluxation. There is complete absence (agenesis) of the odontoid process (arrow, a) and mild dorsal displacement of the body of C2 in relation to C1. (a) Midline parasagittal MPR; bone window; (b–d) transverse images at the level of the C1–C2 joint, from cranial to caudal. Conservative and surgical treatments were offered to the owner, who elected for the former; there was a good long-term outcome.**

is seen when the entire calcified disc is solidly displaced into the vertebral canal.

Delineation of a space-occupying nonperfused tissue against normal or higher perfused tissue can be improved by IV contrast medium administration, which is a great advantage in cases of neoplasms or inflammation. Myelo-CT can be helpful when a soft tissue jet of disc material destroys the subarachnoid border of the cord, which then appear as a hyperdense spot of iodine contrast medium within the intramedullary section (Figure 24.19). Myelo-CT is also diagnostic when a singular disturbance and asymmetry occur, especially in small and toy breed dogs, or to differentiate between multiple compressive segments to find the utmost responsible; but it is less beneficial for non-disc-associated pathology (nerve root swelling and intramedullary neoplasms).

24.5.1.4 Compressive hydrated nucleus pulposus extrusion (HNPE)/Discal cyst

Compressive hydrated nucleus pulposus extrusion (HNPE), also previously described as discal cyst or ventral intraspinal cyst, is an atypical type of disc extrusion in the dog. Overall, HNPE is a relatively rare condition causing a painless acute tetraparesis or tetraplegia in

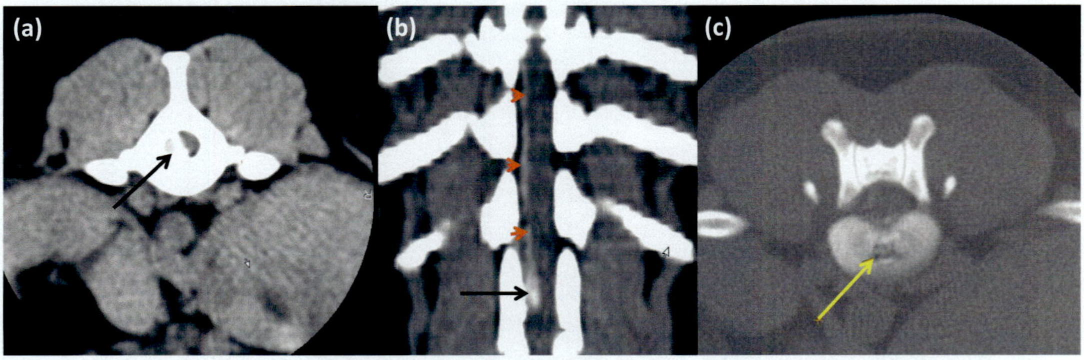

Figure 24.17 Acute thoracolumbar disc protrusion followed by dural haemorrhage cranial to it, which enlightens the dura mater over several segments in a Dachshund, 11 years old (a and c are transverse slices, soft-tissue and bone window), the black arrows show the area with the acute prolapse and the haemorrhage in a dorsal reformation (b). (c) Bone window, shows the collapsed disc space with hypodense remnants of the nucleus (vacuum-phenomenon, yellow arrow).

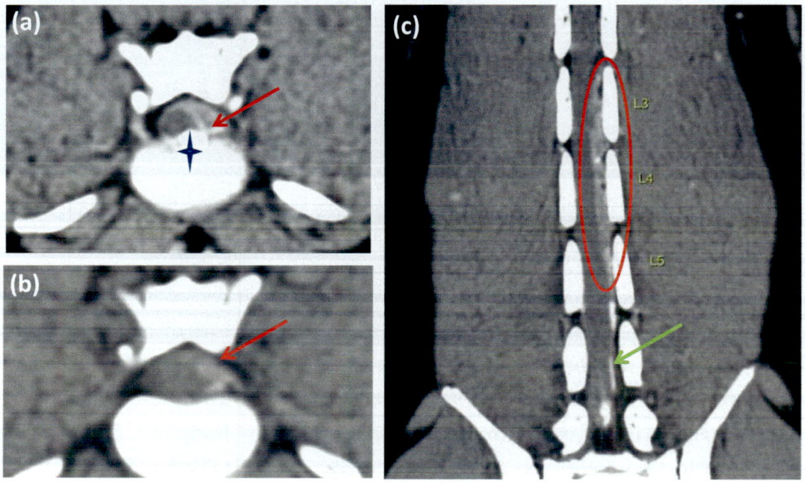

Figure 24.18 Transverse (a, L3–L4; b, L4–L5) and dorsal reformatted (c), CT images in soft-tissue window, of the lumbar spine. L3–L4 shows a mineralized disc (star). A large amount of hyperattenuating (HU 25–40) material of epidural origin is present from the cranial aspect of L3 until the most cranial aspect of L5, causing a secondary severe left-sided compression of the spinal cord on this entire segment (red arrows and oval). At the level of L3–L4 and L4–L5, this material seems to extend till the left foraminal region. Multifocal mineralization of the pachymeninges is seen in the caudal lumbar spine compatible with pachymeningitis ossificans (green arrow). An extended disc herniation L3–L4–L5 with suspected large amount of extruded/migrated disc and severe epidural haemorrhage/inflammation is concluded.

predominantly older small breed dogs. It affects chondrodystrophic and non-chondrodystrophic dog breeds. It is an acute extradural compressive lesion associated with the intervertebral disc, probably consisting of hydrated nucleus pulposus. The exact pathophysiological cause of such an extrusion remains to be elucidated. A HNPE in dogs is typically a cervical disorder, with disc spaces C3–C4 and C4–C5 most commonly affected. Dogs with HNPE most often present with acute (<24 hours) symmetric nonambulatory tetraparesis or tetraplegia. Despite

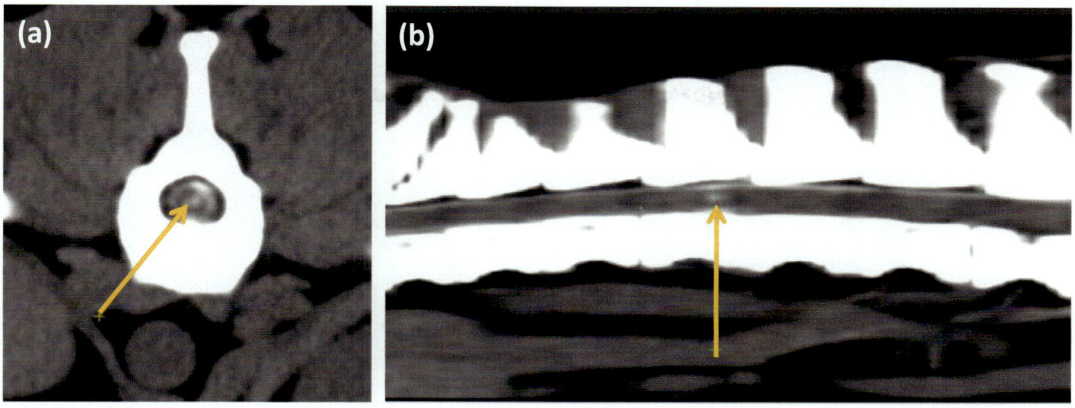

Figure 24.19 Myelo-CT ((a) transverse and (b) sagittal reformatted image in soft-tissue window) of the thoracolumbar spinal cord of a large mixed-breed male dog 10 years old. Intramedullary contrast medium is present due to damage of the subarachnoid barrier suspecting high-velocity disc extrusion (arrows). (Courtesy Dr. Henning Schenk, Tierklinik Lüneburg, Germany.)

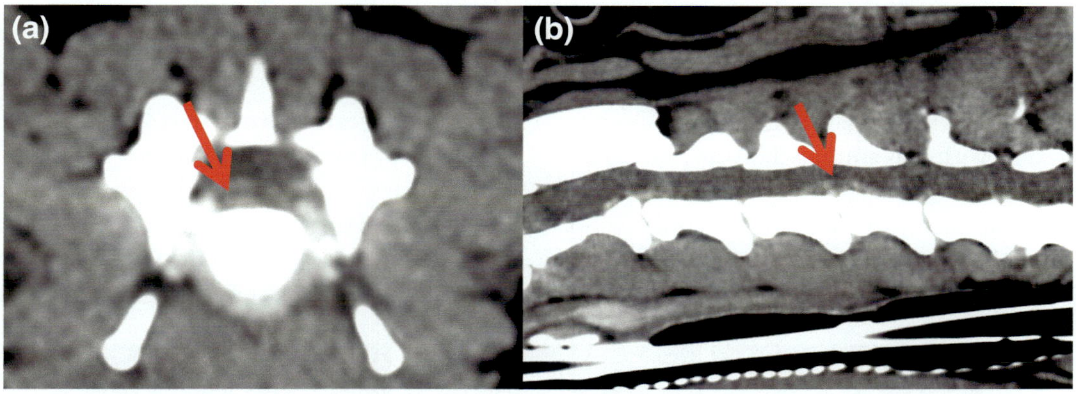

Figure 24.20 Transverse (a) and midsagittal reformatted (b), post-contrast CT images in soft-tissue window, of the cervical spine. (a) Presents the level of the intervertebral disc space of C4–C5. An extradural hypodense, compressive structure surrounded by rim enhancement is present (arrow). On (b) the extent of the HPNE with extradural compression on the spinal cord can be appreciated (arrow).

severe neurological deficits, the clinical outcome of most patients with HNPE is favourable with either medical or surgical management.

On CT, a HNPE appears as an extradural compressive hypodense structure dorsal to the intervertebral disc space, and often rim enhancement is noticed on the post-contrast transverse images and better on sagittal reconstruction (Figure 24.20). CT can be used as a first diagnostic tool for HNPE when MRI is not available.

24.5.1.5 Acute infarct or fibrocartilaginous embolism is not visible on CT

In some cases, a local swelling of the spinal cord is seen. Patients presented with acute asymmetric neurologic symptoms in combination with exclusion of other spinal cord findings on CT.

24.5.16 A haemorrhage within the spinal cord may be not visible on CT

However, it can be identified as a hyperdense lesion (Figure 24.21). Usually, the neurologic symptoms occur with acute onset in these cases.

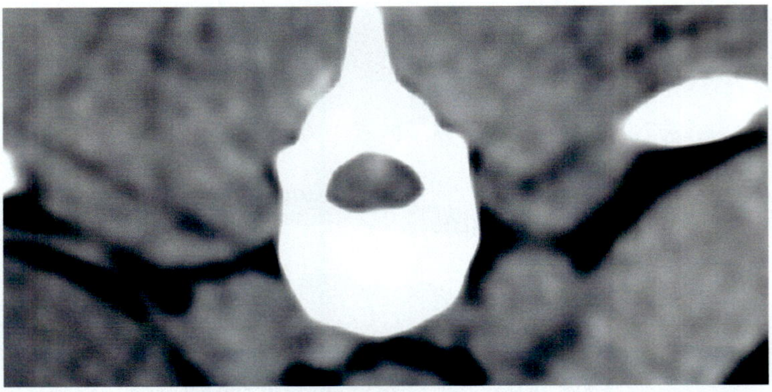

Figure 24.21 Transverse slice of T13 in soft-tissue window of a dog with acute paraplegia. An hyperdensity is present in the dorsal aspect of the spinal cord. Suspicion of spinal cord trauma with intramedullary haemorrhage.

24.5.2 Chronic

24.5.2.1 Chronic disc disease

24.5.2.1.1 Key facts

Bulging of the dorsal longitudinal ligament together with the annulus fibrosus, median or unilaterally.

Displacement of the spinal cord section within the vertebral canal, central dorsally or asymmetric left or right, or into the foraminae.

Rather clear appearance of the bilateral epidural fat.

Mineralization of soft tissues like ligaments, annulus and nucleus or the whole disc.

Various widths of the intervertebral disc spaces, from nearly normal to collapse.

Sclerosis of endplate formation.

Presence of spondylosis and/or spondylarthrosis.

Multiple disc protrusions (mostly in the thoracolumbar area), often slightly different

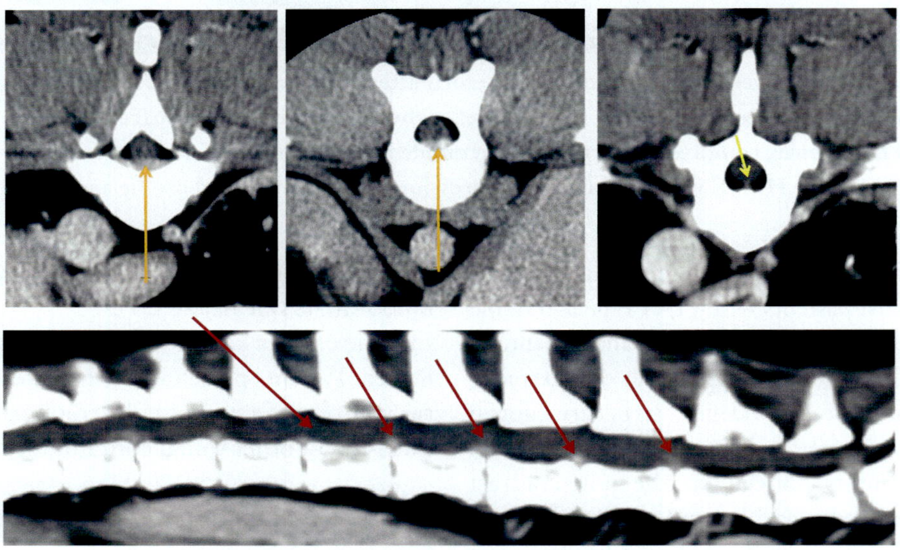

Figure 24.22 Chronic thoracolumbar disc protrusions in the thoracolumbar area in a Labrador Retriever, 10 years old. The difference in chronic protrusion (orange arrows) and thickening of the longitudinal ligament is noticed (yellow arrow). The sagittal reconstruction of the lumbar spine shows multiple protrusions, causing slight-to-mild extradural compression (red arrows).

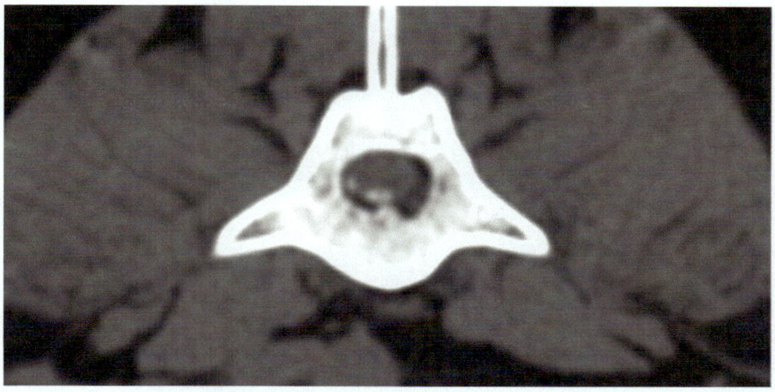

Figure 24.23 **Transverse slice of a lumbar vertebra presenting plaques of bone on the inner dural surface.**

in height (which makes it hard to get the most responsible one).

Only single focal displacement of the spinal cord.

Presence of dural calcifications.

NB: a chronic protrusion also can exacerbate and then will show signs of acute protrusion with unclear margins due to oedema and haemorrhage, displacement and various calcifications, too.

CT images of a singular protrusion will not confirm whether it is acute or chronic (Figure 24.22). In chronic cases, attention should be paid to the asymmetric position of the spinal cord without (acute) thickening of the cord section; varying widths of fatty tissue around the cord are the result of this displacement. Hyperattenuation of the long dorsal ligament, resulting in metaplastic thickening due to continuous mineralization, as well as, occasionally, bony bridging, may also be seen. Mineralized plaques of the dura mater may also be found on the spinal cord in various locations, lateral, dorsal and ventral (Figure 24.23).

Aging discs may protrude in any direction, mainly dorsally median or paramedian, very often with a strong lateral or foraminal shift. Tissue density of the disc may increase due to mineralization, but can also decrease because of the presence of free gas. The vacuum phenomenon arises from sudden or constantly increased pressure on the ruptured chondroid matrix (see also Figure 24.17, showing an acute state). Left or right laterally bent spinous processes are found due to asymmetric musculature forces and scoliosis. IV contrast medium administration leads to slightly improved tissue delineation, but more importantly, it can rule out other inflammatory or neoplastic conditions.

24.5.2.2 Neoplastic conditions

24.5.2.2.1 Key facts

Bone:

Loss of cortical bone, coarse structure of the vertebral body, focal or general loss of spongiosal structure, pathologic fractures and shortening of vertebral body.

Periosteal reactions vary; they can be absent or irregular spikey or broad lamellar (depends mainly on aggressiveness of tissue and time).

Enlargement of neuroforaminae, usually singular and unilateral (mainly brachial and lumbar/sciatic plexus, but also seen along the whole cord; pay attention to high contrast enhancement of a meningioma).

Soft tissue:

Replacement of bone tissue by singular or multiple focal infiltrative tissue within the vertebra but also extraosseous evasion.

Focal compressive soft tissue within the vertebral canal at varying sites, which compresses and obscures the spinal cord section, at the bottom of the vertebral canal sometimes seen over several segments and enhancing in connection with the central midline vein.

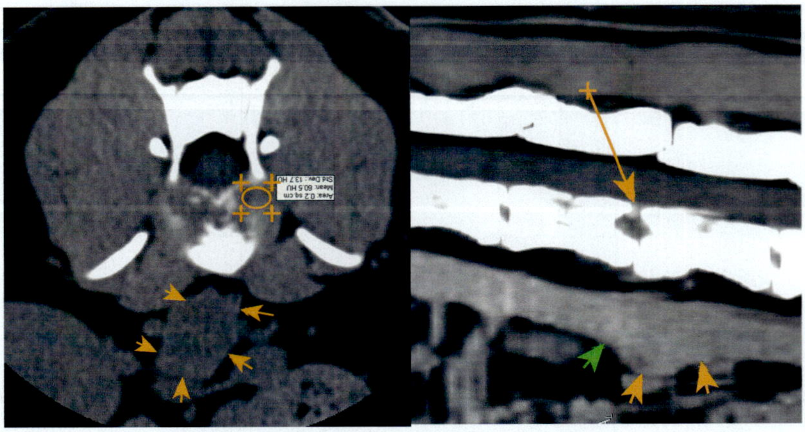

Figure 24.24 **Suspected neoplastic destruction of the intervertebral disc area at L2–L3 due to a soft tissue sarcoma (Great Schnauzer fs, 3 years). The transverse image, native soft tissue window, shows clearly the lysis. On the sagittal reconstruction after IV contrast administration, minimal contrast enhancement is present at the level of the disc of L2–L3 (arrow). Note the enlarged abdominal aorta with an irregularly thickened wall (arrowheads).**

Plain HU measurements of about +40–60 HU and normal-to-moderate contrast uptake of about +70–100HU; a specific tissue diagnosis is not possible and always a biopsy is needed; exceptions to this are meningiomas with a rather high contrast uptake.

Chondroid or fibroid neoplasms occurring from the intervertebral disc area is very rare. The grade of deviation of the spinal cord at the disc level varies and cannot be distinguished from any other protrusion on a plain scan; however, administration of contrast medium markedly enhances the tissue and leads to good delineation. In advanced cases, ruling out an infection can be difficult (Figure 24.24).

Other neoplastic tissues that may arise within the vertebral canal or at the disc level may be primary or of metastatic origin (e.g., lymphoma); the soft-tissue density may be similar to the spinal cord section, which also can obscure the branching nerve. Haemorrhage or even oedematous fluid can cause similar effects. In such a case, IV contrast medium administration can rule out disc disease but cannot differentiate between all other possible causes (Figure 24.25). A follow-up examination, changing the modality (e.g., MRI) or surgical biopsy will further clarify the underlying condition of the pain.

Osteosarcoma or chondrosarcoma that affects the vertebral body, the lamina or the articular facets can be seen clearly, depending on the extent of cortical loss, trabecular destruction or adjacent periosteal reactions. Displacement of the spinal cord can be seen in unilateral disease, although pediculi and lamina involvement leads to overall spinal cord compression, which can be seen again clearly after contrast medium is administered intravenously.

In cases of multiple myeloma or lymphoma or other types of soft tissue sarcoma, the bone marrow is affected (Figure 24.26). Changes in the vertebra's spongy texture are not easily detected, and senile osteoporosis can also have a spongy texture, so every heterogeneous coarse pattern should not indicate a possible severe disease. Focal spots of fatty bone marrow accumulation with loss of spongy architecture are usually benign. Doberman pinschers normally have extraordinary coarse spongious vertebral bone that must not be judged as infiltrative disease. However, special attention should be paid to minor cortical bone losses in small sections of a vertebra, such as articular

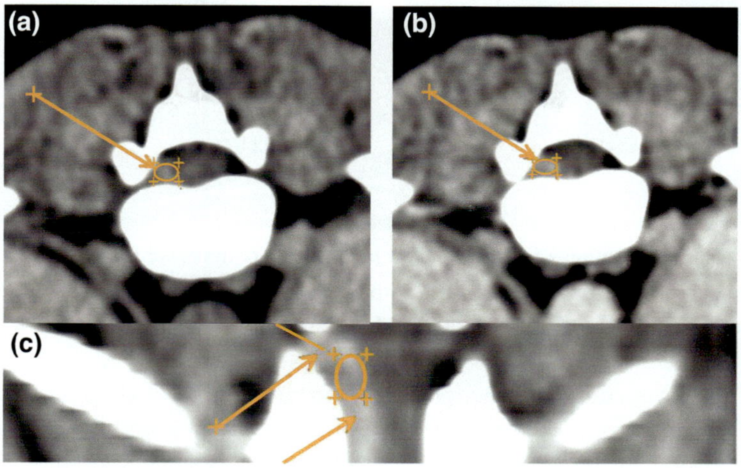

Figure 24.25 **Thoracolumbar right lateral extradural mass revealing lymphoma on explorative surgery (Crossbred, male, 9 years). The space occupying soft tissue intracanalicular mass at the level of T12–T13 shows a relatively low density (46 HU) compared to disc material, which also could be suggestive of focal haemorrhage (a, native soft-tissue window). (b) Transverse and (c) dorsal reconstruction after IV contrast administration reveal the moderate enhancement of the mass.**

facets, spinous processes or adjacent ribs. In such cases, the ribs play an interesting role in detecting infiltrative bone disease in ageing animals; multiple fractures at different levels of the rib body and surrounding small periosteal reactions can accompany these obscured fractures. Abnormal soft-tissue appearance and contrast enhancement can be detected within these altered bony parts.

Neoplasms occurring from an epidural ganglion or the nerve root itself sometimes cannot be distinguished from unilateral protrusion, unless the ROI values of around +40 HU appear similar to the spinal cord section; little-to-moderate enhancement is seen following IV contrast administration, which will be missing in case of disc protrusion. Particular attention should also be paid to the width and shape of the vertebral canal and that of the neuroforaminae; when comparing both sides, one is often slightly more uneven or deepened. In the early state of nerve root thickening, the asymmetry might not be so obvious and may also be caused by inflammation; therefore, follow-up examination and close clinical care will help clarify borderline changes.

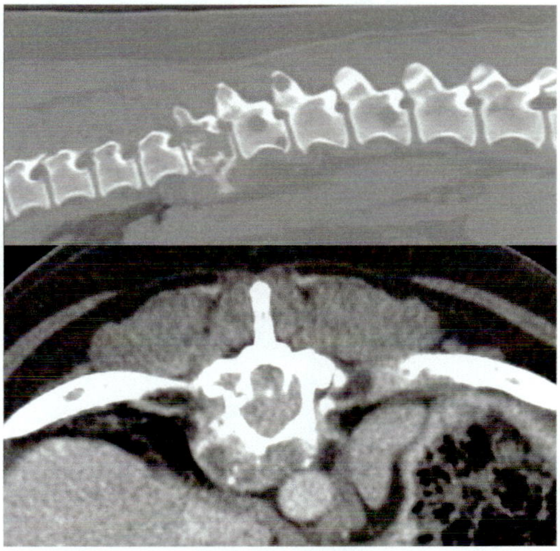

Figure 24.26 **On top, sagittal reformatted image in bone window. Below, transverse post-contrast image of T11 in soft-tissue window. Suspected multiple myeloma in a 5-year-old Rottweiler. Multiple lytic punched out lesions are present in the thoracic and lumbar vertebrae. In T11, lysis of the vertebral spongious body structure and the irregular border of the evading neoplasm are visible. An extradural, extramedullary soft tissue mass is contrast-enhanced and causes spinal cord compression.**

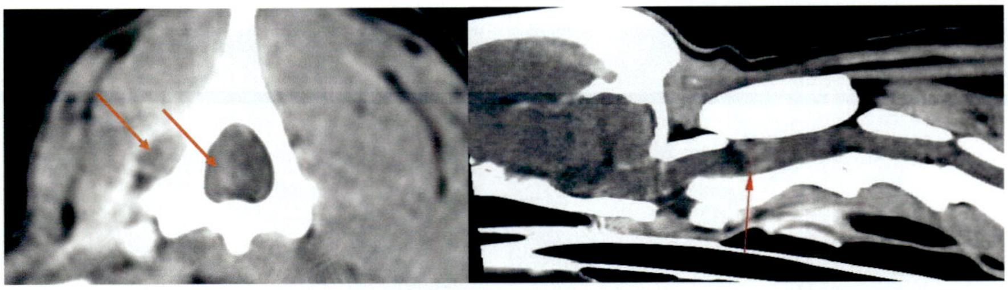

Figure 24.27 IV contrast images of the cervical spine. Left transverse image of C2, right sagittal reformatted image. An intradural extramedullary mass at the level of C2 was histologically confirmed being a meningioma with high malignancy (arrows). There is extension of the mass in the surrounding soft tissues.

An intradural/extramedullary neoplasm cannot be differentiated from the spinal cord in plain images either. If large enough, IV contrast-enhanced studies or myelo-CT are helpful to visualize a space-occupying lesion of this type. Special attention shall be paid to the contrast uptake of the meninges. Focal uptake of the contrast medium up to +100HU can be seen in mass effects arising from the soft tissues around the cord and possibly evading along nerve root canals (Figure 24.27).

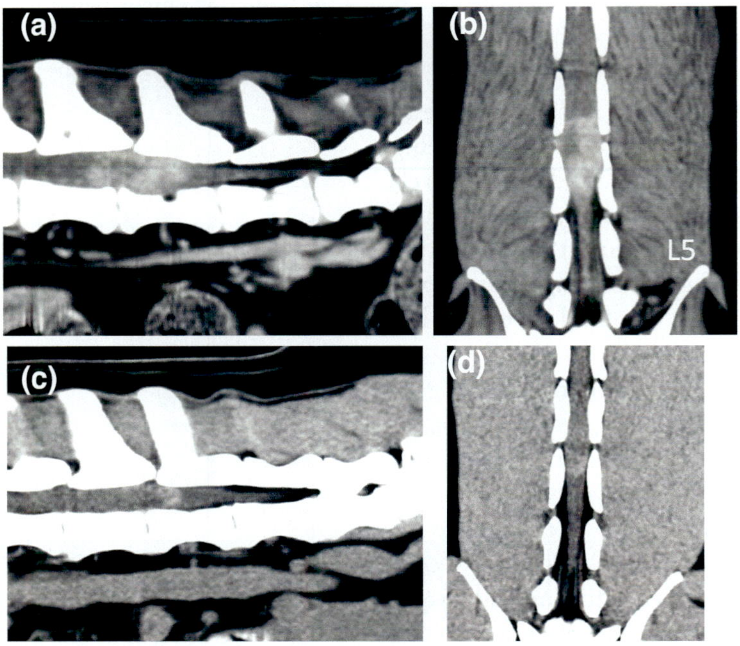

Figure 24.28 IV contrast, sagittal (a) and dorsal (b) reformatted images of the lumbar spine in soft-tissue window. The caudal lumbar spinal cord is focally swollen just cranially to the cauda equina, at the level of L4–L5. An intramedullary mass is demonstrated with a homogeneous enhancement (HU 100) after IV administration of iodinated contrast medium, especially on the left side of the spinal cord. The lumbar spine is deformed with a larger canal at that level; there is thinning of the bone (arch+lumbar process) of L5 with a smooth internal margin. This finding is secondary to chronic pressure atrophy caused by the intramedullary mass. (c) and (d) represent survey sagittal and dorsal reformatted images of the lumbar spine, respectively. At L4–L5, an ill-defined, ovoid hyperintensity (70 HU) is noticed. After IV iodinated contrast administration, no enhancement was seen. The lesion was confirmed as an intramedullary haematoma.

An intramedullary neoplasm has to alter the whole section of the spinal cord and exhibit a perceivable loss of extradural fat; combined with a widening of the vertebral canal; there might be a suspicion on plain scans. IV contrast medium administration may help when this part of the cord undergoes enhancement or a decrease in attenuation, but recognition of tissue changes very much rely on the size and extent of the lesion (Figure 24.28).

24.5.2.3 Spinal infections

24.5.2.3.1 Key facts

Swelling of the spinal cord—ambiguous finding, focal contrast enhancement may help suspect an inflammatory condition and pay attention to possible muscle wasting along nerval involvement.

Discospondylitis—destructive "widening" or collapse of intervertebral disc space, irregularly shaped lucent endplates, usually heavy periosteal reactions, perivertebral/intracanalicular soft tissue thickening and contrast enhancement, haematogenous in origin with possible multifocal appearance and locally also due to foreign material or perforation (bite wound).

Post-surgical complications after treatment of intervertebral disc disease (e.g., ventral slot) or lumbosacral neuroforaminal stenosis include intracanalicular blurring of nerve tissue and fat stranding, focal soft tissue swelling and hypoattenuating perivertebral muscle appearance like abscessation or necrosis; also sinus tracts may be found.

The findings for spinal infections very much rely on the location, stage and duration of disease. An abnormal loss of fat within the vertebral canal due to spinal cord swelling or the thickened peridural tissue (not really seen at the cervical and lumbar intumescence) may be a first sign during a plain scan series. Tissue changes in myelitis cases will be suspected only following IV contrast. This will be seen as a focal hyperattenuation within the spinal cord and, depending on the amount of inflamed tissue, as a more unilateral spot or full enhancement of the entire cord section (Figure 24.29).

At the level of C3, a slightly hyperattenuating spot is visible in the survey transverse image (a) that enhances after IV contrast medium administration (b). The spinal cord shows a mean of 65HU at C3 compared to 46HU at C4 on the survey images. In the sagittal reformatted post-contrast image (c), a mean of 99HU at C3 is measured compared to 40HU at C4.

Spinal infections may be also a sequela long after surgical intervention for treating intervertebral disc disease. Tissue changes may indicate some sort of discospondylitis; in severe cases, a sinus tract will follow. Focal muscle wasting along the disease process may also be seen.

With discospondylitis, soft-tissue proliferation is often seen not only within the vertebral canal but also extracanalicular lateral and ventral

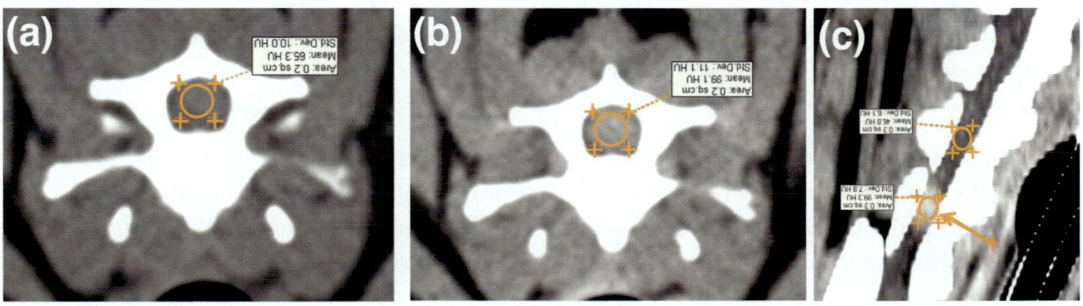

Figure 24.29 **Suspected focal myelitis at C3 in a 4,5 year-old Beagle. At the level of C3 a slightly hyperattenuating spot is visible in the survey transverse image (a) that enhances after iv contrast medium administration (b) the spinal cord shows a mean of 99HU at C3 compared to 65HU on the survey images. In the sagittal reformatted post contrast image (c) a mean of 99HU at C3 is measured compared to 46HU at C4.**

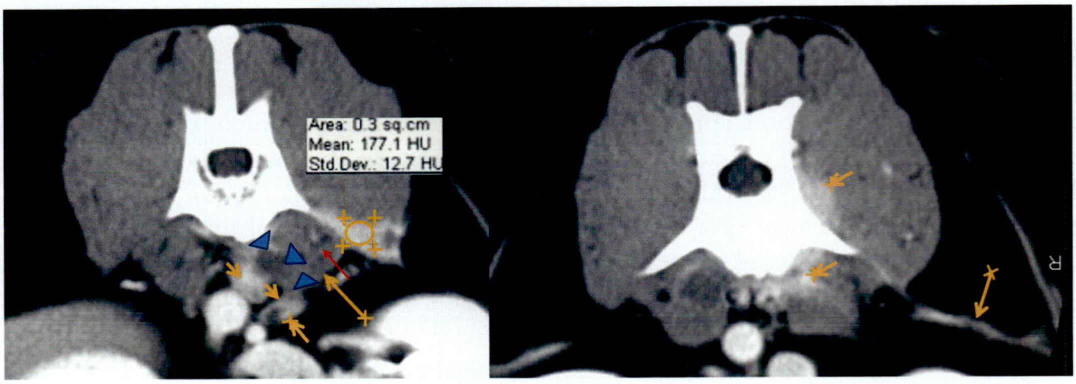

Figure 24.30 Suspected unilateral soft tissue inflammation in a 2-year-old cross-breed. Unilateral soft tissue changes are involving the following: the lateral right side of the second lumbar vertebra and surrounding soft tissues, the pillar of the right diaphragm, the right iliopsoas muscle and the right transverse abdominal fascia. On both contrast-enhanced CT images, the lateral side of L2 (orange arrowheads) is enhanced, and both sections of the iliopsoas muscle appear heterogeneous and swollen (blue arrow). Irregular hyperdense foci (blue arrowheads) appear in the ventral aspect of the vertebra. The transverse abdominal fascia (orange arrow) and muscular borders along the transverse process show 177 HU after contrast administration This is indicative for inflammation; however, an associated foreign body could not be proven.

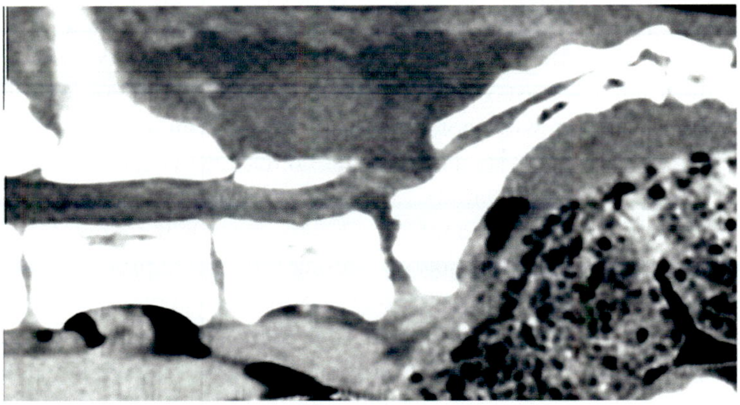

Figure 24.31 Post-operative, sagittal reformatted CT image after IV contrast, 10 days after dorsal laminectomy for the treatment of lumbosacral stenosis. A large abscess is centred on the surgical site, extending from the subcutaneous area until the lumbosacral disc and severely compressing the cauda equina. Signs of active bone changes of L7–S1 are compatible with discospondylitis.

to the vertebral body. Also, symmetrical periosteal reactions cover the cortical bone along the endplates and pediculi. In advanced cases, loss of endplate structure of both adjacent vertebrae, with both appearing moth-eaten and irregular, is common (Figure 24.30). Intervertebral space collapse may even occur. Along the border of the vertebral body's normal spongy tissue, demarcating, hyperdense attenuation, which can be clearly seen in longitudinal multiplanar reformations, is usually present.

The spread of inflammation into, or the involvement of the perivertebral musculature, can be visualized best after IV administration of the contrast medium; therefore, careful screening of all series and images and multiplanar image reformations are strongly recommended. Typically, a bright, soft-tissue ring covers the whole segment and enhances the proliferating soft tissue within the vertebral canal, obscuring the spinal cord and/or the nerve root sections. The HU measurements usually show values of attenuation greater than plus 100 HU (compared with normal enhancement of 20–30 HU after IV contrast administration). In cases of abscessation, spots or irregular formations of hypodense or non-enhancing tissue within the hyperdense soft tissue can be found.

A foreign body of unknown origin may be suspected if unilateral tissue damage is pronounced. In this case, the surrounding tissue should be checked carefully for a possible migrating fistulous tract, which will be only clearly seen after IV contrast because of the highly enhanced inflamed tissue. The lesion's location should also be considered because a migrating, inhaled foreign body will mainly cause unilateral thoracolumbar inflammatory spondyloarthropathy, whereas haematogenously spread discospondylitis can occur at any level of the vertebral column. Other causes of discospondylitis include bite wounds (especially in cats) or sequelae of bullet wounds and other perforating trauma (Figure 24.31).

24.5.2.4 Cervical spondylomyelopathy

Cervical spondylomyelopathy (CSM) or "wobbler syndrome" is a complex and incompletely understood neurological syndrome. It can be considered a collection of disorders, in which a combination of bony and soft-tissue abnormalities results in varying degrees of cervical vertebral canal stenosis and spinal cord compression. This syndrome can affect several large and giant breed dogs. Young adult Great Danes and older Dobermann pinschers are overrepresented in several studies. Many lesions have been attributed to this disorder, leading to a plethora of similar meaning terms to refer to this neurological syndrome. These include spondylolisthesis, cervical spinal subluxation, cervical vertebral instability, cervical stenotic myelopathy, cervical spondylopathy, caudal cervical spondylomyelopathy, wobbler syndrome and cervical malformation/malarticulation syndrome. Clinical signs can range from cervical hyperaesthesia without neurological deficits to non-ambulatory tetraparesis. However, the most common clinical presentation is a gait disturbance predominantly affecting the pelvic limbs. A wide-based ataxia and/or paresis of the pelvic limbs in combination with a short stilted thoracic limb gait can be seen. Based on the characteristic signalment and imaging findings of the affected dogs, more separate syndromes have been recognized over the years. It is unclear if these reflect distinct disease entities or spectra of the same disorder.

- Disk-associated cervical spondylomyelopathy (DA-CSM): this is the most common type of CSM and is also referred to as disk-associated wobbler syndrome. This type of CSM typically affects adult to older large breed dogs, such as the Dobermann pinscher. In DA-CSM, spinal cord compression is mainly caused by one or more type II intervertebral disk protrusions. This disk-associated compression can be seen in combination with dorsal compression resulting from ligamentum flavum hypertrophy and abnormalities in the shape and position of the vertebral bodies. Vertebral malformations are generally mild and can vary from flattening of the cranioventral border to a triangular-shaped vertebral body. An abnormal position of the vertebral body is typically characterized by craniodorsal tilting into the vertebral canal. Other abnormalities that can be seen in dogs with DA-CSM are a collapsed intervertebral disc space, spondylosis deformans and a funnel-shaped vertebral canal with a narrowed cranial orifice. The intervertebral disc spaces between C6–C7 and C5–C6 are most commonly affected. Multiple sites of spinal cord compression can be seen in 25%–52% of affected dogs (Figures 24.32–24.34).

- Osseous-associated cervical spondylomyelopathy: this type of CSM

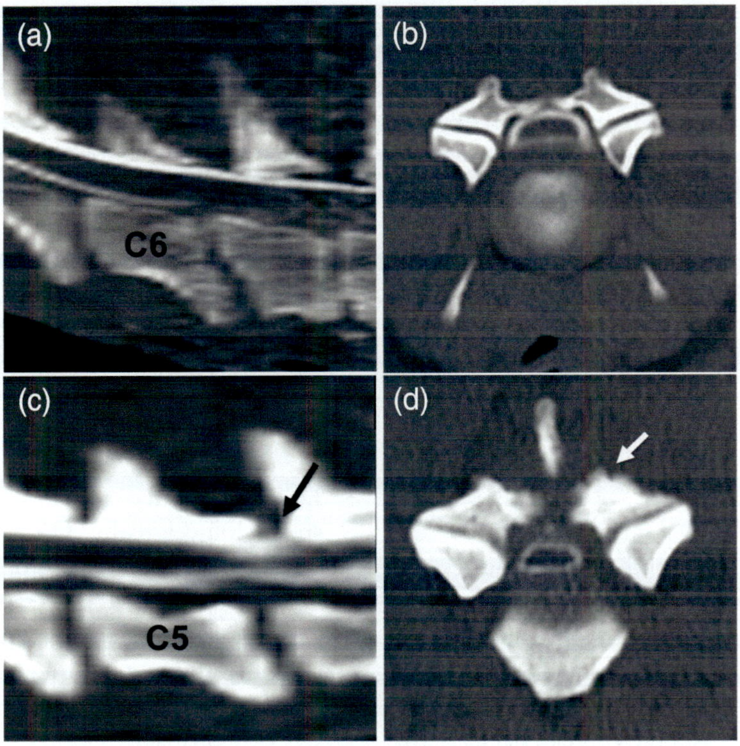

Figure 24.32 CT-myelographic studies of a 6-year-old Doberman pinscher (a and b) and a 9-year-old Whippet (c and d) with DA-CSM. (a) Sagittal reconstruction demonstrates ventral disc-associated compression at the level of C6–C7. A malformed C7 vertebral body with flattening of its ventrocranial border and a collapsed intervertebral disc space can also be seen. (b) Transverse image at the level of C6–C7 demonstrates disk-associated spinal cord compression with flattening of the spinal cord. (c) Sagittal reconstruction demonstrates mild ventral disc-associated spinal cord compression combined with dorsal compression caused by ligamentum flavum hypertrophy at the level of C5–C6 (black arrow). (d) Transverse image at the level of C5–C6 confirms the mainly dorsal spinal cord compression. The dorsal compressive tissue has soft-tissue attenuation characteristics. Mild periatricular new bone formation can also be seen (white arrow).

Figure 24.33 Sagittal-reconstructed CT study of a 7-year-old Doberman pinscher with DA-CSM. Craniodorsal tilting of the C6 vertebral body into the vertebral canal can be seen (white arrow).

typically affects young adult giant breed dogs, such as the Great Dane. In osseous-associated CSM, spinal cord compression is mainly caused by hypertrophied/abnormally shaped articular processes. This dorsolateral to lateral compression is sometimes seen in combination with hypertrophy/proliferation of the dorsal lamina, pedicles and surrounding soft-tissue structures. Vertebral malformations can also be seen in this type of CSM. The intervertebral disc spaces between C6–C7, C5–C6 and C4–C5 are most often affected, but more cranial

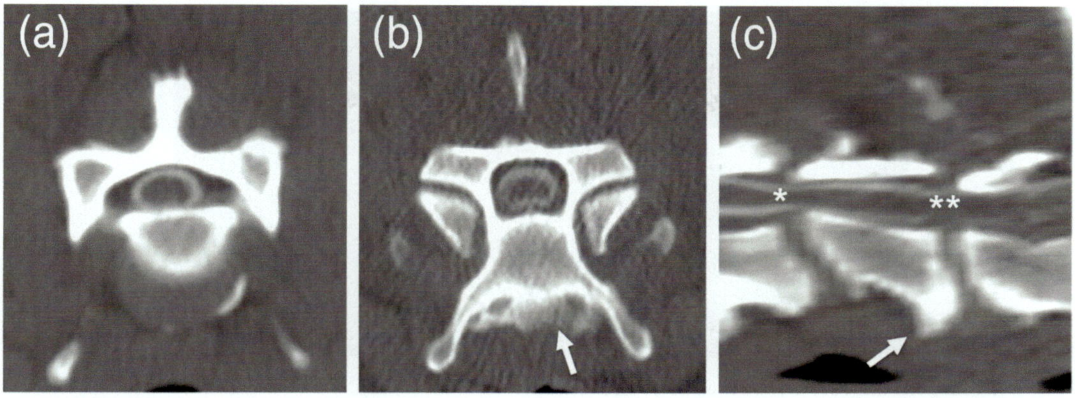

Figure 24.34 CT-myelographic study of a 9-year-old Doberman pinscher with DA-CSM. Transverse images at the cranial (a) and caudal (b) orifices of C6. The vertebral canal at the level of the cranial orifice (a) is narrowed compared to the vertebral canal at the caudal orifice (b). Spondylosis deformans at the caudal aspect of C6 (arrow). (c) Sagittal reconstruction demonstrating a narrowed cranial orifice (*) compared to a wider caudal orifice (**), resulting in a funnel shaped vertebral canal at the level of C6. Spondylosis deformans at the caudal vertebral endplate of C6 (arrow) and a narrowed intervertebral disc space between C6 and C7.

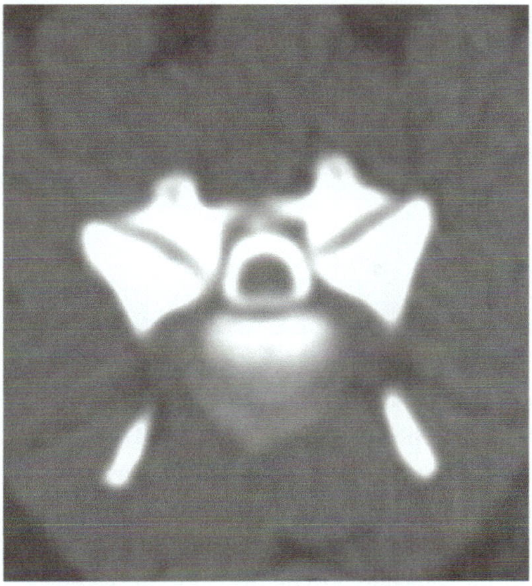

Figure 24.35 Transverse CT-myelography image at the level of C5–C6 in an 11-month-old Great Dane with osseous-associated CSM. Mild-to-moderate bilateral dorsolateral spinal cord compression associated with articular process hypertrophy.

sites can also be involved. Up to 85% of dogs present with more than one site of spinal cord compression (Figures 24.35 and 24.36).

- Other types of CSM, in which spinal cord compression is associated with severe vertebral abnormalities and absolute vertebral canal stenosis, have been described in puppies of large and giant-breed dogs, including Basset Hounds, Dobermann pinschers and Great Danes. Cervical spondylomyelopathy in Bernese Mountain dogs is characterized by multiple cervical spinal cord compressions caused by a combination of both intervertebral disc protrusions and articular process hypertrophy (Figure 24.37).

Obtaining a diagnosis of CSM can be done by using myelography, CT-myelography or MRI. Each of these techniques is associated with specific advantages and disadvantages considering diagnostic potential, availability, safety and costs. From a diagnostic perspective, they should probably be considered complementary to each other for dogs with CSM. The role of applying linear traction during imaging studies has been discussed and debated extensively. Compressive lesions may be categorized based on whether or not compression changes in the "traction" position. Lesions are termed "static"

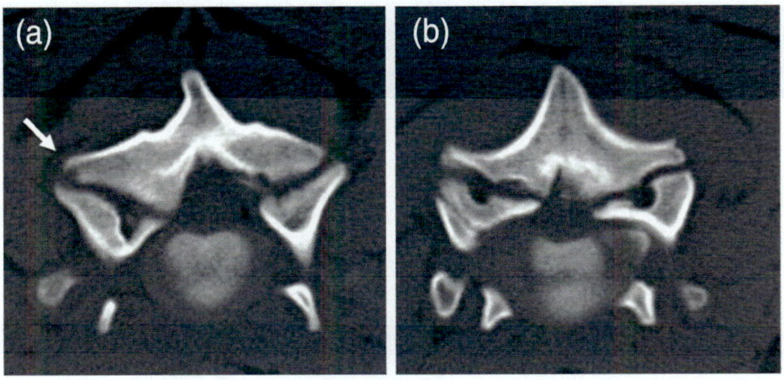

Figure 24.36 Transverse CT images at the levels of C3–C4 (a) and C2–C3 (b) in a 9-month-old English Mastiff with osseous-associated CSM. (a) Severe right-sided articular process and pedicle hypertrophy resulting in vertebral canal stenosis (arrow). Similar, but milder changes on the left side. (b) Severe hypertrophy of the dorsal lamina and bilateral articular process hypertrophy resulting in severe vertebral canal stenosis.

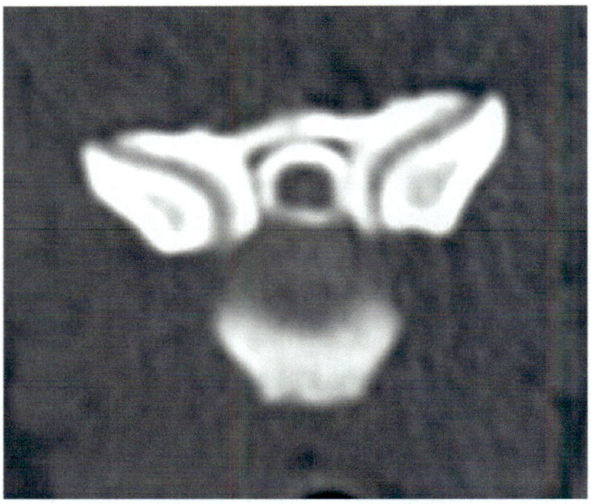

Figure 24.37 Transverse CT-myelography image at the level of C4–C5 in a 6-year-old Bernese mountain dog with CSM. A combination of mild intervertebral disk protrusion and bilateral articular process hypertrophy results in vertebral canal stenosis with mild ventral and bilateral lateral spinal cord compression.

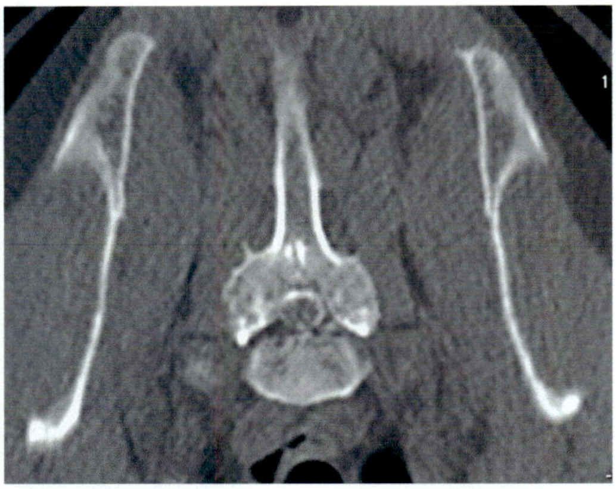

Figure 24.38 Transverse CT-myelography image at the level of T1–T2 in an 8-year-old Doberman pinscher with DA-CSM. Hypertrophy of the articular processes and dorsal arch resulting in vertebral canal stenosis without spinal cord compression. This abnormality was not considered to be the cause of the dog's clinical signs.

when the degree of compression remains the same, whereas "traction-responsive" lesions improve after performing linear traction. Traction studies can be performed during myelography, CT-myelography and MRI studies. Traction usually decreases spinal cord compressions caused by ligamentous structures, while compressions caused by osseous structures do not necessarily change after applying traction. Therefore, most dogs with DA-CSM will show traction-responsive lesions, and most dogs with osseous-associated CSM will demonstrate static lesions. This subdivision of lesion types can guide the surgeon to decide on the best surgical procedure to perform. Traction-responsive lesions have been suggested to benefit from distraction–stabilization surgery, while static lesions would benefit from direct decompressive surgery. However, a recent study has demonstrated promising results after distraction–stabilization surgery in giant-breed dogs with osseous-associated CSM. The degree of compression can also change as the neck is moved between flexed, neutral and gently extended positions. However, performing flexion or extension for a prolonged period of time is not without risk and should ideally be avoided. Relatively few studies have discussed the use of CT and CT-myelography in dogs with CSM. Because of its superior bone detail, CT and CT myelography can readily demonstrate many of the abnormalities seen in dogs with CSM. However, because of its inability to visualize the spinal cord, conventional CT is not recommended to reach a diagnosis of CSM. Although IV contrast-enhanced CT allows assessment of the intervertebral foraminae, it still does not allow detection of spinal cord compression. Therefore, CSM diagnosis can only be made after intrathecal contrast administration, resulting in a CT myelogram. Although an optimal CT myelography study is performed with a lower dose of the contrast medium than a conventional myelographic study, this technique still carries the risk of complications such as post-myelographic seizures and (transient) neurological deterioration. A recent study evaluating the use of CT myelography in dogs with CSM, demonstrated, especially in giant-breed dogs, a considerable number of compressive lesions in the cranial thoracic region. Although these lesions were usually not considered the primary source of spinal course compression, it has been suggested to always include the cranial thoracic region in a CT study of a dog with suspected CSM (Figure 24.38).

A recent study compared the use of myelography, CT myelography and low-field MRI in dogs with DA-CSM. Spondylosis deformans and articular process abnormalities were increasingly detected on CT myelography compared to myelography and low-field MRI, while craniodorsal vertebral tilting was detected less often with CT myelography. In agreement with human studies, CT myelography was considered the most reliable imaging technique to assess articular process abnormalities. In resemblance with people with cervical stenotic myelopathy, it is believed that intervertebral foraminal stenosis can cause or contribute to clinical signs of dogs with CSM (by causing nerve root compression). Unfortunately, assessment of intervertebral foraminal stenosis cannot always reliably be evaluated on CT myelography, MRI or myelography. Obtaining CT images after administration of IV contrast has been suggested to improve the accuracy of evaluating intervertebral foraminal stenosis.

CT can also provide useful morphometric data for the pre-surgical planning of dogs with

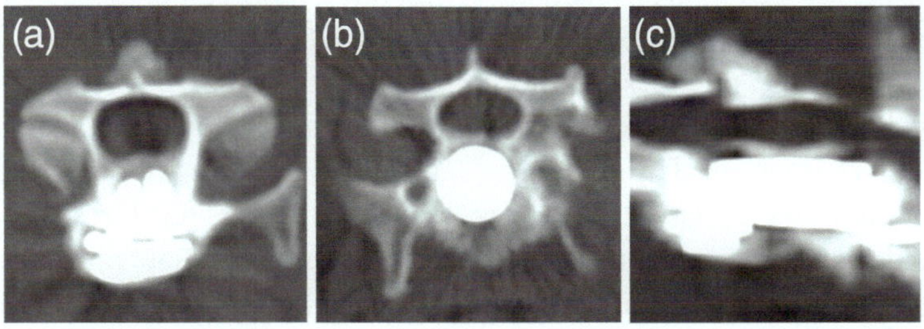

Figure 24.39 CT-study at the caudal aspect of C5 (a), mid-vertebral body of C6 (b), and a sagittal reconstruction at the level of C6. This study was performed 6 months postoperatively in a 7-year-old Doberman pinscher with DA-CSM. Minimal streaking artefact is observed due the titanium nature of the surgical implant and screws. (a) Position of the screws in C5 can be appreciated. (b and c) Position of the implant can be appreciated, and the surgical implant is ventrally and laterally surrounded by tissue with attenuation characteristics of bone.

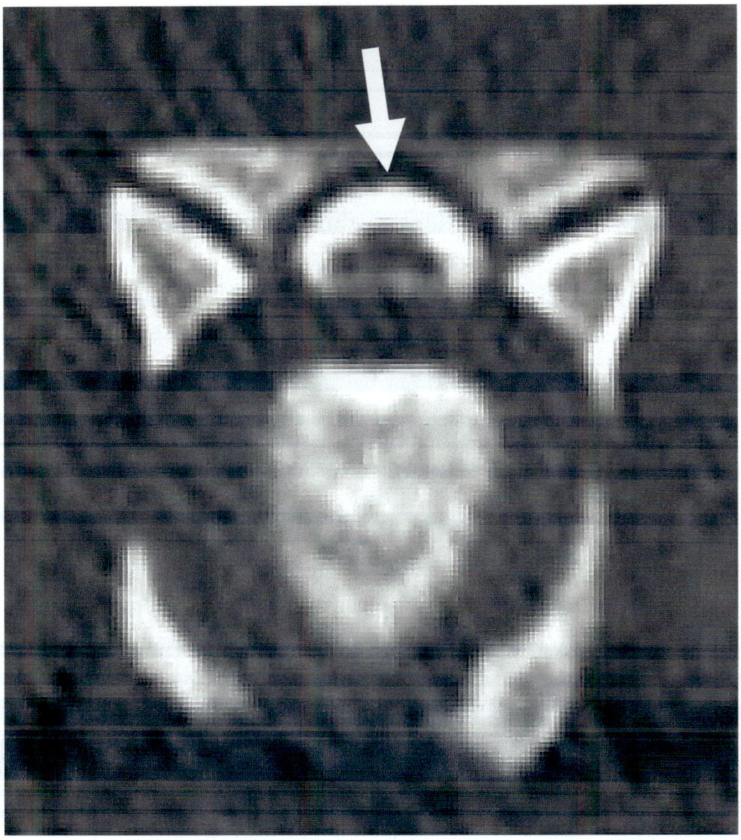

Figure 24.40 Transverse CT-myelography image at the level of C6–C7 in a 10-year-old Doberman pinscher with DA-CSM. Spinal cord atrophy characterized by a triangular-shaped spinal cord and widening of the subarachnoid space relative to the spinal cord parenchyma (arrow).

CSM, evaluate the position of surgical implants, assess postoperative vertebral fusion and eventual bony incorporation of surgical implants. The use of the newer titanium implants probably results in less implant-associated artefacts compared to stainless steel implants (Figure 24.39). In people, CT myelography also provides prognostic information by detecting spinal cord atrophy. Spinal cord atrophy is characterized by a triangularly shaped spinal cord and a relative widening of the subarachnoid space, relative to the spinal cord. A recent study evaluating the use of CT myelography in 58 dogs with CSM demonstrated spinal cord atrophy in 19% of the affected dogs. However, it is currently unclear if this radiographic finding also represents irreversible spinal cord damage in dogs with CSM (Figure 24.40).

24.5.2.5 LS pathology

Lumbosacral diseases in dogs are relatively common entities dealt by radiologists and neurosurgeons. Although many diseases can affect the lumbo-sacral region, degenerative lumbosacral stenosis (DLSS) is the most common disorder affecting the L7–S1 joint. Large breed and middle-aged to older dogs are mainly affected. Labrador Retrievers, Border Collies and German Shepherd Dogs (GSD) seem overrepresented, with a male-to-female ratio being approximately 2:1 in the latter. DLSS is a multifactorial disorder, characterized by many degenerative changes that can occur alone or in combination: spondylosis, disc protrusion, vertebral instability and ligament hypertrophy (i.e., interarcute ligament). In the context of DLSS, these degenerative changes all have in common that they result in various degrees of vertebral canal and/or foraminal stenosis and subsequent compression of the conus medullaris, cauda equina nerve roots or L7 spinal nerve(s). Being the most common cause of compression of the cauda equina, DLSS is also generally (and wrongly) referred to cauda equina syndrome (CES). This syndrome, being the direct consequence of the nerve root compression/damage, is characterized by a variety of symptoms like lumbo-sacral pain, variable degrees of lameness (either unilateral or bilateral), tail paresis and urinary and/or faecal incontinence. Pain is the most common symptom and, not so rarely, the only one. It is important to underline that ataxia, paresis and proprioceptive deficits, when present, should raise the suspicion of a more cranial disease with spinal cord involvement; in fact, it is relatively common for an old, large-breed dog to suffer from a thoracolumbar and lumbosacral disease at the same time. Once a lumbosacral disorder is suspected, the underline cause should be identified by the appropriate imaging modality.

CT and MRI, also due to their constantly increasing access in the veterinary world, represent the best choices to investigate the L–S area, and a good agreement has been demonstrated between these two imaging modalities; however, poor degree of agreement seems to exist between them and surgical findings.

Main objectives of CT and MRI are to confirm diagnosis of DLSS by documenting stenosis of the vertebral canal and/or foramen with associated compression of the neural structures as well as identifying the degenerative changes responsible for that stenosis and the possible associated instability. These findings will ultimately guide therapeutic decision-making, including the choice of the most appropriate surgical technique, when necessary. In fact, DLSS patients are generally considered surgical candidates. Although many surgical techniques have been described to treat patients with L–S diseases and usually the choice relies on the personal preference of the neurosurgeon, the latter wants to know if the L-S joint is stable, if the vertebral canal is stenotic or the nerve roots compressed and the underlying cause. Ideally, the optimal imaging modality should provide the neurosurgeon with all these information at once; however, coupling more than one imaging modality should be considered in order to achieve all the details necessary to offer the best therapeutic option to the patient.

Survey radiography should always be performed first because it is quick and useful to rule out other causes of CES (e.g., discospondylitis, fracture/luxation and vertebral neoplasms), concurrent or predisposing abnormalities which include osteochondrosis (OCD) of the sacral end plate and lumbo-sacral transitional vertebrae (LSTV); moreover, it helps (though controversy exists) to assess possible underlying instability of the LS junction by evaluating potential changes in the lumbo-sacral step during dynamic flexion and extension studies: the L–S step can become evident either on the flexed or on extended view (stress X-rays, Figures 24.41 and 24.42). Instability of the L-S joint has also been referred as retrolisthesis, which consists in ventral displacement of the sacrum in relation to the L7 vertebral body. Despite it not being fully clear what degree of retrolisthesis should be considered clinically

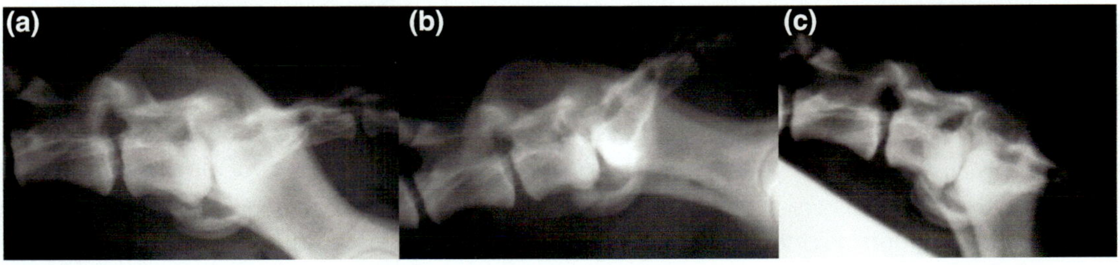

Figure 24.41 Lateral radiographic images in neutral position (a), extension (b), and flexion (c). Note the marked L-S step, evident on the flexed view, suggestive of L-S instability.

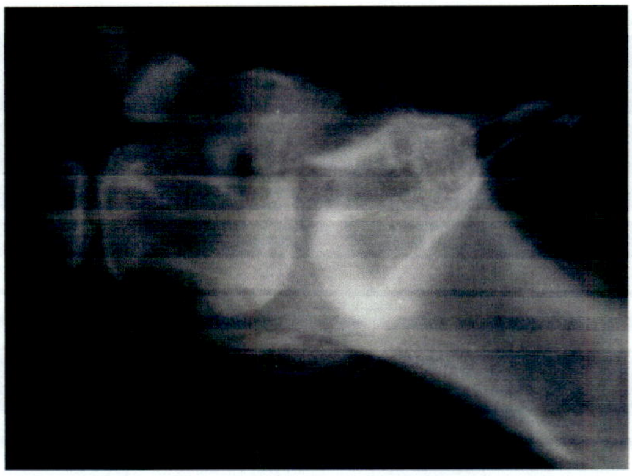

Figure 24.42 **Note the marked L-S step, evident on the extended view, suggestive of L-S instability.**

relevant, it is the author's opinion that it should be suspected and then treated any time it is macroscopically evident on imaging, especially if associated with other changes possibly related to instability (e.g., collapse of intervertebral disc space, sclerosis of vertebral end plates, spondylosis and subluxation of articular facets) and particularly in large and active dogs. Instability of the L-S joint may rarely be the only evident finding but, more commonly, it represents only part of a more complex DLSS, so further cross-sectional imaging modalities like CT or MRI are necessary.

CT, thanks to its great bone resolution, may be complementary to radiography to better define LSTV and OCD lesions or other changes in the bony structures (involving the vertebral bodies and/or the articular facets) like sclerosis, spondylosis, small osteophytes and subtle end-plate lysis (as seen in early discospondylitis). Furthermore, CT has better soft-tissue resolution than conventional radiography, so it can also detect disc protrusion, hypertrophy of the interarcuate ligament, joint capsule and nerve root thickening (Figure 24.43).

First of all, a good and standardized positioning and CT study protocol are essential to achieve good-quality images. The patients are generally placed in dorsal recumbency, with the vertebral column as straight as possible and with the L-S joint in a neutral position; however, the L-S joint is likely to be slightly extended in this position because of the physical weight of the back legs (Figure 24.44). When assessing the L-S joint, it is crucial to know the position of the patient, in order to properly evaluate the potential pathological changes involving the L-S joint itself, as stated more widely below. An optimal CT-study should cover the L4–S3 area, with an ideal slice thickness of 1 mm and an interval of reconstruction of 0.5 mm; tube voltage of 120 kVp, 1 second tube rotation time and tube current of 200 mAs.

In order to properly visualize and differentiate the single nerve roots and eventually the other soft tissues present at the lumbo-sacral level, it is essential that a good amount of epidural fat is present (Figure 24.45); in fact, when stenosis of the spinal canal occurs, either due to disc protrusion or a combination of it and an hypertrophied ligament, the epidural fat starts to disappear, and it can progressively be replaced by fibrous tissue. These changes would finally lead to a loss of the inherent contrast provided by the fat itself and subsequently to a decrease/loss of soft-tissue resolution.

The protruded disc can be directly appreciated or indirectly estimated by evaluating the displacement of the conus medullaris or thecal sac (depending on where the conus medullaris finishes) which, in case of a disc protrusion, is lifted by the floor of the spinal canal, away from the disc itself (Figure 24.46). In order to better localize the compressive soft tissue, CT can be performed after intravenous injection of iodinated contrast medium. In fact, contrast CT can be helpful for predicting the location of compressive soft tissues, especially when there is loss of epidural fat; however, it does not help in differentiating the types of tissues (e.g., inflammatory vs neoplastic tissue).

Having said all of that, it is crucial to bear in mind that a number of middle-/old-age dogs may have some evidence of disc protrusion (and therefore associated loss of epidural fat and some displacement of neural structures) in the absence of clinical indicators: this highlights the importance of performing scanning in only those dogs in which the history and the clinical indicators are strongly suggestive of an L-S disease, in order to avoid "false positive" results.

The articular facets can be nicely depicted on CT images: the bony changes (e.g., osteophytosis, (Figure 24.47), their angle and orientation (i.e., articular process geometry), and then potential subluxation can be identified. The possible association in GSD between facet joint tropism and facet joint orientation with disc degeneration has been investigated by using CT. Facet joint tropism is defined as an asymmetry of left and right facet joint angles, and it has been postulated (despite it is controversial) that facet joint tropism in people could result in disc lesions because of abnormal rotation with increased torsional stress on the intervertebral

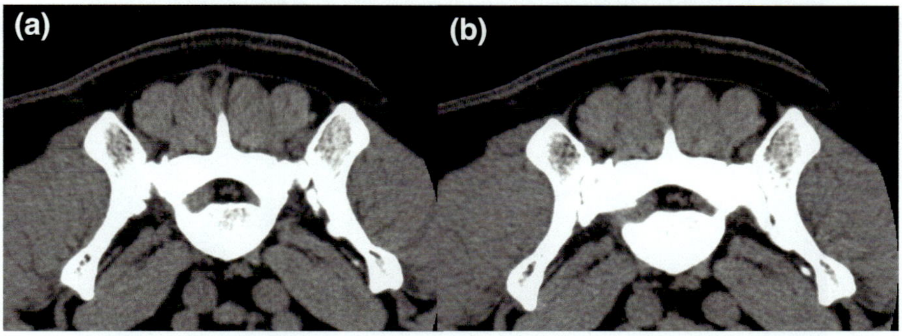

Figure 24.43 (a and b), two consecutive transverse sections (cranial to caudal) showing L7 nerve root thickening on the left side, leaving the vertebral canal through the foramen.

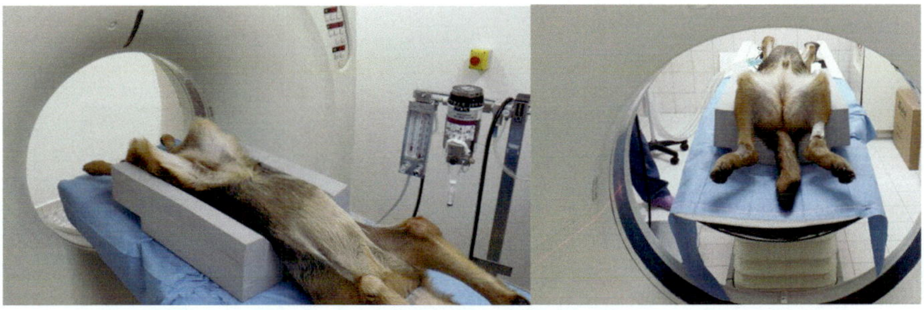

Figure 24.44 A dog positioned in dorsal recumbency for an L-S CT examination.

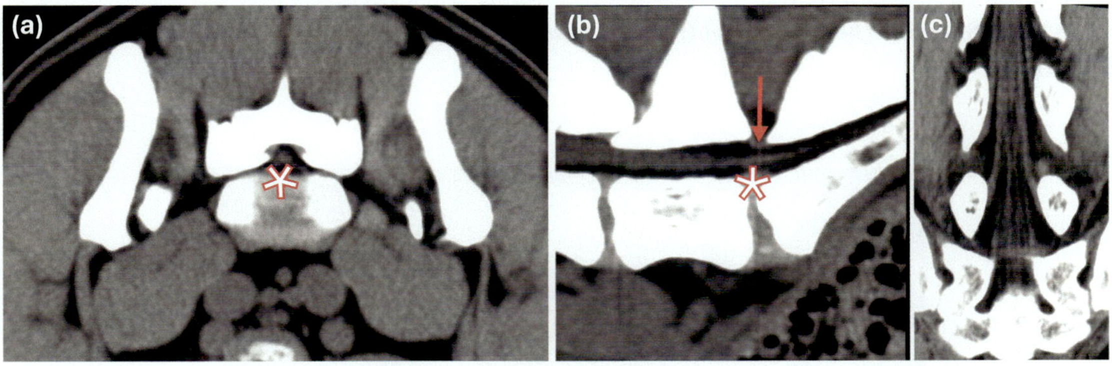

Figure 24.45 (a–c) Transverse, sagittal-MPR and dorsal-MPR of a normal L-S: dorsal aspect of the disc (asterisk) does not protrude into the vertebral canal, interarcuate ligament is not hypertrophied (arrow), no compression of the thecal sac exists and the cauda equina nerve roots are well visualized thanks to the inherent contrast provided by the epidural fat (c).

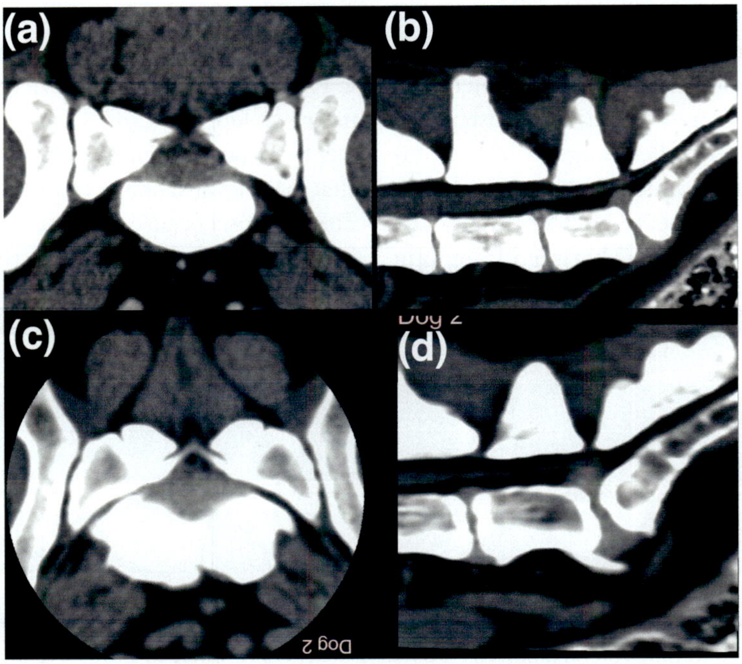

Figure 24.46 (a–d) disc protrusions on transverse and MPR-sagittal planes from two different dogs, with lifted thecal sac due to L7–S1 disc protrusion.

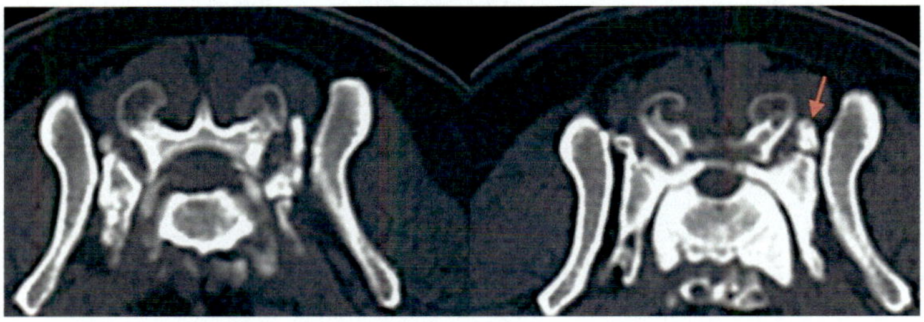

Figure 24.47 Consecutive transverse sections (cranial to caudal) of severe osteophytosis of articular facets of L7–S1; a small bone fragment (possible fracture) is visible on the right sacral articular facet (arrow).

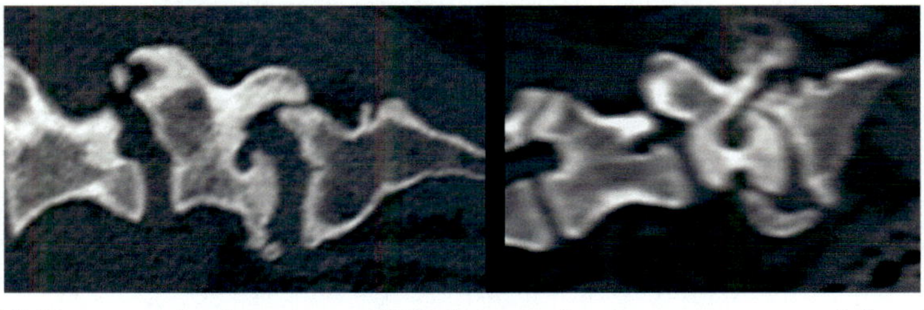

Figure 24.48 MPR-parasagittal sections of L7–S1 articular facets in a normal dog (left) and in a dog with subluxation (right).

disc. Despite the increased facet joint tropism that was identified, no association was found with disc degeneration in GSD (Figure 24.4). Therefore, the role of facet joint tropism in the dog remains open to research, and it may warrant further investigation. Articular process subluxation can also be appreciated on the reformatted images, especially on the sagittal–parasagittal planes (Figure 24.48). Evaluation of the parasagittal planes along with the dorsal reconstructions is useful to evaluate the L7–S1 foramina as well. In fact, thorough assessment of the foramina, with respect to their shape, size and density, is mandatory in all CES patients. Intervertebral foramina are located slightly cranial in relation to the L7–S1 disc space (Figure 24.49) and consist in osseous canals that have been classified into three functional zones: entrance, middle and exit. The entrance zone is located closest to the vertebral canal and is defined as the zone at the medial portion of the L7 vertebral pedicle. The middle zone is located at the centre of the pedicle, and the exit zone is located at the lateral portion of the vertebral pedicle. Positioning of the vertebral column should be taken into consideration when evaluating the L7–S1 foramina. In fact, according to a recent paper, the degree of foraminal stenosis varies by changing the L-S angle, and it may be underestimated when the vertebral column is flexed. LS foraminal stenosis can instead increase, as well as the compression of the nerve roots, during extension of the lumbosacral joint. Dynamic studies (flexion and extension CT) are useful to evaluate L-S instability and changes in the vertebral canal stenosis (Figure 24.50). In case of an L-S instability, the CT exam should also allow the neurosurgeon to plan his stabilization procedure. For this purposes, 3D volume rendering reconstructions are also very useful, giving the chance to pre-operatively assess the implant positioning, screw placement, length, angulation, etc (Figure 24.51).

Post-operative CT control is advised certainly in animals which do not recover as expected (Figure 24.52).

L-S OCD and transitional vertebra are considered to be predisposing factors for DLSS and then CES. The OCD fragment, generally derived from the dorso-cranial border of S1, has got a primary role in promoting the degenerative changes that lead to the stenosis of the vertebral canal or foramina with subsequent compression of L7 nerve root(s), which may become severely thickened (Figure 24.53). Even if lumbosacral OCD can be detected by CT or survey radiography in the first months of life, the onset of clinical signs is in the adult age, being related to the degenerative changes more than to the OCD-fragment itself. Common symptoms are the same as for the other causes of CES and include lumbosacral pain, followed by different degrees of weakness in the rear limbs, manifested as reluctance to rising up or sitting down, running and jumping or as clear lameness in more obvious cases; faecal and/or urinary

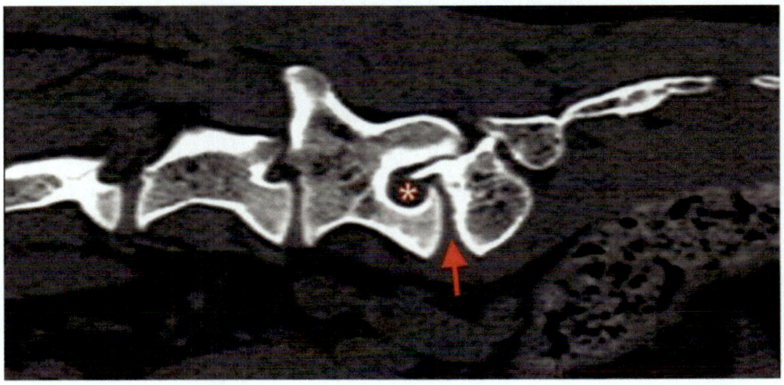

Figure 24.49 The intervertebral foramen (asterisk) is cranial to the intervertebral disc (arrow)

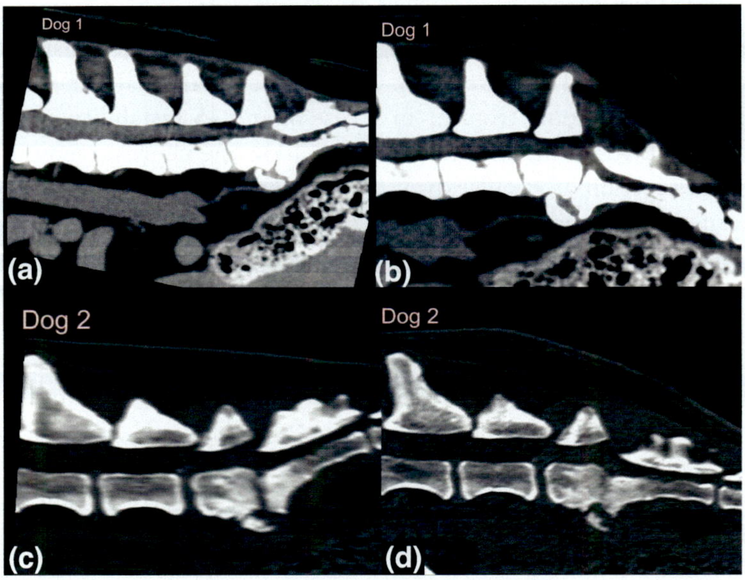

Figure 24.50 Extended (a and c) and flexed (b and d) sagittal-MPRs from two different dogs (dog 1 with DLSS and dog 2 with L-S discospondylitis), showing marked L-S instability (retrolisthesis).

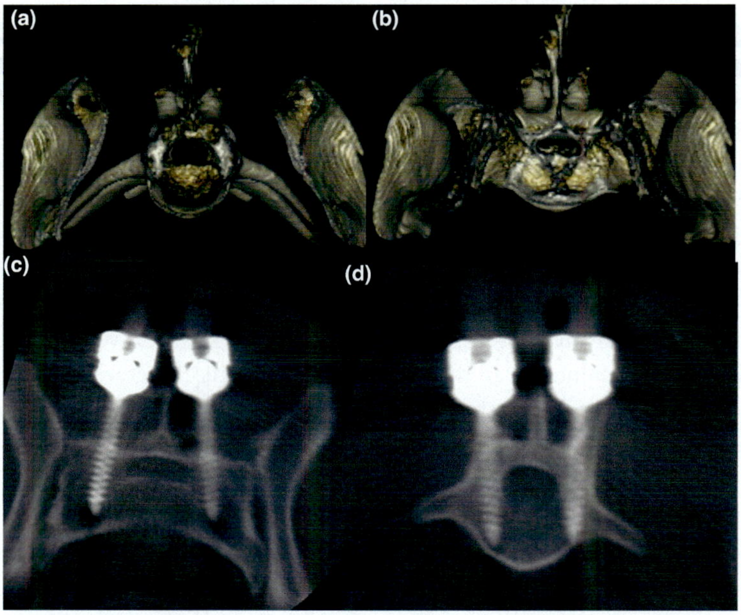

Figure 24.51 3D-volume rendering through L7 (a) and S1 (b) allow pre-operative evaluation of the possible direction of the screws; (c and d). (Courtesy of Dr. Daniele Corlazzoli, transverse sections through L7 and S1, respectively, after implant positioning.)

incontinence can be present. Surgical removal of the bone fragment as well as the other potential compressive tissues (e.g., protruded disc and fibrosis) is considered the treatment of choice; pre-operative CT is then very useful, by allowing the neurosurgeon to localize these alterations: quite commonly the OCD-fragment is lateralized, in proximity to the foramen (Figure 24.54).

L-S transitional vertebra predisposes degeneration and then protrusion of the intervertebral disc cranial to it, with secondary compression of the nerve roots with onset of the clinical signs

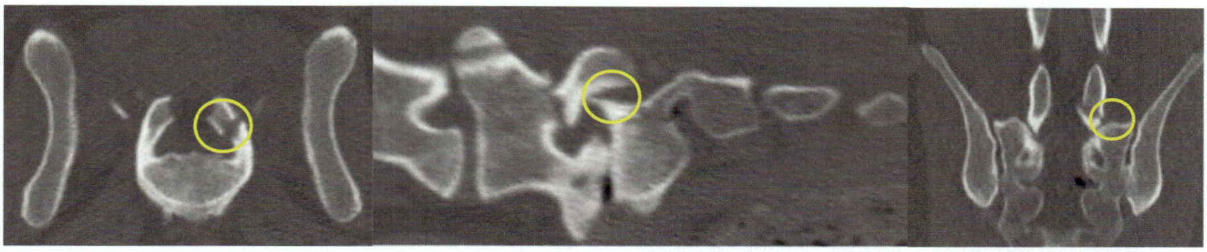

Figure 24.52 Post-operative CT images (a transverse, b sagittal, and c dorsal reconstructions at the level of the articular process) of a 9-year-old Airedale Terrier. An excessive laminectomy in the LS region was performed because of discus hernia. A fracture of the left caudal articular process of L7 is present (yellow circles).

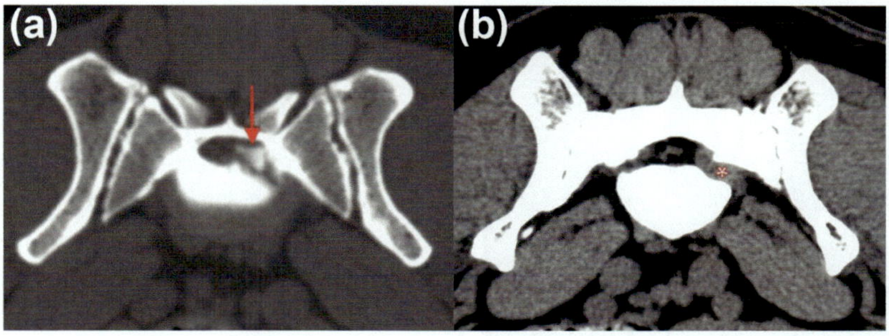

Figure 24.53 (a and b) Two consecutive transverse sections from a German Shepherd dog with CES signs: a bone fragment compatible with an OCD (arrow) and a thickened right L7 nerve root (asterisk) can be appreciated.

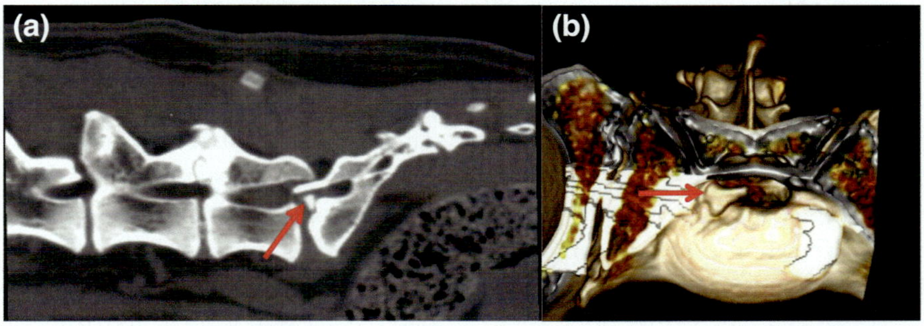

Figure 24.54 (a) Parasagittal MPR and (b) 3D-volume rendering of the same dog in Figure 24.53. The OCD fragment (arrows) lies in proximity of the right foramen.

at a younger age than usually observed in DLSS dogs (Figure 24.55).

Other degenerative changes like juxta-articular cysts are relatively common features in patients suffering from LSTV. Those cysts can only be incidental findings, but sometimes they contribute in determining L-S stenosis and compression/damage of the nervous tissues.

Other diseases than DLSS can affect the L-S joint: exogenous spinal trauma, tumours and infections (i.e., discospondylitis) are in fact relatively frequent at this level.

CT features of discospondylitis of L7–S1 mainly include lysis of the vertebral end plate, which can be detected at a very early stage of the disease, and decrease of the intervertebral disc space (Figure 24.56). When the infected tissue spreads into the vertebral canal (forming spinal epidural empyema in most severe cases), it is also possible to appreciate compression/displacement of the nervous

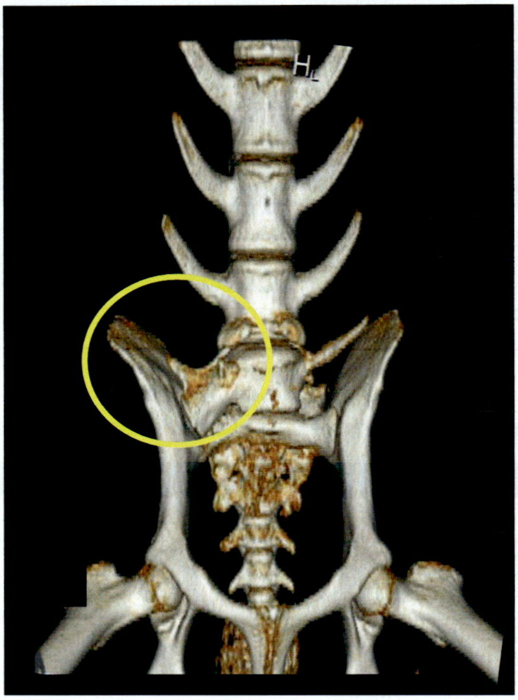

Figure 24.55 **A 3D reconstruction of a transitional vertebra. The seventh lumbar vertebra is sacralized on the right side, forming an iliosacral joint (yellow circle). On the left side, a rather small transversal process can be appreciated.**

structure; contrast CT could be of further help for guiding the neurosurgeon in performing decompression.

Neoplasia can affect either the vertebrae or the nerve roots. Vertebral tumours commonly give a mixture of bone lysis/proliferation, with varying degrees of changes in the bone shape and margins (Figure 24.57). Clinical signs are usually derived from compression or encroachment of the nerve roots. In case the neoplasia primarily arises from the nerve, CT can be of limited value, especially in the early stages of the disease and MRI should be preferred (Figure 24.58).

24.5.2.5.1 Acknowledgement

Author would like to thank Dr. Andrea Arcangeli for his technical support and for supplying some very beautiful CT images.

24.5.2.6 Spinal malformation

Several vertebral malformations have been described in dogs and cats. Probably, the majority of diagnosed vertebral malformations can be considered incidental findings and do not cause clinical signs. Therefore, it is of major importance to correlate imaging findings with the results of thorough general physical and neurological examinations (Figure 24.59). In certain vertebral malformations, clinical signs are not caused by the vertebral malformation itself, but by secondary ligamentous changes. Some dogs with vertebral malformations will, therefore, only demonstrate clinical signs in adulthood. Although most vertebral malformations are congenital in nature, some are acquired. Possible underlying causes include certain lysosomal storage diseases, such as mucolipidosis and mucopolysaccharidosis in cats. Because of its superior detail of bony structures and the possibility of obtaining transverse images, followed by multiplanar and three-dimensional reconstructions, it is obvious that CT can provide

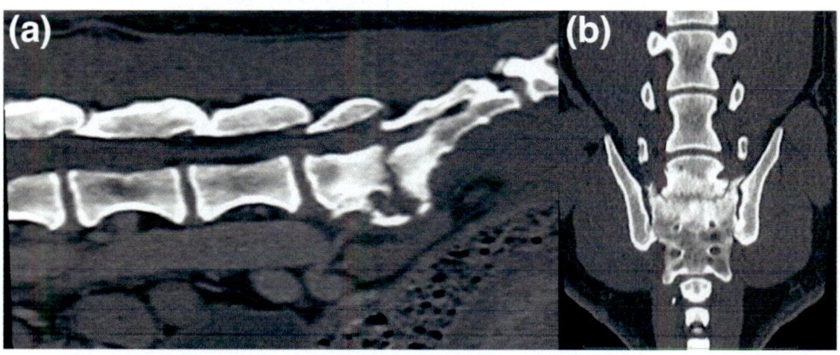

Figure 24.56 **(a) Sagittal and (b) dorsal-MPRs of an L-S discospondylitis showing the lysis of the vertebral end plates and the collapse of the intervertebral disc space.**

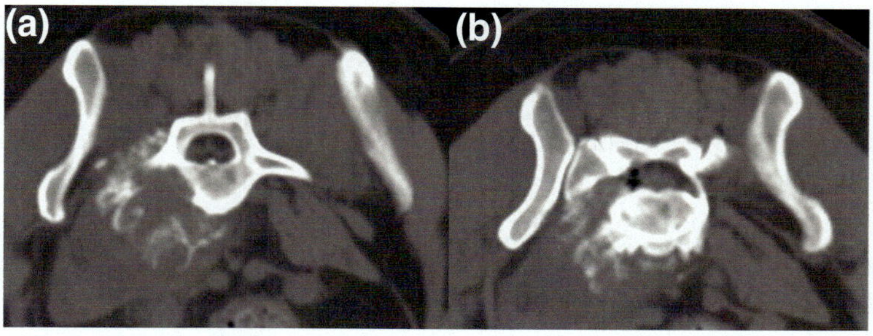

Figure 24.57 (a and b) Consecutive transverse sections (cranial to caudal) showing a vertebral neoplasia involving L7 and the left side of the L-S joint and foramen.

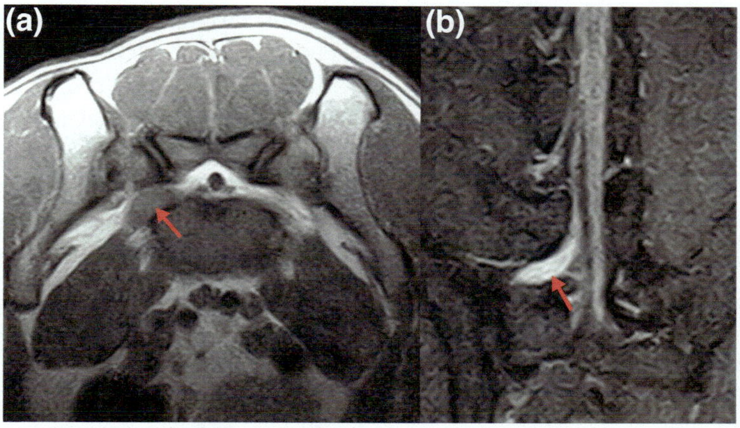

Figure 24.58 (a) T1-weighted transverse section through the L-S foramina and (b) dorsal/oblique STIR image showing evidence of severe changes of the foraminal portion of the right L7 nerve root (arrows), in absence of a foraminal stenosis, compatible with a nerve sheath tumour.

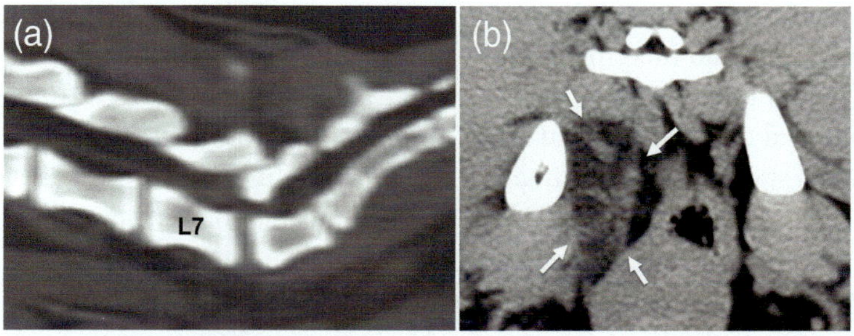

Figure 24.59 (a) Sagittal reconstruction of the lumbosacral region in a 10-month-old bulldog with pain on rectal examination and lumbosacral palpation. Malformed sacrum with extension of the sacral lamina into the vertebral canal, resulting in severe vertebral canal stenosis. Transitional vertebra with lumbardization of S1. (b) Transverse image (soft-tissue window) at the level of the pelvic canal. An abscess can be seen in the pelvic canal (arrows). The dog improved after surgical drainage of the abscess. This case illustrates the importance of correlating imaging with clinical findings and ruling out other potential causes before considering treatment of a vertebral malformation.

detailed information about vertebral anomalies. Because spinal cord compression is not visualized on conventional CT, intrathecal contrast should be administered (CT myelogram) to fully assess vertebral malformations in dogs and cats. Alternatively, conventional CT can be combined with MRI. The most common vertebral malformations are discussed below.

- Abnormalities affecting the vertebral body: Hemivertebra is probably the most common vertebral malformation in dogs and typically affects screw tail breeds, such as English and French Bulldogs, Boston Terriers and Pug dogs. It is characterized by incomplete formation of the vertebral body, often resulting in a wedge-shaped vertebral body. Hemivertebra occur most commonly in the mid-thoracic region, with the vertebral levels between T5 and T8 being most commonly involved. Up to 78% of clinically normal French bulldogs have radiographic evidence of hemivertebra. These malformed vertebrae can lead to varying degrees of vertebral canal stenosis, vertebral instability with subluxation and abnormal angulation of the vertebral column (kyphosis and scoliosis). Hemivertebra may be associated with other spinal disorders, such as intervertebral disc disease, spinal arachnoid diverticula and syringomyelia. For this reason, in combination with the high incidence of hemivertebra in clinically normal dogs, a thorough neurological and imaging study should be performed to reveal the exact cause of neurological deficits. Clinical signs are often mild and stabilize when animals attain skeletal maturity. Therefore, clinical management can be appropriate in many cases. Several surgical techniques have been described, including stabilization with or without decompression and partial realignment of the affected vertebrae. CT can be used for surgical planning and deciding on appropriate surgery for the individual patient (Figure 24.60).
  Other vertebral body malformations include butterfly vertebra and block vertebra. Butterfly vertebrae are caused by a failure of formation of the ventral and central portions of the vertebral body, resulting in two dorsolateral fragments of bone attached to the dorsal arch. On ventrodorsal radiographs, this appears as a sagittal cleft in the vertebral body, hence the name butterfly vertebra. Block vertebrae are characterized by failure of segmentation, leading to a combined vertebra composed of two or more single vertebrae.
- Atlanto-axial instability:
  Atlanto-axial instability with excessive flexion and subluxation of C2 relative to C1 occurs most commonly in young miniature and toy breed dogs. Severity of clinical signs can vary from intermittent cervical hyperaesthesia to non-ambulatory tetraparesis with respiratory difficulties and brainstem signs. Atlanto-axial instability has been associated with hypoplasia, aplasia or an abnormal configuration of the dens. Abnormal ligamentous support of the

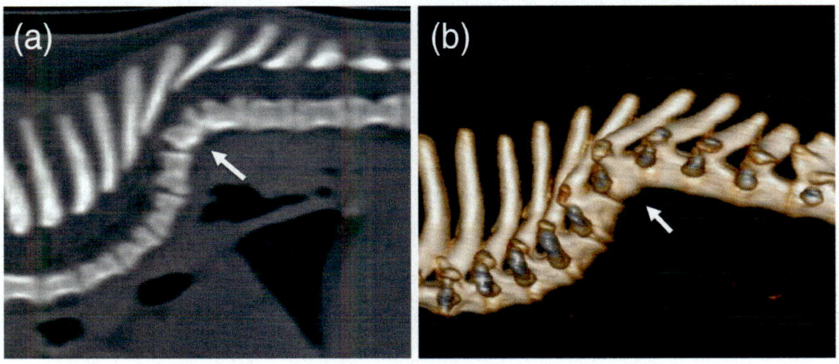

Figure 24.60 **Sagittal (a) and three-dimensional reconstruction (b) of a 9-month-old pug with mild paraparesis. Hemivertebra of T7, resulting in pronounced kyphosis.**

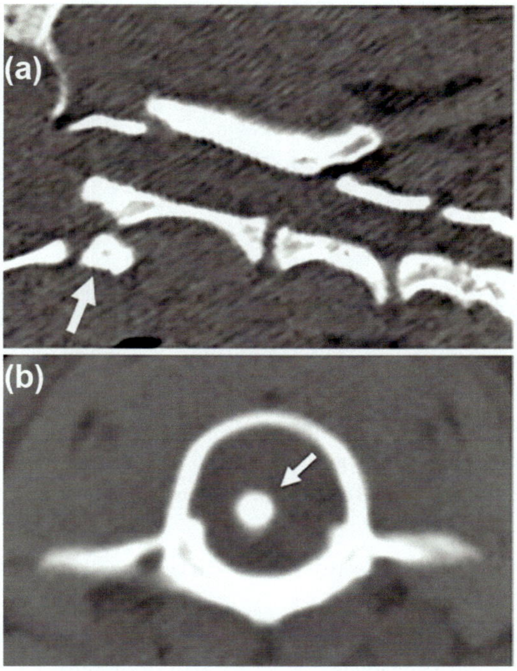

Figure 24.61 (a) Sagittal reconstruction of a 2-year-old Yorkshire terrier with intermittent cervical hyperaesthesia. Atlanto-axial subluxation characterized by cranial and dorsal displacement of the axis relative to the atlas (arrow). (b) Transverse image at the level of the atlas. Abnormal dorsal position (arrow).

dens may also be involved. The body of the axis is displaced dorsally and cranially into the vertebral canal, and the distance between the dorsal arch of the atlas and spinous process of the axis is increased. Demonstration of these abnormalities is diagnostic, and the risks associated with performing a myelogram or CT-myelogram outweighs by far the potential advantage of visualizing spinal cord compression. In contrast to other vertebral malformations, demonstration of atlanto-axial instability should not be considered an incidental finding. Although very brief and gentle flexion of the neck can facilitate a radiographic, or better fluoroscopic, diagnosis of atlanto-axial instability, prolonged or exacerbated flexion can have deleterious and even life-threatening consequences. Therefore, performing a CT study with flexion of the neck is definitely not recommended. Compared with other diagnostic modalities, CT will readily reveal abnormal dens conformation, such as aplasia, hypoplasia or dorsal angulation of the dens. This information is important when selecting the best type of treatment (i.e., necessity to perform and odontoidectomy) in an individual patient (Figure 24.61). Medical treatment can be considered in selected patients, and several surgical techniques have been described for this disorder. CT can assist in surgical planning and evaluation of surgical implant position (Figure 24.62). Other cranial cervical or craniocervical abnormalities, such as atlanto-occipital overlapping syndrome, can occur concurrently with atlanto-axial instability. Atlanto-occipital overlapping syndrome is characterized by a decreased distance between the dorsal arch of the atlas and the occipital bone of the skull. This can result in protrusion of

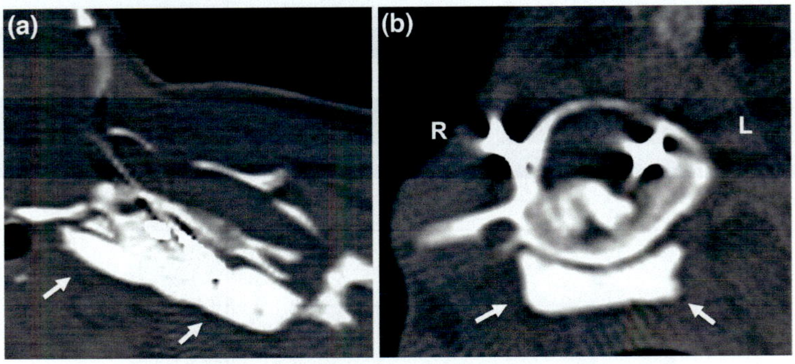

Figure 24.62 Sagittal (a) and transverse (b) postoperative CT study in a 5-year-old Yorkshire terrier with atlanto-axial instability. (a) Realignment of the atlantoaxial junction; position of one of the pins and polymethacrylate bridge (arrows) can be appreciated. (b) Suboptimal placement of the left pin. Despite entering the vertebral canal, the dog made a good recovery. Polymethacrylate (arrows) and streaking artefact due to the stainless steel nature of the implants.

the dorsal arch of the atlas into the foramen magnum (Figure 24.63).

- Cranial thoracic vertebral canal stenosis Osseous stenosis with cranial thoracic spinal cord compression has been described in large and giant-breed dogs. This generally occurs in young male dogs with a Mastiff-type body conformation. Occurrence of multiple compressive sites is usually seen, with the T2–T3 and T3–T4 sites being most commonly affected. Direction of spinal cord compression is typically dorsolateral and is associated with enlargement and abnormal oblique orientation of the articular processes. A variable degree of cranial thoracic stenosis has been demonstrated in clinically normal dogs, and dogs with cranial thoracic stenosis can also be affected by osseous-associated cervical spondylomyelopathy. Therefore, it is important to correlate imaging findings with the results of neurological examination and include a large area of the vertebral column when planning a CT study in dogs suspected of having this disorder. Although no large case-series are available, affected dogs do not necessarily undergo deterioration in condition after clinical management, while decompressive

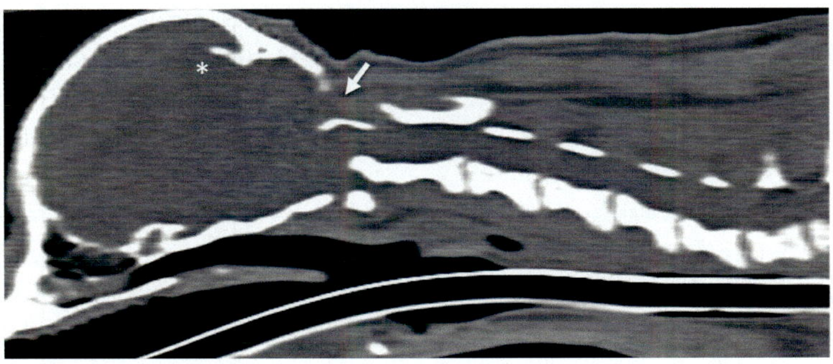

Figure 24.63 Sagittal reconstruction in a 7-month-old Shih Tzu with atlanto-axial subluxation. Additionally, an intracranial quadrigeminal cyst (asterisk)) and atlanto-occipital overlapping syndrome with cranial displacement of the lamina of the atlas into the foramen magnum (arrow) can be appreciated. This case illustrates that multiple congenital abnormalities can occur in one dog, which may complicate treatment and postoperative recovery.

spinal surgery can be associated with considerable postoperative morbidity and mortality (Figure 24.64).

- Transitional vertebrae:
  Transitional vertebrae are anomalies found at the junction between two divisions of the vertebral column. They occur commonly at the thoracolumbar and lumbosacral junction. This can result in variations in the number of vertebrae found within a vertebral segment and abnormal anatomy of the affected vertebra. Transitional vertebrae have symmetrical or asymmetrical morphological characteristics of two different vertebral types. Common examples are lumbardization of the sacrum, sacralization of the last lumbar vertebra and presence of one or two ribs on the first lumbar vertebra or absence of ribs on the last thoracic vertebra with rudimentary transverse processes. Although transitional vertebrae are normally not the cause of spinal cord compression, these anomalies can complicate the surgical approach of routine spinal surgeries. It is of major importance that the neurosurgeon is aware of any transitional vertebrae to avoid performing surgery at the wrong spinal level (Figure 24.65).

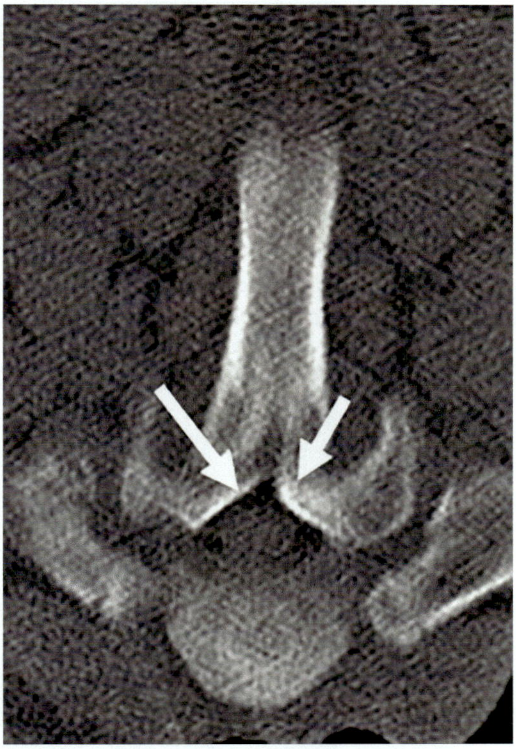

Figure 24.64 CT image at the level of T3 in a 9-month-old English Mastiff. Cranial thoracic stenosis caused by articular process abnormalities (white arrows).

- Abnormalities of the articular processes and vertebral arch:

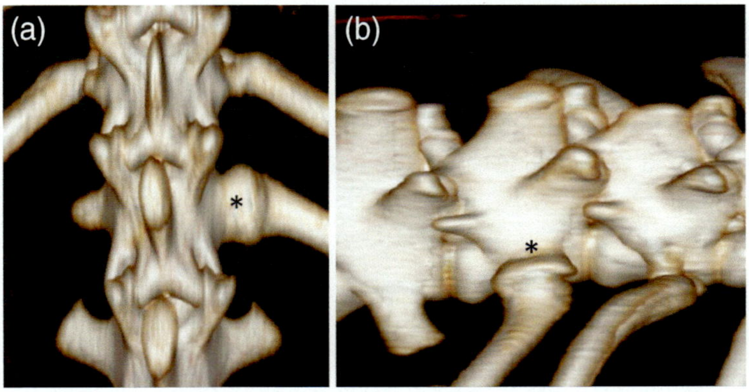

Figure 24.65 Three dimensional reconstruction of a 6-year-old Lhasa Apso with an acute T13–L1 intervertebral disk extrusion. (a) Dorsal view demonstrates a transitional vertebra with an extra rib at L1 (asterisk). (b) Sagittal view. The proximal part of this rib-like structure has characteristics of a transverse process (asterisk). Not recognizing this anomaly could have dramatically complicated the surgical approach.

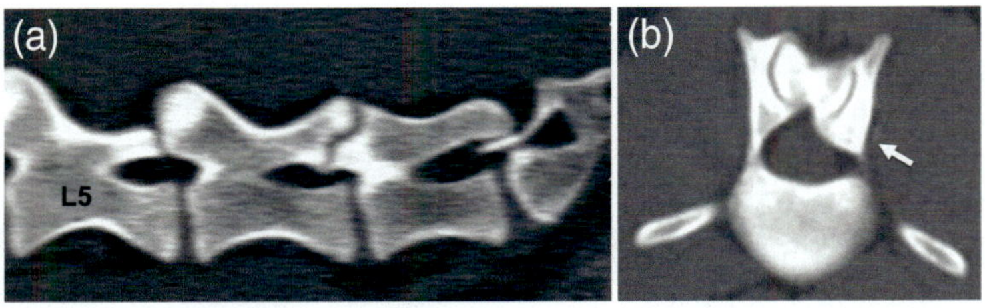

Figure 24.66 Sagittal reconstruction (a) and transverse (b) CT image of an 11-month-old English pointer with articular process hypertrophy at the level of L6–L7 (arrow). This anomaly results in narrowing of the L6–L7 intervertebral foramen with subsequent nerve root compression.

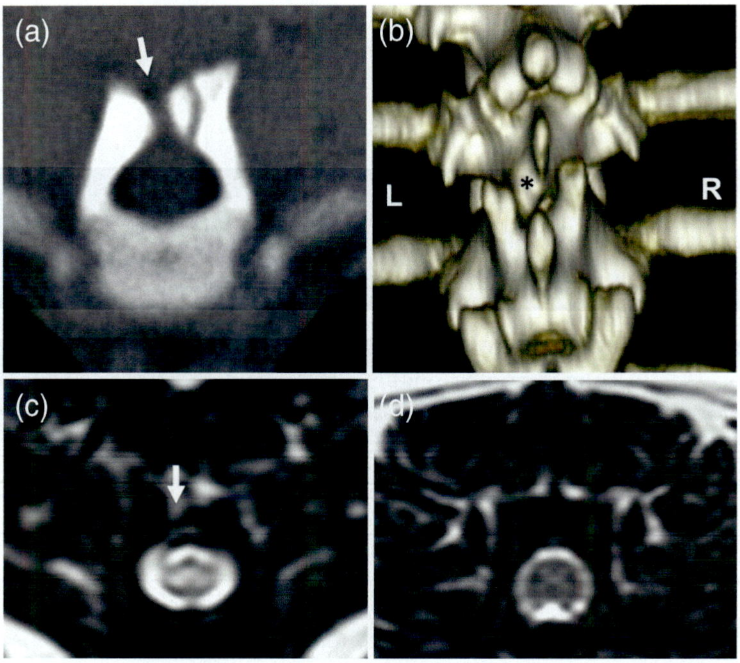

Figure 24.67 (a) Transverse CT image of a 7-year-old Pug with aplasia of the right caudal articular process of T11–T12 (arrow). (b) Three-dimensional reconstruction confirming aplasia of the right caudal articular process. Compare with the normal left-sided caudal articular process (asterisk). (c) Transverse T2-weighted image of the same dog demonstrating constrictive spinal cord compression due to aplasia of the right caudal articular process (arrow). (d) Transverse T2-weighted image at the level of T12–T13 for comparison. The normal dimension of the spinal cord can be appreciated at this level.

Both hypertrophy and hypoplasia of the articular processes may result in spinal cord compression (Figure 24.66). Aplasia and hypoplasia of the caudal articular processes in the thoracolumbar vertebral column of dogs with and without neurological deficits have been described. Secondary changes, such as compensatory hypertrophy of the cranial articular processes and ligamentum flavum, may result in spinal cord compression. Thoracolumbar caudal articular process aplasia/hypoplasia with secondary constrictive myelopathy has been described in pug dogs. Prognosis after surgical decompression and stabilization for this condition are guarded (Figure 24.67).

In contrast to articular process hypoplasia/aplasia, thoracolumbar articular process and vertebral arch proliferation with osteoarthritic changes have been described in related Shiloh Shepherd puppies. Hypertrophied articular processes with spinal cord compression are occasionally seen in other dog breeds. Decompressive surgery combined with spinal stabilization has been recommended for dogs with bilateral articular process abnormalities.

Hypertrophy of the vertebral arch and spinous process has recently been reported in Basset hounds. The articular processes are not a cause of spinal cord compression in this disorder. This bony malformation is associated with secondary ligamentum flavum hypertrophy, resulting in spinal cord compression. The articulation between C4 and C5 is most often affected, and most dogs demonstrate multiple sites of spinal cord compression. Prognosis is good after decompressive surgery (Figure 24.68). Other abnormalities of the vertebral arch include spina bifida, which represents a failure of fusion of the dorsal parts of the vertebra. This may occur alone and cause no clinical signs but can also be associated with protrusion of meningeal and/or spinal cord tissue (Figure 24.69).

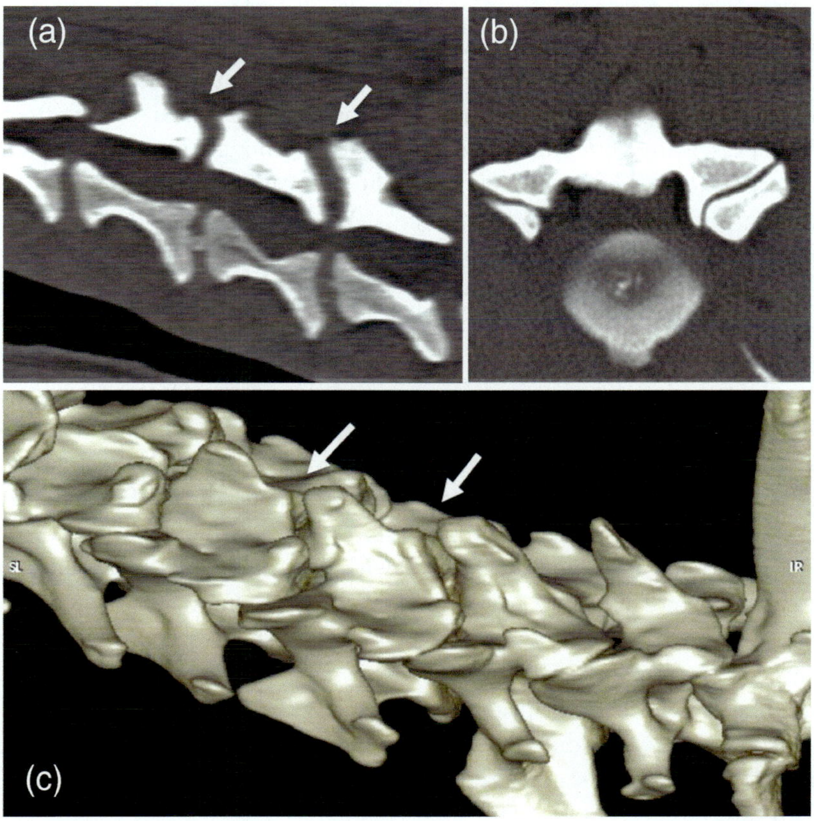

Figure 24.68 **Sagittal-reconstructed (a) and transverse image at the level of C5-C6 (b) and a three-dimensional reconstruction (c) of a 4-year-old Basset hound with dorsal lamina and spinous process hypertrophy at the levels of C4–C5 and C5–C6 (arrows). Pronounced vertebral canal stenosis can be appreciated at the level of C5–C6. The articular processes are not involved in the pathological process. (d) Compare with the normally shaped spinous process of C7.**

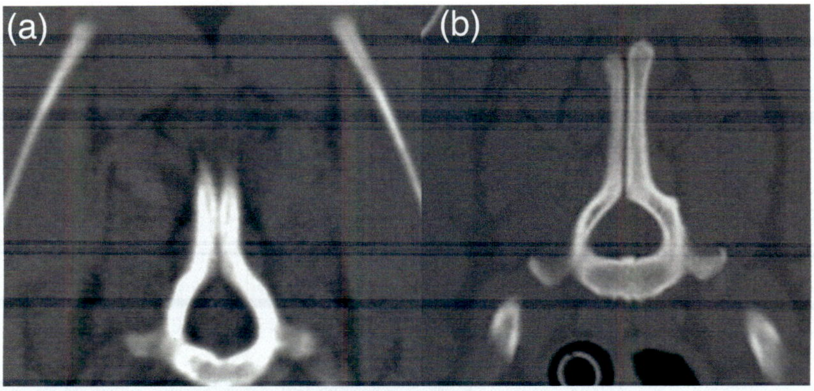

Figure 24.69 **Transverse CT-image at the level of T2 in a 7-month-old Pug (a) and at the level of T1 in a 7-year-old Pug (b) demonstrating clinically irrelevant spina bifida (spina bifida occulta). This anomaly was considered an incidental finding.**

24.5.3 Other

24.5.3.1 Arachnoid diverticula

Arachnoid diverticula, or cysts, are a relatively unusual cause of paresis in dogs but common enough to remain an important differential diagnosis in those presenting with clinical signs of focal spinal cord lesions. The term "cyst" is slightly misleading since histological examination reveals a structure that is not lined by epithelial cells. Arachnoid diverticulum and arachnoid pseudocyst are more appropriate terms, although many clinicians continue to use the term arachnoid cyst.

Arachnoid diverticula fall into different categories: those that occur as a discrete, independent lesion and those associated with a second lesion, such as an intervertebral disc extrusion or congenial vertebral and/or spinal cord anomalies. The former are more common and are typically found in specific locations, with certain breeds predisposed. Rottweilers are over-represented in early descriptions of the condition, with the lesion tending to affect the cervical spinal cord. Recent studies report a less significant over-representation in this breed but reinforce that lesions are seen more commonly at this site in large-breed dogs. Smaller breeds, especially French bulldogs and pugs, are more commonly affected by thoracolumbar lesions. The association with other structural abnormalities, such as congenital vertebral anomalies, may not be causal but may simply reflect the fact that both are common in pugs. Cats appear to be affected less commonly, although cases have been reported in the literature.

MRI studies demonstrate parenchymal spinal cord changes closely associated with the diverticulum, which most likely reflect spinal cord oedema and perhaps develop as a sequela to the changes in CSF flow and pressure around the affected spinal cord, in turn triggered by the fibrous arachnoid adhesions. However, whilst MRI shows these lesions clearly, the presence of an arachnoid diverticulum can be difficult to appreciate, particularly with low-field MRI scanners. In such cases, myelography may be necessary to confirm the presence of an arachnoid diverticulum since this delineates the lesion more clearly in some cases. When there is a high suspicion of the condition, CT myelography is a good imaging modality to use first.

24.5.3.1.1 Aim of the study

As with any surgical spinal cord lesion, the aim is to confirm the precise location and extent of the lesion and to permit surgical planning. The majority of arachnoid diverticula are discrete lesions dorsal to the spinal cord, though lateralized lesions are not uncommon. Arachnoid adhesions can be extensive, in some cases occupying the full circumference of the spinal cord at the affected site. A detailed knowledge of the circumferential and longitudinal extent of the lesion helps establish whether surgery is feasible and also the extent of any laminectomy that may be necessary to fully remove the adhesions.

24.5.3.1.2 CT features

The entire CT study should be systematically examined, including scrutiny of the images in the bone window for the bone algorithm and the soft-tissue window for the soft-tissue algorithm to maximize the study's potential, although the soft-tissue window is of limited use in this condition. Myelography using the contrast medium (either cisternal or lumbar puncture) is essential to confirm the diagnosis. Once the transverse series is obtained (pre- and post-contrast), examining the images in all three planes using multiplanar reconstructions and both the bone and soft-tissue window settings is important. The recognition of the lesion as intradural but extramedullary is also important to allow differentiation from a non-surgical intramedullary lesion.

24.5.3.1.3 Key points in the search for lesions

Plain series

- *Bone window*: search for other possible concurrent anomalies, both congenital (hemivertebra, scoliosis, etc.) and acquired (evidence of previous trauma, recent or old, etc.).
- *Soft-tissue window*: focal dilation of the subarachnoid space may be recognizable with reduction in the diameter/compression of the spinal cord. Intramedullary changes may also be present (e.g. syringiohydromyelia) and may be visible if they are large enough and if the surrounding spinal cord's signal is much less.

Post-myelographic series

- This is better assessed within a bone window. The soft-tissue window could be used if additional intravenous contrast is administered, although this has not been proven to be of any value as spinal cord lesions do not tend to undergo enhancement.

Spinal subarachnoid diverticula are typically classified as intradural, extramedullary lesions.

- Typically are located in the cervical spine (large dogs) and thoracolumbar spine (small/medium dogs).
- The contrast medium accumulates in a bulbous or tear drop-shaped dilation of the subarachnoid space, with associated distortion in the outline of the spinal cord. The cross-section of the spinal cord is triangular at the level of the compression (Figures 24.70 and 27.71).
- Most commonly, a single lesion is located in the dorsal subarachnoid space, either unilateral or bilateral (Figures 24.70 and 24.71). However, diverticula causing dorsal and ventral accumulation of CSF have been described in the cervical spine of Rottweiler dogs.
- The diverticulum is usually oriented with the blind end directed caudally (Figures 24.70 and 24.71), though those in the cervical spine can be directed both caudally and cranially. If a cervical puncture myelogram is performed and the pseudocyst cavity is oriented with the blind end cranially, a golf-tee sign will possibly be seen instead of the more typical appearance.

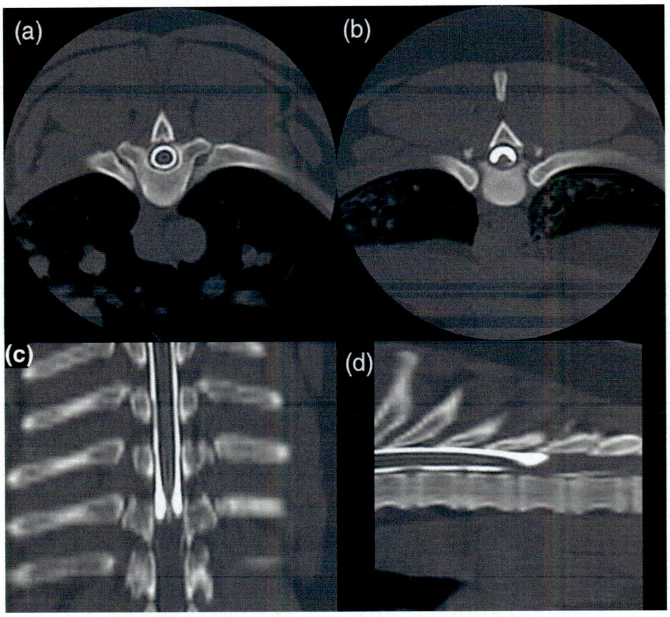

Figure 24.70 Myelographic CT study in a 6-year-old Patterdale Terrier with a 4-week history of progressive tetraparesis. CT images shown here are displayed in a bone window and show transverse images of a normal vertebral column (a) and transverse (b), dorsal (c, MPR), and sagittal plane images (d, MPR) of an arachnoid diverticulum in the caudal thoracic region. Note the dorsal, bilateral dilation of the arachnoid space, which has an abrupt ending and is associated with marked compression of the spinal cord, causing a typical triangular profile in transverse images. The dog underwent decompressive spinal surgery 2 days following his scan, with a dorsal laminectomy of T9–T11 and fenestration of the diverticulum.

In such cases, a lumbar puncture to fill the dilation from the opposite direction will be necessary to fully assess the lesion.

24.5.3.1.4 Practical interpretation

Diagnostic imaging findings must be considered in the context of clinical presentation. If the pseudocyst is small and associated with relatively little intramedullary change, it may be an incidental finding because concurrent anatomical abnormalities are not unusual in dogs with arachnoid diverticula and may be the predominant cause of the clinical signs. On the other hand, even relatively small pseudocyst

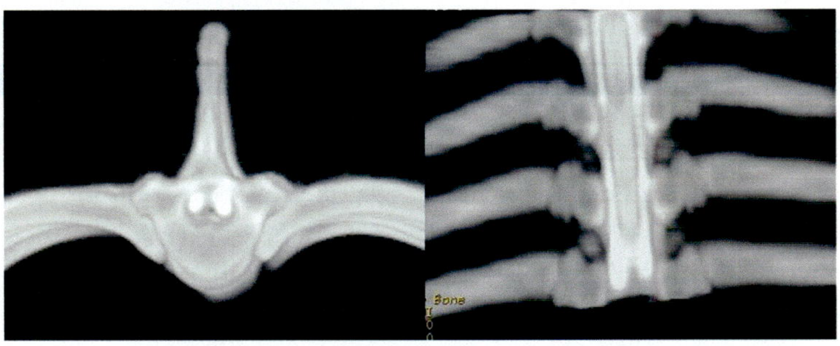

Figure 24.71 Transparent 3D reconstructions in transverse (left) and dorsal planes (right) of the same case as in Figure 24.1, again showing the bilobar dilation of the arachnoid space. The cyst in this case was directed caudally.

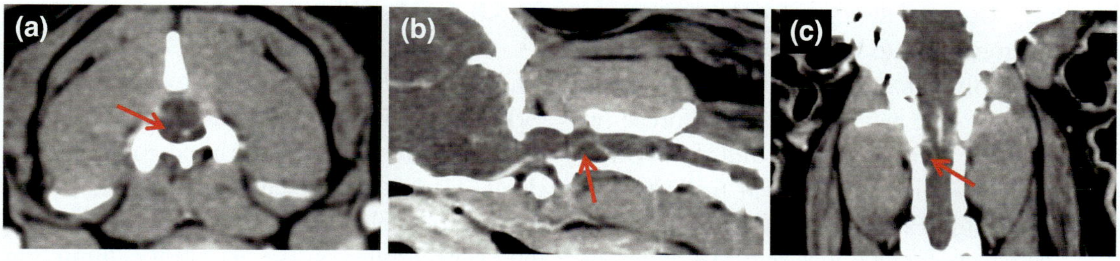

Figure 24.72 Post-contrast transverse (a), sagittal (b), and (c) dorsal reconstructed images of the cranial cervical area in a Cavalier King Charles spaniel of 6 years. Ventral and lateral compression of the spinal cord at the rostral part of C2 by a fluid content structure at the junction of C1–C2 is present, suggesting a synovial cyst or a subarachnoid cyst (arrows). Notice the enhancement of the cystic wall.

cavities can significantly impact spinal cord function.

As a general rule, arachnoid diverticula are clearly delineated by CT myelography, and surgical planning is relatively straightforward. However, as noted above, the lesion can appear as a filling defect in the contrast column, making definitive diagnosis and precise localization more difficult. Clinicians should also be aware that an associated syrinx following contrast medium injection may collapse, widening the arachnoid space and making image interpretation more difficult, which can lead to erroneous localization of the diverticulum.

Once the diverticulum has been identified, a decision must be made whether a conservative or surgical approach is more suitable. Results following surgery tend to be relatively favourable, in the short term at least. However, some diverticula are complex, with both ventral and dorsal components, making the surgical procedure technically challenging and raising the possibility of post-operative deterioration. One study that examined follow-up beyond 1 year following surgery found that the initial improvement noted in some cases was not maintained.

24.5.3.1.5 Synovial cysts

Synovial cysts of the spine are cystic formations connected to the facet joint and containing synovial fluid, represented on CT as hypodense. The cystic wall enhances after IV contrast administration (Figure 24.72).

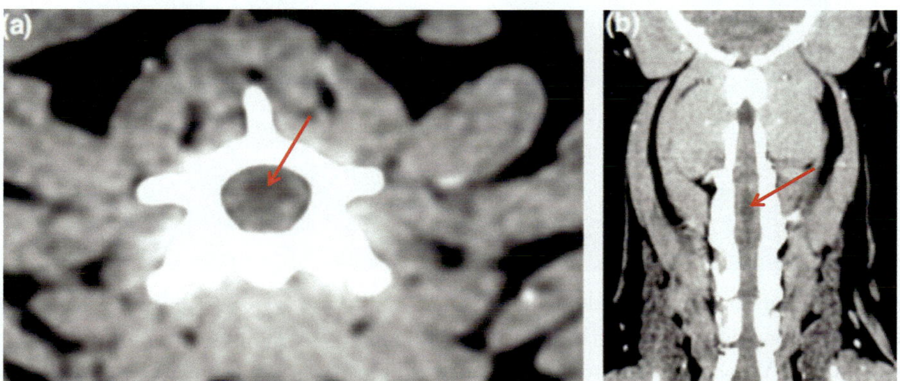

Figure 24.73 Transverse (level of C3) (a) and dorsal-reconstructed (b) CT images of the cranial cervical spine of a Cavalier King Charles spaniel. There is a large tubular hypodense cavity of 3.7 mm diameter in the mid-dorsal aspect of the spinal cord, extending from caudal C2 to cranial C4 (arrows).

24.5.3.2 Cervical syringomyelia

In syringomyelia (SM), fluid-filled cavities develop within the spinal cord. SM is rare in most breeds but has become widespread in Cavalier King Charles spaniels and the Griffon Bruxellois. MRI is the preferred technique to visualize changes in the spinal cord. A recent study demonstrates that low-field MR and multislice CT imaging provide comparable information regarding the presence of SM. The CT images should be reviewed in soft-tissue or brain window. The spinal cord needs to be checked for the presence of an intramedullary fluid-filled cavity with uniform signal intensity and density identical to cerebrospinal fluid, which is hypodense to spinal cord parenchyma (Figure 24.73).

24.5.3.4 Miscellaneous

24.5.3.4.1 Aortal or venous thrombosis

When interpreting CT images of the spine, attention should be paid to the surrounding soft tissues as well. Obtained images should not be limited to a field of view that is too narrow and only concentrates on the vertebral column. On the other hand, a whole-body section may mimic small changes of tissue because of their poor resolution. In the authors' experience, the vertebral column should be displayed with all the adjacent musculature and large draining vessels.

Changes in vessel perfusion, which can be seen only after positive angiography, may occur as a missing enhancement of a certain spinal cord section, missing (unilateral) internal venous

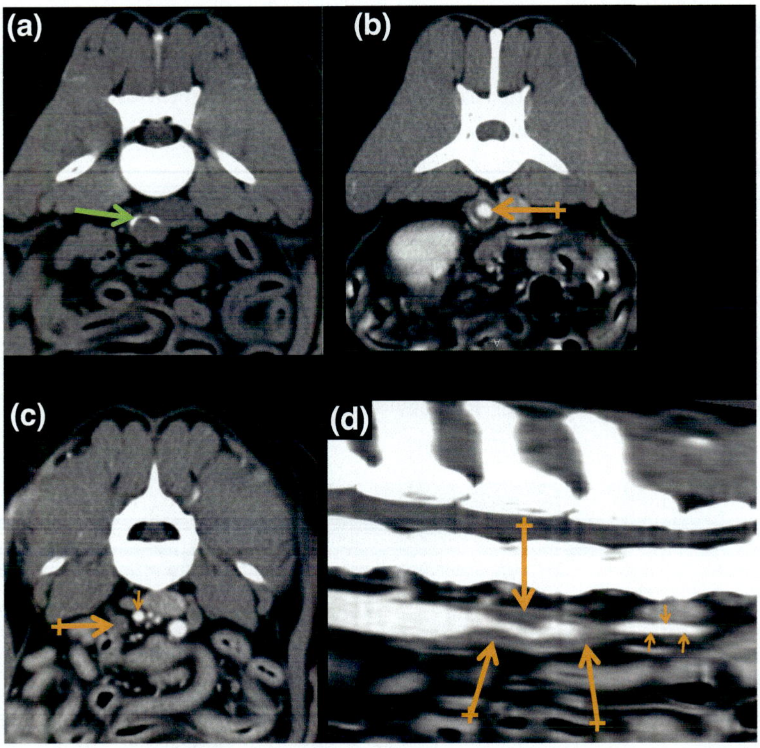

Figure 24.74 **Thrombosis of the abdominal aorta at the level of L4 to L6/left iliac branching in a 7-year-old Rhodesian Ridgeback. In the survey image (a), calcification of the aortic wall is present (green arrow). After IV contrast administration, the reduced diameter of the aorta with thickened wall is shown in image (b) (arrow) and the occlusion of the left external iliac artery in (c) (the arrowhead points at the increased left internal iliac artery due to volume overload). (d) shows the sagittal reformation of the aortic thrombosis cranial to the iliac branching (arrows); also note the thickened left internal iliac artery (arrowheads).**

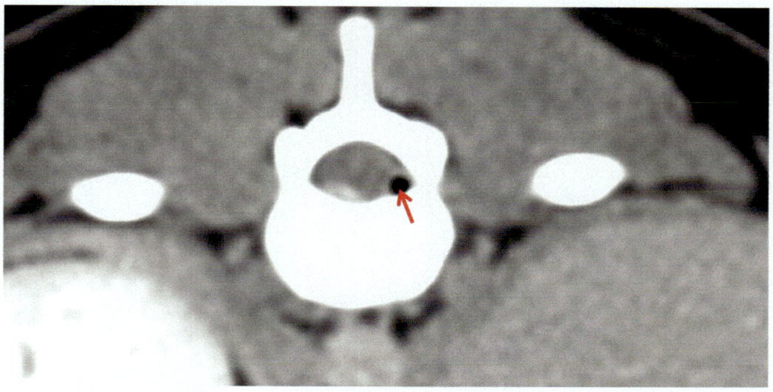

Figure 24.75 **Transverse CT image at the level of the caudal aspect of T13 in soft tissue window after IV contrast administration. Disc herniation is present, causing compression on the spine on the right side. An accumulation of gas within the spinal canal is present (arrow).**

plexus or as increased perfusion with a thickened vessel. Special attention should be paid to the smooth outline of aortal or caval boundaries and the diameter. Thrombosis of the large vessels may be the underlying or influencing cause of disease or treatment in cases of adrenal gland thrombosis affecting the caudal vena cava, aortal/iliac thrombosis, or when haemangiosarcoma is generally suspected (Figure 24.74).

Pneumorrhachis or epidural emphysema is an exceptional finding characterized by the presence of air within the spinal canal (intra- or extradural). Pneumorrhachis can result from disc degeneration, trauma, infection and others. In rare cases, it can cause spinal cord compression, and surgery might be needed (Figure 24.75).

Spinal epidural lipomatosis is also a rare condition. In these cases, hypertrophy of adipose tissue is located in the spinal epidural space (Figure 24.76).

Calcinosis circumscripta is an ectopic deposition of Ca salts in tissues in young large-breed dogs (Figure 24.77). It is most frequently seen in young, large-breed dogs. Calcinosis circumscripta is a solitary lesion found on the cervical spine, metatarsal and phalangeal regions, the tongue, ear pinnae in boxers with cropped ears and on the cheeks of Boston terriers.

Cartilaginous exostosis is a disorder of skeletal development. It has been reported that nodules of cartilaginous tissue, with a variable bony component, arise in the area of growth plates. It can be confused with calcinosis circumscripta. Histopathology is essential for differentiation (Figure 24.78).

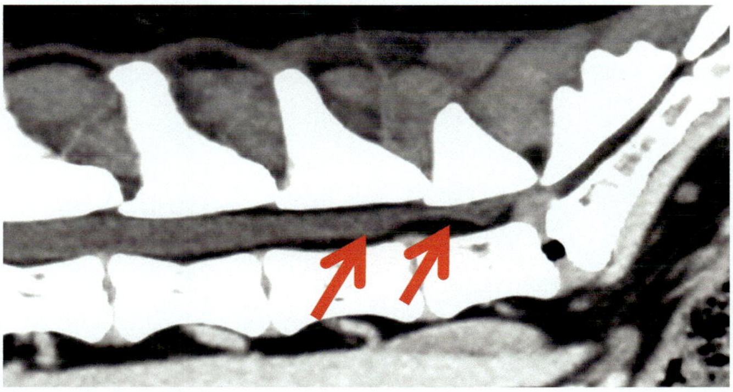

Figure 24.76 **Sagittal-reconstructed image of the lumber and LS region in a Bernese Mountain Dog, post-contrast in soft-tissue window. Cauda equina syndrome and an associated vacuum phenomenon can be noticed. An excessive accumulation of fat is present within the spinal epidural space, creating compression of the nerves of the cauda equina at L6–L7 (arrows).**

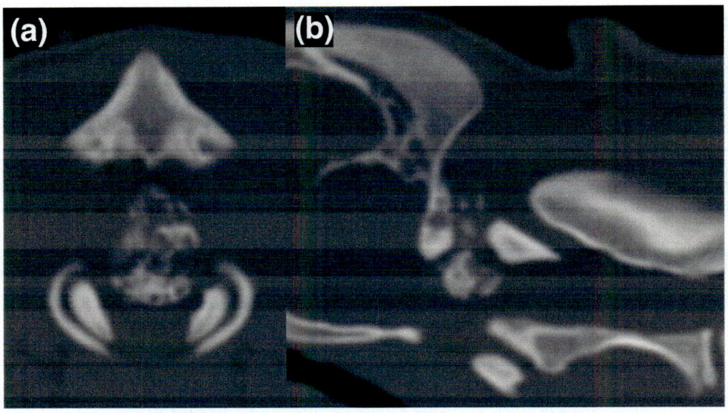

Figure 24.77 Transverse- (a) and sagittal- (b) reconstructed images in bone window, of a 6-month-old Labrador Retriever. Multiple mineral opaque foci are visible within the atlantoaxial space, causing severe compression on the medulla oblongata. The mass was removed and histologically confirmed as calcinosis circumscripta.

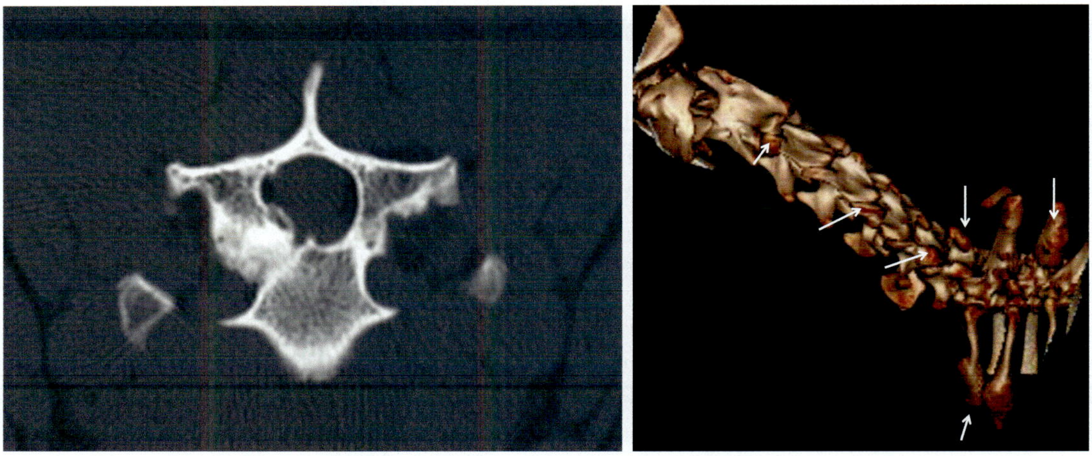

Figure 24.78 Transverse image in bone window at the level of C5 and 3D image of a 7-month-old Doberman. Multiple exostosis are present at the cervical and thoracic vertebrae and ribs (arrows). In C5, the exostosis are causing compression on the spinal cord. These benign proliferative outgrowths of cartilaginous tissue are suggestive of cartilaginous exostosis.

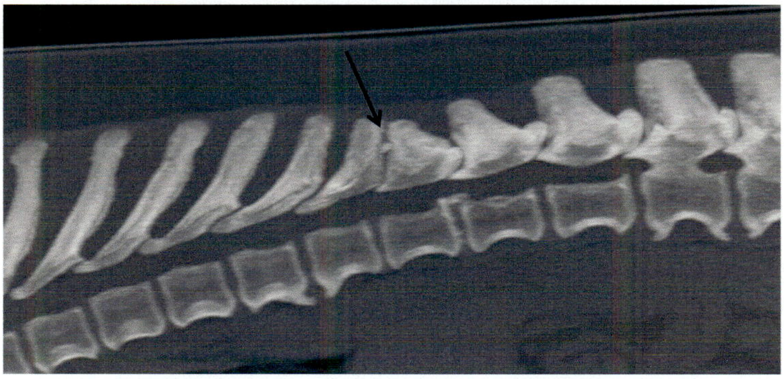

Figure 24.79 Sagittal MPR image of the spine. Close approximation and new bone formation of adjacent spinous processes is noticed (arrow).

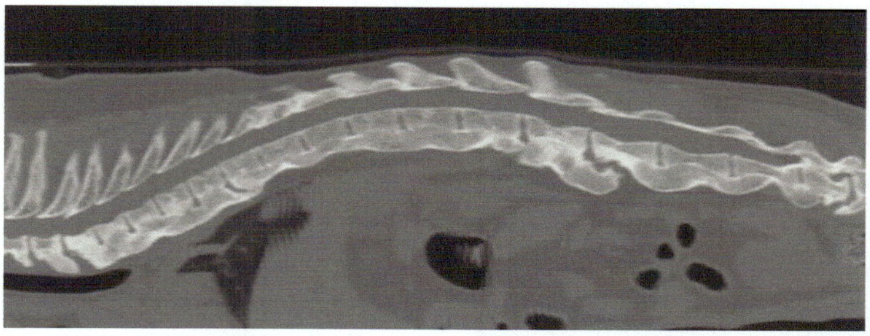

Figure 24.80 Sagittal reformatted CT image in bone window of a cat with diffuse idiopathic skeletal hyperostosis. Smooth proliferative laminar bridging of the ventral aspects of the thoracic and lumbar spine are noted.

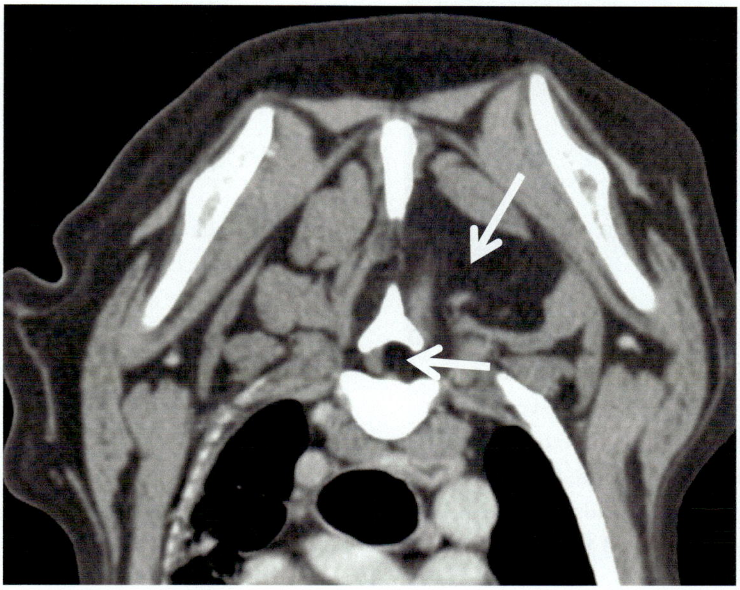

Figure 24.81 Transverse image in soft-tissue window after IV contrast administration at the level of a thoracic vertebra. A large non-enhancing hypodense space occupying mass is present invading the spinal canal (arrows). The HU corresponded to adipose tissue. Histology confirmed a liposarcoma.

Kissing spines (Baastrup's sign) are characterized with an abnormal close contact between the spinal processes and enlargement, flattening and sclerosis of the surfaces of adjacent processes (Figure 24.79).

Diffuse idiopathic skeletal hyperostosis (DISH) is a rare condition in dogs and cats. Boxers are at increased risk. It is a systemic non-inflammatory disease affecting the axial and appendicular skeleton, characterized by ossification of soft tissues, including the spinal ventral longitudinal ligament and enthesis. Criteria for diagnosis are flowing ossification along the ventrolateral aspect of at least four contiguous vertebrae with relative preservation of disc width and absence of facet joint ankyloses (Figure 24.80).

Soft-tissue sarcomas are of mesenchymal origin, arising from connective tissues like muscle, adipose, neurovascular, fascial and fibrous tissue. They could be either benign or malignant. These masses may develop from any anatomical site of the body (Figure 24.81).

24.6 CT-GUIDED PERCUTANEOUS SPINE BIOPSY

Indications for CT-guided percutaneous biopsy are determination of the nature of a bone lesion and investigation for infection and to obtain the sample of organism (Figure 24.82). A CT scan is first performed, with 2.5–5-mm-thick axial images through the affected vertebral body, to locate the best site for needle insertion. The patient is in ventral recumbency, and the point of entry is estimated on the patient's skin.

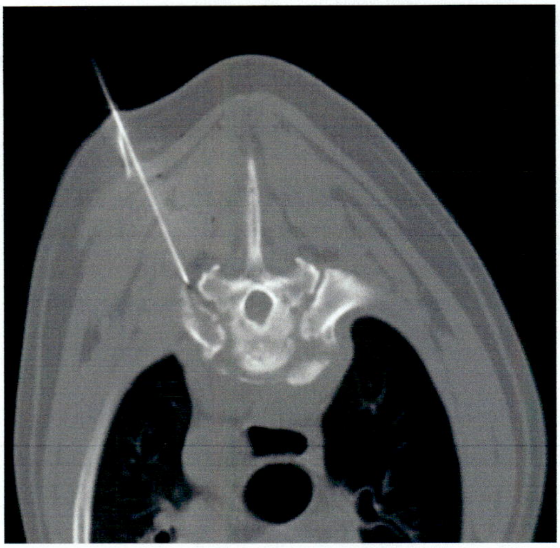

Figure 24.82 **Transverse image in bone window at the level of T4–T5 in a 7-year-Great Dane suspicious of discospondylitis at Th3–Th5. A 20 G needle was inserted, CT-guided, and 2 cc serohaemorrhagic fluid was aspirated. Bacteriologic culture revealed *Staphylococcus (pseudo)intermedius.***

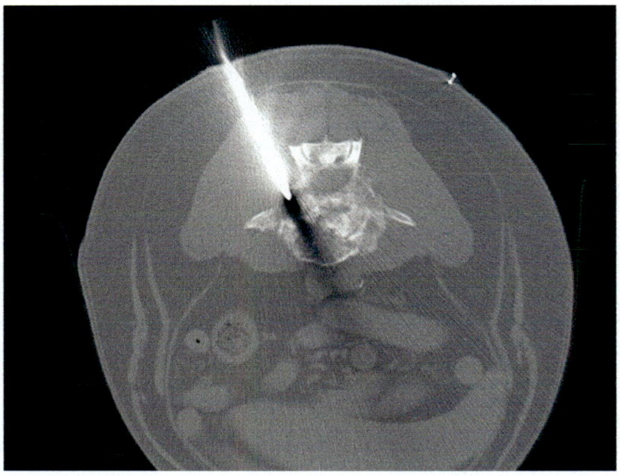

Figure 24.83 **Transverse image in bone window of a 4-year-old Rhodesian Ridgeback showing severe lysis of L5. CT-guided biopsy was performed with histologically confirmed osteosarcoma and osteoblastic productive type.**

The skin is then disinfected, and a strictly sterile condition is mandatory during the hole procedure. The needle, 20 gauge, is inserted through the skin into the body. Another CT scan is performed to confirm that the tip of the needle lies at the desired location. The biopsy needle is introduced through the cortical bone under intermittent CT control to verify correct needle direction (Figure 24.83). When the tip of the needle is seen to be in the proper position, the biopsy specimen is withdrawn through the needle. The needle is removed, and the sample is fixed in 10% formalin. If infection is suspected, the specimens are directly put in a sterile container for microbiological investigation.

24.7 POSTOPERATIVE SPINE CT

CT imaging is useful following surgery to document whether extruded disc material is adequately removed. If only a small amount of material, or none at all, is found. This is described in more detail in the chapter titled "The use of intraoperative- and follow up CT scan" (Chapter 27).

FURTHER READING

Adrega da Silva C, Bernard F, Bardet J-F. Caudal cervical arthrodesis using a distractable fusion cage in a dog. *Veterinary and Comparative Orthopaedics and Traumatology* 2010, 23:209–213.

Aikawa T, Shibata M, Fujita H. Modified ventral stabilization using positively threaded profile pins and polymethylmethacrylate for atlantoaxial instability in 49 dogs. *Veterinary Surgery* 2013, 42(6):683–692.

Argent V, Perillo R, Jeffery N, Freeman P. Recurrence of signs consistent with cervical intervertebral disc extrusion in dogs. *Journal of Small Animal Practice* 2022; 63(6):454–459.

Axlund TW, Hudson JA. Computed tomography of the normal lumbosacral intervertebral disc in 22 dogs. *Veterinary Radiology & Ultrasound* 2003, 44(6):630–634.

Bahr A. The Vertebrae. In: Thrall J (ed) *Textbook of Veterinary Diagnostic Radiology*, Saunders, St Louis, 2007, ch 11. pp. 179.

Bibevski JD, Daye RM, Henrickson TD, Axlund TW. A prospective evaluation of CT in acutely paraparetic chondrodystrophic dogs. *Journal of American Animal Hospital Association* 2013, 49(6):363–369.

Böttcher P, Böttcher IC, Truar K et al. Effect of ventral slot procedure on spinal cord compression in dogs with single static intervertebral disc disease: preliminary findings while evaluating a semiquantitative computed tomographic myelographic score of spinal cord compression. *Veterinary Surgery* 2013, 42:383–391.

Brocal J, De Decker S, José-López R, Manzanilla EG, Penderis J, Stalin C, Bertram S, Schoenebeck JJ, Rusbridge C, Fitzpatrick N, Gutierrez-Quintana R. C7 vertebra homeotic transformation in domestic dogs — are Pug dogs breaking mammalian evolutionary constraints? *Journal of Anatomy* 2018, 233(2):255–265.

Brocal J, De Decker S, José-López R, Guevar J, Ortega M, Parkin T, Ter Haar G, Gutierrez-Quintana R. Evaluation of radiography as a screening method for detection and characterisation of congenital vertebral malformations in dogs. *Veterinary Record* 2018, 82(20):573.

Bruce CW, Brisson BA, Gyselinck K. Spinal fracture and luxation in dogs and cats. A restrospective evaluation of 95 cases. *Veterinary and Comparative Orthopaedics and Traumatology* 2008, 21:280–284.

Caporn TM, Read RA. Osteochondromatosis of the cervical spine causing compressive myelopathy in a dog. *Journal of Small Animal Practice* 1996, 37(3):133–137.

Cerda-Gonzalez S, Dewey CW. Congenital diseases of the of the craniocervical junction in the dog. *Veterinary Clinics of North America Small Animal Practice* 2010, 40:121–141.

Chambers JN, Oliver J, Bjorling DE. Update on ventral decompression for caudal cervical disk herniation in Doberman pinschers. *Journal of the American Animal Hospital Association* 1986, 22/775–778.

Cooper JJ, Young BD, Griffin JF 4th, Fosgate GT, Levine JM. Comparison between noncontrast computed tomography and magnetic resonance imaging for detection and characterization of thoracolumbar myelopathy caused by intervertebral disk herniation in dogs. *Veterinary Radiology & Ultrasound* 2014, 55(2):182–189.

da Costa RC. Cervical spondylomyelopathy. *Veterinary Clinics of North America Small Animal Practice* 2010, 40(5):881–913.

da Costa RC, Echandi RL, Beauchamp D. Computed tomography myelographic findings in dogs with cervical spondylomyelopathy. *Veterinary Radiology and Ultrasound* 2012, 53(1):64–70.

da Costa RC, de Decker S, Lewis MJ et al. Diagnostic imaging in intervertebral disc disease. *Frontiers in Veterinary Science* 2020, 7:588338.

De Decker S, De Risio L, Lowrie M et al. Cervical vertebral stenosis associated with a vertebral arch anomaly in the Basset Hound. *Journal of Veterinary Internal Medicine* 2012, 26(6):1374–1382.

De Decker S, Caemaert J, Tshamala MV et al. Surgical treatment of disk-associated wobbler syndrome by a distractable vertebral titanium cage in seven dogs. *Veterinary Surgery* 2011, 40(5):544–554.

De Decker S, Gielen IM, Duchateau et al. Intraobserver, interobserver, and intermethod agreement for results of myelography, computed tomography-myelography, and low-field magnetic resonance imaging in dogs with disk associated wobbler syndrome. *Journal of the American Veterinary Medical Association* 2011, 238(12):1601–1608.

De Decker S, da Costa RC, Volk HA, Van Ham L. Current insights and controversies in the pathogenesis and diagnosis of disc associated cervical spondylomyelopathy in dogs. *Veterinary Record* 2012, 171(21):531–537.

de Nies KS, Edwards RA, Bergknut N, Beukers M, Meij BP. Caudal lumbar spinal cysts in two French Bulldogs. *Acta Veterinaria Scandinavica* 2018, 60(1):14.

Dennis R, Kirberger RM, Barr F, Wrigley RH. Spine. In: Dennis R, Kirberger RM, Barr F, Wrigley RH (ed) *Handbook of Small Animal Radiology and Ultrasound. Techniques and Differential Diagnoses*, Saunders, 2010; ch 5, pp. 115.

Dewey CW, Marino DJ, Loughin CA. Craniocervical junction abnormalities in dogs. *New Zealand Veterinary Journal* 2013, 61(4):202–211.

Dewey CW, Davies E, Bouma JL. Kyphosis and kyphoscoliosis associated with congenital malformations of the thoracic vertebral bodies in dogs. *Veterinary Clinics of North America: Small Animal Practice* 2016, 46(2):295–306.

Dennison SE, Drees R, Rylander H, Yandell BS, Milovancev M, Pettigrew R, Schwarz T. Evaluation of different computed tomography techniques and myelography for the diagnosis of acute canine myelopathy. *Veterinary Radiology & Ultrasound* 2010, 51(3):254–258.

Drees R, Dennison SE, Keuler NS, Schwarz T. Computed tomographic imaging protocol for the canine cervical and lumbar spine. *Veterinary Radiology & Ultrasound*. 2009, 50(1):74–79.

Dhupa S, Glickman N, Waters DJ. Reoperative neurosurgery in dogs with thoracolumbar disc disease. *Veterinary Surgery* 1999; 28:421–428.

Eagleson JS, Diaz J, Platt SR et al. Cervical vertebral malformation-malarticulation syndrome in the Bernese mountain dog: clinical and magnetic resonance imaging features. *Journal of Small Animal Practice* 2009, 50(4):186–193.

Engel S, Randall EK, Cuddon PA, Webb BT, Aboellail TA. Imaging diagnosis: Multiple cartilaginous exostoses and calcinosis circumscripta occurring simultaneously in the cervical spine of a dog. *Veterinary Radiology & Ultrasound* 2014, 55(3):305–309.

Evans H. Arthrology. In: Evans HE (ed) *Miller's Anatomy of the Dog*, Saunders, 1993; ch 5, pp. 219–258.

Foster E, West N, Butterfield S, Rusbridge C, Crawford A. Respiratory compromise in French bulldogs presented with intervertebral disc extrusion. *Veterinary Record* 2023, 193(12):e3603.

Forterre F, Gorgas D, Dickomeit M et al. Incidence of spinal compressive lesions in chondrodystrophic dogs with abnormal recovery after hemilaminectomy for treatment of thoracolumbar disc disease: a prospective magnetic resonance imaging study. *Veterinary Surgery* 2010, 39:165–172.

Fisher SC, Shores A, Simpson ST. Constrictive myelopathy secondary to hypoplasia or aplasia of the thoracolumbar caudal articular processes in Pugs; 11 cases (1993—2009). *Journal of the American Veterinary Medical Association* 2013, 242(2):223–229.

Flückiger MA, Damur-Djuric N, Hässig M, Morgan JP, Steffen F. A lumbosacral transitional vertebra in the dog predisposes to cauda equina syndrome. *Veterinary Radiology & Ultrasound* 2006, 47(1): 39–44.

Foster A, Morandi F, May E. Prevalence of ear disease in dogs undergoing multidetector thin-slice computed tomography of the head. *Veterinary Radiology & Ultrasound* 2015, 56(1): 18–24.

Galloway AM, Curtis NC, Sommerlad SF, Watt PR. Correlative imaging findings in seven dogs and one cat with spinal arachnoid cysts. *Veterinary Radiology and Ultrasound* 1999, 40(4):445–452.

Geary JC, Oliver JE, Hoerlein BF. Atlantoaxial subluxation in the canine. *Journal of Small Animal Practice* 1967, 8:577–582.

Gnirs K, Ruel Y, Blot S, Begon D, Rault D, Delisle F, Boulouha L, Colle MA, Carozzo C, Moissonnier P. Spinal Subarachniod cysts in 13 dogs. *Veterinary Radiology and Ultrasound* 2003, 44(4):402–408.

Gonzalo-Orden JM, Alonaga JR, Orden A. Magnetic resonance, computed tomographic and radiologic findings in a dog with discospondylitis. *Veterinary Radiology & Ultrasound* 2000, 41(2):142–144.

Hanna FY. Lumbosacral osteochondrosis: radiological features and surgical management in 34 dogs. *Journal of Small Animal Practice* 2001, 42(6):272–278.

Havig ME, Cornell KK, Hawthorne JC, McDonnell JJ, Selcer BA. Evaluation of nonsurgical treatment of atlantoaxial subluxation in dogs: 19 cases (1992—2001). *Journal of the American Veterinary Medical Association* 2005, 227(2):257–262.

Hecht S, Thomas WB, Marioni-Henry K, Echandi RL, Matthews AR, Adams WH. Myelography vs. computed tomography in the evaluation of acute thoracolumbar intervertebral disk extrusion in chondrodystrophic dogs. *Veterinary Radiology & Ultrasound* 2009, 50(4):353–359.

Herzig R, Wang-Leandro A, Steffen F, Matiasek K, Beckmann KM. Imaging and histopathologic features of reversible nerve root and peripheral nerve edema secondary to disc herniation in a cat. *Journal of Veterinary Internal Medicine* 2021, 35(3):1566–1572.

Hettlich BF, Kerwin SC, Levine JM. Early reherniation of disk material in eleven dogs with surgically treated thoracolumbar intervertebral disk extrusion. *Veterinary Surgery* 2012, 41:215–222.

Higgins BM, Cripps PJ, Baker M, Moore L, Penrose FE, McConnell JF. Effects of body position, imaging plane, and observer on computed tomographic measurements of the lumbosacral intervertebral foraminal area in dogs. *American Journal of Veterinary Research* 2011, 72(7):905–917.

Israel SK, Levine JM, Kerwin SC, Levine GJ, Fosgate GT. The relative sensitivity of computed tomography and myelography for identification of thoracolumbar intervertebral disk herniations in dogs. *Veterinary Radiology & Ultrasound* 2009, 50(3):247–252.

Johnson P, De Risio L, Sparkes A, McConnel F, Holloway A. Clinical, morphologic, and morphometric features of cranial thoracic spinal stenosis in large and giant breed dogs. *Veterinary Radiology and Ultrasound* 2012, 53(5):524–534.

Jones JC, Shires PK, Inzana KD, Mosby AD, Sponenberg DP, Lanz OI. Use of computed tomographic densitometry to quantify contrast enhancement of compressive soft tissues in the canine lumbosacral vertebral canal. *American Journal of Veterinary Research* 2002, 63(5): 733–737.

Jones JC, Inzana KD. Subclinical CT abnormalities in the lumbosacral spine of older large-breed dogs. *Veterinary Radiology & Ultrasound* 2000, 41(1):19–26.

Jones JC, Inzana KD, Rossmeisl JH, Bergman RL, Wells T, Butler K. CT myelography of the thoraco-lumbar spine in 8 dogs with degenerative myelopathy. *Journal of Veterinary Science* 2005, 6(4):341.

Kawalilak LT, Tucker RL, Greene SA. Use of contrast-enhanced computed tomography to study the cranial migration of a lumbosacral injectate in cadaver dogs. *Veterinary Radiology & Ultrasound* 2015, 56(5):570–574.

Kerwin SC, Taylor AR. Neurologic causes of thoracic limb lameness. *Veterinary Clinics of North America: Small Animal Practice* 2021, 51(2):357–364.

King JB, Jones JC, Rossmeisl JH, Harper TA, Lanz OI, Werre SR. Effect of multi-planar CT image reformatting on surgeon diagnostic performance for localizing thoracolumbar disc extrusions in dogs. *Journal of Veterinary Science* 2009, 10(3):225–232.

Klang A, Kneissl S, Glänzel R, Fuchs-Baumgartinger A. Imaging diagnosis: fibrodysplasia ossificans progressiva in a cat. *Veterinary Radiology & Ultrasound* 2013, 54(5):532–535.

Kromhout K, van Bree H, Broeckx BJG, Bhatti S, De Decker S, Polis I, Gielen I. Low-field MRI and multislice CT for the detection of cervical syringomyelia in dogs. *Journal of Veterinary Internal Medicine* 2015, 29:1354–1359.

Kuroki K, Vitale CL, Essman SC, Pithua P, Coates JR. Computed tomographic and histological findings of Hansen type I intervertebral disc herniation in dogs. *Veterinary and Comparative Orthopaedics and Traumatology* 2013, 26(5): 379–384.

Lackmann F, Forterre F, Brunnberg L, Loderstedt S. Epidemiological study of congenital malformations of the vertebral column in French bulldogs, English bulldogs and pugs. *Veterinary Record* 2022, 190(1):e509.

Lamb CR, Nicholls A, Targett M, Mannion P. Accuracy of survey radiographic diagnosis of intervertebral disc protrusion in dogs. *Veterinary Radiology & Ultrasound* 2002, 43(3):222–228.

Lavely JS, Vernau KM, LeCouteur RA. Spinal cord empyema in seven dogs. *Journal of Veterinary Internal Medicine* 2002, 16:369.

Leu D, Vidondo B, Stein V, Forterre F. Recurrence rate of intervertebral disc disease in surgically treated French Bulldogs: a retrospective study (2009—2019). *Acta Veterinaria Scandinavica* 2023, 65(1):3.

Lewis M, Olby NJ, Sharp NJH, Early P. Long-term effect of cervical distraction and stabilization on neurological status and imaging findings in giant breeds with cervical stenotic myelopathy. *Veterinary Surgery* 2013, 42(6):701–709.

Lim, Kweon O, Choi M et al. Computed tomographic characteristics of acute thoracolumbar intervertebral disc disease in dogs. *Journal of Veterinary Science* 2012, 11:73–79.

Longo S, Gomes SA, Briola C, Duffy K, Targett M, Jeffery ND, Freeman P. Association of magnetic resonance assessed disc degeneration and late clinical recurrence in dogs treated surgically for thoracolumbar intervertebral disc extrusions. *Journal of Veterinary Internal Medicine* 2021, 35(1):378–387.

Martin-Vaquero P, da Costa RC, Drost WT. Comparison of noncontrast computed tomography and high-field magnetic resonance imaging in the evaluation of Great Danes with cervical spondylomyelopathy. *Veterinary Radiology & Ultrasound* 2014, 55(5):496–505.

Matiasek LA, Platt SR, Dennis R et al. Subfascial seroma causing compressive myelopathy after cervical dorsal laminectomy. *Veterinary Radiology & Ultrasound* 2006, 47:581–584.

Mauler DA, De Decker S, De Risio L, Volk HA, Dennis R, Gielen I, Van der Vekens E, Goethals K, Van Ham L. Signalment, clinical presentation, and diagnostic findings in 122 dogs with spinal arachnoid diverticula. *Journal Veterinary Internal Medicine* 2014, 28:175–181.

McCarthy RJ, Lewis DD, Hosgood G. Atlantoaxial subluxation in dogs. *Compendium of Continuing Education for the Practicing Veterinarian* 1995, 17:215–227.

McConnell F. Imaging of neurological emergencies. In: Platt S, Garosi L (eds) *Small Animal Neurological Emergencies*, Manson Publishing, London, 2012; ch 4, pp. 83.

McDonell JJ, Knowles KE, deLahunta A et al. Thoracolumbar spinal cord compression due to vertebral process degenerative joint disease in a family of Shiloh shepherd dogs. *Journal of Veterinary Internal Medicine* 2003, 17:530–537.

Mclellan AM, Daniel S, Corcuera-Solano I et al. Optimized imaging of the postoperative spine. *Neuroimaging Clinics of North America* 2014, 24:349–364.

McNicholas Jr WT, Wilkens BE, Blevins WE, Snyder PW, McCabe GP, Applewhite AA, Breur GJ. Spontaneous femoral capital physeal fractures in adult cats: 26 cases (1996—2001). *Journal of the American Veterinary Medical Association* 2002, 221(12):1731–1736.

Meij BP, Bergknut N. Degenerative lumbosacral stenosis in dogs. *The Veterinary Clinics of North America: Small Animal Practice* 2010, 40(5): 983–1009.

Moissonier P, Gossot P, Scotti S. Thoracic kyphosis associated with hemivertebra. *Veterinary Surgery* 2011, 40(8):1029–1032.

Moore GE, Mathey WS, Eggers JS, Estep JS. Osteosarcoma in adjacent lumbar vertebrae in a dog. *Journal of the American Veterinary Medical Association* 2000, 217(7):1038–1040.

Morabito S, Auriemma E, Zagarella P, Mercuriali E, Domenech O, Gandini G, Gagliardo T, Zini E, Specchi S. Computed tomographic and angiographic assessment of spinal extradural arteriovenous fistulas in a dog. *Canadian Veterinary Journal* 2017, 58(3):275–279.

Muir P, Johnson KA, Manley PA. Comparison of hemilaminectomy and dorsal laminectomy for thoracolumbar intervertebral disc extrusion in dachshunds. *Journal of Small Animal Practice* 1995, 36:360–367.

Müller MK, Ludewig E, Oechtering G, Scholz M, Flegel T. The vacuum phenomenon in intervertebral disc disease of dogs based on computed tomography images. *Journal of Small Animals Practice* 2013, 54(5):253–257.

Ndikuwera J, Knottenbelt DC, Lawrence J et al. Spinal abscess in a dog. *The Veterinary Record* 1987, 120:554–555.

Newcomb B, Arble J, Rochat M, Pechman R, Payton M. Comparison of computed tomography and myelography to a reference standard of computed tomographic myelography for evaluation of dogs with intervertebral disc disease. *Veterinary Surgery* 2012, 41(2):207–214.

Nykamp SG, Steffey MA, Scrivani PV, Schatzberg SJ. Computed tomographic appearance of epidural empyema in a dog. *The Canadian Veterinary Journal* 2003, 44(9):729.

Olby N. Spinal trauma. In: Platt S, Garosi L (ed) *Small Animal Neurological Emergencies*, Manson Publishing, London, 2012; ch 21, pp. 383.

Olby N. The pathogenesis and treatment of acute spinal cord injuries in dogs. *Veterinary Clinics of North America: Small Animal Practice* 2010, 40(5):791–807.

Owen MC, Davis SH, Worth AJ. Imaging Diagnosis — traumatic myelopathy in a dog with incomplete ossification of the neural arch of the atlas. *Veterinary Radiology and Ultrasound* 2008, 49:570–572.

Parry AT, Upjohn MM, Schlegl K, Kneissl S, Lamb CR. Computed tomography variations in morphology of the canine atlas in dogs with and without atlamtoaxial subluxation. *Veterinary Radiology and Ultrasound* 2010, 51(6):596–600.

Penderis J, Schwarz T, McConnel F, Garosi LS, Thomson CE, Dennis R. Dysplasia of the caudal vertebral articular facets in four dogs: results of radiographic, myelographic and magnetic resonance imaging investigations. *The Veterinary Record* 2005, 156:601–605.

Pfund R, Forward AK, Fentem R, Nagendran A, Fraser AR, Crawford AH. Postoperative outcome of ambulatory dogs with intervertebral disc extrusion causing incontinence and/or tail dysfunction: 18 cases (2010—2020). *Journal of Small Animals Practice* 2022, 63(7):550–558.

Ramirez O 3rd, Thrall DE. A review of imaging techniques for canine cauda equina syndrome. *Veterinary Radiology & Ultrasound* 1998, 39(4):283–296.

Roach WJ, Thomas M, Weh JM et al. Residual herniated disc material following hemilaminectomy in chondrodystrophic dogs with thoracolumbar intervertebral disc disease. *Veterinary and Comparative Orthopaedics and Traumatology* 2012, 25:109–115.

Robertson I, Thrall DE. Imaging dogs with suspected disc herniation: pros and cons of myelography, computed tomography, and magnetic resonance. *Veterinary Radiology & Ultrasound* 2011, 52(1 Suppl 1):S81–S84.

Rodiño-Tilve V, Guevar J, Hammond G, Penderis J. Incomplete ossification of the atlas in a dog: surgical stabilisation using a SOP plate. *Journal of Small Animal Practice* 2017, 58(5):301.

Roudebush P, Maslin WR, Cooper RC. Canine tumoral calcinosis. *Compendium: Continuing Education for Veterinarians* 1988, 10:1162–1164.

Ruoff CM, Kervin SC, Taylor AR. Diagnostic imaging of discospondylitis. *Veterinary Clinics of North American Small Animal Practice* 2018, 48:85–94.

Rylander H, Lipsitz D, Berry WL, Sturges BK, Vernau KM, Dickinson PJ, Añor SA, Higgins RJ, LeCouteur RA. Retrospective analysis of spinal arachnoid cysts in 14 dogs. *Journal of Veterinary Internal Medicine* 2002, 16(6):690–696.

Sande RD. Radiography, myelography, computed tomography, and magnetic resonance imaging of the spine. *Veterinary Clinics of North America: Small Animal Practice* 1992, 22:811–831.

Sanders WP, Truumees E. Imaging of the postoperative spine. *Seminars in Ultrasound, CT and MRI* 2004, 25:523–535.

Santifort KM, Gutierrez-Quintana R, Bernardini M, Kortz GD, Gomes SA, Lowrie M, De Decker S, Glass EN, Bergknut N. Vertebral vascular canal dysplasia in French and English Bulldogs: clinical, CT, and MRI characteristics and prevalence. *Veterinary Radiology & Ultrasound* 2022, 63(3):281–291.

Shamir MH, Tavor N, Aizenberg T. Radiographic findings during recovery from discospondylitis. *Veterinary Radiology & Ultrasound* 2001, 42(6):496–503.

Schroeder R, Pelsue DH, Park RD, Gasso D, Bruecker KA. Contrast-enhanced CT for localizing compressive thoracolumbar intervertebral disc extrusion. *Journal of American Animal Hospital Assocication* 2011, 47(3):203–209.

Schwarz T, Owen MR, Long S, Sullivan M. Vacuum disk and facet phenomenon in a dog with cauda equina syndrome. *Journal of the American Veterinary Medical Association* 2000, 217(6):862–864.

Schulz K, Walker M, Moon M et al. Correlation of clinical, radiographic, and surgical localization of intervertebral disc extrusion in small-breed dogs: a prospective study of 50 cases. *Veterinary Surgery* 1998, 27/105–111.

Seiler GS, Häni H, Busato AR, Lang J. Facet joint geometry and intervertebral disk degeneration in the L5-S1 region of the vertebral column in German Shepherd dogs. *American Journal of Veterinary Research* 2002, 63(1):86–90.

Seiler G, Kinns J, Dennison S, Saunders J, Schwarz T. Vertebral column and spinal cord. In: Schwarz T, Saunders J (eds) *Veterinary Computed Tomography*, Wiley-Blackwell, Chichester, 2011; ch 13, pp. 209.

Seim HB, Prata RG. Ventral decompression for the treatment of cervical disk disease in the dog: a review of 54 cases. *Journal of the American Animal Hospital Association* 2000, 36:68–74.

Seo E, Choi J, Choi M, Yoon J. Computed tomographic evaluation of cervical vertebral canal and spinal cord morphometry in normal dogs. *Journal of Veterinary Science* 2014,15(2):187–193.

Sharp NJH, Cofone M, Robertson ID et al. Computed tomography in the evaluation of caudal cervical spondylomyelopathy of the Doberman pinscher. *Veterinary Radiology and Ultrasound* 1995, 36:100–108.

Shimizu J, Koga M, Kishimoto M et al. Effect of an autogenous free fat graf ton hemilaminectomy defects in dogs. *Journal of Veterinary Medical Science* 2009, 71:1291–1294.

Shores A. Spinal trauma. Pathophysiology and management of traumatic spinal injuries. *Veterinary Clinics of North America: Small Animal Practice* 1992, 22(4):859–888.

Signoret M, Gros L, Musso C, Hahn H, Le Boedec K, Cauzinille L. IV versus myelography computed tomography for thoracolumbar intervertebral disc extrusion surgical planning in French Bulldogs. *Journal of American Animal Hospital Association* 2023, 59(5):207–213.

Smith CJ, Guevar J. Spinal subarachnoid diverticula in dogs: a review. *Canadian Veterinary Journal* 2020, 61(11):1162–1169.

Smith BA, Hosgood G, Kerwin SC. Ventral slot decompression for cervicalintervertebral disc disease in 112 dogs. *Australian Veterinary Practitioner* 1997, 27:58–64.

Skeen TM, Olby NJ, Muñana KR, Sharp NJ. Spinal arachnoid cysts in 17 dogs. *Journal of the American Animal Hospital Association* 2003, 39:271–282.

Suiter E, Grapes N, Martin-Garcia L, De Decker S, Gutierrez-Quintana R, Wessmann A. MRI and clinical findings in 133 dogs with recurrent deficits following intervertebral disc extrusion surgery. *Veterinary Record* 2023, 193(5):e2992.

Suwankong N, Voorhout G, Hazewinkel HA, Meij BP. Agreement between computed tomography, magnetic resonance imaging, and surgical findings in dogs with degenerative lumbosacral stenosis. *Journal of the American Veterinary Medical Association* 2006, 229(12):1924–1929.

Tamura S, Doi S, Tamura Y, Takahashi K, Enomoto H, Ozawa T, Uchida K. Thoracolumbar intradural disc herniation in eight dogs: clinical, low-field magnetic resonance imaging, and computed tomographic myelography findings. *Veterinary Radiology & Ultrasound* 2015, 56(2):160–167.

Thatcher H, Targett M, Alcoverro E, Stee K, Schofield I, Lowrie M, Gomes SA. Incidence and clinical characterisation of thoracolumbar intervertebral disc extrusions in Basset Hounds compared with Dachshunds. *Veterinary Record* 2024, 194(1):e3212.

Thomas WB, Sorjonen DC, Simpson ST. Surgical management of atlantoaxial subluxation in 23 dogs. *Veterinary Surgery* 1991, 20(6):409–412.

Thrall D. *Textbook of Veterinary Diagnostic Radiology*. Elsevier Health Sciences 2007, pp. 179–219.

Tomlinson J. Tetraparesis following cervical disc fenestration in two dogs. *Journal of the American Veterinary Medical Association* 1985, 187:76–80.

Van Ham LMV, van Bree H, Tshamala M, Thoonen H. Use of computed tomography and computed tomographic myelography for assessment of spinal tumoral calcinosis in a dog. *Veterinary Radiology & Ultrasound* 1995, 36(2):115–118.

Warren-Smith CMR, Kneissl S, Benigni L, Kenny P, Lamb CR. Incomplete ossification of the atlas in 5 dogs with cervical spinal signs. *Veterinary Radiology and Ultrasound* 2009, 50(6):635–638.

Watson AG, De Lahunta A. Atlantoaxial subluxation and absence of transverse ligament on the atlas in a dog. *Journal of American Veterinary Medicine Association* 1989, 195:35–37.

Westworth DR, Sturges BK. Congenital spinal malformations in small animals. *Veterinary Clinics of North America Small Animal Practice* 2010, 40:951–981.

Widmer WR, Thrall DE. Canine and feline intervertebral disc disease, myelography, and spinal cord disease. In: Thrall J (ed) *Textbook of Veterinary Diagnostic Radiology*, Saunders, St Louis, 2007; ch 12. pp. 194.

Wilson ER, Aron DN, Roberts RE. Observation of a secondary compressive lesion after treatment of caudal cervical spondylomyelopathy in a dog. *Journal of the American Veterinary Medical Association* 1994, 2:153–154.

Wrzosek M, Płonek M, Zeira O, Bieżyński J, Kinda W, Guziński M. Congenital bipartite atlas with hypodactyly in a dog: clinical, radiographic and CT findings. *Journal of Small Animal Practice* 2014, 55(7):375–378.

CHAPTER 25

CT OF THE PERIPHERAL NERVES, BRACHIAL, AND LUMBOSACRAL PLEXUS

Stephan Kaiser

25.1 INTRODUCTION

Symptoms of peripheral neuropathy are often difficult to distinguish from those of an orthopaedic problem and often lead to delayed diagnosis. If not caused by an acute trauma, animals often show very slowly progressive lameness with noticeable muscle atrophy. When neurological deficits are already present, differentiation with regard to orthopaedic causes is easier.

25.2 INDICATIONS FOR USING CT

CT is a useful diagnostic method for brachial and lumbosacral plexus lesions. In the clinical evaluation prior to a decision for CT, significant muscle atrophy is a key finding. A CT scan of the brachial and lumbosacral plexus and related spinal cord areas is thus recommended for dogs with severe muscle atrophy and no convincing orthopaedic cause for the problem. Detection, however, depends upon noticeable soft tissue asymmetry or contrast enhancement as these lesions often have properties similar to muscle. Small or diffuse lesions may be subtle and hard to detect. With therapy-resistant lameness and missing lesions on the CT scan, CT should be repeated after a short duration.

25.3 TECHNICAL ASPECTS

25.3.1 Positioning

Dorsal recumbency with symmetrically positioned extended forelimbs and hindlimbs. Non-oblique body positioning allows comparison with the contralateral unaffected brachial and lumbosacral plexus.

25.3.2 Technical parameters

The technical parameters should include a field of view (FOV) large enough to allow examination of the entire axillary or pelvic region and the vertebral canal. Use of a large FOV is also important as specific clinical anatomic localization can be difficult. A thin slice and a small reconstruction interval provide a more thorough visualization of the vertebral canal, intervertebral foramina, individual spinal nerve roots, and peripheral nerves.

25.3.3 Contrast

Contrast-enhanced CT images are obtained routinely in patients with clinically documented peripheral neuropathy. Contrast medium is administered as a bolus injection of 600 mg of iodine/kg body weight with an injection rate of 2 mL/s immediately prior to the scanning sequence. For standard CT image acquisition to detect peripheral neuropathy, contrast agents can be administered by hand or using a power injector. Regardless of whether contrast medium is administered, the use of saline flush must be considered.

CT myelography is best for showing intradural lesions. During CT myelography, contrast medium SolutrastTM (iopamidol 200M, 0.2 mL/kg body weight) is injected atlanto-occipitally or lumbally into the subarachnoid space. After atlanto-occipital injection of contrast medium, the animal is placed in an oblique position for a few minutes, with its head elevated.

DOI: 10.1201/9781315118130-27

25.4 NORMAL CT ANATOMY AND/OR ANATOMIC VARIATIONS

25.4.1 Brachial plexus

The ventral branches of the sixth, seventh, and eighth cervical spinal nerves (C6, C7, and C8) and the first and second thoracic spinal nerves (T1 and T2) merge in various combinations to form the brachial plexus and peripheral nerves of the thoracic limb. Occasionally, the ventral branch of the fifth cervical nerve (C5) also contributes to the brachial plexus. The ventral branches may be identified on the CT scan from the intervertebral foramina distally, together with the subclavian vein and artery cranially located to the first rib, ventrally of the scalenus muscle, until they form the brachial plexus and the peripheral nerves. Nerves appear as isodense soft tissue tubular structures surrounded by fat opacity mass. Normal nerves do not accumulate in contrast medium. The contrast scans are used to distinguish blood vessels from nerves. The musculocutaneous, ulnar, and median nerves are identified on the medial aspect of the mid-humerus and followed proximally to the axillary region and distally to the elbow. The radial nerve, formed by multiple nerve components, is seen on the mediocaudal aspect of the humerus.

25.4.2 Lumbar Plexus

In dogs beginning from the third and in cats from the fourth, the lumbar nerves unite caudally with the strong ventral branches from the intervertebral foramina distally, until they form the lumbosacral plexus. The nerve roots of the fourth and fifth lumbar nerve unite in the iliopsoas muscle to form the femoral nerve. The femoral nerve passes through the iliopsoas muscle. Both leave the abdominal cavity. As a branch of the femoral nerve, the saphenous nerve descends medially and accompanies the femoral artery and distally the saphenous artery. The sixth and seventh lumbar nerves unite with both of the sacral nerves medial to the body of the ilium to form the strong sciatic nerve. It is accompanied by the caudal gluteal vein and artery caudal to the hip joint.

25.5 PATHOLOGICAL FINDINGS

25.5.1 Trauma

Peripheral nerve injuries can occur as a result of trauma, including fractures, gunshots, bite wounds, laceration, or surgical complications. Avulsion of the brachial plexus occurs as a result of an extreme abduction of the shoulder or an extreme caudally directed traction of the forelimbs. In this way, tractive and extension forces are built up, which may lead to partial or complete avulsion of the perineuriumless nerve roots. In order to make a prognosis, the severity of the lesion as well as the identification of the affected nerves are important criteria. The dorsal and ventral nerve roots do not appear on a CT scan. A CT myelographic scan may show the nerve roots as small ventral and dorsal gaps in the contrast-enhanced subarachnoid space (Figure 25.1). CT myelography should be used if a partial or complete avulsion of the intradural nerve root from the spinal cord is

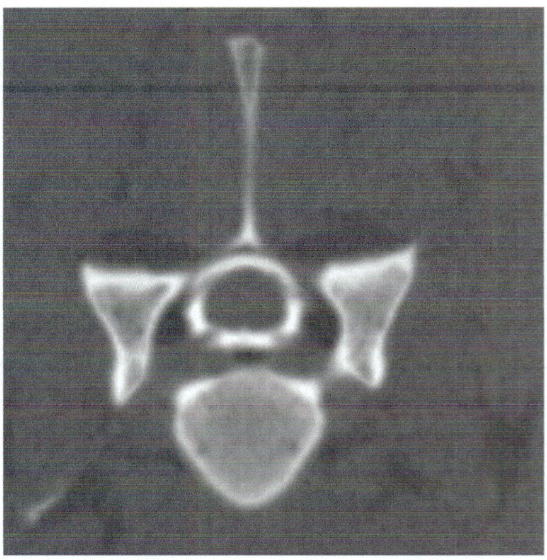

Figure 25.1 This transverse CT myelogram of a normal adult dog at C7 shows the contrast medium within the subarachnoid space. The filling defects show the ventral and dorsal nerve roots.

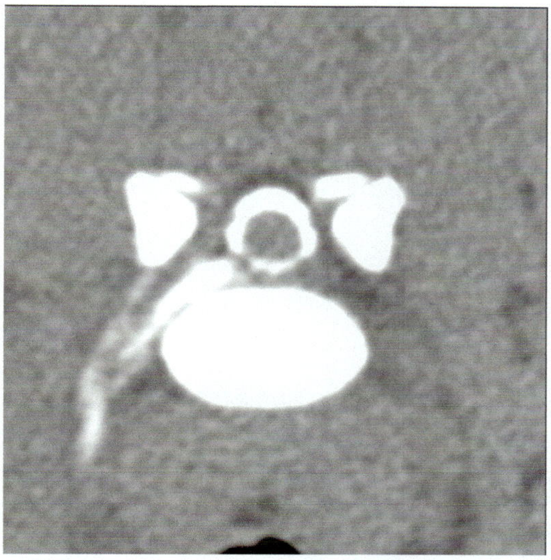

Figure 25.2 Nerve root (C7) avulsion in an adult dog that was hit by a car. Transverse CT myelogram shows leakage of contrast medium from the subarachnoid space into the extradural space.

assumed. The avulsion of the nerve root and its localization can be seen on a CT myelogram. The scan shows extradural leakage of contrast from the subarachnoid space at the site of the nerve root avulsion (Figure 25.2). The femoral nerve is almost completely embedded within the iliopsoas muscle. Thus, dogs and cats with iliopsoas muscle injuries can have varying degrees of monoparesis and pelvic limb lameness. Clinically, these patients develop atrophy of the quadriceps muscle, absence of cutaneous sensation in the medial aspect of the limb, and a depressed patellar reflex. CT abnormalities with regard to the iliopsoas muscle are explained in a previous chapter.

Injuries of the sciatic nerve are often caused by fractures of the body of the ilium or complex pelvic fractures (Figure 25.3a and b).

25.5.2 Neoplastic

Most of the nerve tumours originate from the structures surrounding the nerve. They are referred to as peripheral nerve sheath tumours. The majority of canine nerve sheath tumours arise in the caudal cervical area. They originate peripherally and spread proximally towards the

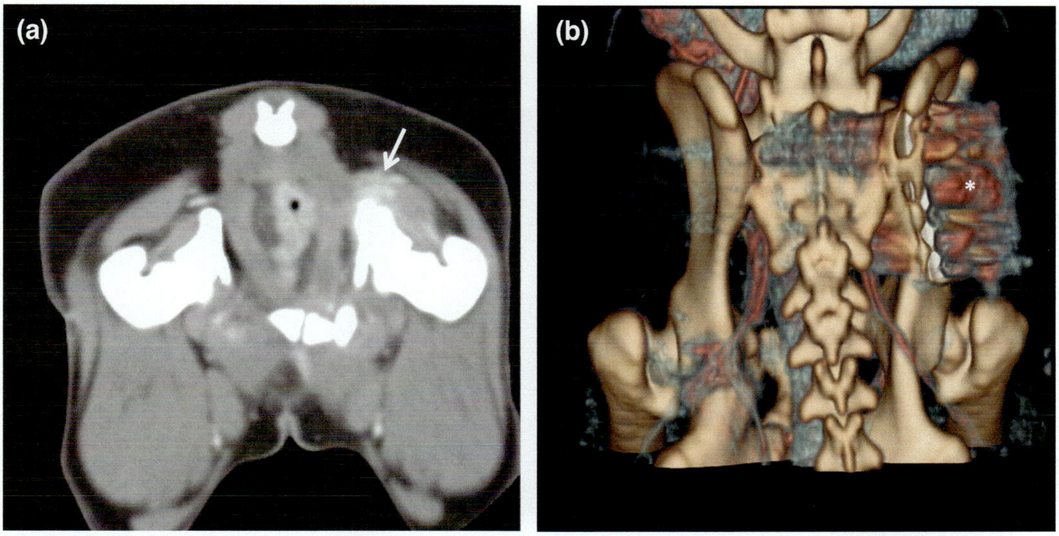

Figure 25.3 (a) Transverse post-contrast CT image of a dog 4 weeks after osteosynthesis of the right ilium fracture. The caudal fracture of the left acetabulum was not surgically treated. Note the non-uniform enhancement of the area dorsal to the left acetabulum (arrow). Compare with the contralateral sciatic nerve. The abnormalities are considered a neuritis and perineuritis of the sciatic nerve. (b) Volume-rendered image of a pelvic fracture in a dog 4 weeks after surgery. A significant contrast enhancement caudally of the left acetabulum can be identified. Note the artefacts because of the metallic implants on the right ilium (asterisk).

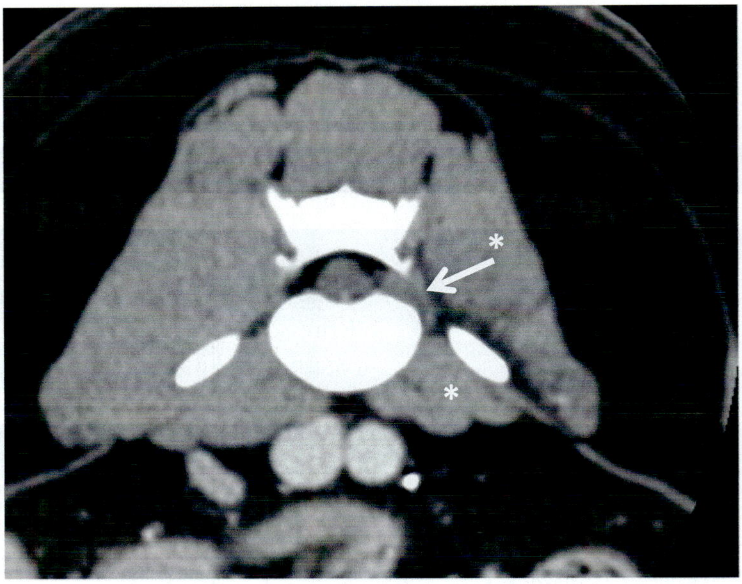

Figure 25.4 Transverse post-contrast CT image of a dog at L5/6 with a peripheral nerve sheath tumour of the left L6 nerve. A mass with ring enhancement within the left intervertebral foramen is identifiable (arrow). The left paraspinal and iliopsoas muscles are significantly atrophied (asterisk).

spinal cord along the nerve roots. Tumours of the peripheral nerves, the nerve roots, and the lumbosacral plexus are rare. Surgical treatment and prognosis depend on the spread of the tumour and the degree of infiltration. CT used for diagnosis of peripheral nerve sheath tumours provides a visualization of masses in the area of the plexus and can show if the tumour has spread proximally into the intervertebral foramen or the spinal canal (Figure 25.4). The

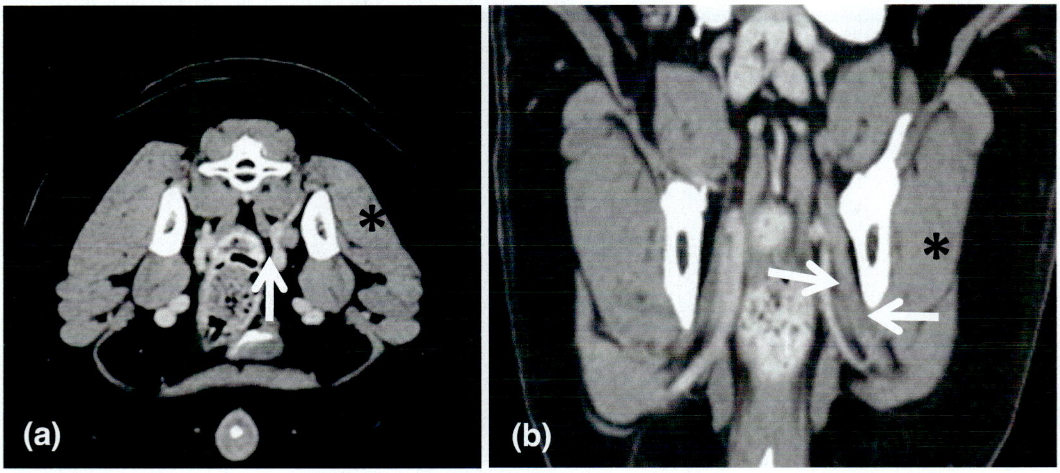

Figure 25.5 (a) Transverse post-contrast CT image of a dog with a peripheral nerve sheath tumour of the left lumbosacral plexus nerves. A well-marginated contrast-enhanced mass (arrow) is identifiable in the pelvic canal. The left gluteal muscles are significantly atrophied (asterisk). (b) Dorsal reconstructed post-contrast CT image shows the mass extending in a linear thickening that follows the course of the left lumbosacral plexus nerves (arrows). Compare with the opposite normal lumbosacral plexus nerves. Note the significant atrophy of the left gluteal muscles (asterisk).

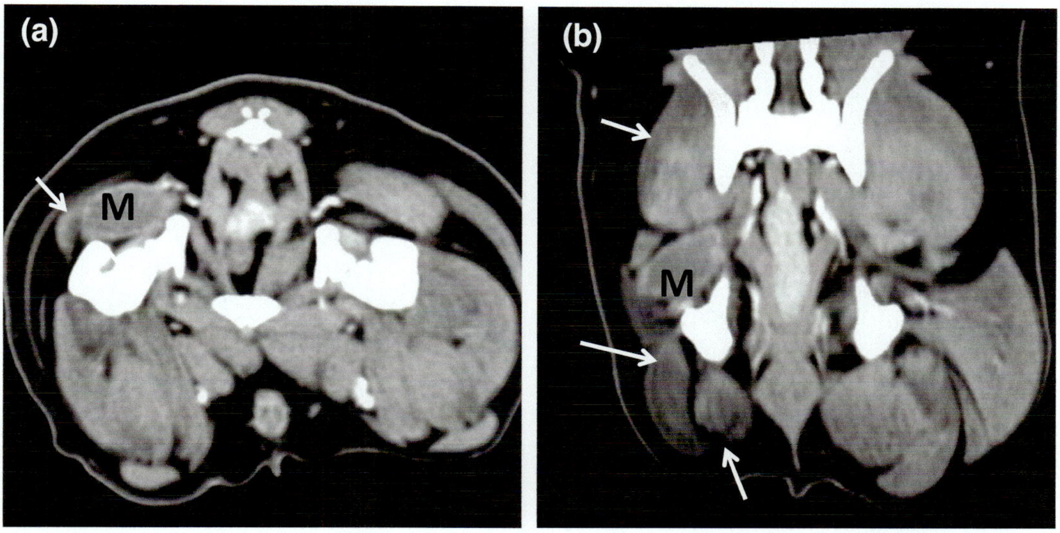

Figure 25.6 (a) Transverse post-contrast CT image of a dog with a peripheral nerve sheath tumour of the right sciatic nerve. A well-marginated mass with a ring enhancement (M) is identifiable. The right gluteal muscles are significantly atrophied (arrow). Compare with the opposite normal sciatic nerve. (b) Dorsal reconstructed post-contrast CT image shows the mass with a ring enhancement (M). The right gluteal muscles and muscles caudally of the femur are markedly atrophied (arrows). Compare with the opposite normal sciatic nerve and muscles.

tumours present as diffuse thickening of the nerve (Figure 25.5a and b) or as nodular masses (Figure 25.6a and b). At early stages, however, peripheral nerve sheath tumours cannot be distinguished from neuritis or fibrosis. In such cases, a biopsy should be performed prior to choosing a therapy. Most masses of the brachial and lumbosacral plexus are characterized by

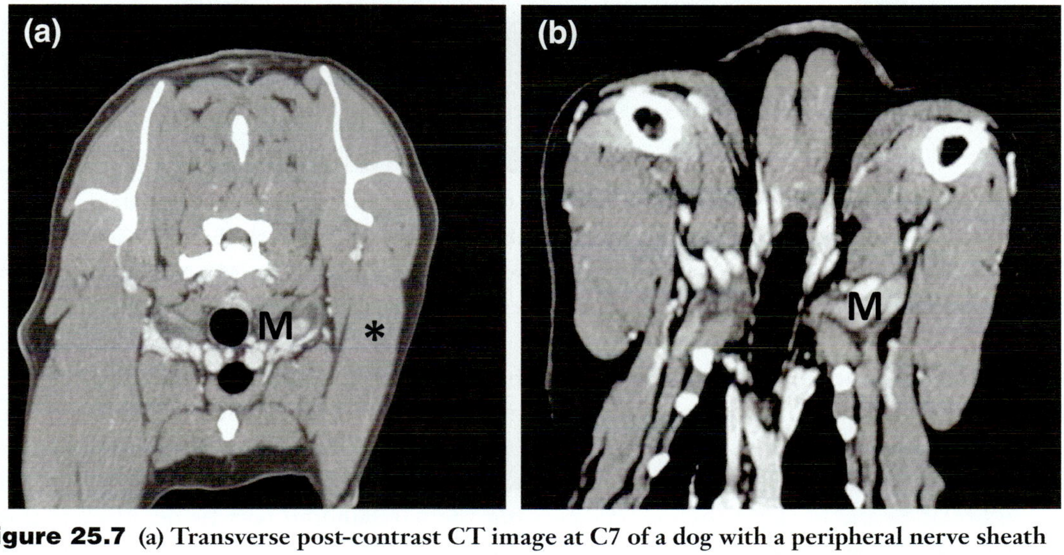

Figure 25.7 (a) Transverse post-contrast CT image at C7 of a dog with a peripheral nerve sheath tumour. A well-marginated contrast-enhanced mass (M) is identifiable in the left axilla. The right triceps brachii muscle is significantly atrophied (asterisk). (b) Dorsal reconstructed post-contrast CT image shows the mass (M) extending in a linear thickening that follows the course of the brachial plexus nerves. Compare with the opposite normal brachial plexus nerves.

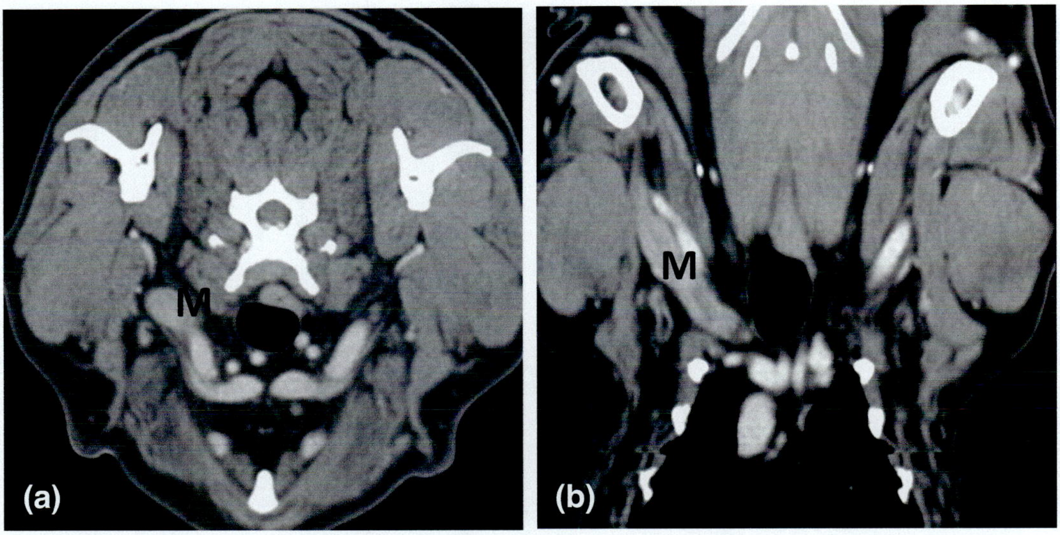

Figure 25.8 (a) Transverse post-contrast CT image at C6 of a dog with a peripheral nerve sheath tumour. A well-marginated contrast-enhanced mass (M) is identifiable in the right axilla. Compare with the opposite normal brachial plexus nerves. (b) Dorsal reconstructed post-contrast CT image shows the mass (M) extending in a linear thickening that follows the course of the brachial plexus nerves.

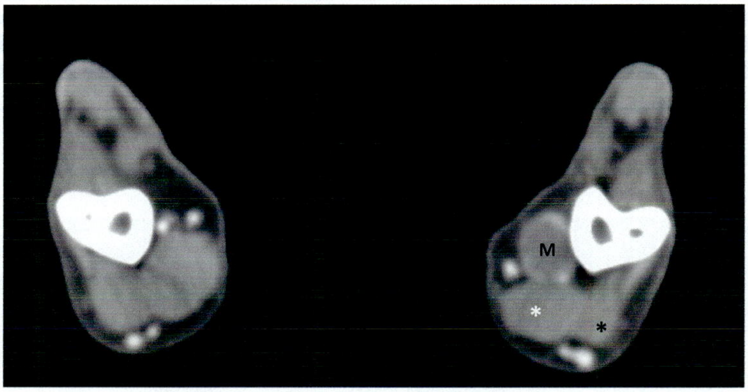

Figure 25.9 Transverse post-contrast CT image at the level of the distal humerus of a dog with a peripheral nerve sheath tumour from the left ulnaris nerve. Medial at the humerus is a well-marginated mild hypodense mass in the muscle with a ring enhancement (M). Note the atrophy of the biceps brachii (white asterisk) and brachiocephalicus muscles (black asterisk).

contrast enhancement. In some patients, this enhancement is fairly uniform (Figure 25.7a and b). In others, there is heterogeneous enhancement, frequently with a hypoechoic central area (Figure 25.8a and b). These hypoattenuating areas may indicate necrosis. In one study, a prominent rim enhancement was seen in 54% ($n=13/24$). Rim enhancement is noted frequently, especially in well circumscribed masses (Figure 25.9). Seventy-nine percent ($n=19/24$) of masses had well-defined margins, while the remainder were ill-defined, infiltrating the surrounding soft tissues. In addition, the affected limb shows severe muscle atrophy (Figure 25.10).

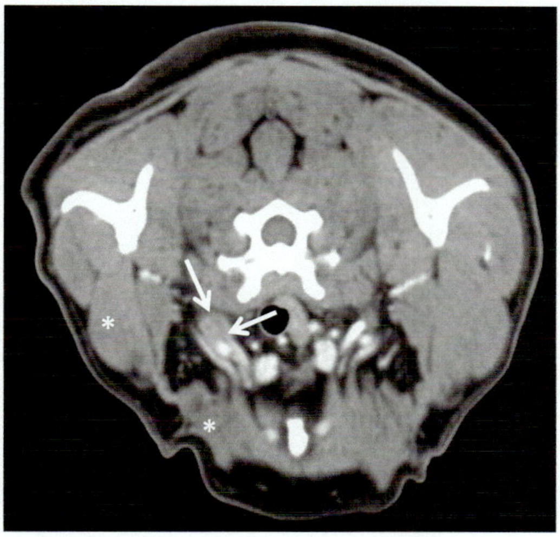

Figure 25.10 Transverse post-contrast CT image of a dog at C6. A small contrast-enhanced mass with ring enhancement (arrows) is identifiable in the right axilla by absence of a similar structure in slice images of the contralateral axilla. The right triceps brachii and pectoralis muscles are significantly atrophied (asterisk).

FURTHER READING

Breur G, Blevins W. 1997. Traumatic injury of the canine iliopsoas muscle. *Journal of the American Veterinary Medical Association* 210:1631–1634.

Forterre F. 2013. Tumouren der Kopf- und peripheren Nerven. In: Kessler M (ed) *Keintieronkologie*, Enke Verlag, Stuttgart, ch 29, pp. 474–477.

Forterre F, Matiasek K, Schmahl W, Brunnberg L. 2003. Periphere Nervenerkrankungen: Monoparese, -plegie bei Hund und Katze: Retrospektive Studie über 94 Fälle. *Kleintierpraxis* 48(3):141–150.

Forterre F, Gutmannsbauer B, Schmahl W, Matis U. 1998. CT myelography for diagnosis of brachial plexus avulsion in small animals. *Tierarztl Prax Ausg K Kleintiere Heimtiere* 1998:322–329.

Frewein J. 1994. Peripheres nervensystem. In: Frewein J (ed) *Anatomie von und Katze*, Blackwell Wissenschafts-Verlag, Berlin, ch 16, pp. 381–393.

Rossmeisl JH, Rohleder JJ, Hancock R, Lanz OI. 2004. Computed tomographic features of suspected traumatic injury to the iliopsoas and pelvic limb musculature of a dog. *Veterinary Radiology & Ultrasound* 45(5):388–392.

Rudich SR, Feeney DA, Anderson KL, Walter PA. 2004. Computed tomography of masses of the brachial plexus and contributing nerve roots in dogs. *Veterinary Radiology & Ultrasound* 45(1): 46–50.

CHAPTER 26

CT OF THE BRAIN

Ingrid Gielen, Nuria Corzo-Menendez and Peter Smith (Head trauma) and Pierre Moisonnier (CT-guided brain biopsy)

26.1 INTRODUCTION AND PATHOLOGY

Computed tomography and magnetic resonance imaging (MRI) are both widely used in the detections of various brain diseases in animals. The first reports of CT scans of small animals were published in the 1980s and dealt with the normal brain anatomy as seen on CT and various types of tumours in dogs and cats. MRI is considered superior to CT for visualizing pathological changes in the brain because its analytic sensitivity for allowing identification of soft-tissue alterations is greater. In the past few years, there has been a huge increase in the literature about MR and brain anatomy and disorders. Because MR offers better soft-tissue contrast, it is more suitable than CT to diagnose brain tumours and to visualize secondary tumour features, such as oedema, cyst formation, and necrosis. Other subtle changes, such as displacement of the ventricular system, depth of the gyri, and a hernia of the temporal lobe and cerebellum are more easily diagnosed with MRI. However, most space-occupying processes can be detected with CT because masses create a displacement of the normal anatomical structures (mass effect) and disturb the normal symmetry. CT cannot display clearly any lesions within the brainstem, cerebellum, and the piriform lobe because these areas are surrounded by highly opaque structures. Bony structures, such as the petrous temporal bones of the skull base, create an artefact, called beam hardening, that makes demonstration of lesions in the caudal fossa of the skull difficult, or even impossible. Lesions in these anatomic locations commonly fail to be identified when CT is used and, in these cases, MRI is indicated.

Unfortunately, different kinds of masses can create almost identical images, so it is not always possible to distinguish between neoplastic and non-neoplastic diseases. Certain tumours, such as meningiomas and pituitary tumours, can be easily recognized, but in most cases, biopsies are necessary to determine the exact nature of the lesions. CT has a place in brain imaging. There needs to be an awareness of the possibilities and advantages and the limitations of brain CT.

26.2 INDICATIONS FOR USING CT

Brain CT is performed to detect

- brain tumours and metastasis, abscess
- enlarged brain cavities (ventricles)
- infarct or haemorrhage
- head injury and skull fractures
- guidance for brain surgery or biopsy
- for planning radiotherapy

26.3 TECHNICAL ASPECTS

26.3.1 Positioning

General anaesthesia or sedation should be obtained to provide optimal and symmetrical positioning. In trauma cases and patients in bad condition, anaesthesia is contraindicated. The animal may be positioned in dorsal or sternal positioning with the head first in the gantry, although dorsal positioning with the fore limbs fixed parallel to the thorax provides an easy

DOI: 10.1201/9781315118130-28

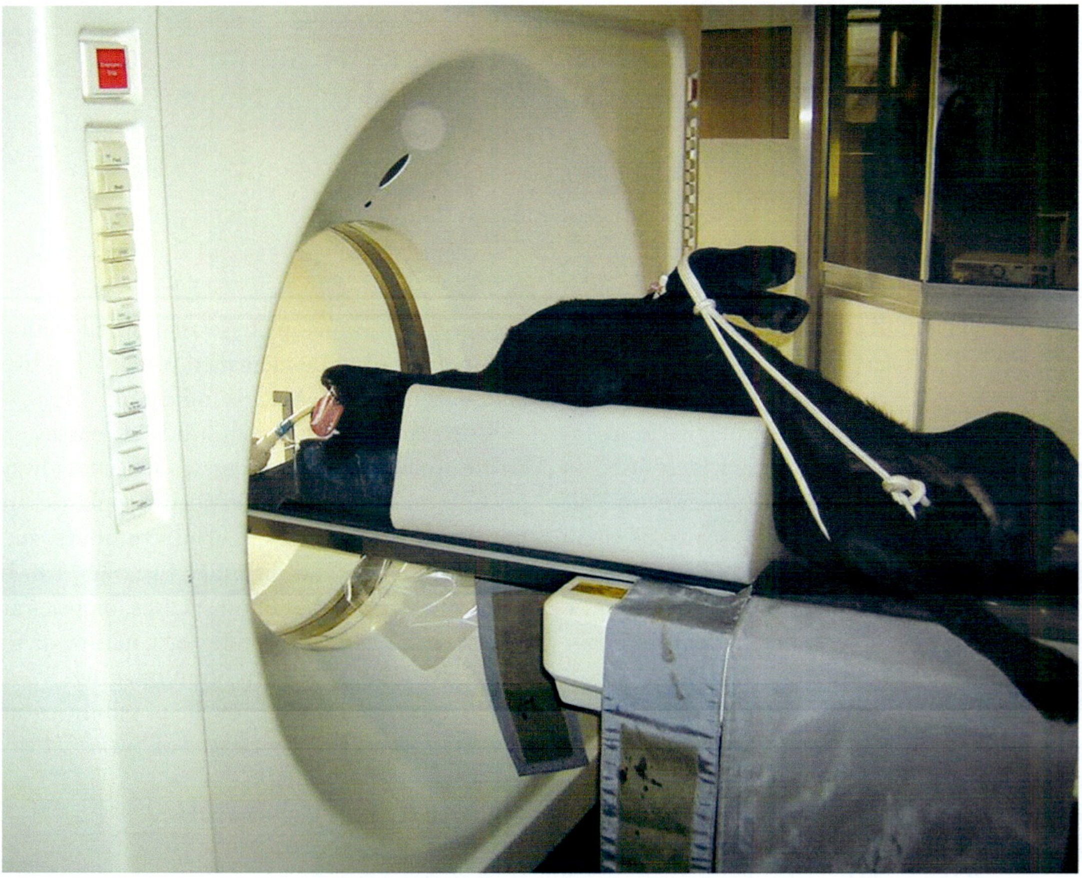

Figure 26.1 **Dorsal positioning with the fore limbs fixed parallel to the thorax provides an easy access to administer IV contrast to the cephalic vein.**

access to the cephalic vein in order to administer IV contrast (Figure 26.1). Sponge wedges are helpful to position the head and neck in a straight and symmetrical way. Abrupt motion and panting should be avoided.

Positioning depends on the structures to be visualized and personal preferences:

- For evaluation of brain pathology, dorsal recumbency; in case of brain biopsy, ventral recumbency.
- For visualization of the bulla tympanica, dorsal recumbency
- For detection of nasal pathology, ventral recumbency is handier if biopsies need to be taken.
- For evaluation of retrobulbar processes, ventral recumbency is preferred.

A problem with head CT in animals compared to humans is that they have a substantially thicker skull, producing significantly more beam-hardening artefacts, resulting in streaking artefacts in the caudal fossa (Figure 26.2). This makes the determination of lesions in the caudal fossa difficult and impossible.

26.3.2 Technical parameters

Transverse slices parallel to the hard palate from the cribriform plate till the atlas in standard or soft tissue and in bone algorithm are obtained (Figure 26.3). A tube voltage of 120 kVp with mA 100–140 is used with the field of view (FOV) (a minimum of 12 cm is recommended) adapted.

Thin slices (1–2 mm) in bone algorithm can be reformatted to form thicker slices (2–4 mm) in soft-tissue algorithm for interpretation.

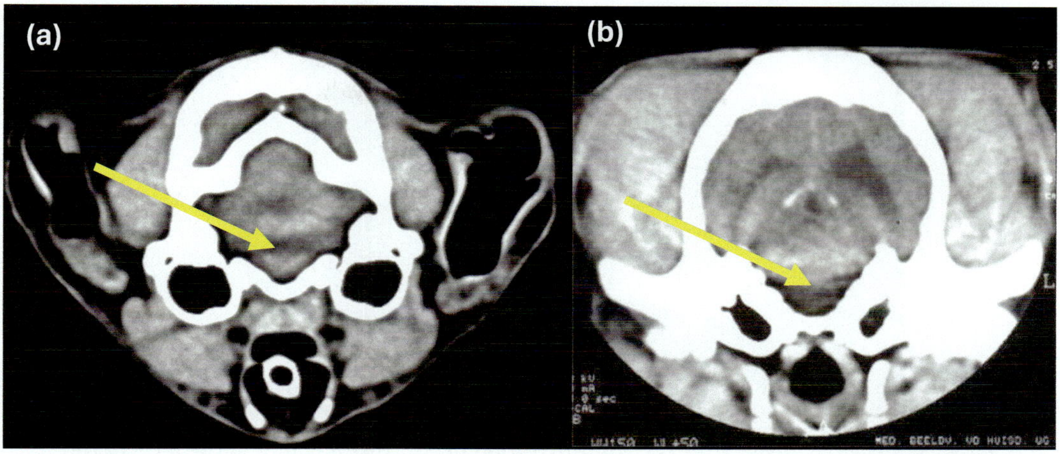

Figure 26.2 Transverse images of a cat (a) and dog (b) at the level of the caudal fossa of the skull and the tympanic bulla. The yellow arrows points to a black horizontal area where the diagnosis becomes impossible. This artefact is present when low-energy protons are absorbed by radiopaque structures (e.g., skull base and temporal bone).

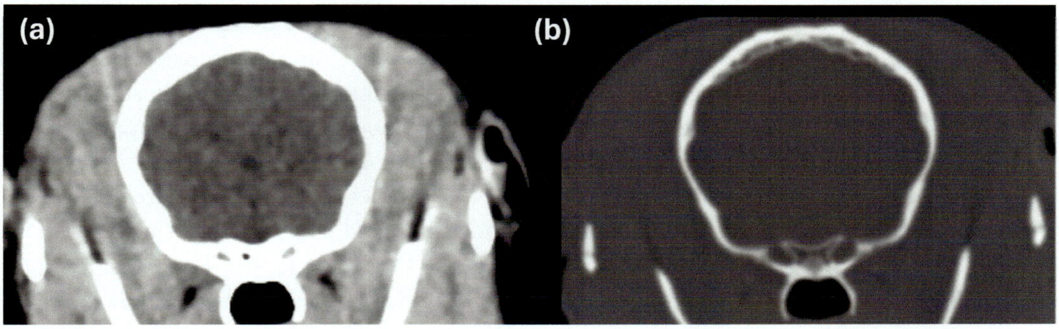

Figure 26.3 Transverse CT images in soft-tissue algorithm (a) and bone algorithm (b). For a CT examination of the brain, standard or soft tissue and bone algorithm are obtained, allowing the visualization of all tissue densities.

This results in better-quality soft-tissue images showing less noise and grittiness. Additionally, the reduction in artefacts will less hinder the evaluation of the caudal fossa. Scanning interleaved or with 50% overlap has several benefits. Partial volume artefacts and the chance of missing a small lesion at the edge of a slice are reduced. Overlapped images result in minimizing motion artefacts and reconstructions of better quality. Sagittal and dorsal reconstructions result in long-axis images, giving the best assessment to evaluate the entire lesion and its relationship with the surrounding structures. The multiplanar reformatting (MPR) imaging is proven to be valuable regarding the accurate localization of lesions.

26.3.3 Contrast (IV)

The application of contrast agents determines the perfusion of the tissues and distinguish between normal and abnormal tissue. IV iodine contrast agent (600–900 mg iodine/kg) should be the standard for the evaluation of intracranial lesions. Contrast is administered manually or via a power injector. A scan delay of 1–1.5 minutes is observed after the administration of the bolus. If there is suspicion of pathology, however, which is not well visible, the study should be repeated after 1 minute as some lesions have a delayed wash in and wash out. In the pituitary gland, the choroid plexus, and the meninges, contrast enhancement is considered

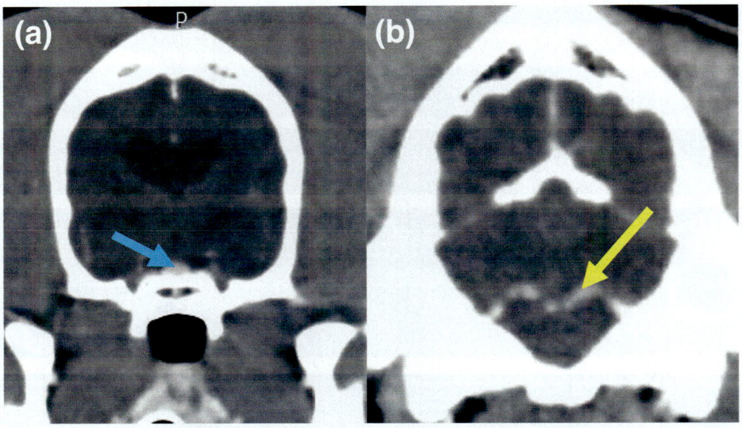

Figure 26.4 **Transverse image after the administration of contrast medium at the level of the pituitary gland at the level of the sella turcica (a) and at the level of the caudal fossa (b). In normal physiological circumstances, the pituitary gland homogenously enhances after contrast administration (blue arrow). Contrast enhancement of the plexus choroideus of the fourth ventricle shows a specific shape and is enhancing in healthy animals (yellow arrow).**

physiological (Figure 26.4). Contrast enhancement is regarded as pathologic in other brain regions and occurs from hypervascularization or disintegration of the blood–brain barrier.

Normally, water-soluble, iodinated contrast media do not cross the intact blood–brain barrier, though where this barrier is damaged by a lesion like a tumour or inflammation, the contrast medium will cross the blood–brain barrier and will enter the interstitium and enhance the radio absorption.

Malignant intracranial lesions can be distinguished from non-malignant lesions through their contrast absorption and flushing characteristics. The normal brain enhances only 2–4 HU after the administration of an intravenous contrast medium injection (caused by small vascular structures), but the density of a lesion will increase by 20–40 HU. This enhancement simply reflects vascularity, and this phenomenon can be useful for delineating areas of necrosis.

Abnormal brain tissue accumulates the vascular contrast medium due to the following:

- Disruption the blood–brain barrier
- Hypervascularity of neoplasia and inflammation
- Leaky tumour vessels

26.3.4 Interpretation

To make sure no significant anomalies are missed, the interpretation of a head CT must be done systematically. First compare the left and right hemispheres for symmetry.

CT studies of the brain are evaluated in both bone and soft tissue settings. Brain window (WW 80, WL 40); bone window (WW 3,000, WL 500); soft tissue window (WW 260, WL 80); stroke window (WW 40, WL 40). The images show low-density tissues as darker (blacker) and high-density structures as brighter (whiter). The density of the parenchyma needs to be evaluated for hypo-iso or hyperdense regions. White matter appears more dark than grey matter. Cerebrospinal fluid, grey matter, and white matter show a different density and can be distinguished, (cerebrospinal fluid ±5–10 HU; white matter ±25–30 HU; grey matter ±35–50 HU).

Low-density lesions are oedema and necrosis. Images can be displayed in different window settings to allow evaluation of injury to different tissues. The brain, cerebrospinal fluid (CSF) spaces, and surrounding bone are normally symmetrical structures. A brain CT should be inspected for normal symmetry as deviation often is correlated to pathology. The term "mass effect" refers to the compression or displacement of normal brain structures (including sulci

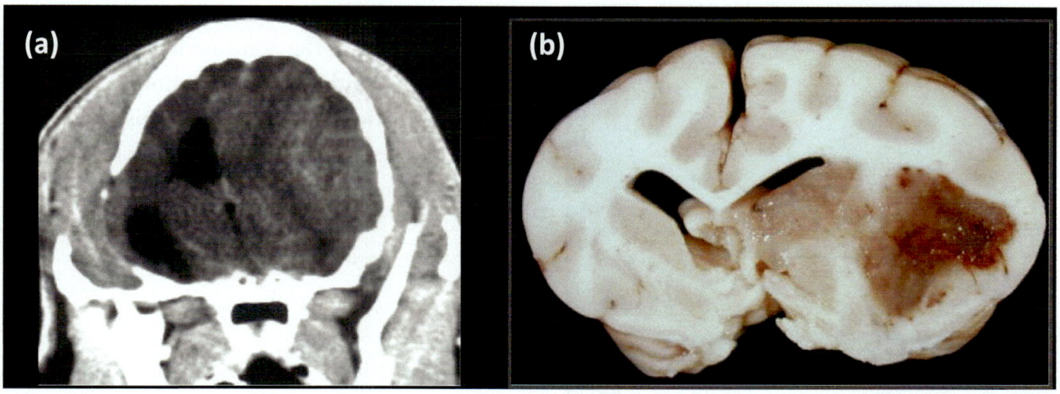

Figure 26.5 **Transverse CT image of the brain in soft-tissue algorithm (a) and anatomic specimen (b). Midline shift is an extreme displacement of the brain tissues crossing the midline of the skull. It causes increased pressure on parts of the brain following trauma or in cases of a space-occupying lesion.**

and ventricles) by adjacent masses like a tumour, haemorrhage, oedema, or obstruction of cerebrospinal fluid flow (see neoplasia). Midline shift occurs when this effect is extreme and crosses the midline of the skull (Figure 26.5). Studies should be examined for artefacts as these can hinder the interpretation. Masses may contain calcifications and are often surrounded by oedema. Vasogenic oedema affects the white matter, shows a finger like pattern, and develops in the presence of abnormally leaky blood arteries following a breakdown in the blood–brain barrier (such as those found in the context of neoplasms) (Figures 26.6 and 26.7). Cytotoxic oedema affects the grey matter and adjacent white matter. This type of oedema shows a wedge like pattern and occurs in ischaemia (see Figure 26.8). Oedema causes tissues to become less dense until they reach water's density (zero on the Hounsfield scale), which causes them to appear "hypodense" (darker/blacker) on a CT scan. Oedema may be visible as an area of decreased density in the brain parenchyma.

Ensure that the four ventricles are in their expected positions and are not enlarged. The lateral ventricles should be symmetric. Even though asymmetry in the size of the lateral ventricles is frequently linked to pathology, it may also be seen in healthy dogs as an incidental finding and must always be compared with the clinical symptoms. There is some information available about ventricle size and typical variations in various breeds.

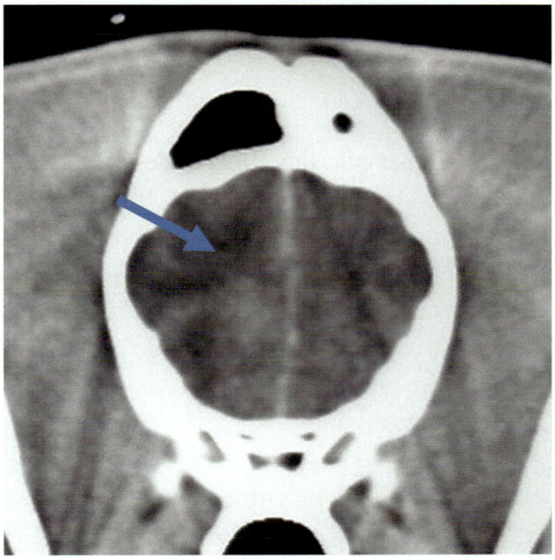

Figure 26.6 **Transverse post contrast image at the level of the frontal lobes. A finger-like pattern, hypodense, present in the white matter represents vasogenic oedema (arrow).**

In the setting of trauma, images should be obtained and reviewed at window settings appropriate for demonstrating brain and bone abnormalities as well as small subdural haematomas and soft tissue lesions or for 3D reconstruction.

26.3.4.1 CT interpretation of the acquired images

- Looking for a mass-effect or midline shift=shift in the position of normal brain structures (Figure 26.7):
  - Distortion lateral ventricles
  - Displacement falx cerebri

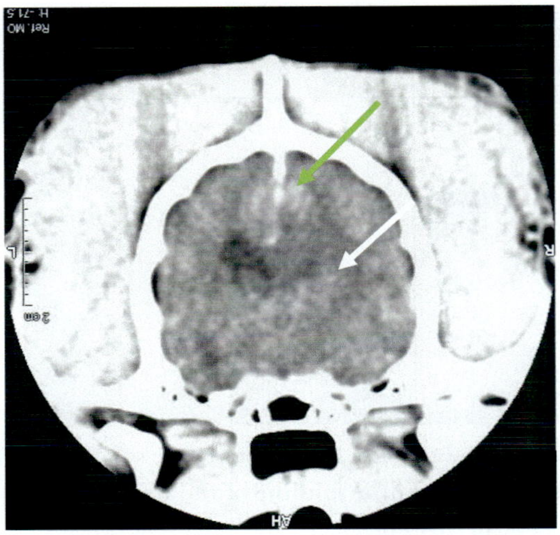

Figure 26.7 Transverse image after the administration of the contrast medium at the level of the frontal lobes. Deviation of the falx cerebri is noticed (green arrow). Distortion of the lateral ventricles is present. The left lateral ventricle is not visible. Vasogenic oedema is appreciated in the white matter of the left frontal lobe (white arrow).

- Changes in density (Figures 26.8 and 26.9):
    - Hyperdense
    - Isodense
    - Hypodense
- Other signs: calcifications, oedema, necrosis, and fluid (Figures 26.10 and 26.11)
- Looking for enhancement of lesions after IV administration of iodinated contrast medium: presents evidence of disruption of the blood–brain barrier. (Figures 26.10 and 26.11)

CT examination

- is a morphologic examination
- gives you no idea about functionality
- gives you no idea about histology.
- to reach a definite diagnosis, a brain biopsy is required.

26.4 NORMAL CT ANATOMY AND/OR ANATOMIC VARIATION(S)

For an accurate interpretation of CT scans of the brain, a thorough knowledge of the normal brain anatomy is required. Normal CT anatomy of the canine brain is described in literature, and comprehensive atlases of the normal canine brain are available (Figure 26.12). CT does not clearly show the anatomical borders of the brain lobes.

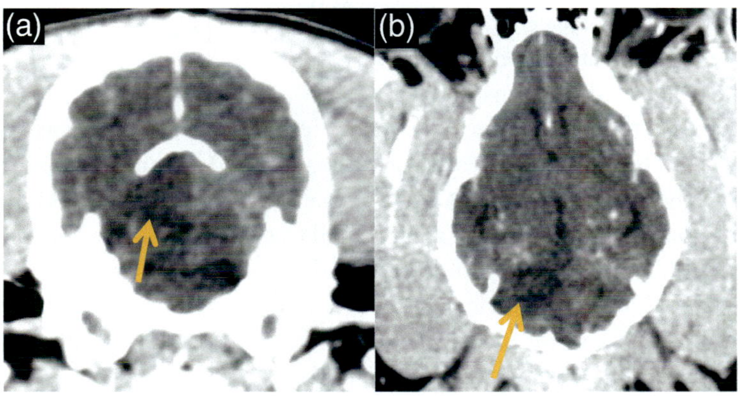

Figure 26.8 Transverse- (a) and dorsal-reconstructed (b) images after the administration of contrast medium in a dog with a cerebellar infarct. There is a wedge-shaped lesion in the dorsal cerebellum that is hypointense on pre and post contrast images and shows no contrast enhancement (arrows).

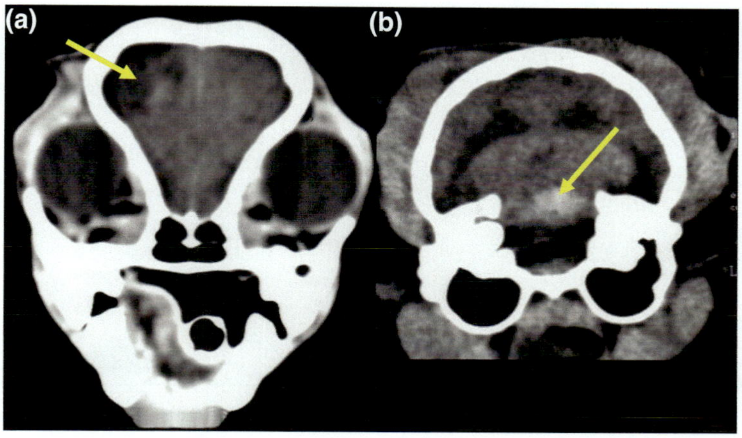

Figure 26.9 Transverse slices at the level of the frontal lobes in a dog (a) and at the temporal lobes in a cat (b). The dog brain shows a hypodense area in the dorsal aspect of the right frontal lobe. There is a hyperdense area present in between the petrous bones, on top of the beam-hardening artefact (arrows).

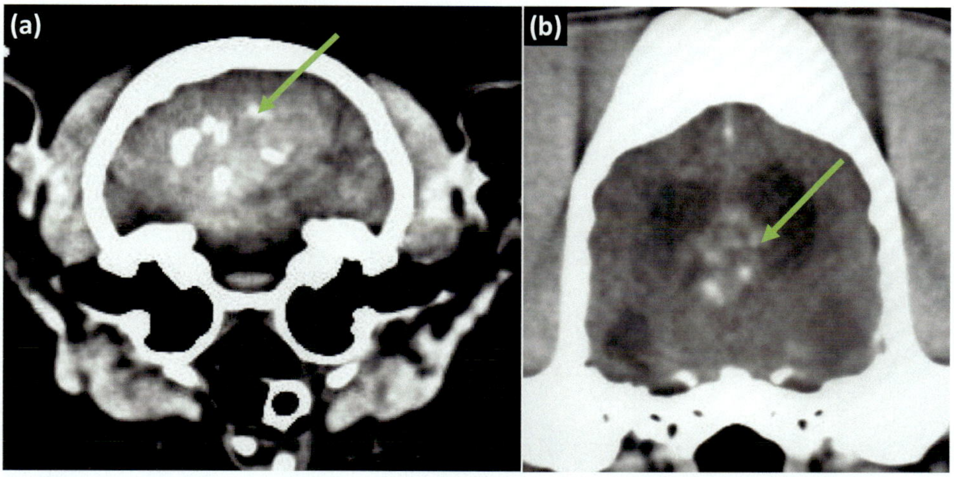

Figure 26.10 Transverse survey images of a cat (a) and a dog (b). Multiple opacities are present, representing calcifications (arrows). In both cases, a confirmed meningioma was present.

Numerous bony landmarks present on the CT images such as the osseous tentorium cerebelli, falx cerebri, dorsum sellae turcicae, and chiasmatic sulcus along with bony foramina and canals facilitated the identification of important intracranial structures such as the cerebral peduncles and lobes, the transverse and longitudinal cerebral fissures, the hypothalamus and hypophysis, the olfactory peduncle and tract, the optic chiasm and nerve, and the pons.

The brain is divided into supra- and infratentorial regions. The supratentorial region contains the cerebrum. Below the tentorium cerebelli is the infratentorial region containing the cerebellum. In brachycephalic dogs and Labrador retrievers, enlarged ventricles have been described as a common normal morphologic variation and also linked to normal ageing.

In toy breeds like chihuahuas, Shih Tzus, and Yorkshire terriers, persistent open fontanelles are a common genetic condition in normal developmental skull closure (Figure 26.13).

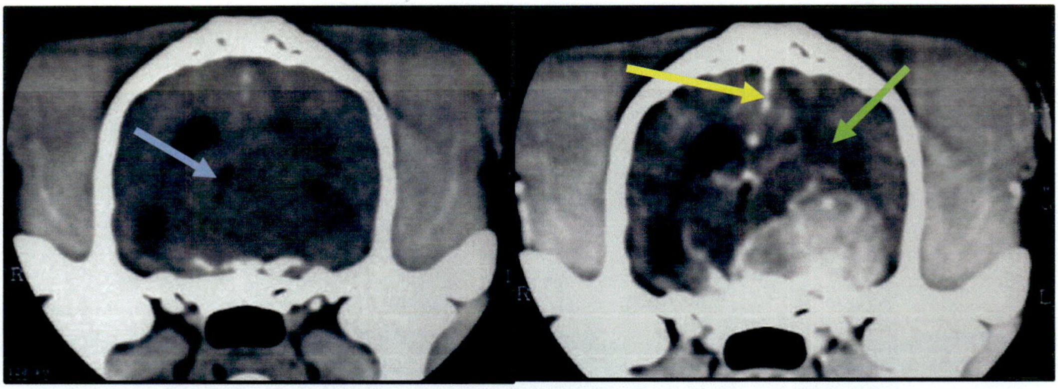

Figure 26.11 Transverse images (a) of survey and (b) after the administration of contrast medium at the level of the temporal lobes. The lateral and third ventricles show distortion (blue arrows) with deviation of the falx cerebri (yellow arrow) and a midline shift towards the right side. There is a homogenous contrast-enhancing, sharply delineated broad-based mass at the base of the skull. Vasogenic oedema is present (green arrow).

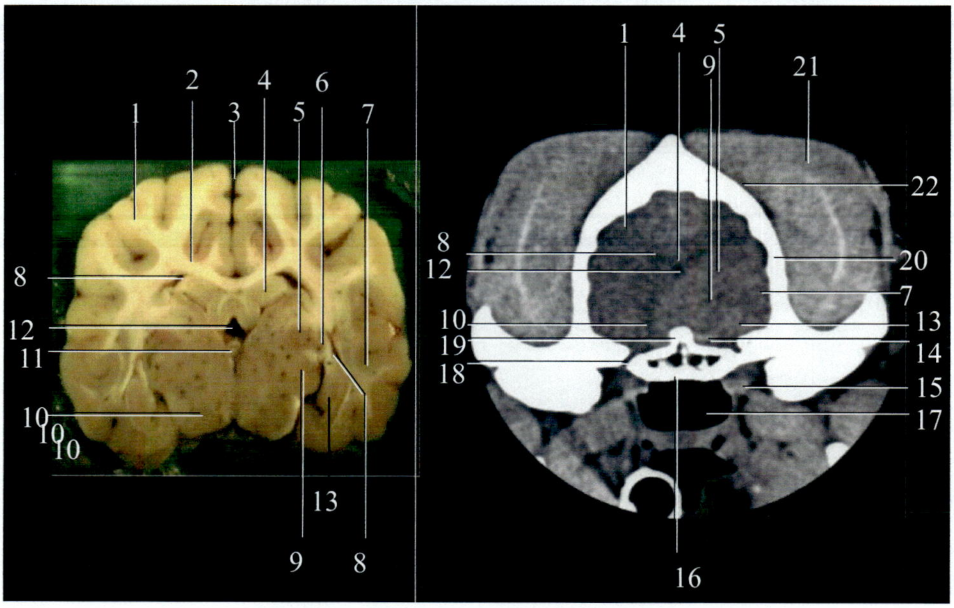

Figure 26.12 Anatomic specimen (left) and corresponding CT image in the soft-tissue algorithm at the level of the temporal lobes of a clinically normal dog. (Adapted from De Rycke et al., 2005.) 1. Parietal lobe; 2. Corpus callosum; 3. Longitudinal cerebral fissure; 4. Hippocampus; 5. Lateral geniculate body; 6. Optic tract; 7. Temporal lobe; 8. Lateral ventricle; 9. Medial geniculate body; 10. Cerebral peduncle; 11. Pineal gland; 12. Third ventricle; 13. Hippocampus; 14. Internal carotid artery (in the cavernous sinus); 15. Mandibular nerve; 16. Basisphenoid bone; 17. Nasopharyngeal cavity; 18. Oval foramen; 19. Dorsum sella turcica; 20. Temporal bone; 21. Temporal muscle; 22. Parietal bone.

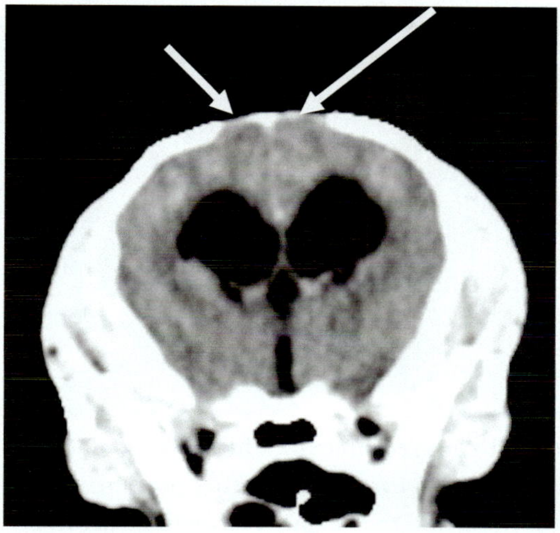

Figure 26.13 CT image of the head of a chihuahua with congenital hydrocephalus. Note the open fontanelles (white arrows).

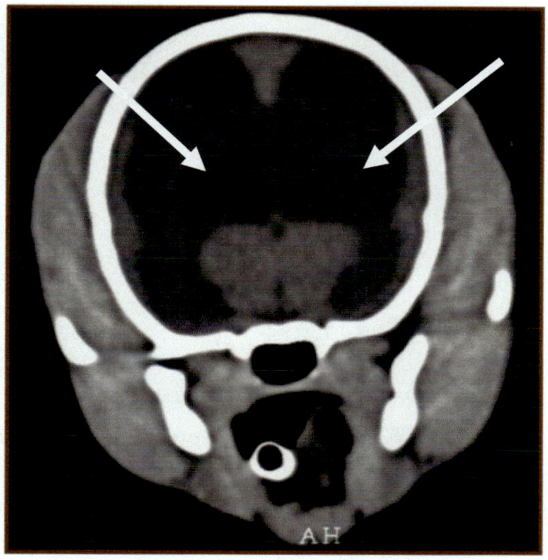

Figure 26.14 Transverse CT image of the head of a young toy breed with congenital hydrocephalus. There is marked bilateral enlargement of both lateral ventricles (white arrows).

26.5 PATHOLOGIC FINDINGS

Encephalitis (brain inflammation or infection), cerebral degeneration (death of brain cells), and brain tumours (growth of a mass inside the dog's brain) are the three main types of canine brain disorders. Canines of any age can develop brain tumours, although older dogs are more likely to do so due to weakened immune systems. In cats, the most common intracranial tumour is meningioma, with lymphoma being the second most common.

26.5.1 Congenital/developmental brain disease

26.5.1.1 Hydrocephalus

Hydrocephalus can occur as a congenital condition as well as a result of trauma or brain tumour. It is a condition in which excessive fluid is found within and around the brain (Figures 26.14 and 26.15). The body may form too much fluid or, as occurs in most cases, the fluid that is produced cannot drain from the central nervous system, as it normally does. In a hydrocephalic dog, the ventricles are fluid-filled. They become swollen, and the increased pressure damages and/or prevents development of brain tissue. Toy breeds such as Maltese, Yorkshire terriers, Pomeranians, and chihuahuas are commonly affected. Hydrocephalus occurs in other breeds as well.

- Generalized (non-obstructive) or localized (obstructive) ventricular enlargement.
- Fluid-filled.
- Cranial sutures and/or open fontanelles may be evident after the normal age for closure and skull ossification.

However, imaging diagnoses of hydrocephalus may be challenging. It has been reported that lateral ventricle dilatation and asymmetry is observed in physiological and pathological cases. Symmetric or asymmetric enlargement of the lateral ventricles is relatively common in normal adult dogs and puppies. Moreover, minimal or mild ventriculomegaly as well as asymmetry may be correlated with breed, gender, or body weight among healthy dogs (Figure 26.16).

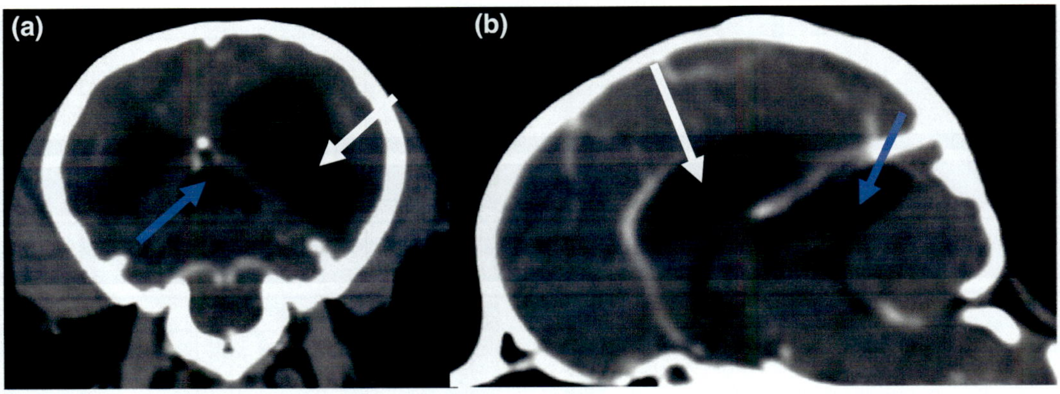

Figure 26.15 Transverse- (a) and sagittal-reconstructed (b) images after the administration of contrast agent of a Pinscher showing ataxia and circling. There is a cyst-like structure present extending from the third ventricle and between the cerebellum and the occipital lobes (blue arrows). Severe hydrocephalus is present (white arrows). This quadrigeminal arachnoid cyst is causing compression on the cerebellum.

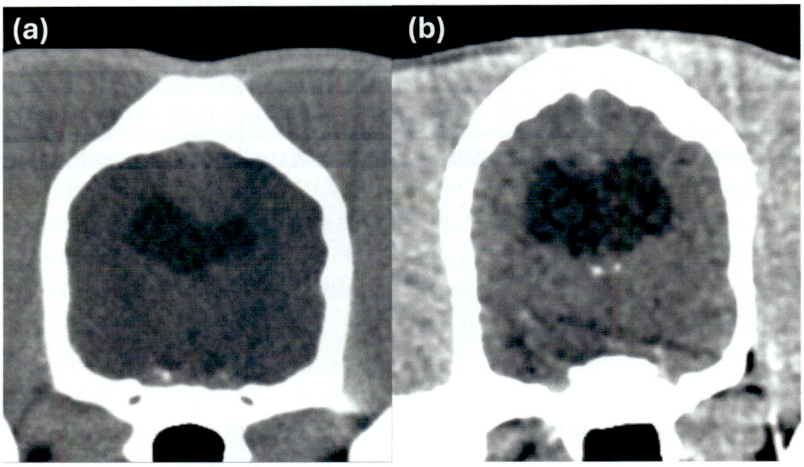

Figure 26.16 Transverse images of a mixed-breed dog (a) and a Boxer (b). Both were referred for symptoms not related to neurologic disorders. There is a prominent asymmetry of the lateral ventricles present as an incidental finding. The appearance of large ventricles is normal in brachycephalic dogs, like this Boxer.

26.5.1.2 Intracranial cysts

Intracranial arachnoid cyst (IAC) is an infrequently reported developmental disorder seen primarily in small-breed dogs, and in cats. It usually occurs in the caudal fossa, in the region of the quadrigeminal cistern. It can be associated with the third ventricle. IAC, also called intracranial intra-arachnoid cyst and quadrigeminal cyst, is a developmental brain disorder in which cerebrospinal fluid (CSF) accumulates within a split of the arachnoid membrane. Most reported cases of IAC are found in small-breed dogs, with a predominance of brachycephalic animals. Shih Tzu's may be overrepresented. Male sex also appears to be a predisposing factor. Similar to IAC in people, it is thought that canine IAC is often an incidental finding.

The typical appearance of IAC on CT is a well-demarcated, cystic-appearing structure that is non-contrast-enhancing with IV contrast administration (Figure 26.15).

- accumulation of CSF in a cavity surrounded by a membrane, often causing compression on the cerebellum and occipital lobes.
- causes:
  - congenital
  - acquired: trauma, infection, haemorrhage, increase of intracranial pressure....

Other intracranial cysts are arachnoid diverticula, epidermoid cysts, dermoid cysts, ependymal cysts, porencephalic cysts, and choroid plexus cysts (Figures 26.17 and 26.18). They cause compression on the brain, resulting in clinical signs. On CT, they appear fluid-like and are rare lesions in dogs. Epidermoid cysts need to be included in the differential list for compressive intracranial lesions certainly if localized to the cerebellopontine angle or the fourth ventricle. The described choroid plexus cysts in dogs were in the fourth ventricle.

26.5.1.3 Dandy–Walker

Dandy–Walker syndrome refers to a complex of developmental anomalies characterized by the morphological triad of aplasia or hypoplasia of the cerebellar vermis, cyst-like dilatation of the fourth ventricle, and hydrocephalus. It is a rare malformation and usually represented in young animals (Figure 26.19).

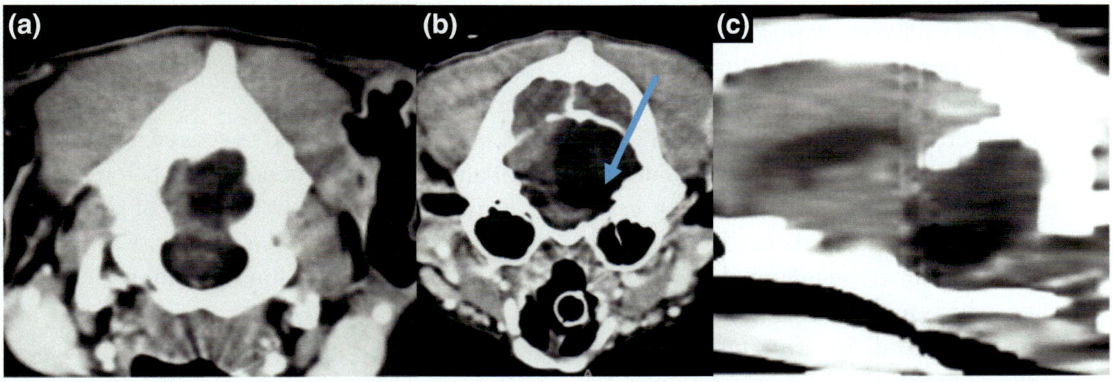

Figure 26.17 Transverse- (a, b) and sagittal-reconstructed (c) images showing a cerebellar cyst. The cyst is present in the cerebellopontine angle (arrow), causing prominent compression on the left and dorsal aspects of the cerebellum.

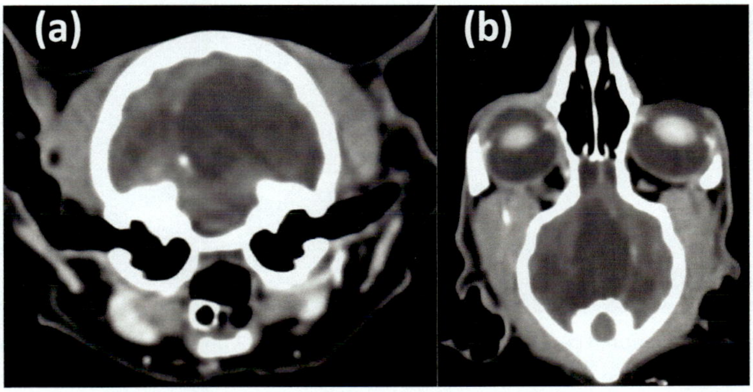

Figure 26.18 Transverse- (a) and dorsal-reconstructed (b) images after administration of the contrast medium in a cat. Hydrocephalus with dilation of the lateral ventricles is present, causing left thalamic, brainstem, and cerebellar compressions. The cyst is closely associated with the left choroid plexus. Minimal wall cyst contrast enhancement is noted. Differential diagnoses of the lesion include choroid plexus cyst, arachnoid diverticula, or ependymal cyst.

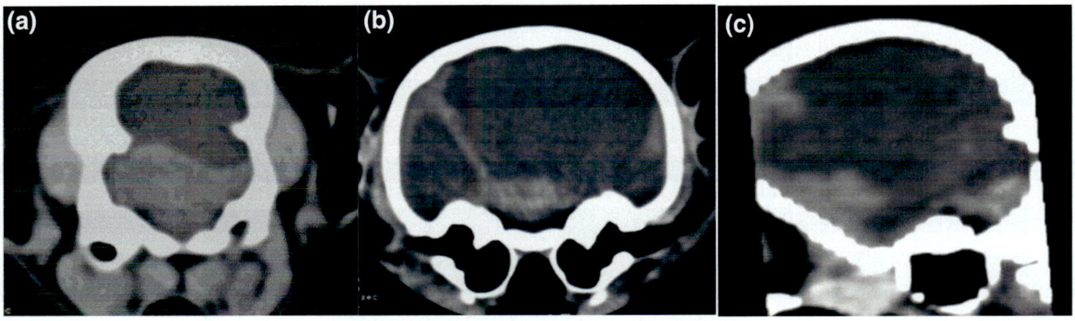

Figure 26.19 Transverse- (a, b) and sagittal-reconstructed (c) images of a 4-month-old British short-haired Pointer with an abnormally shaped head, bulging eyes, and cerebral symptoms. Generalized hydrocephalus is present along with dilation of the third ventricle, the aquaductus, and lateral ventricles. A large cyst is communicating with the fourth ventricle. The cerebellar vermis and left hemisphere are compressed by the cyst. The cerebellar vermis does not appear to have hypoplasia. Based on the CT findings, a preliminary diagnosis of Dandy–Walker deformity was made.

26.5.1.4 Chiari-like malformation and secondary syringomyelia

In Chiari-like malformation, seen in brachycephalic toy breeds, the size of the skull's caudal fossa is not in proportion to the volume of the cerebellum and brainstem. In these cases, the cerebellum may protrude caudally into and through the foramen magnum, blocking the CSF flow, which may cause fluid-filled cavities inside the spinal cord. So, Chiari malformation is the underlying issue of fluid cavitation of the spinal cord (syrinx), syringomyelia. The standard screening method for this disease is MRI. However, it has been described that CT can be used as an alternative for diagnosing cerebellar herniation and fluid cavitation of the spinal cord (Figure 26.20).

Cerebellar hypoplasia is a congenital condition with underdevelopment occurring before birth.

It is characterized by absence of the caudal portions of the cerebellar vermis. Continuous large retrocerebellar fluid accumulations are present (Figure 26.21).

Meningocele (meninges) and encephalocele (brain tissue) and meningoencephalocele (both brain and meninges) are congenital deformations where tissue herniates through a defect in the calvarium (Figure 26.22).

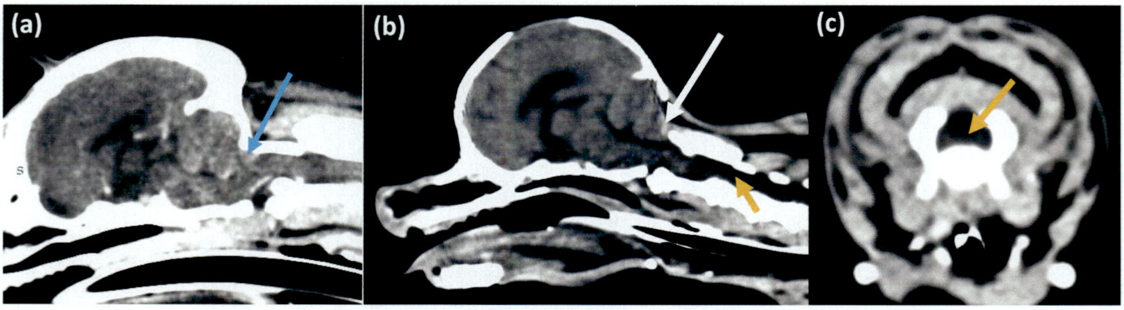

Figure 26.20 Sagittal-reconstructed images of the brain of a Cavalier King Charles (a) and a chihuahua (b). Transverse slice of the cervical spine at the level of C4 (c). There is a herniation of the cerebellum through the foramen magnum (blue arrow). Chiari-like malformation but no evidence of syringomyelia. Caudal displacement of the cerebellum into the foramen magnum is noticed (white arrow); this is typical for Chiari malformation. There is a dilation of the central canal, with a hypodense aspect, fluid accumulation within the spinal cord, a typical finding in dogs with syringomyelia (orange arrows). Note the open fontanelles.

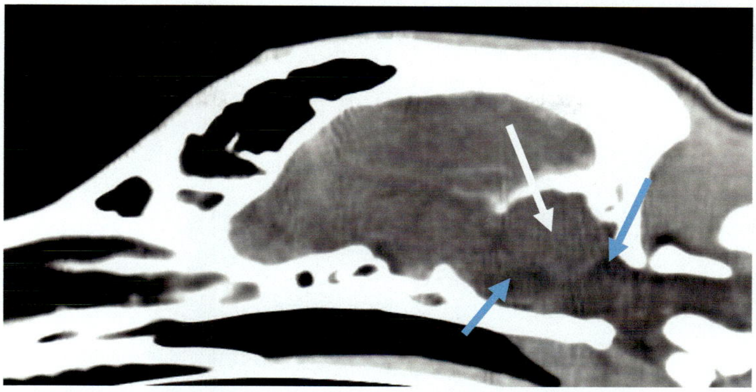

Figure 26.21 Sagittal-reconstructed image of a 6-month-old Labrador Retriever with cerebellar hypoplasia. The dog showed neurologic symptoms at the fore limbs since birth and now since recently ataxia. The cerebellum (white arrow) appears much smaller and with an increased amount of fluid around it (blue arrows).

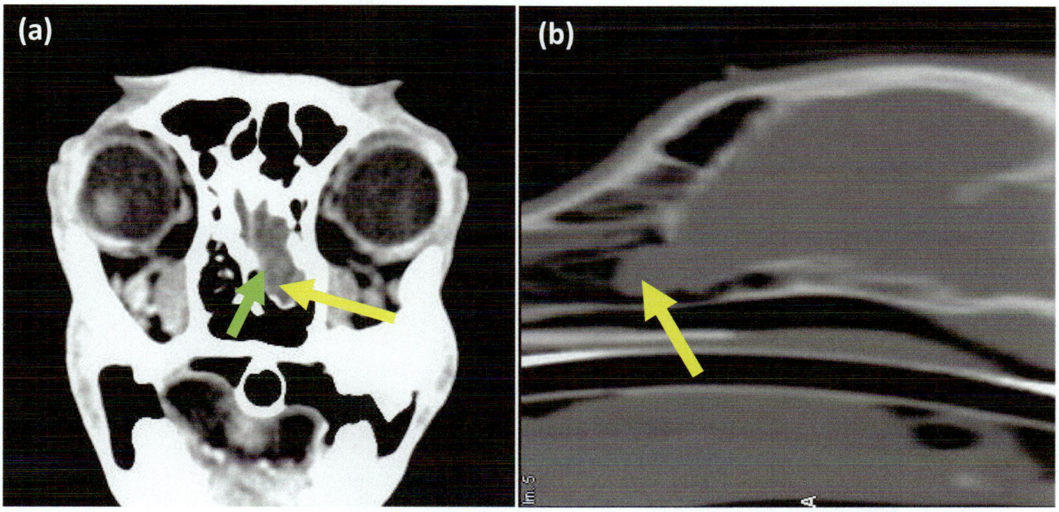

Figure 26.22 Transverse post-contrast images in soft-tissue window, at the level of the ethmoid complex (a) and sagittal-reconstructed image (b) in bone window, showing the caudal nasal cavity of a 5-month-old Border with epileptic seizures. The bony ventral part of the cribriform plate is absent (green arrow). The extension from a intracranial mass through the ethmoidal labyrinth is noticed (yellow arrows). These images correspond with a confirmed intranasal meningoencephalocele.

26.5.2 Degenerative

Can be an incidental finding in older dogs and cats. The criteria for brain atrophy in dogs have not yet been established because of the wide variation in the morphology of the ventricles and sulci of the brain depending on the breed and size. Brain atrophy corresponds with widened and enlarged sulci and symmetrical lateral ventricular enlargement and is often degenerative disease-related (Figure 26.23).

Cerebellar abiotrophy is a post-natal cerebellar degeneration. The *c*erebellum appears malformed, much smaller and with an increased amount of CSF around it (Figure 26.24)

26.5.3 Neoplastic

26.5.3.1 Brain tumours

A brain tumour is any intracranial tumour defined as an abnormal uncontrolled growth of

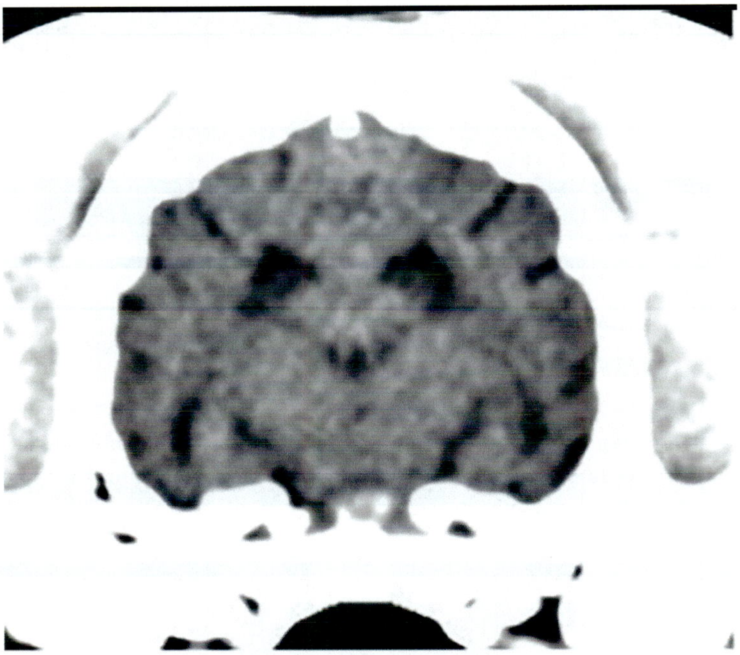

Figure 26.23 Transverse survey image showing brain atrophy. There is irregularity of the gyri and widening of the sulci. The lateral ventricles are dilated.

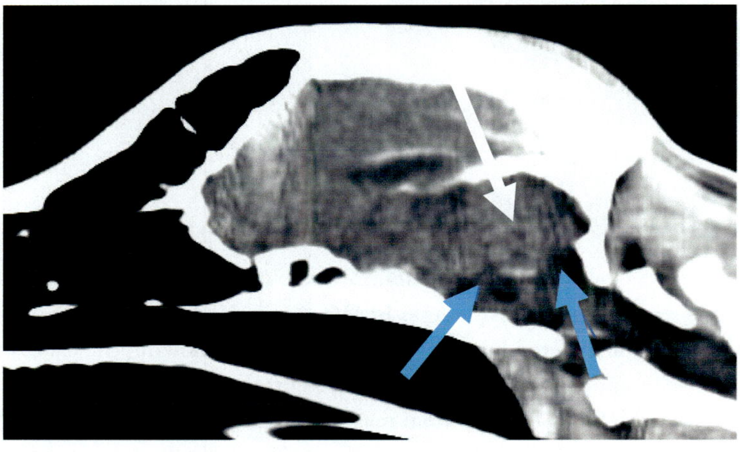

Figure 26.24 Sagittal-reconstructed image of the brain of a 6-month-old Golden retriever with recent vestibular ataxia. The cerebellum is malformed (white arrow) with absence of the caudal portions of the cerebellar vermis and large caudal fossa fluid accumulations. The fourth ventricle appears enlarged and continuous with large retrocerebellar fluid accumulations (blue arrows). There is a presumptive cerebellar hypoplasia or atrophy/abiotrophy.

cells, normally either found in the brain itself, in the cranial nerves, in the brain envelopes (meninges), skull, or pituitary and pineal gland. They may be classified as primary or secondary. Primary brain tumours (those arising from the cells of the brain and its lining) in dogs include meningioma, glioma, choroid plexus papilloma, pituitary adenoma, or adenocarcinoma.

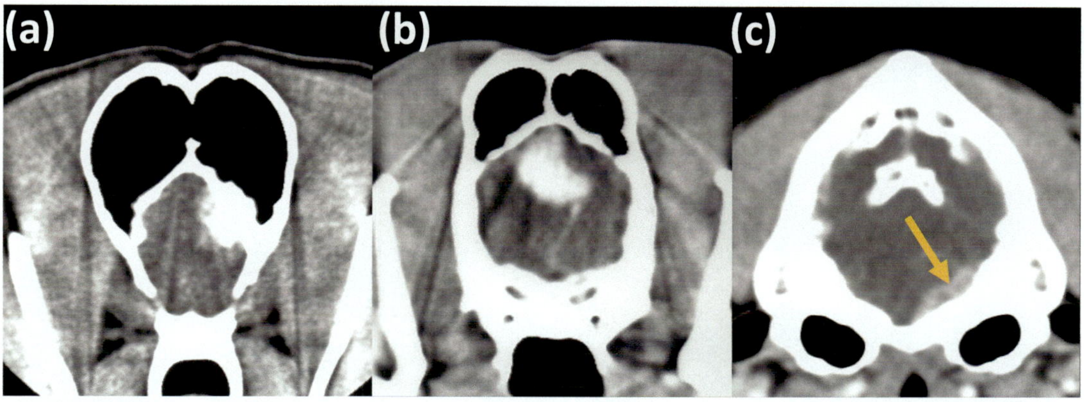

Figure 26.25 (a–c) Transverse post-contrast images of confirmed meningiomas. Note the different occurrence of the location. From left to right: against the ventral border of the left frontal sinus, at the right side of the falx cerebri, and in the left cerebellopontine angle (orange arrow).

Secondary or metastatic brain tumours originate from a primary tumour elsewhere in the body (from breast, lung or prostate cancer, haemangiosarcoma, or extension of a pituitary gland tumour)

and metastasize to the brain or is a tumour that affects the brain by extending into brain tissue from an adjacent non-nervous system tissue, such as bone or the nasal cavity.

Most mass lesions are observed within the cerebrum. Few lesions occur within the intraventricular system and within the brain stem.

26.5.3.1.1 Primary brain tumours

- Meningioma
- Glioma
    - Astrocytoma

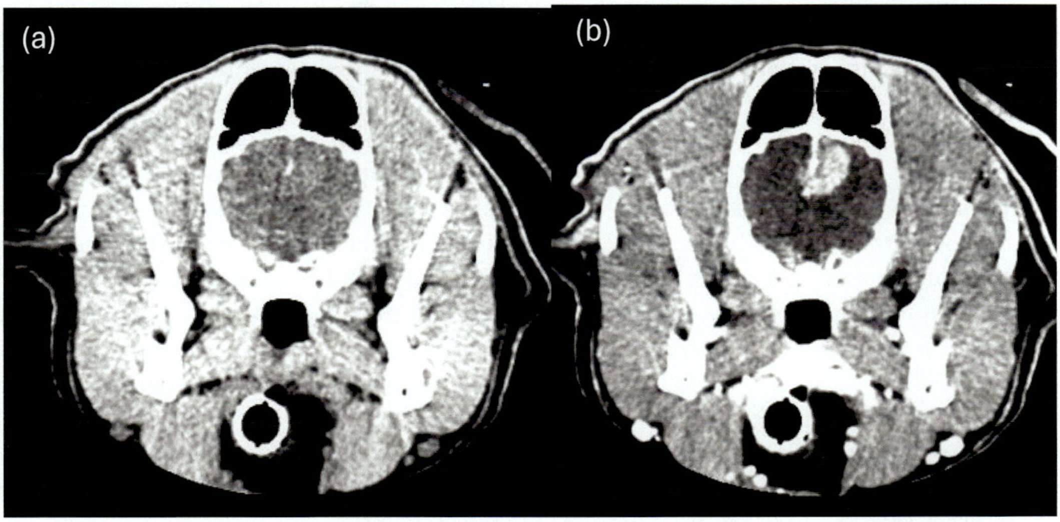

Figure 26.26 A confirmed papillary meningioma can be very aggressive – similar as seen in humans, in a 14-year-old Golden Retriever. Transverse images (a) of survey and (b) after the administration of contrast medium. There is no mass effect nor oedema seen on the survey image. There is homogenous contrast-enhancing, sharply delineated mass at the level of the frontal lobes and the falx cerebri mainly on the left side. There is slight deviation of the falx towards the right side. The process is surrounded by a rim of oedema.

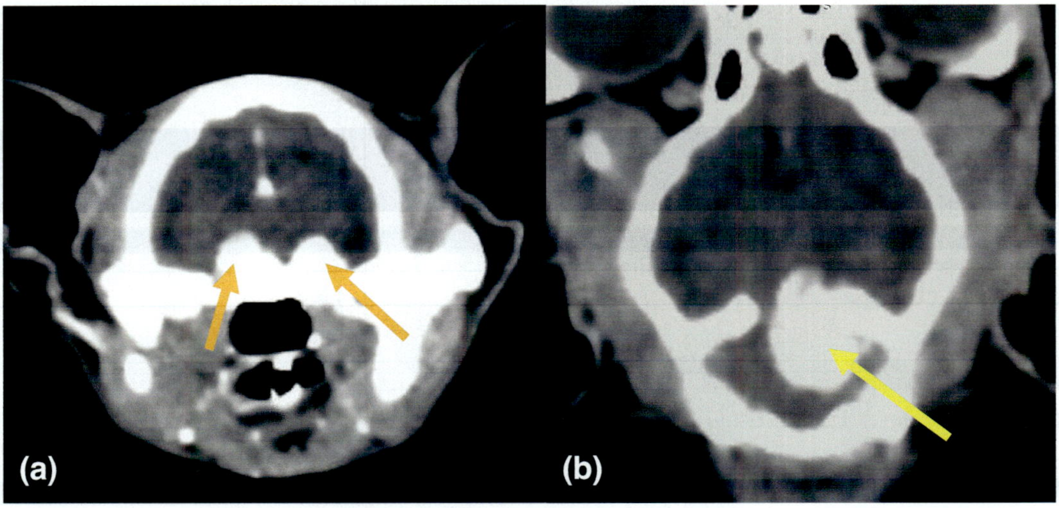

Figure 26.27 **Multiple meningiomas in a 12-year-old cat. Transverse- (a) and dorsal-reconstructed (b) images after the administration of contrast medium. Two round, well-defined extra-axial masses showing homogenous contrast enhancement are present at the base of the skull (orange arrows). A third larger mass showing the same appearance is present at the tentorium cerebellum (yellow arrow).**

- *Diffuse astrocytoma
- *Anaplastic astrocytoma
- *Glioblastoma multiforme
- *Pilocytic astrocytoma
- *......
- Oligodendroglioma
- Plexus choroïdeus papilloma/sarcoma
- Ependymoma
- (Pituitary) hypophysis tumour

Brain tumours are not uncommon in older dogs; however, there is an increasing incidence of brain tumours occurring in younger dogs. Brain tumours vary widely in their level of malignancy and some can be treated quite effectively.

Age appears to be a risk factor, with 95% of canine meningiomas occurring in dogs over 7 years of age.

Dogs older than 5 years are more susceptible to developing brain tumours; the median age of affected pets is 9 years. Certain breeds of dogs are at higher risk for developing primary brain tumours than others. Brain tumours that originate from the membranes covering the brain (meningiomas) are found more frequently in dolichocephalic breeds of dogs, which are characterized by long heads and noses, such as the Collie. Conversely, brachycephalic breeds of dogs, which are characterized by their short-nosed and flat-faced appearance, are more likely to develop gliomas, which are tumours of the interstitial tissue of the central nervous system.

Meningiomas are usually well delineated and show a homogenous contrast enhancement. Calcifications may be present in this mass, and in half of the cases, hyperostosis is present (Figure 25-30). Often, however not pathognomonic for this type of tumour, a dural tail sign is seen. Thickening and enhancement of the dura adjacent to a meningioma is considered a dural tail sign

Meningioma is most common in cats and 30%–40% in dogs. CT features of meningioma are as follows:

- Extra-axial, they do not arise in the braincells, but occur in the arachnoid mater of the meninges (Figure 26.25)
- Broad-based (Figure 26.11)
- Sharply marginated

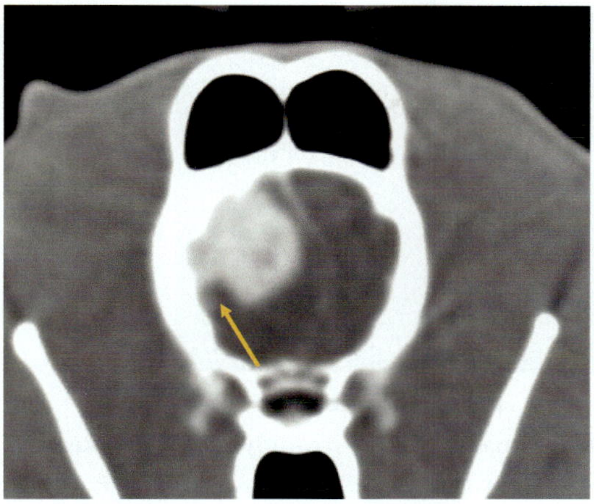

Figure 26.28 Transverse post-contrast image at the level of the frontal lobes of a dog with a meningioma. There is a prominent midline shift present. Extending from the mass is a tail, a dural tail present as a thickening of the enhanced dura mater (orange arrow).

- Markedly enhancing (Figures 26.26 and 26.27)
- Dural tail (Figure 26.28)
- May be associated by oedema
- May show a cystic component
- (calcifications) (Figure 26.29)
- (bone remodelling) (Figure 26.30)

Gliomas, which comprise astrocytomas, oligodendrogliomas, ependymomas, and choroid plexus tumours, are the most prevalent type of brain tumour (Figure 26.31). Concerning the majority of supratentorial tumours in all age categories, astrocytomas make up about 80% of all gliomas. Astrocytomas are frequently classified into infiltrating and confined tumours. The restricted category includes pilocytic astrocytomas and subependymal giant-cell astrocytomas since they do not penetrate and frequently preserve anatomical borders. Grade II, III, and IV astrocytomas are invading due to their propensity to insinuate and invade. Tumour cells are frequently present outside of mass.

Oligodendroglioma and astrocytoma are intra-axial lesions that typically have a very irregular pattern. Following intravenous contrast delivery, heterogeneous, non-uniform contrast uptake can be noticed, as well as peripheral contrast uptake with a central hypo-dense zone. Peripheral contrast uptake is a relatively vague symptom that could point to an inflammation or brain abscess.

Astrocytoma (benign or malignant)

- Diffuse astrocytoma
- Anaplastic astrocytoma
- Glioblastoma multiforme

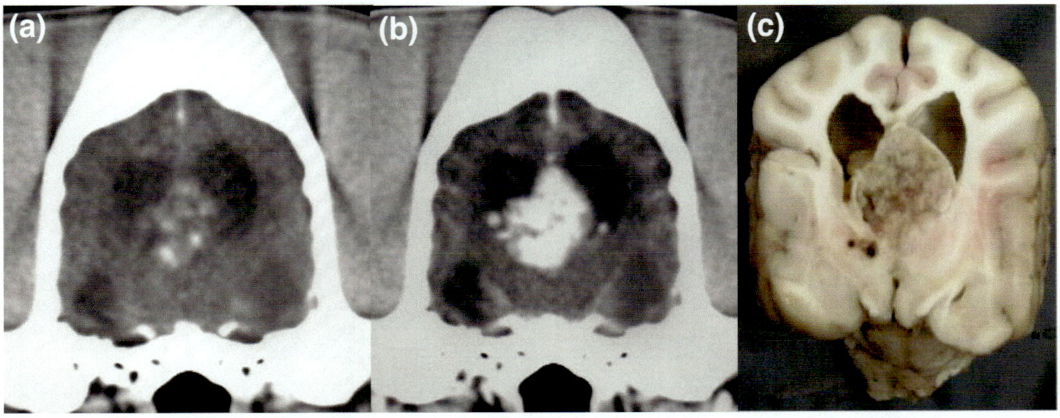

Figure 26.29 Transverse images (a) of survey and (b) after the administration of contrast medium and (c) corresponding anatomic specimen of a confirmed meningioma in between both hemispheres. There are calcifications seen on the survey image. The post-contrast images show homogenous marked contrast enhancement. The shape of the tumour on CT corresponds with the anatomic specimen.

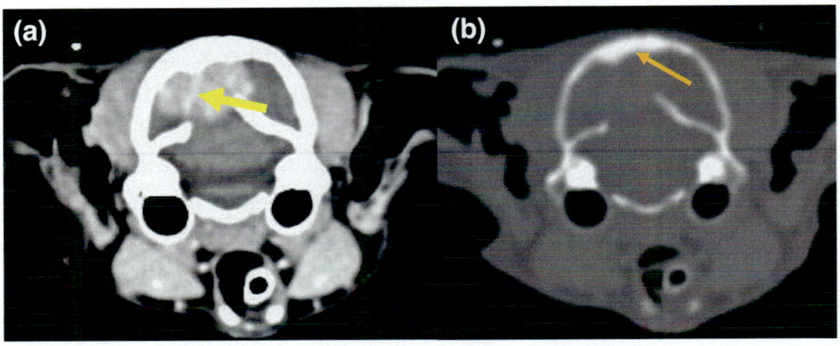

Figure 26.30 Transverse images (a) in soft-tissue window after the administration of contrast medium and (b) survey in bone window, showing a meningioma in a cat. A strongly homogeneously contrast-enhancing extraaxial mass with a broad base is present in the occipital lobe (yellow arrow). There is adjacent calvarial hyperostosis present (orange arrow).

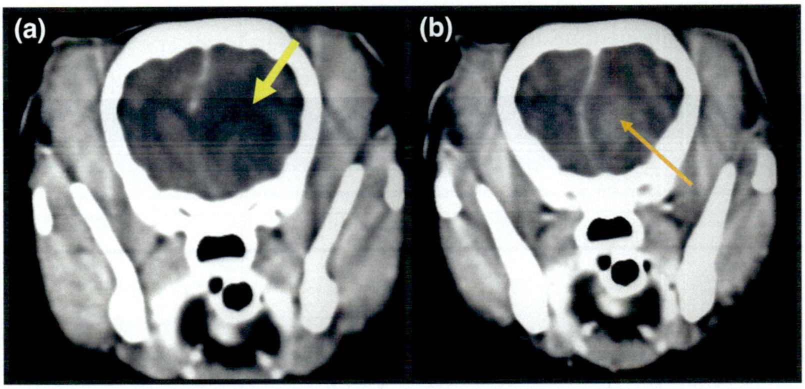

Figure 26.31 Transverse images (a) of survey and (b) after the administration of contrast medium of a confirmed glioma. There is a midline shift and vasogenic oedema seen (yellow arrow) in the left frontal lobe. There is mild contrast enhancement of the mass with ill-defined margins (orange arrow).

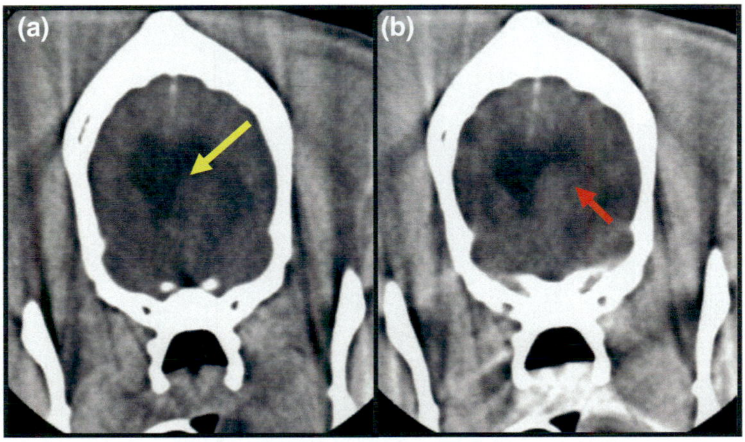

Figure 26.32 Transverse survey (a) and post-contrast (b) images of a confirmed oligodendroglioma in a Boxer. There is a mass effect present on the left lateral ventricle (yellow arrow). The mass shows minimal contrast enhancement (red arrow).

- Pilocytic astrocytoma
- ……

Oligodendroglioma

CT features of astrocytoma:

- Intra-axial, little mass-effect
- Infiltrating
- Surrounded by oedema
- Rarely contrast-enhancing

CT features of oligodendroglioma:

- Intra-axial, large hemispheres
- May invade the ventricles
- Various degrees of contrast enhancement, may not enhance (Figure 26.32)
- Peripherally enhancing (ring) (Figure 26.33)

Plexus choroïdeus papilloma/carcinoma are well-defined masses showing uniform contrast enhancement after IV contrast administration. They arise from the choroid plexus epithelium and are usually found in the lateral, third, or fourth ventricle and/or in the lateral aperture at the cerebellomedullary pontine angle.

Within or adjacent to ventricles: (Figure 26.34)

- Hyperdense usually
- Marked contrast enhancement
- Associated with dilated ventricles

Ependymoma (Figure 26.35)

This is a benign neoplasm arising from the ependymal cells. These cells normally line the ventricles; the tumour is within and expanding the ventricle. The CT findings are often non-specific.

- Well-circumscribed, good vascular formation
- may invade the adjacent brain parenchyma
- intraventricular
- 50% show calcifications
- occur most commonly in lateral/third ventricles
- cause obstructive hydrocephalus, particularly when located in the fourth ventricle

Hypophysis (Pituitary) tumour (Figure 26.36)

The pituitary gland is usually well-visible on CT at the level of the sella turcica.

In Cushing's disease, a pituitary tumour, macroadenoma may be expected.

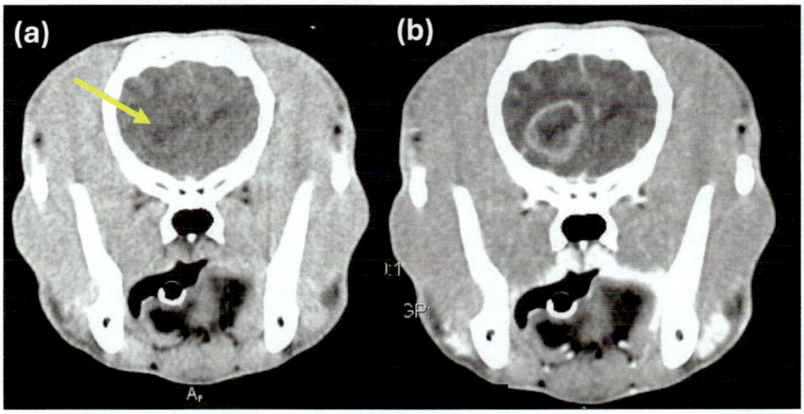

Figure 26.33 **Transverse image (a) survey and (b) after the administration of contrast medium of a confirmed oligodendroglioma. There is a midline shift and vasogenic oedema seen (yellow arrow) in the right frontal lobe. There is a peripheral contrast uptake with a central hypo-dense zone.**

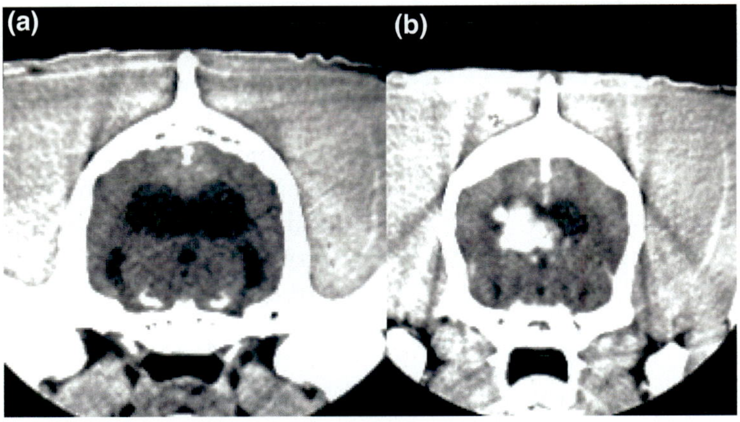

Figure 26.34 Transverse images (a) survey and (b) after the administration of contrast medium of a confirmed choroid plexus tumour. The brain parenchyma shows enlarged lateral ventricles. After IV contrast, a marked contrast-enhancing mass is present within the right lateral ventricle. This mass exerts a "mass" effect and is blocking the CSF flow.

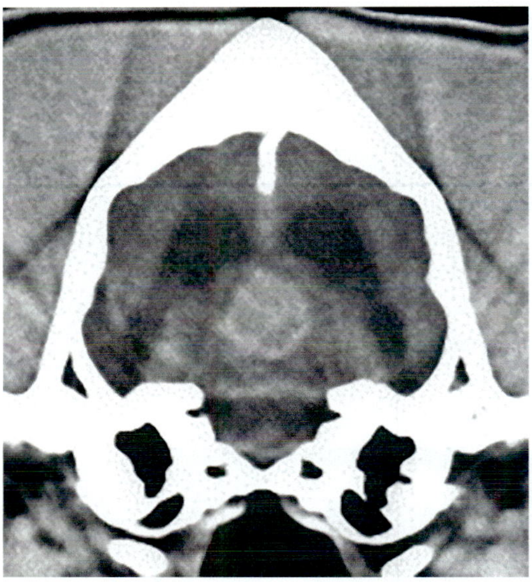

Figure 26.35 Transverse image after the administration of contrast medium. The brain parenchyma shows enlarged lateral ventricles. Post contrast is a round shaped mass present within the third ventricle. This neoplasia is most likely a choroid plexus tumour or an ependymoma.

- Extra-axial
- Located in the sella turcica
- Displaces the third ventricle dorsally
- Enhances uniformly, sharply marginated
- "Mushroom appearance"

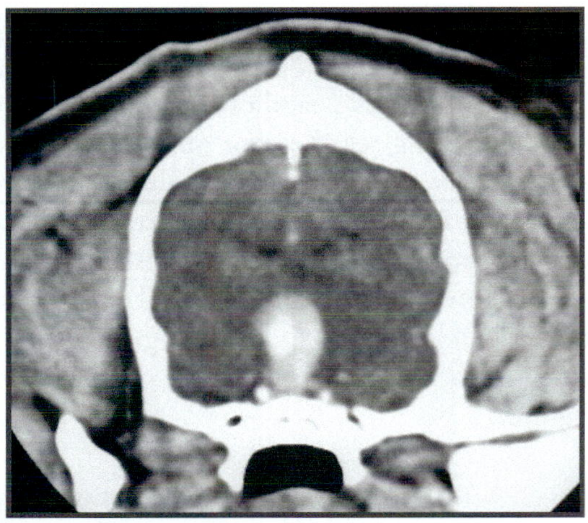

Figure 26.36 Transverse post-contrast CT image of a pituitary macroadenoma in a dog. A large, smoothly delineated mass with homogeneous contrast enhancement is visible, extending dorsally from the pituitary fossa.

- (microadenomas usually indistinguishable because of isoattenuation, and dynamic contrast CT is required)

How can you know that the hypophysis is enlarged?

- Measure the height of the hypophysis (mm) = H.

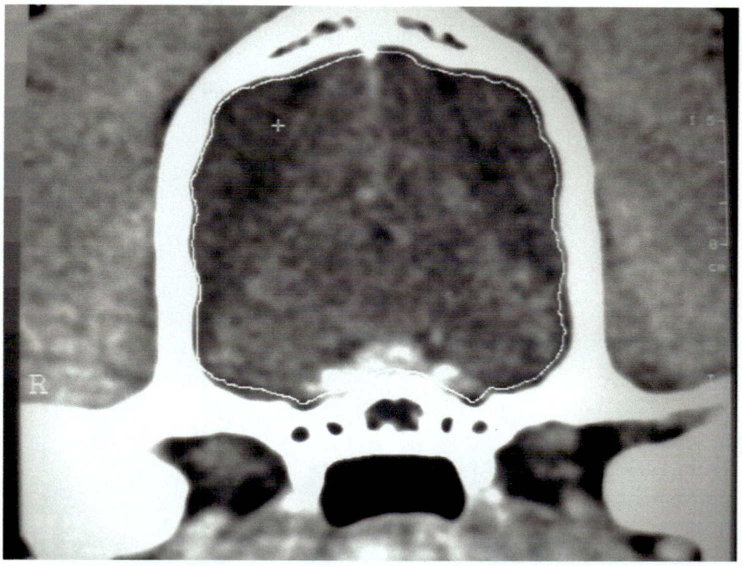

Figure 26.37 **The transverse image that contains the largest cross-section of the pituitary gland is selected. The surface of the brain including the hypophysis is measured in (mm^2).**

- Select the transverse image that contains the largest cross-section of the pituitary gland
- Measure the surface of the brain section (including the hypophysis) (mm^2)=B. (Figure 26.37)
- Make the calculation: $H/B \times 100$
- If the outcome is <0.32: the hypophysis is not enlarged
- If the outcome is >0.32: the hypophysis is enlarged

Cranial nerve V lesions, trigeminal neuropathy, is usually visible on CT certainly when chronic, progressive unilateral masticatory muscle atrophy is present (Figures 26.38 and 26.39)

Secondary brain tumours (metastases).

Brain metastases are estimated to account for approximately 25%–50% of intracranial tumours in human patients. In dogs, the reported incidence is less probably because brain scanning is not routinely performed in tumour patients.

CT features: (Figure 26.40)

- Brain metastases may be solitary or multiple.
- Brain metastases generally do not undergo calcification.
- On pre-contrast imaging, the mass may be isodense, hypodense, or hyperdense (classically melanoma) compared to normal brain parenchyma with variable amounts of surrounding oedema.
- Following administration of contrast, enhancement is also variable and can be intense, punctate, nodular, or ring-enhancing if the tumour has outgrown its blood supply.

26.5.4 Inflammation

Encephalitis is the state of inflammation in a dog's brain. Encephalitis may occur alone or in conjunction with meningitis (meningoencephalitis), which is recognized as the inflammation of the meninges. The meninges comprise the membranes surrounding the brain and spinal cord, and/or myelitis, or the inflammation of the spinal cord itself. Encephalitis is most likely to be found in a young adult or adult dog and has a higher occurrence in the following breeds: chihuahua, Yorkshire terrier, Maltese, pug, and German Short-Haired Pointers.

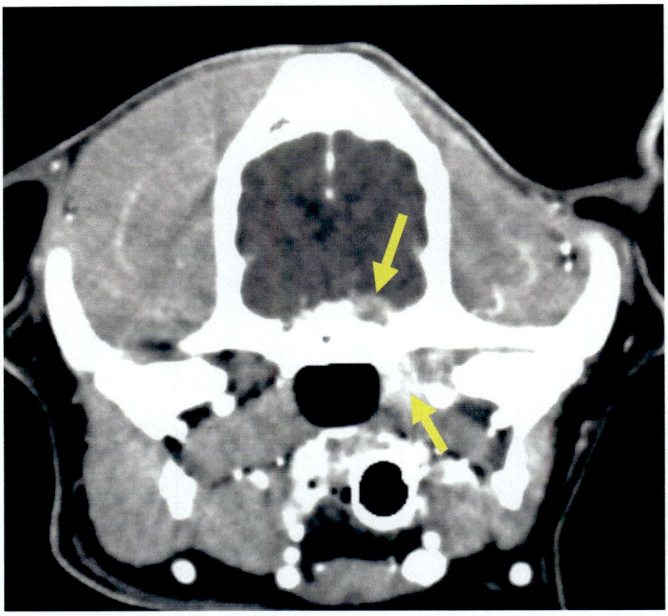

Figure 26.38 Transverse image after the administration of contrast medium, of an 8-year Akita. The left temporal and masseter muscles show prominent muscle atrophy. There is an extra-axial contrast-enhancing mass at the origin of the right trigeminal nerve with involvement of the extracranial trigeminal nerve extending into the right mandibular branch (yellow arrows).

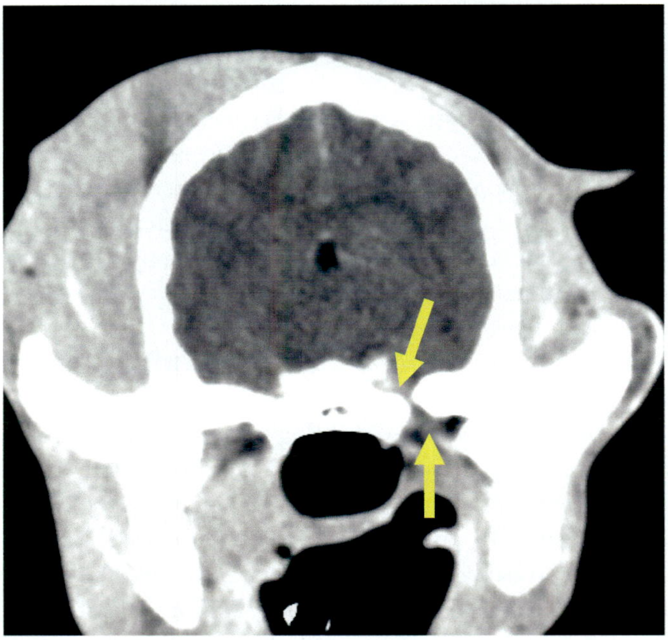

Figure 26.39 Transverse image after the administration of contrast medium, of a 2-year-old Jack Russel Terrier. The oval foramen on the right side is enlarged. There is prominent atrophy of the left masticatory muscles. The left trigeminal nerve is thickened and contrast-enhancing (yellow arrows). The lesion resolved on follow-up examinations. The presumptive diagnosis was unilateral trigeminal neuritis.

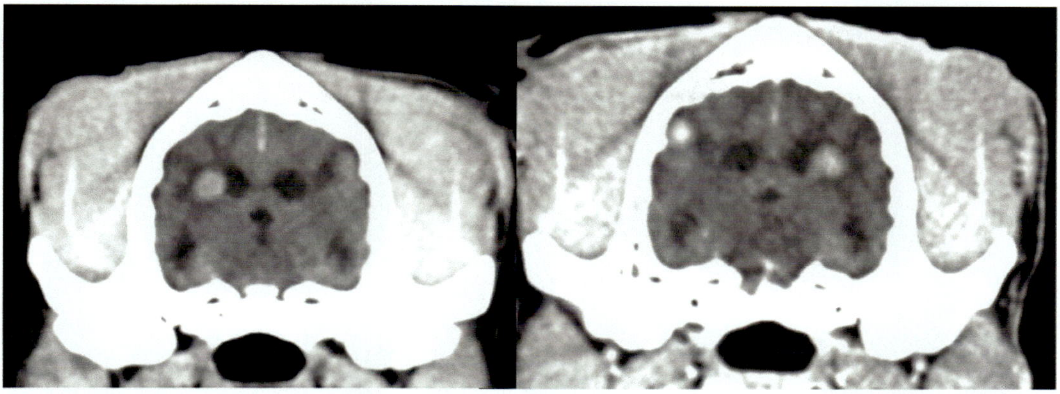

Figure 26.40 **Transverse images after the administration of contrast medium showing multiple round contrast-enhancing nodules. The dog was suffering from mammary neoplasm.**

Some cases of inflammatory brain disease may be detected with CT, although MRI still remains the gold standard. Both CT and MRI can visualize non-neoplastic processes, such as infection and inflammation, but even with MRI, it is not always easy to distinguish between neoplastic versus non-neoplastic space-occupying processes. Some patterns of CT abnormalities of inflammatory brain disorders are similar as those noticed in brain tumours. These comprise ventricle asymmetry, deviation of the falx, focal enhancing or non-enhancing lesions, and ring enhancement. Inflammatory diseases often do not show visible abnormalities on CT.

The combination of CT imaging with CSF analysis may help provide the diagnosis.

26.5.4.1 Idiopathic inflammatory CNS disease

Non-infectious causes of brain inflammation are significantly more common. Some are idiopathic, meaning there is no known cause, and all are autoimmune diseases. They are a group of sporadic inflammatory diseases that affect the brain and/or spinal cord in the absence of an infectious cause (i.e., pathogen-free). The most important ones are granulomatous meningoencephalitis(GME) and necrotizing {Meningo} encephalitis (NE) and meningoencephalomyelitis of unknown origin (MUO) (Figure 26.41). Primary inflammatory illnesses like granulomatous meningo-encephalitis are distinguished by the presence of multifocal, contrast enhancing, intra axial, granulomatous lesions in several regions of the brain.

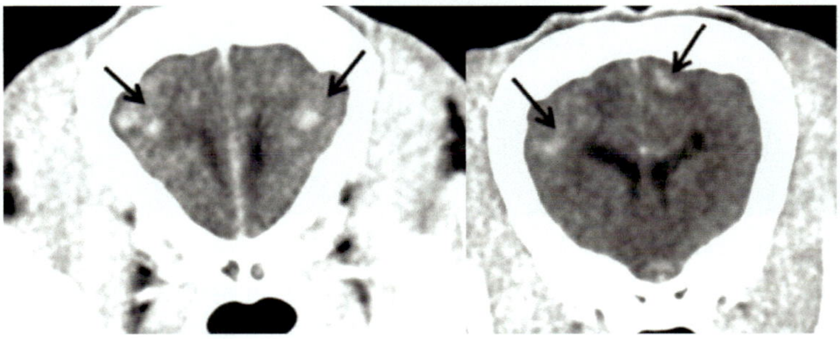

Figure 26.41 **CT image administration of IV contrast at the level of the frontal lobes of an 8-month-old female entire French Bulldog with meningoencephalomyelitis of unknown origin. Multiple hyperdense, rounded, intraaxial lesions are visible within the frontal and parieto-temporal cerebral cortex of the dog (arrows).**

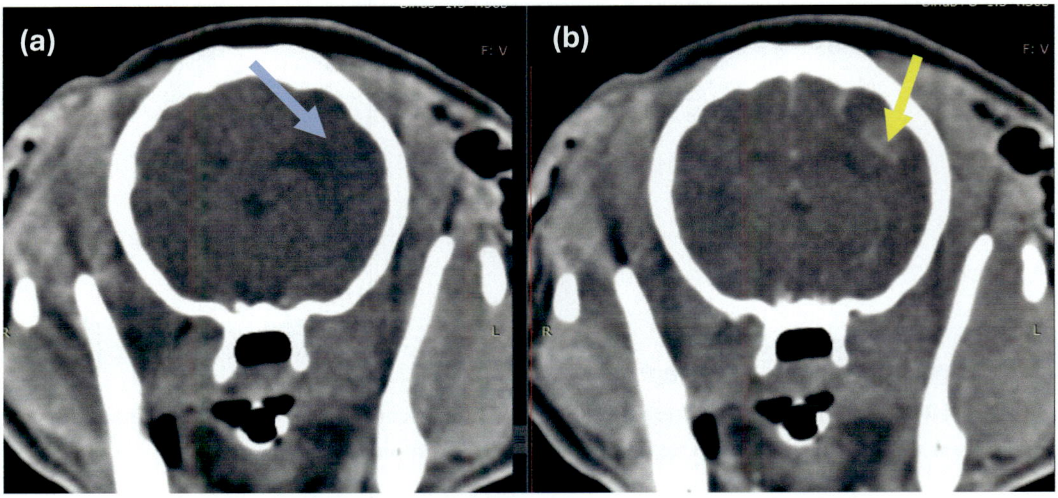

Figure 26.42 **Transverse images (a) of survey and (b) after the administration of contrast medium of a confirmed granulomatous meningoencephalitis. In the left frontal lobe are hypodense areas present, suggestive of oedema. (blue arrow). IV contrast shows diffuse hyperdensities intraaxially in the left frontal lobe (yellow arrow). Asymmetric ventricles, with left larger than right, are present.**

Some lesions may be associated with vasogenic oedema and mass effect (Figure 26.42). The only way to definitively diagnose GME is through a biopsy of the brain, which is rarely done in living patients. Necrotizing encephalitis is characterized by multifocal areas of decreased opacity, with necrosis or cystic changes, which may or not enhance with contrast (Figure 26.43).

26.5.4.2 Infectious inflammatory CNS disease

Causes of Brain Inflammation in Dogs: Infectious causes of brain inflammation are less common. They include bacteria, viruses (such as distemper, rabies, and parvovirus), fungal infections (such as aspergillosis, histoplasmosis, and blastomycosis), protozoa, rickettsia, complications of vaccines, and parasitic infections (such as Rocky Mountain spotted fever and ehrlichiosis) (Figure 26.44)

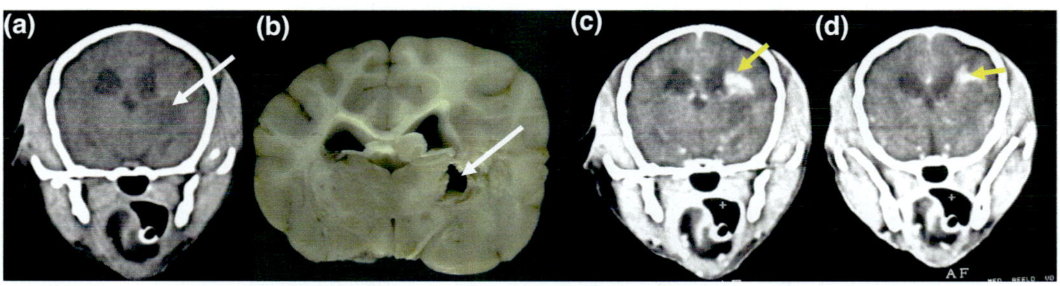

Figure 26.43 **(a)Transverse survey CT image, (b) macroscopic anatomical section at the same level as a, (c, d) transverse images at different levels after IV contrast injection, of the brain of a Yorkshire terrier with confirmed necrotizing encephalitis. A hypodense, intraaxial, zone can be seen in the left cerebrum (white arrow), asymmetric and dilated lateral ventricles are seen, as well as a mild mass effect to the right. The area of necrosis on the macroscopic section (white arrow) matches with the hypodense zone seen on the survey CT image. Several contrast enhancing, intraaxial, areas are seen at other levels of the brain (yellow arrows).**

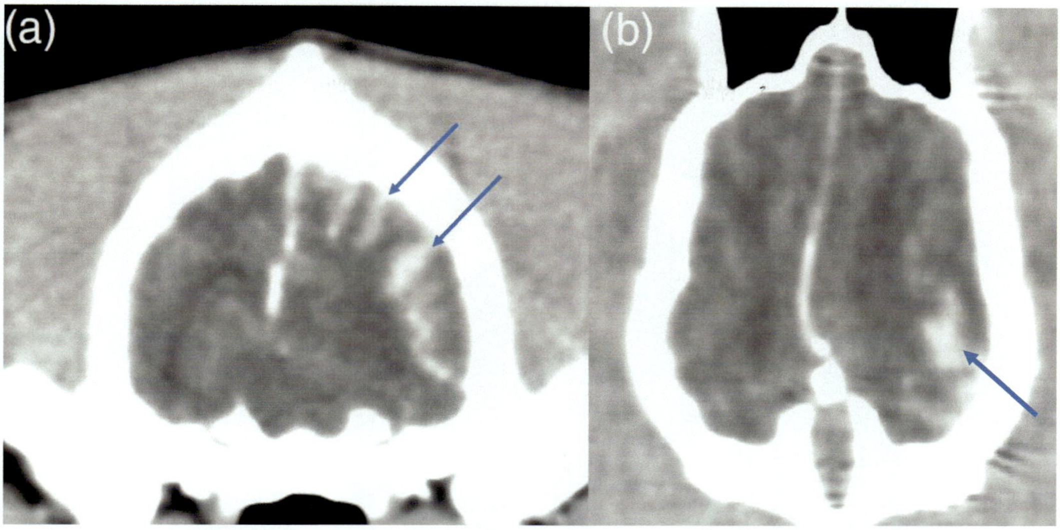

Figure 26.44 Transverse- (a) and dorsal-reconstructed (b) images after administration of contrast medium of a 2-year-old mongrel dog with confirmed Ehrlichia canis infection. The grey matter is mainly affected, causing mass effect characterized by sulcal effacement. The cerebral sulci and meninges of the left temporal lobe show prominent contrast enhancement (arrows).

CT features include the following:

- Asymmetric, bilateral (often multifocal) lesions in the forebrain, brainstem, and/or spinal cord.
- These lesions variably enhance with intravenous contrast administration.
- Cystic areas may be seen in necrotic brain areas.
- Sometimes, enlargement of the ventricles due to obstructive lesions.

26.5.5 Vascular lesions

In contrast to the high incidence in man, intracerebral haemorrhage resulting from spontaneous rupture of vessels is rare in dogs. Usually, it may occur in association with skull fractures. Haemorrhagic stroke: when a blood vessel ruptures and blood is leaking, leaking into the brain. Ischaemic stroke: when a blood clot blocks the blood vessels to the brain.

The cause might be as follows: trauma (subarachnoid or intracranial haemorrhage), rupture of congenital vascular abnormalities, primary or secondary brain tumours, vasculitis, systemic hypertension, intravascular lymphoma, haemorrhagic infarction, and impaired coagulation.

As in men, CT is a vital tool in the assessment of small animal patients with serious head injury (see b. Trauma head). CT is the best procedure to detect acute stroke, which appears as homogeneous, hyper-dense areas associated with a mass effect during the first 72 hours; the lesions then become isodense and, after a week, hypodense.

CT features of haemorrhage: (Figure 26.45)

- Acute (24–72 hours)
  - Region of increased opacity
- Chronic (>72 hours)
  - Usually decreased opacity

The location can be as follows:

- Epidural
- Subdural
- Intracerebral

CT features of infarct or ischaemic stroke: (Figure 26.8)

- Seldom!
- Depending on the duration.

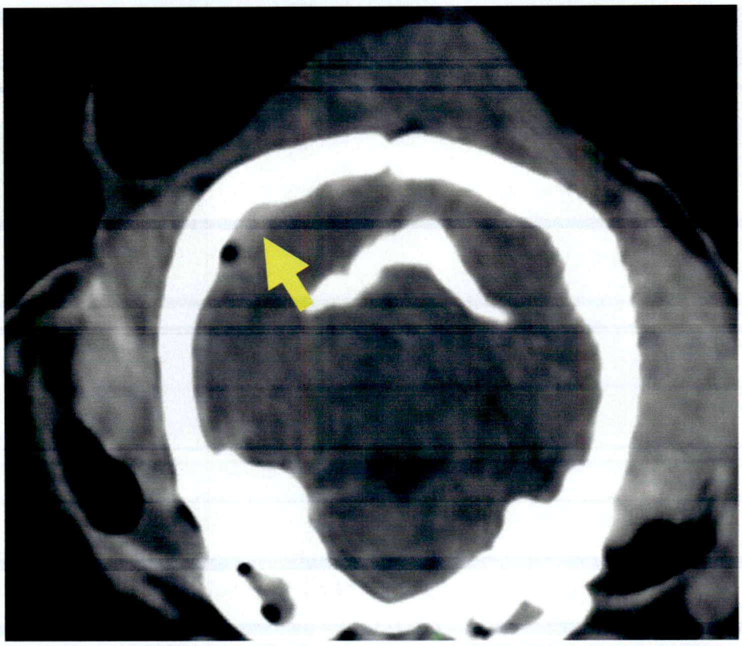

Figure 26.45 **Transverse survey image at the level of the occipital lobes of a chihuahua after head trauma. A subdural haematoma appears hyperdense, concave towards the brain in the right side (arrow).**

- Acute (8–24 hours): Focal area of hypodensity (necrosis and oedema).
- After several days: Hypodensity increases (more oedema). Mass effect.
- 7–10 days: Less oedema. Irregular increase in density.
- After several weeks: Decrease in density.

26.6 HEAD TRAUMA-CRANIOFACIAL AND BRAIN INJURIES

Head trauma results from a variety of causes, with traffic accidents the most common. The clinical signs reflect soft-tissue trauma and severe bone fractures; however, brain damage is the more critical component of any head injury.

The effect of trauma on the nervous system is complicated. The traumatized neurons and glial cells at the centre of the injury die immediately and create an area of tissue necrosis, followed by a wave of secondary cell death during which the peripheral neurons undergo a number of metabolic changes. This wave is driven by excitotoxicity and results in a progressive increase in the size of the necrotic region following the trauma. Excitotoxicity is a complicated process during which the traumatized cells depolarize and release excitatory neurotransmitters, particularly glutamate. These neurotransmitters cause depolarization of the adjacent surviving cells, whose metabolism is deranged, leading to a critical increase in intracellular calcium concentration. This increase in turn activates a variety of cellular enzymes and ultimately leads to the destruction of cellular macromolecules and cell death. The intracellular calcium increase during excitotoxicity is key, and experiments with neurons in tissue culture show that the whole process can be inhibited by depleting calcium levels.

Along with these cellular changes, the traumatized brain tends to swell, in turn increasing the intracranial pressure (ICP). Moreover, blood flow autoregulation, the process by which blood flow to the brain is maintained despite fluctuations in systemic blood pressure, is disturbed. These factors together impair the delivery of blood and oxygen to neurons, which exacerbates the metabolic derangements inflicted on damaged neurons and can lead to a vicious cycle that further increases ICP.

26.6.1 Consequences of Raised Intracranial Pressure

ICP depends on the volume of the three main constituents of the intracranial tissues – brain, blood, and CSF. When the brain swells, ICP can be maintained within normal limits by a reduction in the volume of blood and/or CSF within the skull. Unfortunately, this compliance is limited, and when exhausted, the ICP rises precipitously and the swollen brain becomes compressed against various bony prominences of the skull, ultimately herniating. Herniation can occur at various sites, but the tentorium – the bony shelf that separates the forebrain and cerebellum – and the foramen magnum are the most common. In humans with forebrain swelling, herniation across the tentorium results in pressure on the occulomotor nerves, producing dilated, unresponsive pupils – a bad prognostic indicator. A similar phenomenon has been cited in dogs, although dogs with sub-tentorial herniation rarely have occulomotor paralysis, and their clinical signs of increased ICP tend to become obvious only when herniation at the foramen magnum occurs, leading to pressure on the caudal brainstem and signs of vestibular dysfunction. Moreover, almost three-quarters of the dogs that show evidence of herniation on diagnostic imaging show no clinical signs of herniation.

26.6.2 Presentation

Head trauma cases present with a variety of common injuries. Various skull bones can be fractured, and significant brain injury is a risk. However, the basic principles of emergency assessment (airway, breathing, and circulatory system) should take priority and a neurological assessment delayed until the patient is stable. Neurological evaluation should then be repeated frequently to monitor for signs of deterioration.

Neurologic signs frequently involve a depressed level of consciousness as a result of direct trauma to the cerebral cortex and damage to the brainstem systems that provide generalized arousal of the cerebral cortex. Close attention should be paid to the pupils, with miosis common contralateral to the more severely contused cerebral hemisphere. Miosis may also occur as a result of Horner's syndrome, which can occur when various structures in the neuraxis, neck, and head are damaged. Bilateral mydriasis with an absent pupillary light reflex, which can occur with caudotentorial herniation of the cerebral hemispheres, should be considered a poor prognosis. However, as noted previously, this is a relatively rare finding in dogs. Other brainstem reflexes should also be examined, although care should be taken when manipulating the head and neck to evaluate the oculocephalic reflex because of the risk of fractures or subluxations of the head and/or neck. The oral cavity should also be examined cautiously since fractures of the maxilla, mandible, and sphenoid bones are not unusual in head trauma patients.

26.6.3 Indications

Diagnostic imaging is used to determine the nature, number, and extent of lesions affecting the entire head and brain; assess their consequences; and allow effective treatment planning. Additionally, injuries elsewhere in the body should be thoroughly assessed, with a particular focus on thoracic and abdominal injuries that may impact patient outcomes.

CT is the fastest, most effective way to get the necessary information because although MRI provides more detail about the brain parenchyma, CT provides superior bony detail.

Specifically, use CT to assess these possibilities:

- Rib fractures, pulmonary contusion or pneumothorax, which can all impair breathing and potentially worsen brain injury.

- Significant thoracic or abdominal haemorrhage.
- Significant intracranial haemorrhage that may require surgery.
- Any skull fractures that would require immediate attention, such as depressed fractures of the cranial vault that may require removal of bone fragments, or mandibular fractures that may preclude normal prehension of food. Severe nasal trauma may cause breathing difficulties, and the combination of bone and soft-tissue damage may necessitate use of a nasal oxygen tube and/or feeding tube.
- The cosmetic impact of the trauma.

CT, combined with the results of the clinical and neurologic examinations, should help establish a reliable prognosis.

26.6.4 CT features

The entire study should be systematically examined, including scrutiny of the images in a bone window for the bone algorithm and a soft-tissue window and brain window for the soft-tissue algorithm to maximize the CT study's potential. After obtaining the transverse series, it is important to look at the images in all three planes by using multiplanar reconstructions (MPR) and both the bone- and soft-tissue window settings. Reconstructed images in multiple planes and 3D are particularly useful in craniofacial trauma assessment. IV contrast media may be required to improve visualization of the soft tissues and brain parenchyma.

vThe bone window is indicated to evaluate all bone lesions, whereas the soft-tissue/brain windows are used to examine, in parallel, all the soft tissues and brain parenchyma, with a specific focus on the presence of haemorrhage/contusion.

26.6.5 CT features: craniofacial injuries

The complex anatomy of the skull makes interpretation difficult. Also, multiple concurrent injuries are common, which distorts the normal structure and further complicates interpretation. Some fractures can be very small, but their presence must still be recognized. Initial examination of all the images, one by one from rostral to caudal in the transverse plane and using both bone and soft-tissue windows, is a good start. Next, MRP (again using both windows) should be examined; 3D reconstructions will allow further understanding of most of these lesions. If possible, the fractures should be described, including the bone affected, shape (transverse, oblique, spiral, and comminuted), degree of displacement, direction of the fragments, and whether they are intraarticular and open.

The injuries most commonly seen are (Figures 26.46–26.48):

- Fractures of the incisive bone, symphyseal separation.
- Hard palate fractures, including maxillary midline separation, separation of the nasomaxillary suture, fracture of the palatine process of the maxillary bone, fracture of the maxillary bone, fracture of the horizontal lamina of the palatine bone, and fracture of the palatine bone.
- Mandible fractures, including the mandibular body and ramus.
- Nasal cavity fractures, where fractures of the nasal bone and vomer are possible. The integrity of the turbinates should be examined carefully, and the nasal contents should be inspected – if they are hyperdense using a soft-tissue window, haemorrhage is likely.
- Fractures of the sphenoid bones, pterygoid bone, ethmoid bone, and frontal bone.
- Zygomatic bone fractures usually associated with temporomandibular joint fractures/luxation.
- Temporomandibular joint (TMJ) injuries. With TMJ luxation, the mandibular fossa appears empty – the condyle is usually displaced rostrally, causing dental malocclusion, and mandibular shift may be present. If the TMJ is subluxated, there may be an uneven joint space with

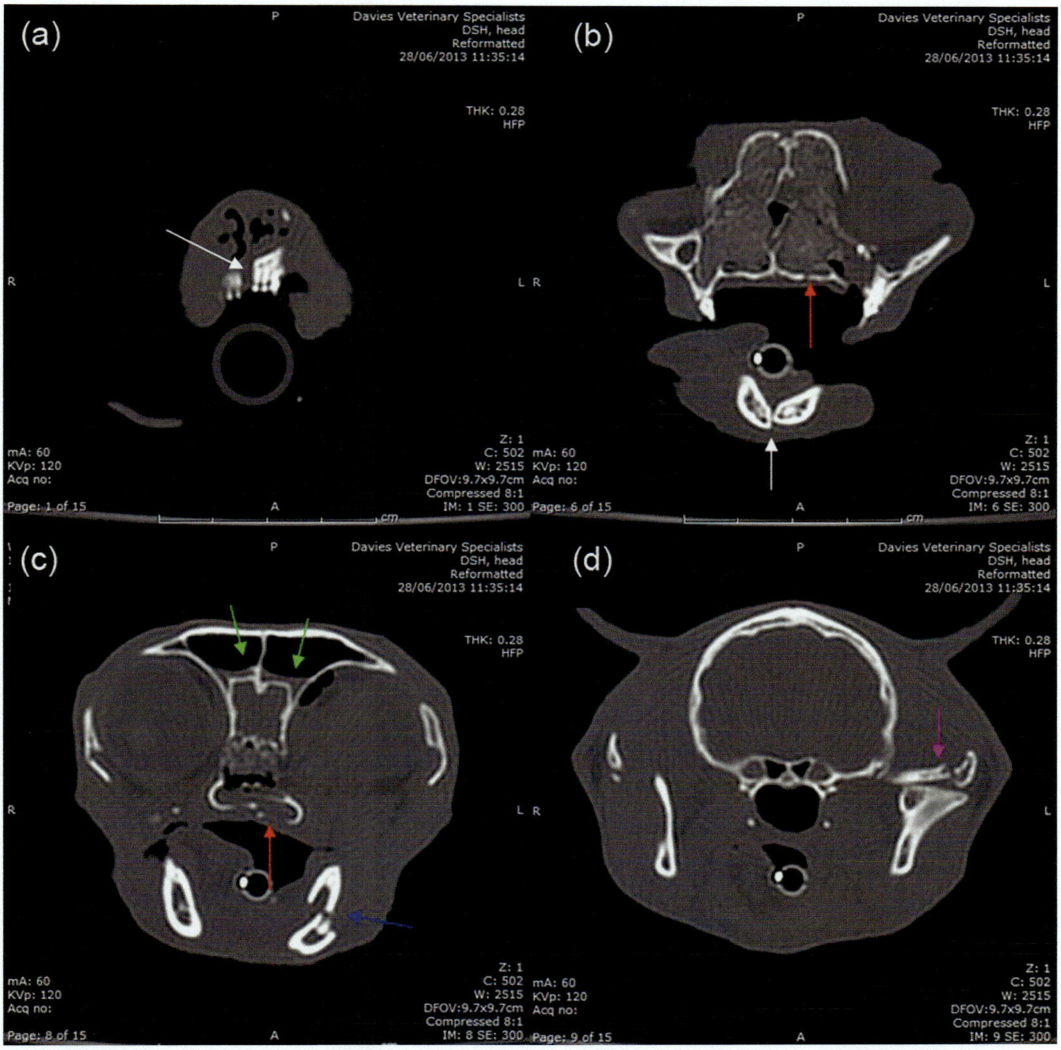

Figure 26.46 **Transverse CT images (bone window) of the head of a cat following a road traffic accident, at the level of the rostral nasal cavity/incisor bone (a), mid-nasal cavity (b), frontal sinuses (c), and temporomandibular joints (d). Note the symphyseal separation of the incisor bone and rostral mandible (white arrow), multiple maxillary and hard palate fractures, with misshapen and occluded nasopharynx due to accumulation of fluid/haemorrhage (red arrows); this can cause severe respiratory problems. There is also fluid in both frontal sinuses (green arrows). There is also a fracture of the left mid-mandibular body/tooth (blue arrow) and a comminuted fracture of the left retroarticular process of the zygomatic arch (intra-articular fracture), with subluxation of the TMJ (pink arrow).**

misalignment of the condyle (compared with the contralateral side). Concurrent fractures, sometimes intraarticular, are common – these can affect the coronoid process, condylar process, mandibular fossa, and/or retroarticular process.

- Tympanic bullae fractures, where hyperdense contents using a soft-tissue window are compatible with haemorrhage. Bear in mind that external/ middle ear disease/changes may have already been present
- Hyoid bone fractures
- Malocclusion

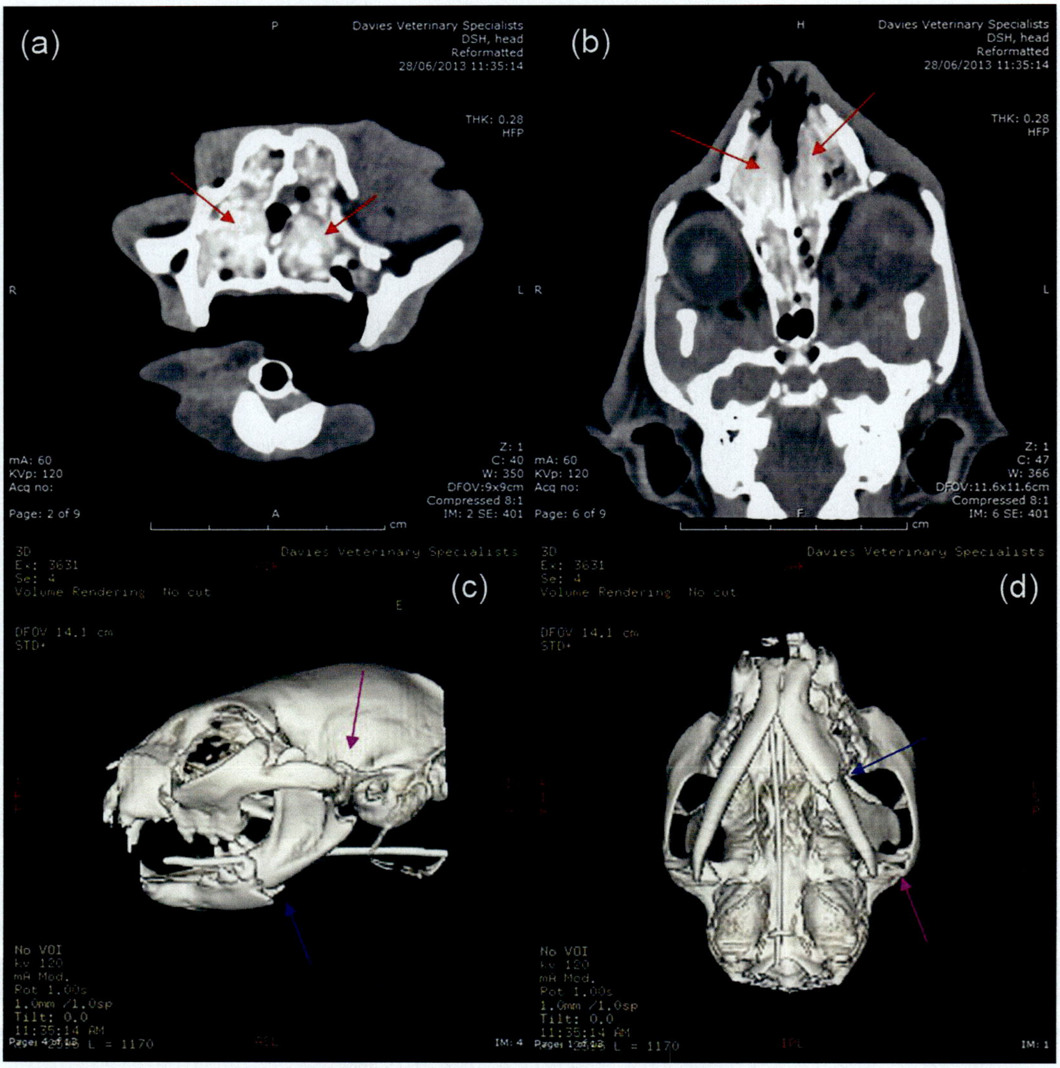

Figure 26.47 Transverse (a) and dorsal plane (b) CT images of the same cat as in Figure 26.45 using a soft-tissue window, before contrast administration, showing a large amount of hyperdense material in both nasal cavities; this is likely to represent haemorrhage (red arrows). (c and d) show 3D reconstructions, permitting the transverse mandibular fracture (blue arrows), TMJ subluxation/fracture (pink arrows) and mild malocclusion to be easily evaluated.

- Pharyngeal/nasopharyngeal occlusion due to, for example, haemorrhage (soft-tissue window).
- Temporal bone and other skull fractures (see head injuries).
- Other soft-tissue injuries that can affect ocular and periocular structures, the skin, and subcutaneous and muscular structures in any region of the skull.

26.6.6 CT features: brain injuries

Brain/skull injuries are not as common as craniofacial injuries, but their presence can make a big difference in case management.

- Skull fractures, when the temporal bone is most commonly affected.
  If a depressed fracture with a fragment compresses the brain parenchyma, brain contusion/haemorrhage can occur.

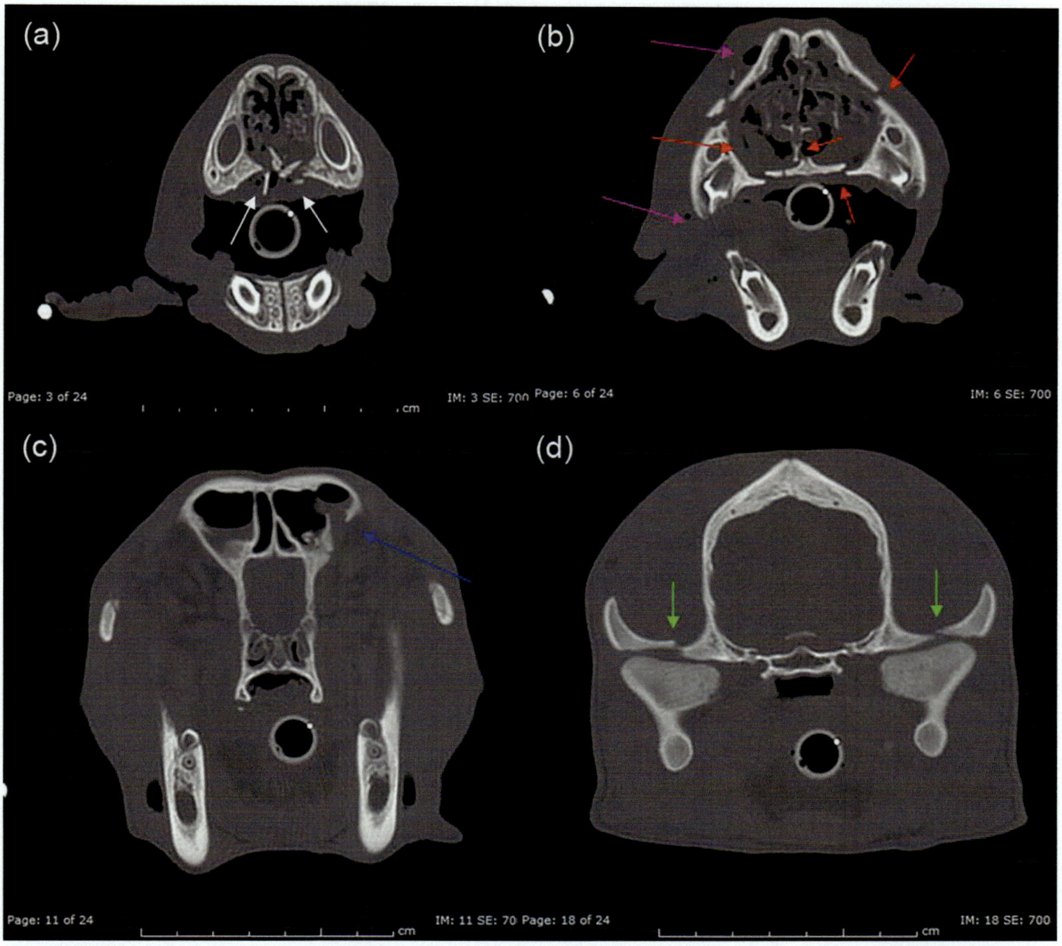

Figure 26.48 **Transverse CT images (bone window) of the head of a dog after being hit by a car, at the level of the rostral nasal cavity (a), mid-nasal cavity (b), frontal sinuses (c), and temporomandibular joints (d). There is a comminuted incisor bone fracture, with free bone fragments displaced into the nasal cavity; this has caused an oronasal communication (white arrows, a). There are also multiple nasal, hard palate, and vomer fractures (red arrows, b) and a comminuted, depressed fracture of the left frontal bone is present, with fluid/blood accumulation in both sinuses (blue arrow, c). Note also the subcutaneous emphysema, most likely caused by air tracking from the nasal cavity (pink arrows, b). There are bilateral intraarticular TMJ fractures due to minimally displaced zygomatic process fractures with no TMJ subluxation (green arrows, d).**

If major displacement of the brain outward is present, herniation can occur. If the fracture is open, pneumocephalus can develop (air accumulation within the cranial vault). See Figures 26.49 and 26.50.

- Acute haemorrhage/haematoma, where CT is highly indicated due to the high signal caused by the globin portion of the blood, which is hyperdense in the acute and subacute phases but decreases to become isodense at 4 weeks. Between 6 days and 6 weeks following the initial injury, peripheral contrast enhancement may be present, representing revascularization. Haemorrhage can occur at several sites: intraparenchymal (within the substance of the brain), which is intraaxial; epidural (between the dura and inner skull

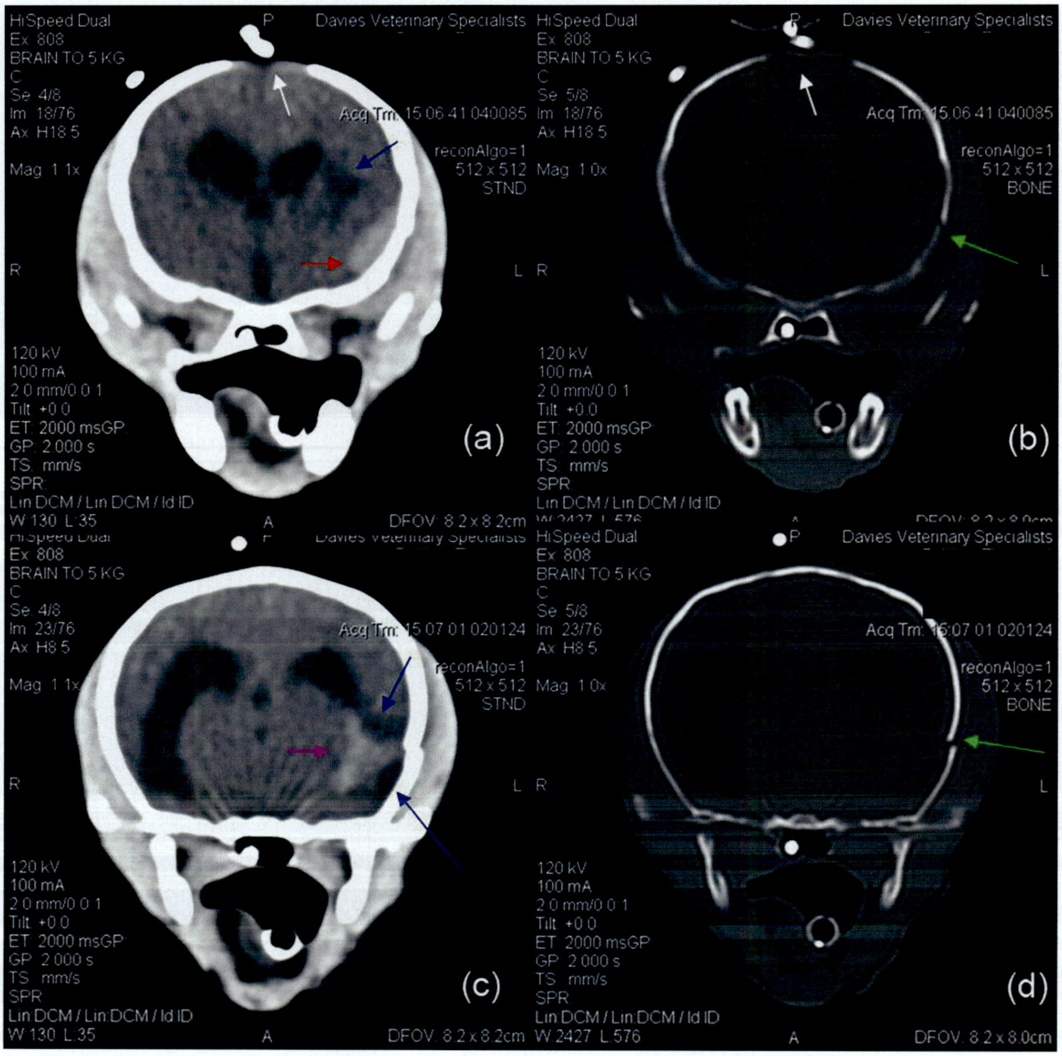

Figure 26.49 **Transverse CT images from a 5-month-old chihuahua that suffered head trauma after a domestic accident, displayed in brain window (a and c) and bone window (b and d) at the level of the temporal lobes. There is an epidural (red arrow, a) and extensive intraparenchymal haemorrhage (pink arrow, c) in the left temporal lobe associated with hypodense intraparenchymal changes/contusion (blue arrows, a and c) and minimally displaced temporal bone fractures (green arrows, b and d). Note the open fontanelle (white arrow, a and b).**

surface), which is typically biconvex and well-defined; subdural (between the dura and pia/subarachnoid), most commonly crescent-shaped; subarachnoid (within the arachnoid space), which causes an increased signal in the sulci and fissures as the blood mixes with the CSF; and intraventricular, seen as hyperdense material in the ventricular system. Despite these characteristic CT findings associated with haemorrhage at various sites, the precise location can be difficult to determine. Irrespective of the site, assessing whether there is a mass effect on the surrounding structures is important (Figures 26.49 and 26.50)

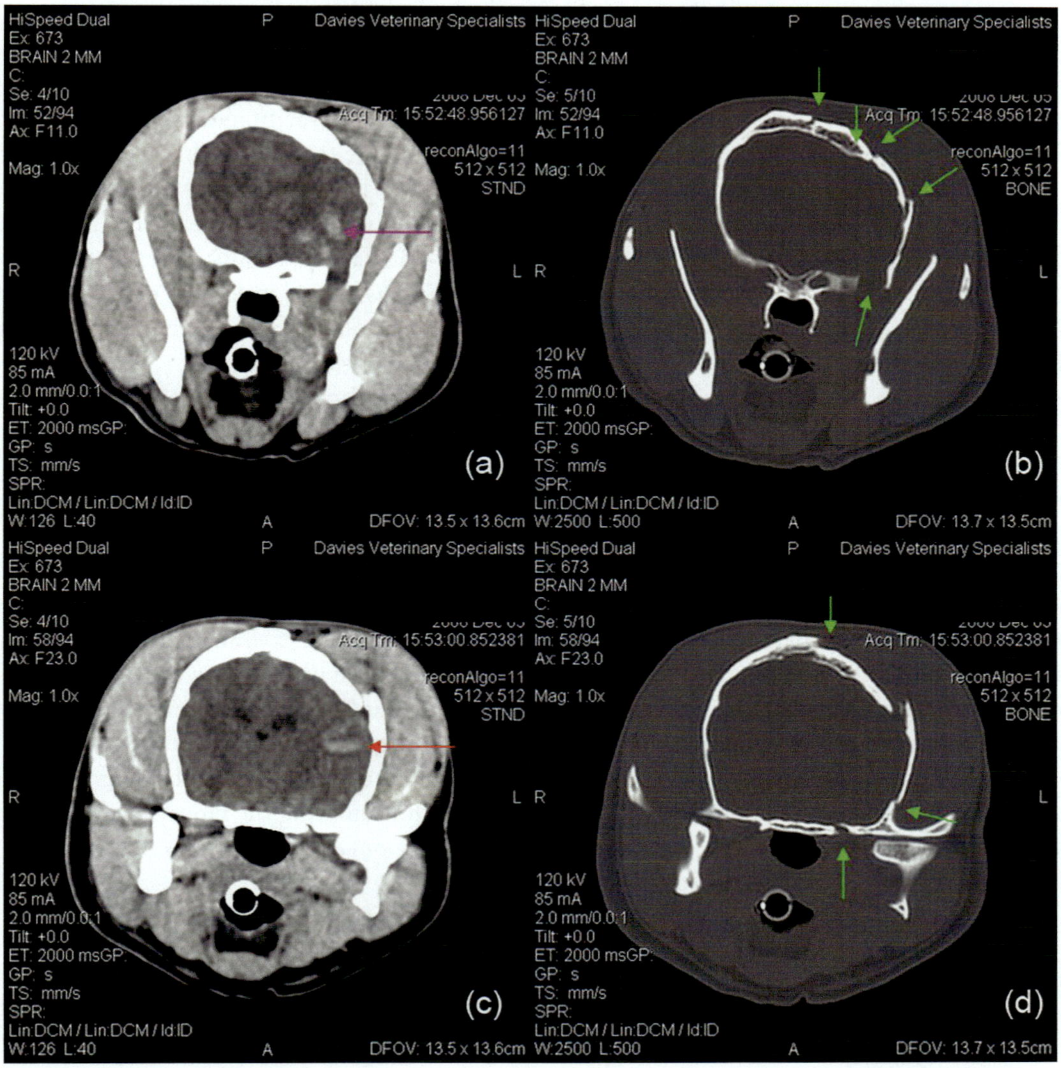

Figure 26.50 Transverse CT images from a 4-year-old Border terrier, which was kicked by a horse, displayed in brain window (a and c) and bone window (b and d) at the level of the frontal lobes (a and b) and parietal/temporal lobes (c and d). There is extensive intraparenchymal haemorrhage in the left frontal lobe (pink arrow, a) and temporoparietal lobes (red arrow, c), closely associated with multiple complex depressed and displaced fractures of the overlying skull (green arrows, b and d).

- Brain contusion, which is seen as hypodense areas within the brain parenchyma, indicating either oedema or necrosis. In some cases, a mixed signal with hyperdense areas may be caused by intra-parenchymal haemorrhage (Figure 26.50).
- *Brain herniation*: CT may be less sensitive than MRI for detecting structural abnormalities in the brain parenchyma, but it should allow evaluation of herniation at the foramen magnum and tentorium, which will be better identified in sagittal reconstructions.
- *Soft-tissue changes*: Changes in the signal of the muscles/soft tissues surrounding the traumatized skull are common. In areas of superficial trauma, surface abnormalities

without any changes in the deeper tissue may occur. With penetrating injuries such as bite wounds, open/tracking lesions may be present. Intravenous contrast administration helps identify and classify soft-tissue injuries.

26.7 CT-GUIDED BRAIN BIOPSY IN THE DOG

26.7.1 Introduction

Recent imaging techniques can provide detailed anatomic information about the brain and valuable information about various pathologies. In one report, MRI had a sensitivity and specificity exceeding 90% for differentiating neoplastic and inflammatory brain diseases in dogs. Despite this, the ability to distinguish some conditions in animals is still limited. In the same report, the sensitivities for detecting meningioma, glioma, and granulomatous meningoencephalitis were reported as 60%, 84%, and 50%, respectively. In a retrospective review of 300 human patients, preoperative diagnosis based on clinical signs and imaging differed from histopathologic diagnosis from a biopsy in as many as 53% of cases. Moreover, CT or MRI characteristics have not proven reliable for grading brain tumours in the dog. The differential diagnosis of brain lesions thus remains problematic, and choosing the appropriate therapy remains difficult, particularly when deciding whether surgery is indicated for veterinary patients.

A tissue sample must be obtained for histopathologic examination to make a specific diagnosis. A variety of techniques for biopsy removal have been reported, including the following:

- Open surgical biopsy, which is more likely a conventional brain surgery procedure than a "true" mini-invasive biopsy. This technique is used only for the histologic evaluation of a superficial tumour or to evaluate a diffuse disease because exposing the subcortical areas necessitates extensive operations.
- The "free-hand needle biopsy" can be performed through a burr hole with a biopsy needle, taking into account the CT anatomy of the skull in relation to the abnormal tissue inside the brain. This technique is suitable when the lesion of the cortex is diffuse (i.e., a storage disease) but may lead to unsatisfactory results; for example, the target was missed in 12% of patients and in six of eight dogs in which the technique was used in another study.
- Ultrasound-guided biopsy, which needs a specific material, increases the operating time (compared to freehand) but improves the sensitivity of the biopsy in human or animal patients. Ultrasound (US)-guided brain biopsy was indicated in human patients for supratentorial lesions exceeding 15 mm in diameter. US examination of the brain has also been described in the dog. The author has limited experience with this technique in animals, but it does allow the tumour's exact location to be determined before attempting the biopsy for gliomas or meningiomas in dogs.

- In laboratory animals, stereotactic instruments and a brain atlas with stereotaxic coordinates of brain structures have been developed. Orthogonal radiographs have also been used *in situ* to measure the coordinates of the lesion and determine the place of trepanation with a high error level. However, dogs' heads are so variable that reliance on coordinates of a standard atlas has proven impractical in the clinic.
- Many CT-guided and MRI-guided stereotaxic brain biopsy devices were developed in both humans and animals. CT is considered superior in imaging bony structures and MRI in visualizing soft tissues but, for practical reasons, CT-guided stereotactic brain biopsy (CTGSBB) is still relevant because it is faster and easier to perform, without the

interference between the biopsy device and the antenna that MRI requires. In humans, application accuracy of CTGSBB systems is typically within 2–3 mm, complications are uncommon (0%–13%), and diagnostic yield exceeds 90% in most reports. The most common systems used in humans are the Talairach, Reichert–Mundinger, Brown–Roberts–Wells, Leksell, and the Laitinen. In veterinary patients, various CT-guided or MRI-guided devices have been used. These studies concluded that stereotactic brain biopsy was accurate and safe, but did not conclude whether CT-guided or MRI-guided biopsy was the superior procedure.

26.7.2 General indications

According to the National Comprehensive Cancer Network (NCCN), brain biopsy must be considered when maximal safe resection of the brain tumour is not feasible or if MRI suggests primary CNS lymphosarcoma. CT-guided stereotaxic brain biopsy (CTGSBB) remains a method of choice in human patients and offers neurosurgeons multiple options:

- A precise point of tissue retrieval for histologic or microbiologic tissue assay
- Cystic structure or abscess cavity aspiration
- Installation of permanent or temporary drainage conduits
- Installation of multiple catheters for point source of brachytherapy
- A precise direction for cerebroscopic instrumentation
- Intraoperative point localization or guidance.

These indications are likely more common in veterinary practice because our knowledge of brain pathology and imaging is less accurate.

26.7.3 CTGSBB devices components

Devices may be divided into two groups: (1) those used only within the scanner to perform the biopsy and (2) those that allow the translation of scanning data to an operating room setting. In practical terms, scanner access is time-limited at most institutions, and the ability to perform neurosurgical procedures in a standard operating room probably limits the complication rate. Moreover, CT and MRI correctly identify many lesions, which can then be treated, so using stereotactic devices during the initial diagnostic imaging procedure is not reasonable.

Several components make up the current CTGSBB systems documented for animal use: (1) a restraint, which is designed to accurately secure the head of the patient in a standard and reproducible position; (2) the tridimensional coordinate generator; and (3) the biopsy piece that allows the needle to penetrate into the predefined target.

All the components used in the devices under CT must combine CT-compatibility (no artefact production) and rigidity to secure the head of the dog. They ensure a reproducible head restraint and allow the production of extracranial reference points on CT slices, which give the tridimensional coordinates of intracranial targets.

26.7.4 Restraint

Usually, the bony parts of the ear canal, teeth and nose are used to secure the head. The part of the device used for restraint is graduated, enabling the surgeons to reproduce the same positioning. When the biopsy is not performed within the CT, the dog's head must be secured in a reproducible position in the device twice. Frame-base models were designed to ensure a very stiff and reproducible restraint (Figures 26.51 and 26.52) and were initially the model used most commonly. In human patients, they have been replaced by more flexible frameless systems,

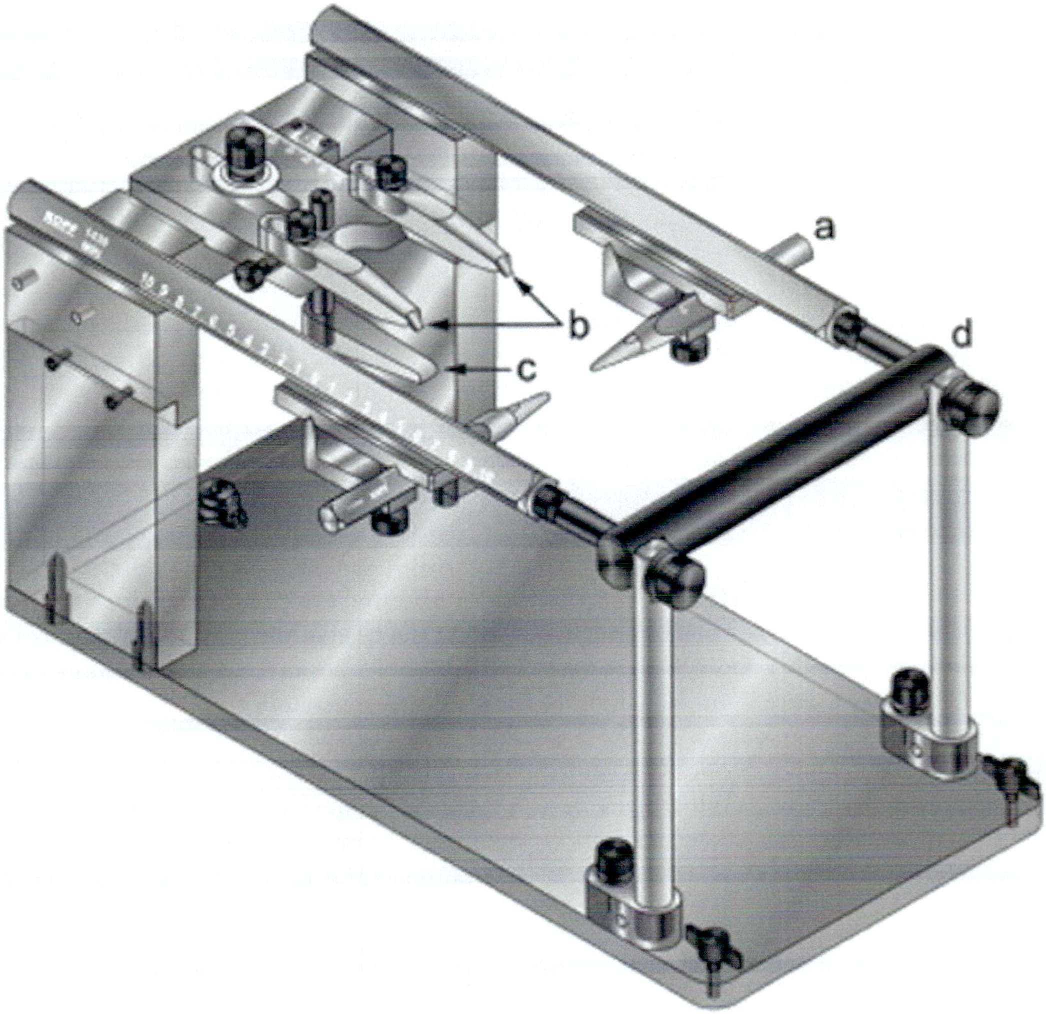

Figure 26.51 Frame-base CT-guided brain biopsy device (Kopf instrumentation). The frame is made of Lexan® and is adapted for the use in CT. The arm bars of the frame have an embedded ruler to allow for precise rostrocaudal positioning of stereotactic instruments. The restraint is obtained with ear bars inserted in the ear canal, tooth bar (used as a bite plate), and two eye bars (inserted on the ventral aspect of the orbit).

do not require placement of bone-grasping head frames, and yield similar accuracy and diagnostic results. The frameless systems also adapt more easily to the wide variations in head shapes, an advantage with veterinary patients.

A bite-plate can be used to secure the dog's head and conduct CTGBB in a frameless CT-guided device (Figure 26.53). A maxillary dental impression using dental acrylic material can be placed in a mould identical in shape to the head-holding fixture, ensuring that the head is properly secured, and can be placed in the same position during the different steps of the procedure.

26.7.5 Three-dimensional coordinate generator

In the modified Laitinen system, the localizer unit is composed of two triangular localizers fixed to the frame bars (Figure 26.54a).

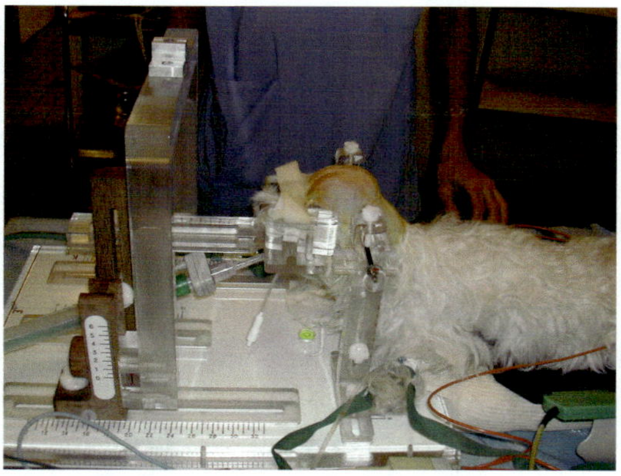

Figure 26.52 Frame-base CT-guided brain biopsy device (Modified Laitinen Instrumentation). The bony part of the ear canal (ear bars), the teeth (maxilla support), and the nose (nose constrictor) are used to secure the head in a reproducible way (millimetric scale on each component).

The rostral extremity of the triangle includes radiopaque reference points (Figure 26.54b). The localizers allow the production of extra-cranial reference points on CT slices that may be especially related to intracranial targets. Fiducial markers (Figure 26.53) connected to tridimensional analysis software are used in the new devices.

26.7.6 Biopsy component

The biopsy component in the device leads the biopsy needle to the centre of the biopsy target. This is generally an arc system that allows the setting of multiple angles to achieve the appropriate trajectory to the target point in rigid fixation (Figures 26.55a and b). The extremity of the arc system has a needle holder that can be moved in various directions and lowered towards the skull. The biopsy needle can be a 25G-spinal needle, a Hamilton or a Cedan, which is a 2-mm-diameter side-holed brain cannula that includes outer and inner biopsy needles (Figure 26.55b). Both needles show a lateral window. The outer needle has a blunt non-bevelled end. The windows demonstrate the cutting edges to ensure the brain parenchyma is cut properly. The needle is generally connected to a small-volume syringe (10 μL–2 mL) to limit the depression in the brain parenchyma (Figure 26.55b).

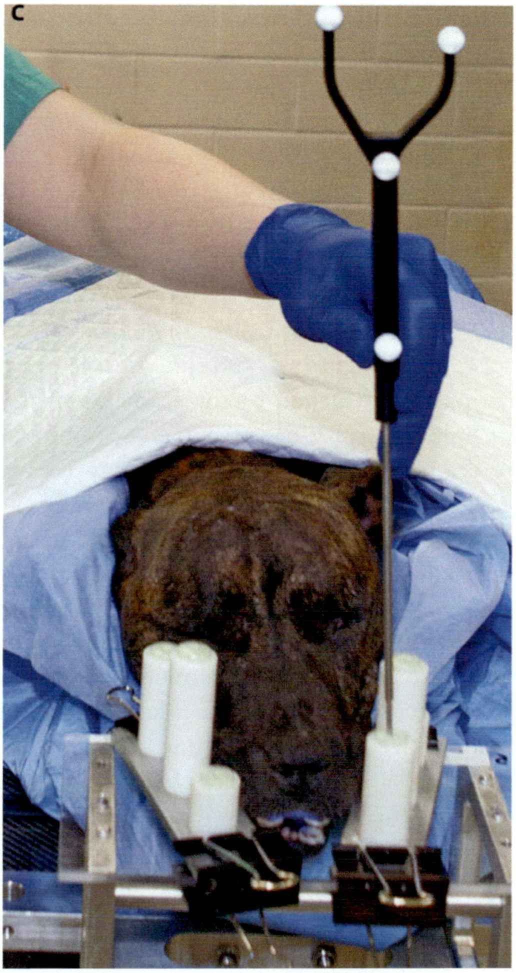

Figure 26.53 Frameless stereotactic CT-guided biopsy device. (Reprint with permission from Taylor et al. 2013.) A bite plate is connected to the maxilla by a dental putty and secured to fiducial markers that allow stereotactic tridimensional localization of intracranial structures.

26.7.7 Biopsy using the different devices

26.7.7.1 Kopf stereotactic instrumentation

The Kopf stereotactic instrumentation, initially used for experimental purposes on laboratory animals, has been modified and made compatible with CT–target localization. This restraint system (Model 1430MRI Kopf stereotactic instrument) is commercially available and does not require additional modification for use in small animals. The frame is made of Lexan® and has been adapted for CT use. The arm bars of the frame have an embedded ruler to allow for precise rostrocaudal positioning of the stereotactic instruments. A stereotactic electrode manipulator attached to the frame's arm bar can hold a variety of instruments, including the craniotomy drill and the biopsy needle. The base of the electrode manipulator is calibrated in two planes and can direct biopsy needles through an oblique trajectory, which penetrates the least amount of normal tissue. The manipulator's electrode vertical and lateral rods have built-in adjustment guides that allow instruments to be placed within 0.1 mm of the desired target. The end of the lateral arm has a "V" groove and locking thread that attach to the craniotomy drills or the electrode holder. The electrode holder holds the needle for biopsy. Troxel et al. recommend a side-cutting biopsy needle with an inner and outer cannula.

The head is secured within the frame with two horizontally adjustable ear bars placed in the external ear canals, with the tip resting against the petrous temporal bone. The maxilla is placed on a vertically adjustable tooth bar that is used as a bite plate, and two adjustable eye bars are placed against the ventral portion of the orbit. A small Vacu-Positioner sand bag is placed within the frame under the head and a vacuum pump is used to withdraw air from the bag, creating a firm conforming mould that limits the ventral displacement of the head.

Continuous transverse CT images of the brain are acquired with the arm bars of the head holder in the field of view. The three-dimensional coordinates are determined using the on-screen measuring tools of the CT workstation. The rostrocaudal coordinate (z) is calculated using the distance between the target location and the centre of the ear bar. Lateral coordinates (x) are calculated by measuring the lateral distance between the top

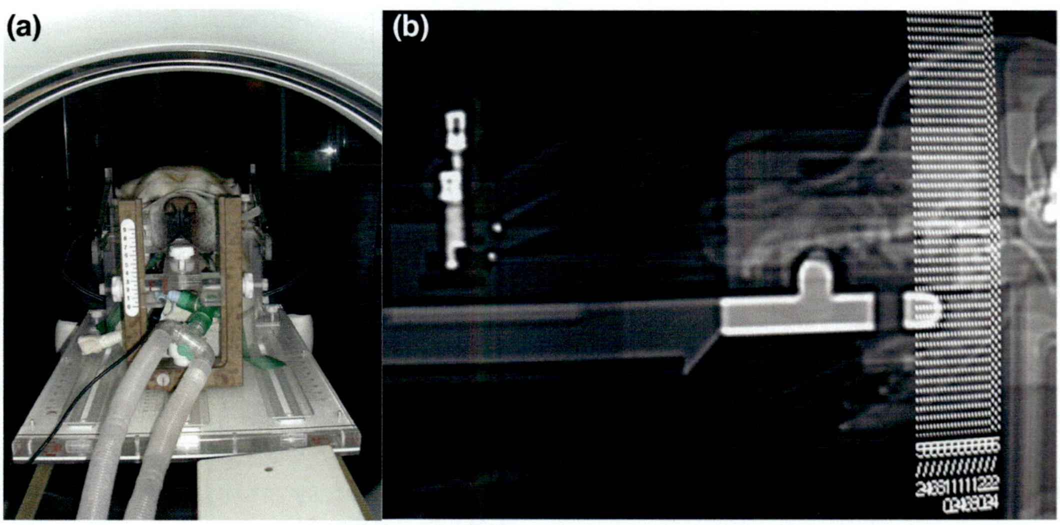

Figure 26.54 (a and b) Head restriction of a Labrador with a frame-base device (Laitinen Modified) for CT examination. The localizer unit is composed by to two triangular localizers fixed to the frame bars (a) that show radiopaque reference points in the rostral extremities of the triangles visible in the scout view before CT (b).

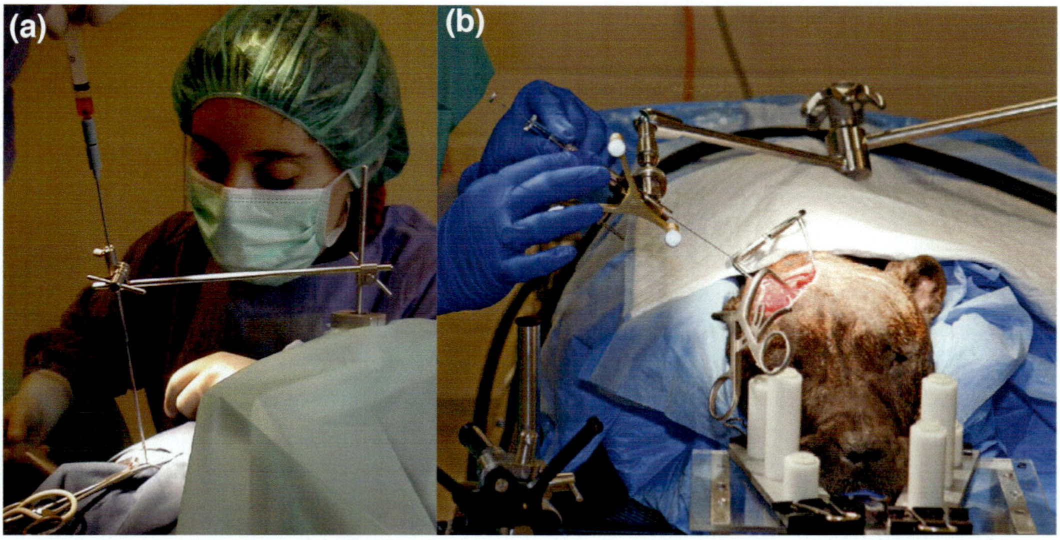

Figure 26.55 **(a and b) Biopsy part of a frame-base device (a) and a frameless device (b Reprint with permission from Taylor et al. 2013.) The arc system allows multiple angles of setting for an appropriate trajectory to the target point. The needle holder secures a 25G-spinal needle (b) or a Cedan needle (a), which can be directed to the targeted parenchyma.**

of the arm bar and the centre of the region of interest. The depth of the biopsy (y) is determined by measuring the vertical distance from the top of the arm bar to the centre of the region of interest.

A small skin incision and a craniotomy are made in reference to the needle trajectory, and the biopsy needle is then pushed into the centre of the suspected lesion.

26.7.7.2 Pelorus Mark III instrumentation (from Koblik et al., 1999a,b)

The modified Pelorus Mark III stereotactic system was one of the first CT-guided devices available for veterinary surgery. Initially designed for human patients, modifications were necessary to accommodate a 90-degree shift in orientation of the canine head compared with the human head during the imaging phase and to facilitate other phases of the biopsy procedure that are affected by the uneven and variable topography of the canine skull.

The Pelorus system utilizes a 5-cm-diameter aluminium skull ring that is secured by four plastic screws to the dog's skull on the caudal aspect of the frontal sinuses and used as a platform to attach various imaging and biopsy devices to the patient. A plastic adaptor also is used to ensure a good fit on the dog's head. AO-spiked washers are placed between the plastic adaptor and the shin to reduce pressure ischaemia. A double-transfer plate with two openings is secured to the aluminium ring.

One opening is used to secure the plastic swivel socket ball socket and guide rod assembly during imaging and while the biopsy needle trajectory is generated. The other opening is used to secure the target swivel socket. The guide rod assembly, which contains a 1-mm radiopaque marker, is attached to the appropriate opening in the double-transfer plate and to the table mounting assembly. In dogs, the skull ring is approximately parallel to the CT patient table.

A lateral scout view image is used to prescribe a set of contiguous 1-mm slices through the guide rod assembly and a set of contiguous 3-mm slices through the brain lesion. It is important to keep the field of view, centring point, and magnification factors the same for both sets of images. The x, y, and z coordinates of the reference marker and the desired biopsy target site are determined using standard software available on the scanner's display console and recorded on a calculation sheet.

Following the CT study, the guide rod assembly is disconnected from the table mounting assembly and the double-transfer plate is detached from the skull ring. The guide rod assembly is then attached to the alignment rod of the phantom frame. The position of the guide rod assembly's reference marker to the coordinate system of the phantom frame is fixed, and the position of the guide rod relative to the transfer plate is not altered. Therefore, the location of the reference marker and biopsy target site in CT space can be translated to the coordinate system of the phantom frame. The x, y, and z offset from the reference marker to the target site are calculated. The z offset is used to set the position of a movable cross-bar, the x offset to set the location of the carrier assembly, and the y offset to set the height of a calibrated pointer within the carrier assembly. The tip of the calibrated pointer represents the target site. A 3-mm drill bit is passed through the guide sleeve, an 18G spinal needle is used to create a small rent in the dura mater, and a side-cutting aspirator biopsy needle is advanced to the prescribed depth. An additional series of 1-mm-thick CT images through the brain lesion area is acquired to confirm the needle tip's position prior to biopsy.

26.7.7.3 Modified Laitinen instrumentation (from Moissonnier 2002)

The Laitinen system, initially designed for human patients, was modified to accept the various shapes and volume of dogs' heads. The frame supports two triangular localizer units, including radiopaque reference points that allow the production of extracranial reference points on CT slices that may be especially related to intracranial targets. CT-guided stereotaxic biopsy is a four-step procedure.

The anaesthetized dog is positioned in ventral recumbency, with the head elevated and secured in the device and submitted for a CT examination (Figure 26.54a).

On the images, three-dimensional coordinates of the mass are measured (x, y, and z) in millimetres. The x coordinate is measured from the outer left triangular localizer unit, the y ventro-dorsally from the x axis, and the z from the rostral reference point of the localizer units (Figure 26.56).

The head is removed from the device and the phantom base positioned between the localizer units. The biopsy needle is pointed to the intracranial target (x, y, and z).

The dog is secured again in the device in the same position, and a burr hole craniotomy is performed. The biopsy needle is introduced through the arc system and lowered into the nervous tissue to the predetermined depth until the needle block-locking device reaches the arc system guide (Figure 26.55a).

Proper positioning of the head in the device is very important for choosing the craniotomy site and the trajectory for biopsy.

26.7.7.4 Biopsy under CT scan (from Giroux 2002)

The stereotactic device is positioned and fixed on the CT table. The head is immobilized in the maxillary support (in the maxillary dental impression) with the maxilla aligned in a sagittal position. A 14-gauge, 6-inch biopsy needle is placed in the needle stabilization fixture so that the needle is level with the base of the ear. Scans of the brain are then performed with a CT-generated radiograph (pilot image) of the head, using a fixture to stabilize the needle. Four to six transverse slices are posted on the pilot image at each of three gantry tilt angles, and the angle that provides the best alignment between the CT slice plane and the needle is recorded and used for subsequent needle

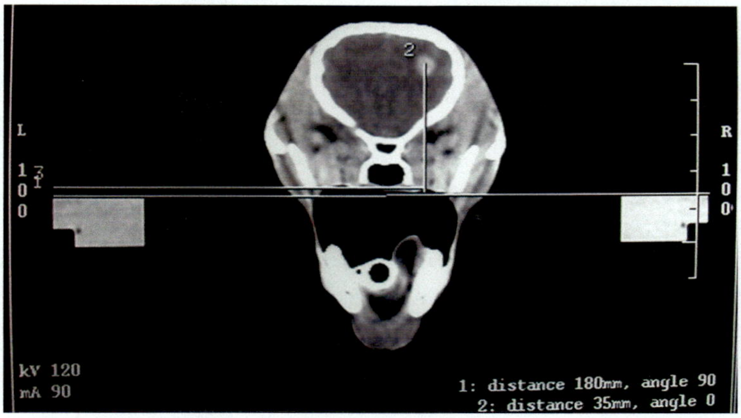

Figure 26.56 **CT scan examination of the head of a Bichon frisé, demonstrating a parenchymal mass submitted to CTGSBB. The three-dimensional coordinates of the mass are measured (x, y, and z) in millimetres. The x coordinate (distance 1) is measured from the outer left triangular localizer unit. The y coordinate (distance 2) is measured ventro-dorsally from the X axis. The Z coordinate is measured from the rostral reference point of the localizer units (Figure 26.54b).**

placement. Best alignment is defined as maximum visualization of the needle lumen and tip in a bone window setting. The true needle tip is identified based on an abrupt and distinct appearance on CT images. Using the chosen gantry tilt angle, additional transverse images of the target region of the brain are obtained. The couch position of the needle is recorded and used as a reference point. The targets are identified from transverse CT images and their couch positions recorded. The difference between the target and needle couch positions is calculated. The x-axis motor of the stereotactic device is used to move the head this distance to centre the target under the needle. Once the target is centred, optimal trajectory and target depth measurements are recorded from images displayed on the CT monitor. The head is rotated using the b-axis motor of the stereotactic device until the optimal trajectory is determined using the criteria of head rotation that is no greater than 30 degrees, minimal surface-to-target distance, and minimal involvement of the lateral ventricles. Window and level settings are adjusted as needed to maximize visualization of the target and needle. Surface-to-target measurements are performed using an electronic cursor and the CT computer's software for distance measurement. For each target, the needle is advanced to the external surface of the skull, and the desired site for craniotomy is marked with a felt tip maker. The head is moved to a couch position outside the gantry, and the craniotomy is performed. The head is repositioned at the previous couch location and the needle advanced to the brain surface. The distance measurement from the exposed brain surface to the target is repeated and recorded as the target depth, which is marked on the needle using a high-precision calliper and fine-tip felt marker. Transverse CT scans of the brain are repeated twice during needle placement to monitor the trajectory of the needle relative to the target. The brain parenchyma can be retrieved at one-half the target depth.

26.7.7.5 Biopsy procedure with a frameless device (from Taylor 2013)

The dog is secured to a stereotactic table using predrilled acrylic bite plates of thermoplastic putty conformed to the dentition. Six fiducial markers are secured to the bite plates. The fiducial images obtained by CT are imported into the stereotactic software. Using the referenced image, the surgeon selects a trajectory for targeting the phantom lesion that will give the most direct dorsal access to the lesion's centre in three planes (x, y, and z) through the least amount of normal brain tissue and will avoid the ventricular system and venous sinuses.

26.7.8 Accuracy of the different systems

The accuracy of the CTGSBB devices currently being used was documented. Overall, significant progress has been made with CT-guided stereotactic brain biopsy in veterinary medicine, but direct comparison shows little variability, usually close to 3 mm, as summarized in Table 26.1.

CTGSBB leads to the removal of satisfactory tissue samples in both human and animal patients (Figure 26.57), although the target is sometimes missed. The major factors leading to poor results from a biopsy procedure include (1) the placement of the needle outside the mass or at its boundaries; (2) targeting an area of the tumour that shows non-specific features of the overall mass; and (3) the tumour's characteristics (size, location, or texture). The first factor correlates to the device's accuracy. The second correlates to the choice of the biopsy target on CT images and the pathologist's ability to differentiate between normal and pathologic brain tissue, and the third depends on the tumour itself. The end of the biopsy needle is either smooth (Cedan), enabling it to penetrate the neural tissue without damaging its vascular component, or sharp (Hamilton), which makes penetration easy but carries a higher risk of a secondary haemorrhage. In the authors' experience, the Cedan needle is safe for most tumours (only one case of severe haemorrhage was seen), but it does not penetrate very stiff tumours such as meningiomas well.

We believe that early assessment of the brain sample by cytological examination is particularly important because it allows the surgeon to perform another biopsy while the dog is still in the operating room. Moreover, when a final diagnosis is reached, the dog's owner can be informed immediately of the neoplasm's histology and the related survival period that will help determine

Table 26.1 **The accuracy of CTGSBB devices in different studies is shown in this overview**

CRITERIA (REF)	FRAME-BASED	FRAMELESS	BIOPSY PERFORMED IN	
			OPERATING ROOM	CT-SCAN ROOM
Needle placement error (Troxel et al., 2008) ten dog and five cat cadaver heads.	(X, Y, and Z errors) 0.9–1.0–1.7 mm DV trajectory error: 0.9±0.9 mm Oblique trajectory error: 1.7±1.6 mm			X
Accuracy in 50 live animals (Koblik 1999a)	3.5±1.6 mm		X	
Target (Koblik 1999b)	49/50 biopsy (98%)			
Ten cadaver head target (Moissonnier 2000),	2.9±1.08 mm		X	
23 owner dogs (Moissonnier 2002)	95% of biopsy			
Sixteen cadaver head target (Giroux 2002).	Caudate nucleus (75%) Pituitary gland (98.6%)			X
Two dogs (Flegel et al., 2002)	Not quantified		X	
LeCouteur (2011)	Centre of the arc system		X	
Ten cadaver heads (Taylor 2013)		4.3 then 2.9 mm (learning)		

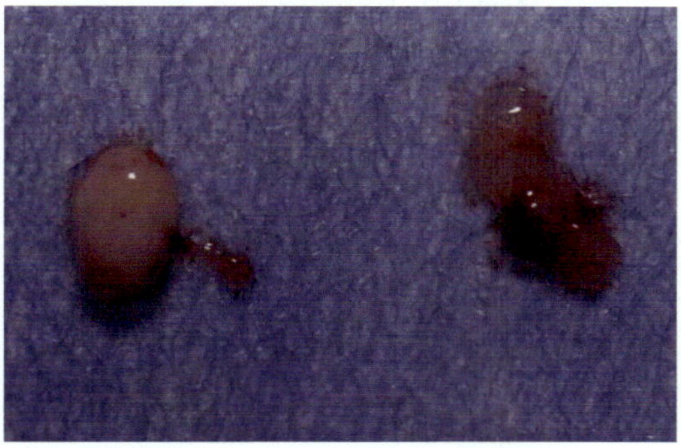

Figure 26.57 **Tissue samples (normal brain left, tumoral tissue right) removed after CTGSBB with a Cedan needle. Early cytological examination of the sample gives the surgeon the opportunity to perform another biopsy while the dog is still in the operating theatre.**

the most appropriate treatment. Because the structural features of brain tumours vary, the characteristics observed in biopsy samples may not represent the whole neoplasm. The reliability of the cytological assessment depends on the cytopathologist's (or the clinician's) ability to make a correct diagnosis from the biopsy sample, usually between 69% and 83% in veterinary patients and 92.2% in human patients. Accurate tumour typing and distinction between reactive and neoplastic astrocytic proliferation are two broad categories that cause difficulties in making a definitive diagnosis. Glial tumour biopsy in human patients leads to a definitive diagnosis in 89% of cases.

26.7.9 Complications associated with CTGSBB

The preoperative condition of dogs with intracranial tumours may be poor, and those who do not receive treatment will likely not survive long. Complications following CTGSBB include haemorrhage, temporary worsening of neurologic deficits, and death. The rates of mortality (7%–8%) and morbidity (approximately 25%) related to the procedure are high compared with those described in retrospective studies of human patients, whose mortality rate varies between 0%, 0.2% or 0.6% and whose morbidity rate was estimated at 3%. The needle diameter (Cedan vs. Hamilton) may be one cause. Also, CTGSBB is performed in the dog without precisely evaluating tumour vascularization, which is only qualitatively assessed by contrast medium affinity on CT examination, whereas in human patients, angiographical data and visualization of the surgical trajectory are incorporated, which enhances the safety and accuracy of the CTGSBB. Increased neurologic deficits and perioperative seizures have also been described in human patients as potential complications resulting from brain biopsy. These transitory complications can be treated in both veterinary and human patients, but biopsy-induced seizures may become uncontrollable and require euthanasia in the dog.

26.7.10 Conclusion

CT-guided stereotactic brain biopsy is a well-standardized procedure in veterinary surgery. Both simple and complex devices are available and are accurate enough to correspond well to clinical situations where large tumours may be submitted for biopsy.

The procedure is accurate, with a low complication rate that is superior to the rate in human patients. The new frameless devices provide similarly accurate information for dogs as the frame-based devices; however, the MRI stereotactic system can more clearly show the exact location of the abnormal brain tissue, particularly when the brain lesions are not easily visible with CT.

SUGGESTED READING

Adamantos S, Garosi L: Head trauma in the cat. 1. Assessment and management of craniofacial injury. *J Feline Med Surg*. 2011; 13: 806–814.

Amengual-Batle P, José-López R, Durand A, et al.: Traumatic skull fractures in dogs and cats: a comparative analysis of neurological and computed tomographic features. *J Vet Intern Med*. 2020; 1–11: 1975–1985.

Apuzzo MLJ, Chandrasoma PT, Cohen D, Zee C-S, Zelman V: Computed imaging stereotaxy: experience and perspective related to 500 procedures applied to brain masses. *Neurosurg*. 1987, 20: 930–937.

Arnold SA, Platt SR, Gendron KP, West FD: Imaging ischemic and hemorrhagic disease of the brain in dogs. *Front Vet Sci*. 2020, 7: 279.

Auriemma E, Barthez PY, Van Der Vlugt-Meijer Rh, et al.: Computed tomography and lowfield magnetic resonance imaging of the pituitary gland in dogs with pituitary-dependent hyperadrenocorticism: 11 cases (2001—2003). *J Am Vet Med Assoc*. 2009; 235: 409–414.

Bar-Am Y, Pollard RE, Kass PH, Verstraete FJM: The diagnostic yield of conventional radiographs and computed tomography in dogs and cats with maxillofacial trauma. *Vet Surg*. 2008; 37: 294–299.

Berger MS: Ultrasound-guided stereotactic biopsy using a new apparatus. *J Neurosurg*. 1986, 65: 550–554.

Brown RA: Simplified CT-guided stereotactic biopsy. *Am J Neurosurg Neuroradiol*. 1981, 2: 181–184.

Carapella CM, Mastrostefano R, Raus L, Ricco A: Methodology of stereotactic biopsy. *J Neurosurg Sci*. 1989; 33: 5–10.

Cerda-Gonzales S, Olby NJ, McCullough S, et al.: Morphology of the caudal fossa in Cavalier King Charles Spaniels. *Vet Radiol Ultrasound*. 2009; 50: 37–46.

Chandrasoma PT, Smith MM, Apuzzo MLJ: Stereotactic biopsy in the diagnostic of brain masses: comparison of results of biopsy and resected surgical specimen. *Neurosurg*. 1989; 24: 160–165.

Chen AV, Wininger FA, Frey S, Comeau RM, Bagley RS, Tucker RL, Schneider AR, Gay JM: Description and validation of a magnetic resonance imaging-guided stereotactic brain biopsy device in the dog. *Vet Radio & Ultrasound*. 2012; 53: 150–156.

Cherubini GB, Mantis P, Martinez TA, et al.: Utility of magnetic resonance imaging for distinguishing neoplastic from non-neoplastic brain lesions in dogs and cats. *Vet Radiol Ultrasound*. 2005; 46: 384–387.

Coffey RJ, Lunsford LD: Animal research stereotactic instrument modified for computed tomographic guidance. *Appl Neurophysiol*. 1987; 50: 81–86.

Cornelis I, Van Ham L, Gielen I, De Decker S, Bhatti SFM: Clinical presentation, diagnostic findings, prognostic factors, treatment and outcome in dogs with meningoencephalomyelitis of unknown origin: a review. *Vet J*. 2019; 244: 37–44.

Dammers R, Haitsma IK, Schouten JW, et al.: Safety and efficacy of frameless and frame-based intracranial biopsy techniques. *Acta Neurochir*. 2008; 150: 23–29.

Davis DH, Kelly PJ, Marsh WR: Computer-assisted biopsy of intracranial lesions. *Appl Neurophysiol*. 1987; 50: 172–177.

De Divitiis E, Spaziante R, Cappabianca P: Reliability of stereotactic biopsy-a model to test the value of diagnoses obtained from small fragments of nervous system tumors. *Appl Neurophysiol*. 1983; 46: 295–303.

De Rycke LM, Gielen IM, Van Meervenne SA, Simoens PJ, van Bree HJ: Computed tomography and cross-sectional anatomy of the brain in clinically normal dogs. *Am J Vet Res*. 2005; 66(10): 1743–1756.

Dickson LR, Dopfmer I, Dalefield RR, Graydon RY, Jolly RD: Method of cerebro-cortical biopsy in lambs. *New Zealand Vet J*. 1989; 37: 22–23.

Dorward NL, Paleologos TS, Alberti O, et al.: The advantages of frameless stereotactic biopsy over frame-based biopsy. *Br J Neurosurg*. 2002; 16: 110–118.

Doward NL, Olaf A, Dijkstra A, et al: Clinical introduction of an adjustable rigid instrument holder for frameless stereotactic interventions. *Comp Aid Surg*. 1997; 2: 180–185.

Eich CS, Whitehair JG, Moroff SD, Heeb LA: The accuracy of intraoperative cytopathological diagnosis compared with conventional histological diagnosis. *J Am Anim Hosp Assoc*. 2000, 36: 16–18.

Fike JR, Cann CE, Turowski K, et al.: Differentiation of neoplastic from non-neoplastic lesions in dog brain using quantitative CT. *Vet Radiol*. 1986; 27: 121–128.

Flegel T, Podell M, March PA, et al: Use of a disposable real-time CT stereotactic navigator device for minimally invasive dog brain biopsy through a mini-burr hole. *Am J Neuroradiol*. 2002; 23: 1160–1163.

Forrest LJ, Schwarz T: Oral cavity, mandible, maxilla and dental apparatus. In: Schwarz T & Saunders J (eds): *Veterinary Computed Tomography*, Chichester: Wiley-Blackwell, 2011, ch 12, p. 111.

Freeman C, Platt S: Head trauma. In: Platt S & Garosi L (eds): *Small Animal Neurological Emergencies*, London: Manson Publishing, 2012, ch 20, p. 363.

Fuchs C, Meyer-Lindenberg A, Wohlsein P, et al.: Computer tomographic characteristics of primary brain tumors in dogs and cats. *Berl Munch Tierarztl.* 2003; 116: 436–442.

Gallagher JG, Penninck D, Bourdrieau RJ: Ultrasonography of the brain and vertebral canal in dogs and cats: 15 cases (1988–1993). *J Am Vet Med Assoc.* 1995; 207: 1320–1324.

Gandini G, Gentilini F, Cimatti L, et al.: Evaluation of the clinical signs and computed tomographic findings in 27 dogs with intracranial space-occupying lesions (1999–2000). *Vet Res Commun.* 2003; 27: 399–401.

Garosi L, Adamantos S: Head trauma in the cat. 2. Assessment and management of traumatic brain injury. *J Feline Med Surg.* 2011; 13: 815–823.

Gielen I, Kromhout K, Gavin P, Van Ham L, Polis I, van Bree H: Agreement between low-field MRI and CT for the detection of suspected intracranial lesions in dogs and cats. *J Am Vet Med Assoc.* 2013; 243: 367–375.

Giroux, A; Jones, JC; Bohn, JH; Duncan, RB; Waldron, DR; Inzana, KR A NEW DEVICE FOR STEREOTACTIC CT-GUIDED BIOPSY OF THE CANINE BRAIN: DESIGN, CONSTRUCTION, AND NEEDLE PLACEMENT ACCURACY. Veterinary Radiology & Ultrasound, 2002; 43; 3; 229-236

Gomes E, Degueurce C, Ruel Y, et al.: Anatomic study of cranial nerve emergence and associated skull foramina in cats using CT and MRI. *Vet Radiol Ultrasound.* 2009; 50: 398–403.

Hall WA: The safety and efficacy of stereotactic biopsy for intracranial lesions. *Cancer*. 1998; 82: 1749–1755.

Harari J, Moore, MM, Leathers CW, Roberts GD, Gavin PR: Computed tomographic-guided, free-hand needle biopsy of the brain tumors in dogs. *Prog Vet Neurol.* 1993; 4: 41–44.

Hariz MI, Berghenheim AT, DeSalles AA, Rabow L, Trojanowski T: Percutaneous stereotaxis brain tumour biopsy and cyst aspiration with a non-invasive frame. *British J Neurosurg.* 1990; 4: 397–406.

Heath RG, John S, Foss O: Stereotactic biopsy. A method for the study of discrete brain regions of animal and man. *Arch Neurol.* 1961; 4: 291–300.

Hecht S, Anderson KM, Castel A, Griffin JFt, Hespel AM, Nelson N, et al.: Agreement of magnetic resonance imaging with computed tomography in the assessment for acute skull fractures in a canine and feline cadaver model. *Front Vet Sci.* 2021; 8: 603775. doi: 10.3389/fvets.2021.603775

Hecht S: Brain. In: Schwarz T & Saunders J (eds): *Veterinary Computed Tomography*. Ames, IA: Wiley-Blackwell, 2011; pp. 197–204.

Heidner GL, Kornegay JN, Page RL, Dodge RK, Thrall DE: Analysis of survival in a retrospective study of 86 dogs with brain tumors. *J Vet Int Med.* 1991; 5: 219–226.

Hikenbottom MD, Grotta J: Neuroprotective therapy. *Semin Neurol.* 1998; 18(4): 485–492.

Jeffrey ND, Brearley MJ: Brain tumours in the dog: treatment of 10 cases and review of recent literature. *J Small Anim Pract.* 1993; 34: 367–372.

Jones TR, Kaplan RT, Lane B, et al.: Single- versus multi-detector row CT of the brain: quality assessment. *Radiol.* 2001; 219: 750–755.

Kim JE, Kim DG, Paek SH, et al: Stereotactic biopsy for intracranial lesions: reliability and its impact on the planning of treatment. *Acta Neurochir.* 2003; 145: 547–555.

Knight R, Meeson RL: Feline head trauma: a CT analysis of skull fractures and their management in 75 cats. *J Feline Med Surg.* 2019; 21(12): 1120–1126.

Koblik PD, LeCouteur RA, Higgins RJ: Modification and application of a Pelorus Mark III stereotactic system for CT-guided brain biopsy in 50 dogs. *Vet Rad Ultrasound*: 1999a; 40: 424–433.

Koblik PD, LeCouteur RA, Higgins RJ: CT-guided brain biopsy using a modified Pelorus Mark III stereotactic system: experience in 50 dogs. *Vet Rad Ultrasound.* 1999b; 40: 434–440.

Kraft SL, Gavin PR, DeHaan C, Moore M, Wendling LR, Leathers CW: Retrospective review of 50 canine intracranial tumors evaluated by magnetic resonance imaging. *J Vet Intern Med.* 1997; 11: 218–225.

Kraft SL, Gavin PR: Intracranial neoplasia. *Clin Tech Small Anim Pract.* 1999; 14: 112–123.

Kromhout K, van Bree H, Broeckx B, Bhatti S, De Decker S, Polis I, Gielen I: Low-field magnetic resonance imaging and multislice computed tomography for the detection of cervical syringomyelia in dogs. *J Vet Intern Med.* 2015; 29(5): 1354–1359.

Kromhout K, van Bree H, Broeckx B, Bhatti S, Van Ham L, Polis I, Gielen IKromhout K, van Bree H, Broeckx BJG, Bhatti S, Van Ham L, Gielen I: Low-field MRI and multislice CT for the detection of cerebellar (foramen magnum) herniation in Cavalier King Charles Spaniels. *J Vet Intern Med.* 2015; 29: 238–241.

Laitinen LV, Liliequist B, Fagerlund M, Eriksson AT: An adaptated for computed tomography-guided stereotaxis. *Surg Neurol.* 1985; 32: 559–566.

LeCouteur RA: Brain biopsy, CT-guided stereotactic. In: Cote E (ed): *Clinical Veterinary Advisor Dogs and Cats*, 2nd ed. St Louis: Mosby Inc., 2011; pp. 1214–1215.

LeCouteur RA: Tumors of the nervous system. In: Withrow SJ & MacEwen EG (eds): *Clinical Veterinary Oncology*, Philadelphia, PA: JB Lippincott, 1989; pp. 325–349.

Leksell L: Stereotactic apparatus in cerebral surgery. *Acta Chir Scand.* 1949; 99: 229–233.

Lim RKS, Liu C, Moffitt RL: *A Stereotaxic Atlas of the Dog's Brain*. Springfield: Thomas, 1960.

Maciunas RJ, Galloway RL: Magnetic resonance and computed tomographic image-directed stereotaxy for animal research. *Stereotact Funct Neurosurg.* 1989; 53: 197–201.

Marino DJ, Loughin CA, Dewey CW, et al.: Morphometric features of the craniocervical junction region in dogs with suspected Chiari-like malformation determined by combined use of magnetic resonance imaging and computed tomography. *Am J Vet Res.* 2012; 73: 105–111.

Marshall LF, Jenett B, Langfitt TW: Needle biopsy for the diagnosis of malignant glioma. *J Am Anim Hosp Assoc.* 1974; 228: 1417–1418.

McConnell F: Imaging of neurological emergencies. In: Platt S & Garosi L (eds) *Small Animal Neurological Emergencies*, London: Manson Publishing, 2012, ch 4, p. 83

Mihara F, Hirakata R, Hasuo K, et al.: Gd-DTPA administered MR imaging of intracranial mass lesions: a comparison with CT and precontrast MR. *Radiat Med.*1989; 7: 227–235.

Moissonnier P, Blot S, Devauchelle P, et al.: Stereotactic CT-guided brain biopsy in the dog. *J Small Anim Pract.* 2002; 43: 115–123.

Moissonnier P, Bordeau W, Delisle F, et al: Accuracy testing of a new stereotactic CT-guided brain biopsy device in the dog. *Res Vet Sci.* 2000; 68: 243–247.

Morandi F: Calvarium and zygomatic arch. In: Schwarz T & Saunders J (eds) *Veterinary Computed Tomography*, Chichester: Wiley-Blackwell, 2011, ch 16, p. 161.

Motta L, Mandara MT, Skerritt GC: Canine and feline intracranial meningiomas: an updated review. *Vet J.* 2012; 192: 153–165.

Mundinger F: CT stereotactic biopsy for optimizing the therapy of intracranial processes. *Acta Neurochir Suppl.* 1985; 35: 70–74.

National Comprehensive Cancer Network (NCCN): Clinical practice guidelines in oncology: central nervous system cancers. 2009. Available at https://www.nccn.org (accessed January 2010).

Niebauer GW, Habil MV, Dayrell-Dayrell BL, Speciale J: Evaluation of craniotomy in dogs and cats. *J Am Vet Med Assoc.* 1991; 198: 89–95.

Ohlerth S, Scharf G: Computed tomography in small animals — basic principles and state of the art applications. *Vet J.* 2007; 173: 254–271.

Parizel PM, Makkat S, Van Miert E: Intracranial hemorrhage: principles of CT and MRI interpretation. *Eur Radiol.* 2001; 11: 1770–1783.

Paul AE, Lenard Z, Mansfield CS: Computed tomography diagnosis of eight dogs with brain infarction. *Aust Vet J.* 2010; 88(10): 374–380.

Platt S, Olby N: Neurological emergencies In: Platt SR & Olby NJ (eds): *BSAVA Manual of Canine and Feline Neurology*, Quedgeley Gloucester, UK: British Small Animal Veterinary Association2013, ch 20, pp. 388–408.

Plummer SB, Wheeler SJ, Thrall DE, Kornegay JN: Computed tomography of primary inflammatory brain disorders in dogs and cats. *Vet Radiol Ultrasound.* 1992; 33: 307–312.

Polizopoulou ZS, Koutinas AF, Souftas VD, Kaldrymidou E, Kazakos G, Papadopoulos G: Diagnostic correlation of CT-MRI and histopathology in 10 dogs with brain neoplasms. *J Vet Med Ser A.* 2004; 51: 226–231.

Pollard RE, Reilly CM, Uerling MR: Cross-sectional imaging characteristics of pituitary adenomas, invasive adenomas and adenocarcinomas in dogs: 33 cases (1988–2006). *J Vet Int Med.* 2010; 24: 160–165.

Porat-Mosenco Y, Schwarz T, Kass PH: Thick-section reformatting of thinly collimated computed tomography for reduction of skull-base-related artifacts in dogs and horses. *Vet Radiol Ultrasound.* 2004; 45: 131–135.

Rangel-Castillo L, Gopinath S, Robertson CS: Management of intracranial hypertension. *Neurol Clin.* 2008; 26(2): 521–541.

Reichert T, Mundinger F: Beschreibung und anwendung eines zeilgerätes für stereotaktische hirnoperationene. *Acta Neurochir (Wien)* 1955; Suppl 3: 308–337.

Robbins PD, Yu LL, Stockes BAR: Steretaxic biopsy of 100 intracerebral mass at Sir Charles Gairdner hospital. *Pathology.* 1995; 26: 410–413.

Schroder H, Meyer-Lindenberg A, Nolte I: Comparative examination of the lateral cerebral ventricles of different dog breeds using quantitative computer tomography. *Berl Munch Tierarztl.* 2006; 119: 506–511.

Schwarz T: Temporomandibular joint and masticatory apparatus. In: Schwarz T & Saunders J (eds): *Veterinary Computed Tomography*, Chichester: Wiley-Blackwell, 2011, ch 13, p. 125.

Snyder JM, Shofer FS, Van Winkle TJ, et al.: Canine intracranial primary neoplasia: 173 cases (1986–2003). *J Vet Intern Med.* 2006; 20: 669–675.

Spiegel EA, Wycis HT, Marks M, Lee AJ: Stereotactic apparatus for operation of the human brain. *Science.* 1947; 106: 349–350.

Sturges BK, Dickinson PJ, Bollen AW, et al.: Magnetic resonance imaging and histological classification of intracranial meningiomas in 112 dogs. *J Vet Intern Med.* 2008; 22: 586–595.

Sutcliffe JC, Battersby RDE: Intraoperative ultrasound-guided biopsy of intracranial lesions: comparison with freehand biopsy. *British J Neurosurg.* 1991; 5: 163–168.

Talairach J, Hecaen H, David M, Mornier M, De Ajuriaguerra J: Recherches sur la coagulation thérapeutique des structures sous-corticales chez l'homme. *Rev Neurol.* 1949; 81: 4–29.

Taylor AR, Cohen ND, Fletcher S, Griffin JF, Levine JM: Computed tomography-guided stereotactic brain biopsy system in dogs. *Vet Radiol Ultrasound.* 2013; 54: 332–342.

Thomas WB, Sorjonen DC, Hudson JA, Cox NR: Ultrasound-guided brain biopsy in dogs. *Am J Vet Res.* 1993; 54: 1942–1947.

Tidwell CS, Chalela JA, SAaver JL: Comparison of MRI and CT for detection of acute intracerebral hemorrhage. *J Am Med Assoc.* 2004; 292: 1823–1830.

Tipold, A: Diagnosis of inflammatory and infectious diseases of the central nervous system in dogs: a retrospective study. *J Vet Intern Med.*1995; 9: 304–314.

Tokgoz N, Oner YA, Kaymaz M, et al.: Primary intraosseous meningioma: CT and MRI appearance. *AJNR Am J Neuroradiol.* 2005; 26: 2053–2056.

Troxel MT, Vite CH, Massicotte C, McLear RC, Van Winkle TJ, Glass EN, Tiches D, Dayrell-Hart B: Magnetic resonance imaging features of feline intracranial neoplasia: retrospective analysis of 46 cats. *J Vet Intern Med.* 2004; 18: 176–189.

Troxel MT, Vite CH: CT-guided stereotactic brain biopsy using the Kopf stereotactic system. *Vet Rad Ultrasound.* 2008; 49: 438–443.

Vali Y, Gielen I, Soroori S, Ludewig E: The diagnostic value of intravenous contrast computed tomography in addition to plain computed tomography in dogs with head trauma. *Bmc Vet Res.* 2021; 17(1). https://doi.org/10.1186/s12917-021-02764-6

Van Der Vlugt-Meijer RH, Meij BP, Van Den Ingh TS, et al.: Dynamic computed tomography of the pituitary gland in dogs with pituitary-dependent hyperadrenocorticism. *J Vet Int Med.* 2003; 17: 773–780.

Vernau KM, Kortz GD, Koblik PD, LeCouteur RA, Bailey CS, Pedroia V: Magnetic resonance imaging and computed tomography characteristics of intracranial intra-arachnoid cyst in 6 dogs. *Veti Radiol Ultrasound.* 1997; 38: 171–176.

Willis C, Lybrand S, Bellamy N: Excitatory amino acid inhibitors for traumatic brain injury. *Cochrane Database Syst Rev.* 2004; CD003986.

Wolff C, Holmes SP, Young BD, et al.: Magnetic resonance imaging for the differentiation of neoplastic, inflammatory and cerebrovascular brain disease in dogs. *J Vet Intern Med* 2012; 26: 589–597.

Young BD, Levine JM, Porter BF, et al.: Magnetic resonance imaging features of intracranial astrocytomas and oligodendrogliomas in dogs. *Vet Radiol Ultrasound.* 2011; 52:132–141.

Zarelli M, Schwarz T, Puggioni A, Pinilla M, O'Doherty JV, McAllister H: An optimized protocol for multislice computed tomography of the canine brain. *Vet Radiol Ultrasound*. 2014; 55(4): 387–392.

Zarelli M, Shiel R, Gallagher B, Skelly C, Cahalan S, McAllister H: Imaging diagnosis: CT findings in a dog with intracranial hemorrhage secondary to angiostrongylosis. *Vet Radiol Ultrasound*. 2012; 53(4): 420–423.

CHAPTER 27

THE USE OF INTRAOPERATIVE- AND FOLLOW-UP CT SCAN

Ingrid Gielen and Henri van Bree

The use of *intraoperative CT* in taking biopsies under CT guidance was already mentioned in the chapter dealing with this issue. *Follow-up CT* can be useful in the postoperative evaluation of surgical interventions including spinal and orthopaedic surgery as well as in the follow-up of several diseases. In certain cases, CT is used to determine healing and if a certain bone has healed completely.

27.1 ORTHOPAEDIC SURGERY

In orthopaedic surgery, the main objectives of intraoperative CT are the confirmation of anatomic reduction and accurate placement of implants. Metallic implants and instruments cause artefacts that may obscure the area of interest (Figure 27.1). There is sufficient evidence that the application of intraoperative imaging leads to precise implant positioning and improves the radiological outcomes (Figure 27.2). In horses, it has been reported that CT is used before surgery for planning, during the procedure to guide surgical steps, and afterwards to review if the results of the surgery are useful.

Follow-up CT can be used in the postoperative evaluation of OCD lesions. The diagnosis of elbow dysplasia and fragmented medial coronoid process is based on clinical examination and early osteoarthritic changes seen on x-rays. Should changes not been seen on x-rays, then either CT scan or arthroscopy may be required. In case of deterioration or persistence of lameness, post-operative CT can be useful, especially in these cases where fragmentation

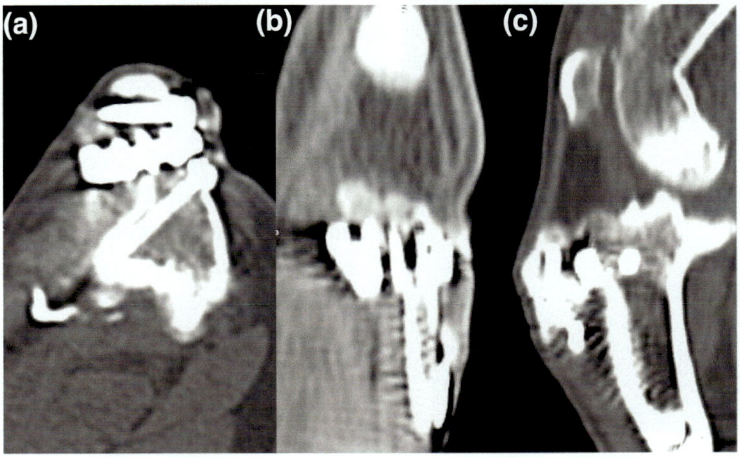

Figure 27.1 Transverse image (a) and dorsal (b) and sagittal (c) reconstructed images in bone algorithm of a postoperative study after a tibial tuberosity advancement surgery procedure of a stifle joint of a dog with cranial cruciate ligament rupture. The dog was responding favourably after the procedure. Metallic implants cause artefacts obscuring the area of interest.

DOI: 10.1201/9781315118130-29

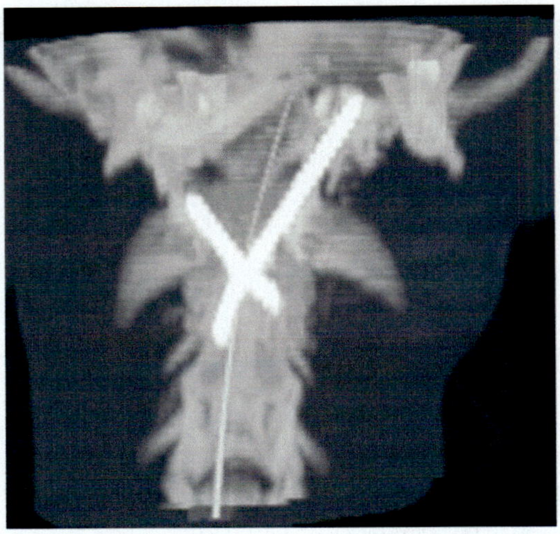

Figure 27.2 Maximum intensity projection (MIP) CT-image after atlanto-axial stabilization surgery. Precise implant positioning can be appreciated immediately after the surgery.

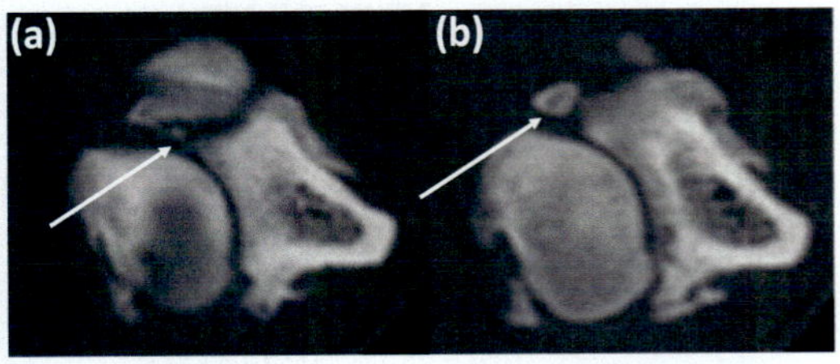

Figure 27.4 Transverse images in bone algorithm of an 11-month-old male Labrador. (a) shows a small displaced fragment of the medial coronoid process (arrow). This fragment was arthroscopically removed. Already considerable DJD is present at this age. Three years postoperatively, the dog was presented again with lameness, and a new larger fragment was noticed (arrow) (b). This new fragment was also arthroscopically removed, and the dog responded well to the surgery. Notice that after 3 years, the DJD shows practically no progression.

has been undetectable during the initial elbow arthroscopy. Also, in cases with multiple fragments, CT post-op can be useful to evaluate if all fragments have been removed (Figure 27.3). Here, one should differentiate between scar tissue and real fragments (Figure 27.4). In cases of recurrent lameness after a surgical procedure, follow-up CT can be useful in determining its cause (Figure 27.5).

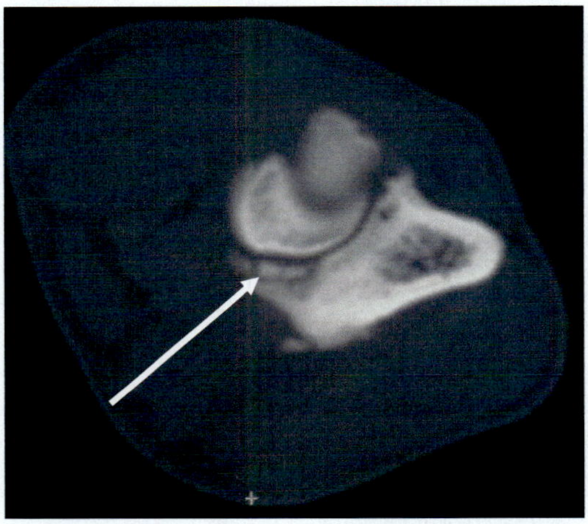

Figure 27.3 Transverse image in bone algorithm of an elbow with a fragmented medial coronoid process. Three small fragments can be appreciated in between the radius and ulna (arrow). In these cases, postop CT can be useful to evaluate if all fragments have been removed.

Patients admitted with tarso-crural OCD often present with multiple fragments. In these patients, a post-op CT can be useful to check if all fragments have been removed (Figure 27.6).

27.2 SPINAL SURGERY

Dorsal laminectomy and partial discectomy of the lumbosacral disc are common surgical treatments for decompression of the cauda equina and spinal nerves within the vertebral canal. Studies in dogs that report postoperative outcomes include complications such as recurrent lumbosacral pain due to scar tissue-induced neural compression, recurrent disc herniation, and fractures involving an intervertebral joint (Figure 27.7). Fractures of the articular processes as a post-operative complication after laminectomy are reported in a few cases in veterinary medicine (Figure 27.8). It is possible that the extension of the laminectomy can result in weakening of the articular process or instability of the intervertebral joint and therefore predisposed to fracture development.

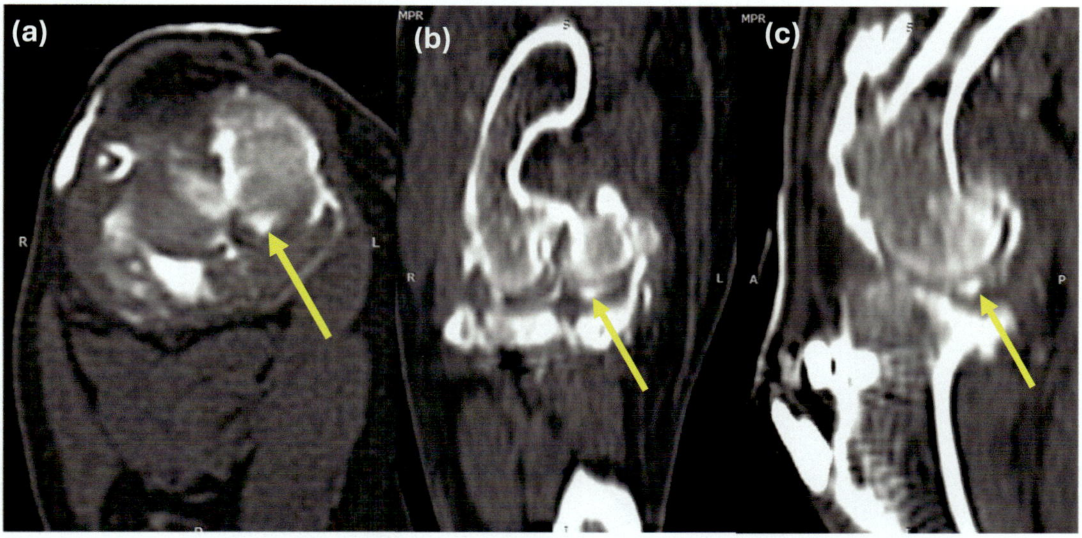

Figure 27.5 Transverse (a), dorsal (b), and sagittal (c) reconstructions of an arthro-CT study, postoperatively, after a TTA procedure of a stifle joint of dog with a cranial cruciate ligament rupture. This dog showed recurrent lameness. The arthro-CT images display a minimal amount of contrast medium extending into the meniscal defect (arrows). A tear in the caudal pole of the medial meniscus is observed.

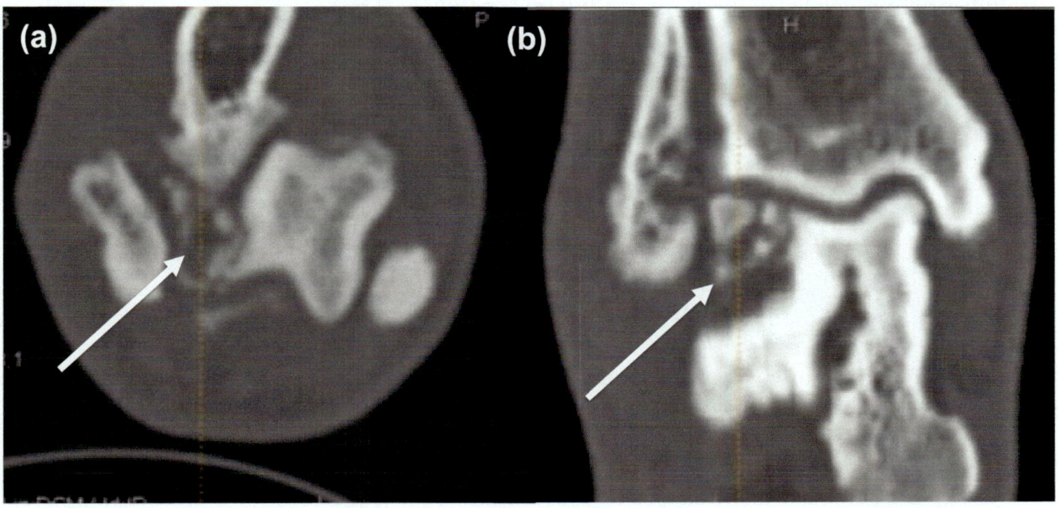

Figure 27.6 Transverse (a) and dorsal (b) reconstructions of a tarso-crural OCD with multiple fragments of the lateral ridge (arrows). In these cases, a post-op CT can be useful to check if all fragments have been removed.

One recent study indicates that imaging signs of neural compression are common after degenerative lumbosacral stenosis (DLSS) surgery, even in dogs showing clinical improvement.

Computed tomography imaging can be used to evaluate various surgical techniques and for follow-up; for example, when a ventral slot procedure is performed, CT can evaluate parameters such as slot depth, width, and angulation, which can be useful for detecting early complications (Figure 27.9). In case of disc herniation, recurrent pain can be due to insufficient removal of disc material, recurrent disc herniation, or abundant scar tissue formation (Figure 27.10).

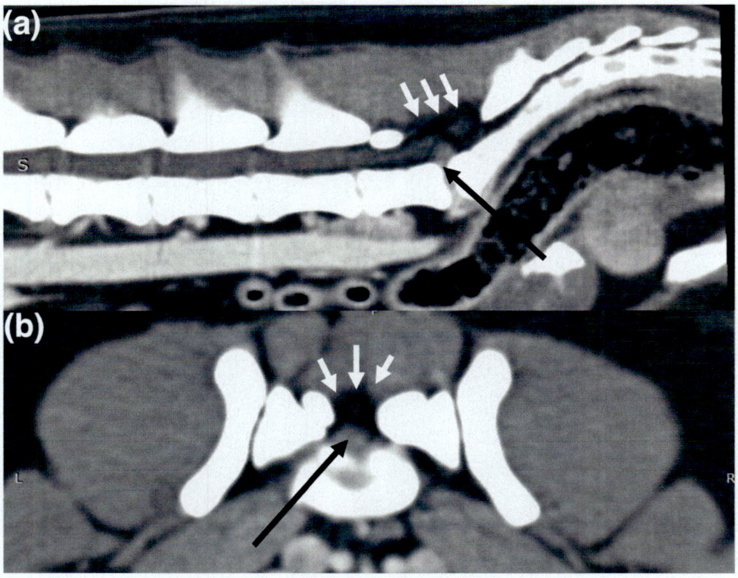

Figure 27.7 **Sagittal reconstruction (a) and transverse image (b) of a post-operative dorsal laminectomy. The dog was admitted with recurrent lumbosacral pain. The surgical site where a dorsal laminectomy has been performed can be appreciated (small white arrows). A huge protrusion of the LS disc compressing the cauda equina at the sagittal on the right side is present (large black arrows).**

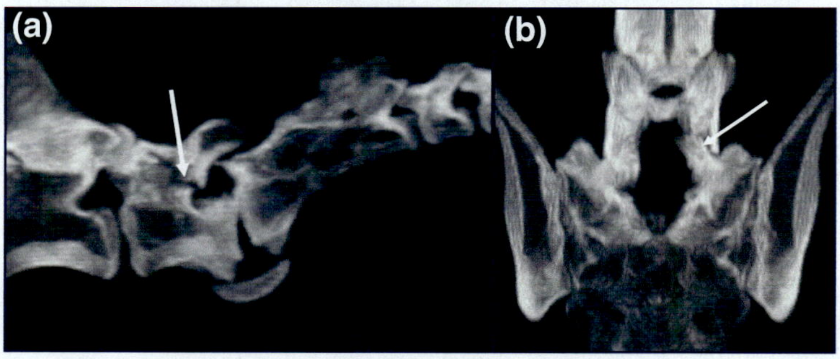

Figure 27.8 **Maximum intensity projection (MIP) CT images, sagittal (a) and dorsal (b) of a fracture of the left articular process (arrows) as a post-operative complication after laminectomy.**

CT imaging is useful following surgery to document whether the extruded disc material is adequately removed. If only a small amount of the material, or none at all, is found, CT scanning can show if the correct site was decompressed, more disc material was present on the opposite side of the spinal cord, or if the material was missed during surgery (Figure 27.11).

Differentiation between recurrent disc herniation and epidural fibrosis is important.

Reoperation to remove epidural fibrosis is considered contraindicated in humans because surgery would cause formation of more scar tissue, whereas surgery is indicated in cases of significant recurrent disc herniation. Besides the remaining disc material, compression post-operatively could be caused by haemorrhage or fat graft (Figures 27.12 and 27.13). The surgical site infection (SSI) rate in dogs undergoing hemilaminectomies or laminectomies for thoracolumbar disc herniation or lumbosacral disease, without use of perioperative antibiotics, is 0.6%, while the expected SSI rate in clean operative wounds in dogs and cats is 2.0%–4.8% (Figure 27.14). The SSI rate in human spinal surgery is 0.7%–4.3%.

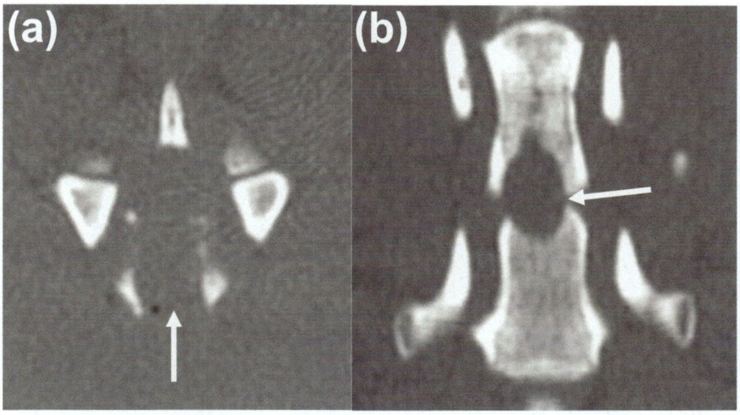

Figure 27.9 Early post-operative CT images, transverse (a) and dorsal reconstruction (b), of a ventral slot in bone algorithm at the level of C4–C5 (white arrows). Parameters such as slot depth, width, and angulation can be evaluated.

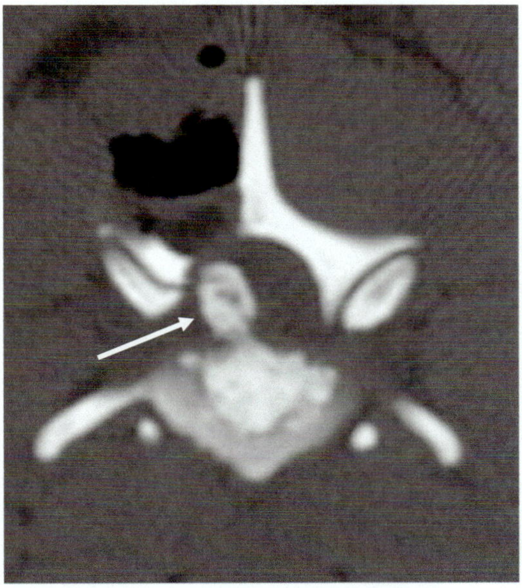

Figure 27.10 Transverse image in bone algorithm at the level of C2–C3 after mini-hemilaminectomy. On this early post-operative image, residual disc material is still present at the right parasagittal side (arrow).

27.3 BRAIN LESIONS

In dogs with intracranial, space-occupying lesions, CT can be used for the follow-up evaluation of chemotherapy treatment. Mass size and accompanying oedema can be evaluated (Figure 27.15). In case of surgical removal of the space-occupying mass, a follow-up CT is helpful to check if the process is completely removed or not (Figure 27.16).

Head trauma with resultant traumatic brain injury (TBI) is associated with a high risk of morbidity and mortality in small animals. CT is the imaging modality of choice for initial assessment of TBI in humans, and scoring systems have been proposed in order to guide prognosis. CT has previously been suggested as a valuable tool in all cases of canine head trauma. Imaging features recorded include presence of midline shift, intracranial haemorrhage, brain herniation, skull fractures, and percentage of total brain parenchyma affected (Figure 27.17). Although CT is a valuable tool in the detection of structural abnormalities, a recent study failed to identify any early CT imaging features with prognostic significance in canine TBI patients.

In humans, the use of follow-up CT in cases of chronic subdural haematoma remains controversial. The radiologic evidence of a second haematoma often leads to reoperation, even if the patient has no or minimal symptoms. Some people argue that a follow-up with routine CT scans may be harmful due to additional and maybe unnecessary surgeries and hospital days in a patient population

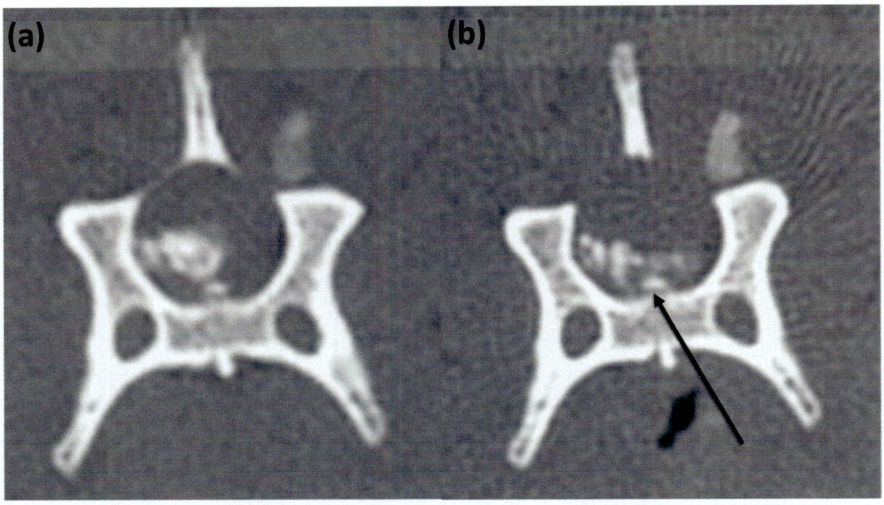

Figure 27.11 Pre- (a) and post -(b) operative transverse images in bone algorithm at the C4–C5. After the surgery, new disc material or displacement of old material is noticed (black arrow).

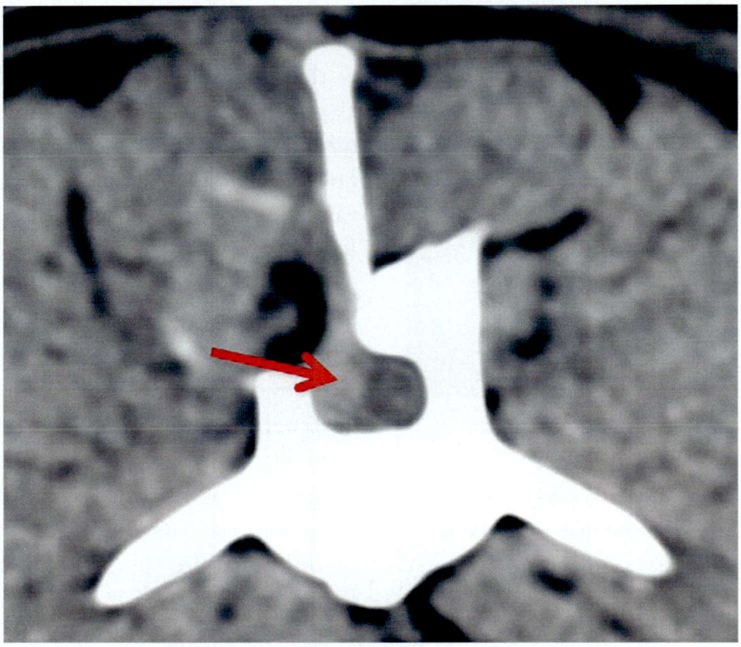

Figure 27.12 Transverse post-contrast image in soft tissue algorithm at the L2–L3 after hemilaminectomy. This early postoperative image shows a hyperdense aspect at the right lateral side (red arrow), consistent with an acute bleeding, causing extradural compression on the spinal cord.

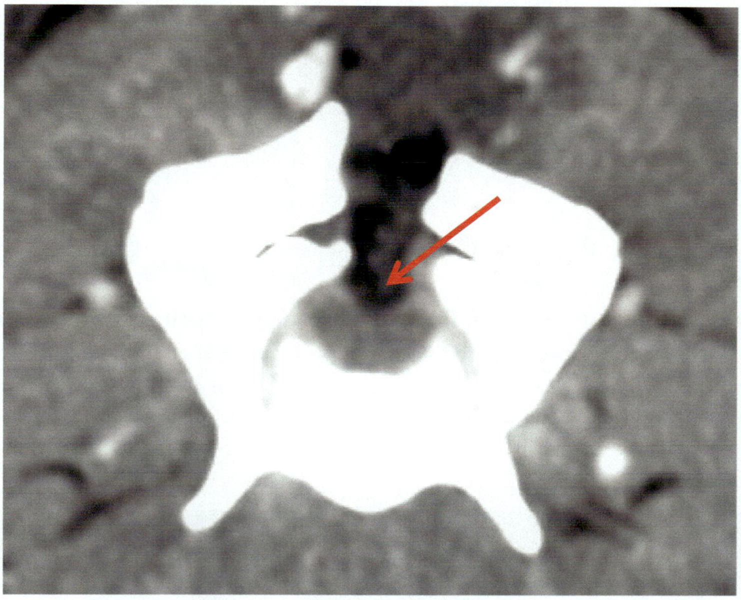

Figure 27.13 Transverse post-contrast CT-image in soft tissue algorithm at the level of C2–C3 early after laminectomy. A hypodense aspect (red arrow) with fat density (fat graft) causes dorsal compression on the spinal cord.

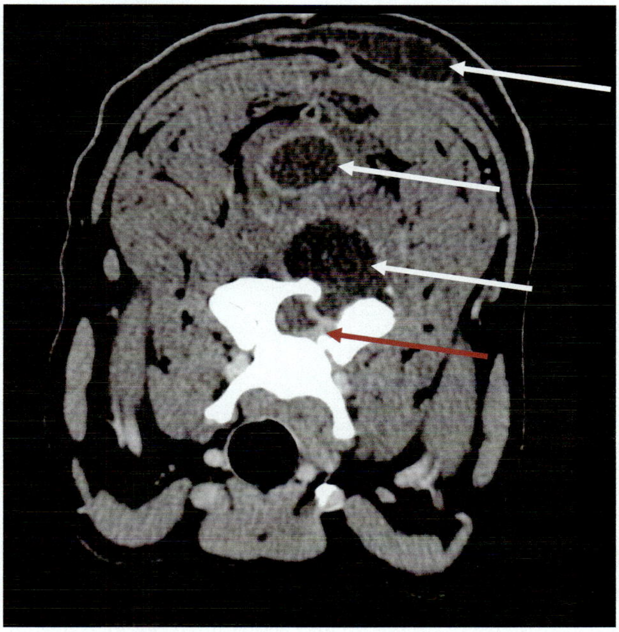

Figure 27.14 Transverse post-contrast image in soft tissue algorithm at the level of C6–C7 at a longer postoperative period after hemilaminectomy. Fluid-filled pockets are present in the surrounding muscles (white arrows). There is left dorsolateral compression on the spinal cord visible on the surgical site (red arrow).

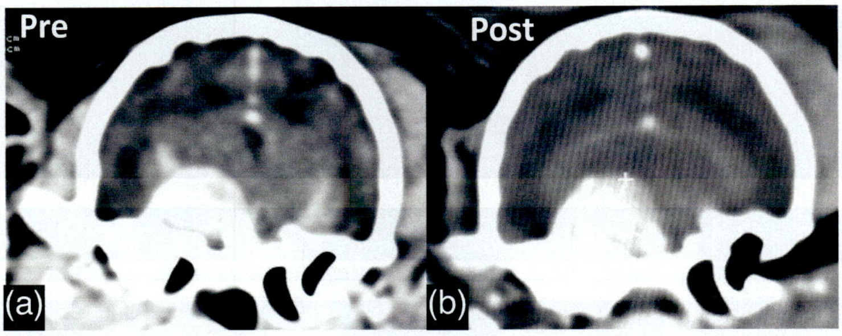

Figure 27.15 Pre- (a) and post-treatment (b) transverse images, both post contrast, in soft tissue algorithm of a cat with a meningioma treated with symptomatic and chemotherapy. There is no reduction in tumour size visible. The oedema and mass effect seen are reduced post-treatment. The cat showed clinical improvement after treatment.

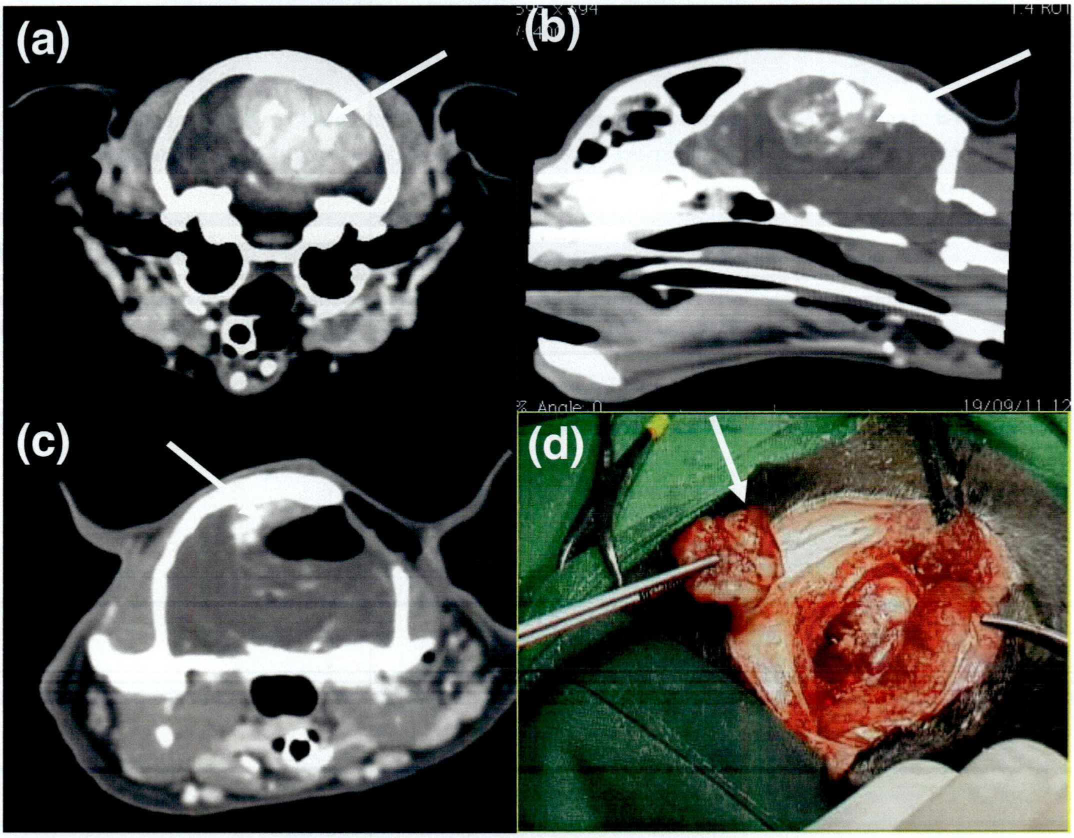

Figure 27.16 Pre surgical transverse image (a) and sagittal (b) reconstruction in post-contrast soft tissue algorithm of a cat with a large meningioma. On the post-surgical transverse image (c), it can be appreciated that the mass was not completely removed (arrows). (d) shows the size of the mass *in situ*.

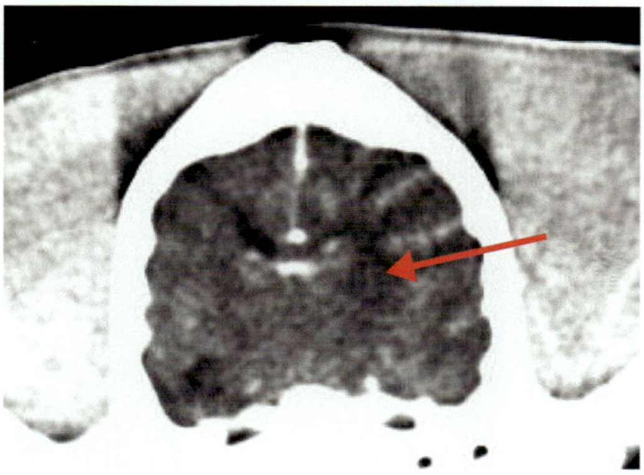

Figure 27.17 Transverse post-contrast image in soft tissue algorithm of a dog with brain contusion after head trauma. An important midline shift toward the right side is observed. There is no fracture nor cerebral bleeding noticed. Oedema and high intracranial pressure are present (arrow).

characterized by advanced age and fragility. In veterinary medicine, information on this issue is scarce.

FURTHER READING

R. Bergman et al. Post-operative computed tomography in two dogs with cerebral meningioma. *Veterinary Radiology & Ultrasound*. 2000;41(5):425–432.

B. A. R. Dyall, H. G. Schmökel Surgical site infection rate after hemilaminectomy and laminectomy in dogs without perioperative antibiotic therapy. *Veterinary and Comparative Orthopedics and Traumatology*. 2018;31:202–213.

K. O'Shea et al. The usefulness of computed tomography following open reduction and internal fixation of acetabular fractures. *Journal of Orthopaedic Surgery*. 2006;14(2):127–132.

R. A. R. Perrin et al. The use of computed tomography to assist orthopaedic surgery in 86 horses (2002—2010). *Equine Veterinary Education*. 2011;23(6):306–313.

K. L. Perry. A retrospective study of the short-term complication rate following 750 elective elbow arthroscopies. *Veterinary and Comparative Orthopedics and Traumatology* 2014;27:68–73.

M. Rapp et al. Postoperative computed tomography and low-field magnetic resonance imaging findings in dogs with degenerative lumbosacral stenosis treated by dorsal laminectomy. *Veterinary and Comparative Orthopedics and Traumatology* 2017;30:143–152.

A. D. Schweitzer et al. Traumatic brain injury: Imaging patterns and complications. *RadioGraphics*. 2019;39:1571–1595.

J. G. Teplick et al. Computed tomography of the postoperative lumbar spine. *American Journal of Neuroradiology*. 1983;4:1053–1072.

S. Van Meervenne et al. Comparison between symptomatic treatment and lomustine supplementation in 71 dogs with intracranial, space-occupying lesions. *Veterinary and Comparative Oncology*. 2014;12(1):67–77.

K. Vishal et al. Does the use of intraoperative CT scan improve outcomes in Orthopaedic surgery? A systematic review and meta-analysis of 871 cases. *Journal of Clinical Orthopaedics and Trauma*. 2021;18:216–223.

S. Wyatt et al. Early CT in dogs following traumatic brain injury has limited value in predicting short-term prognosis. *Veterinary Radiology and Ultrasound*. 2020, 62;1–9.

COMPUTED TOMOGRAPHY PROTOCOL, SCAN PARAMETERS & POSITIONING

CT scan positioning and protocols differ based on the body region being examined and the clinical purpose of the scan. Patients are positioned in dorsal recumbency and enter the scanner head first. However, other body region examinations may require ventral positioning or feet-first entry to optimize image quality or patient comfort. In veterinary CT, patient positioning is the most commonly overlooked factor in CT protocols, although it has a great impact on the image quality. Iso-centring will help you achieve the best results. Good iso-centring will result in consistent image quality and lower doses of radiation. When positioning a patient, it's imperative that the body is not rotated or skewed in either direction. Ensure the scanned region of the patient is placed straight and symmetrically positioned on the midline of the CT table. Changes in obliquity may make structures appear abnormal when they are not. Use positional aids, sponges, or troughs if needed, and ensure that all metallic or other unnecessary objects are removed. Place the patient either head-first or hindlimb-first into the gantry, depending on which will be closest to the region for imaging.

A1 CT BRAIN

Patient positioning should be in dorsal or sternal recumbency, with the head extended and the hard palate parallel to the CT table. Head elevation is recommended to ensure continued extension (Figure A1). Dorsal positioning with

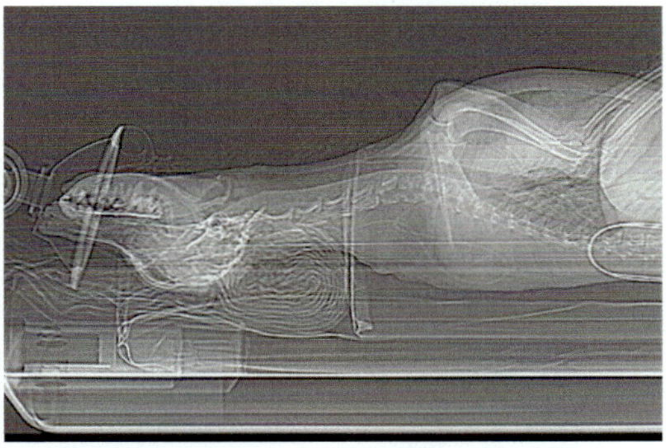

Figure A1 **Lateral scout view for head and brain examination with the head extended and elevated and the hard palate parallel to the CT table.**

the fore limbs fixed parallel to the thorax provides easy access to the cephalic vein in order to administer IV contrast (Figure A2). Studies should extend from the cribriform plate to the second to third cervical vertebra. The incidence of metal objects in monitoring devices in the scan field can lead to artifacts

Transverse slices perpendicular to the hard palate in standard or soft tissue and in bone algorithm are obtained. A tube voltage of 120 kVp with mA 120–150 is used with the FOV (a minimum of 12 cm is recommended) adapted. Thin slices (1–2 mm) in bone algorithm can be reformatted to thicker slices (2–4 mm) in soft tissue/standard algorithm for interpretation. These soft tissue or standard images result in better quality showing less noise and grittiness. Additionally, the reduction in artefacts will less hinder the evaluation of the caudal fossa. Scanning interleaved or with 50% overlap has several benefits. The multiplanar reformatting

PARAMETERS						
KVP	MA	THICKNESS (MM)	INTERVAL (MM)	ALGORITHM/ KERNEL	POSITIONING	CONTRAST
120 120	140 140	1.25 mm 2–4 mm	0.6 50% overlap or increment	Bone Soft tissue/ standard	Ventral or dorsal recumbency Dorsal recumbency: forelimbs symmetrically extended caudally. Symmetrical and non-oblique positioning of the head.	IV 2 mL/kg of 300–350 mg I/mL in bolus. Repeat the contrast scan in case of delayed enhancement.

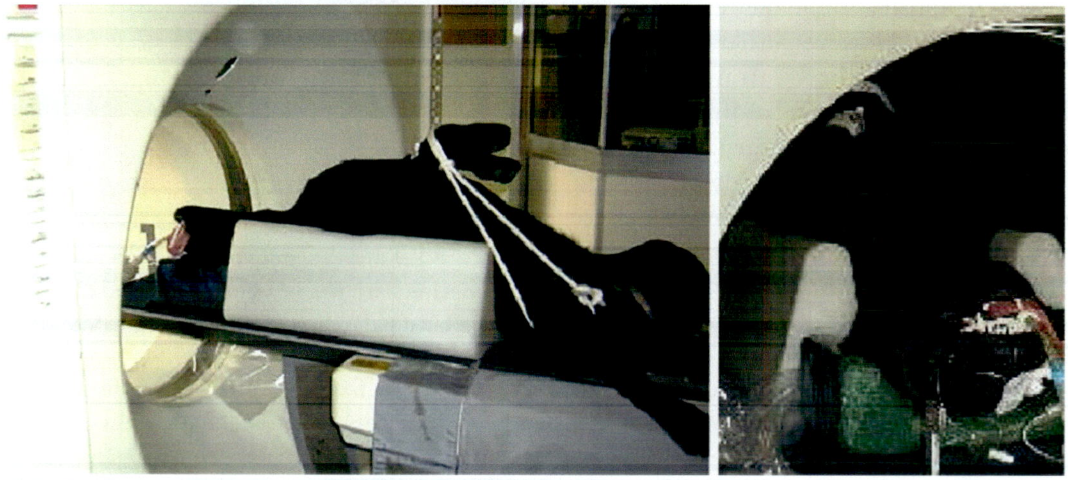

Figure A2 **Positioning in head and brain CT. Dorsal positioning with the forelimbs fixed parallel to the thorax provides easy access to administer IV contrast to the cephalic vein. Position aids are helpful to elevate the head and non-oblique positioning.**

(MPR) imaging is proven to be valuable regarding the accurate localization of lesions. IV contrast is administered at 2 mL/kg of 300–350 mg I/mL in bolus. Scan delay is 1–1.5 minutes after the administration of the bolus. If there is suspicion of pathology, however, not well visible, the study should be repeated after 1 minute as some lesions have a delayed wash-in and wash-out. In the pituitary gland, the choroid plexus, and the meninges, contrast enhancement is considered physiological.

A2 CT SPINE

The patient should be positioned in dorsal recumbency. It is advised to pay attention to straight non-oblique positioning or the vertebral column.

Cervical spine

- Position the patient with a slight elevation of the head to maintain physiological cervical extension (Figure A3).
- Secure the forelimbs caudally using non-elastic tape to minimize movement and ensure alignment (Figure A4).

Thoracolumbar spine

- Position the patient with forelimbs extended cranially and tied using tape to stabilize the thoracic region (Figures A5 and A6).

	PARAMETERS						
	KVP	MA	THICKNESS (MM)	INTERVAL (MM)	ALGORITHM/ KERNEL	POSITIONING	CONTRAST
Small (<10 kg)	100	80–120	1–1.25 mm	0.6	Bone	Dorsal recumbency	IV 2 mL/kg of 300–350 mg I/mL in bolus. Repeat the contrast scan in case of delayed enhancement.
Medium (10–25 kg)	120	120	2–3 mm	50% overlap or increment	Soft tissue/ standard		
Large (>25 kg)	140	140					

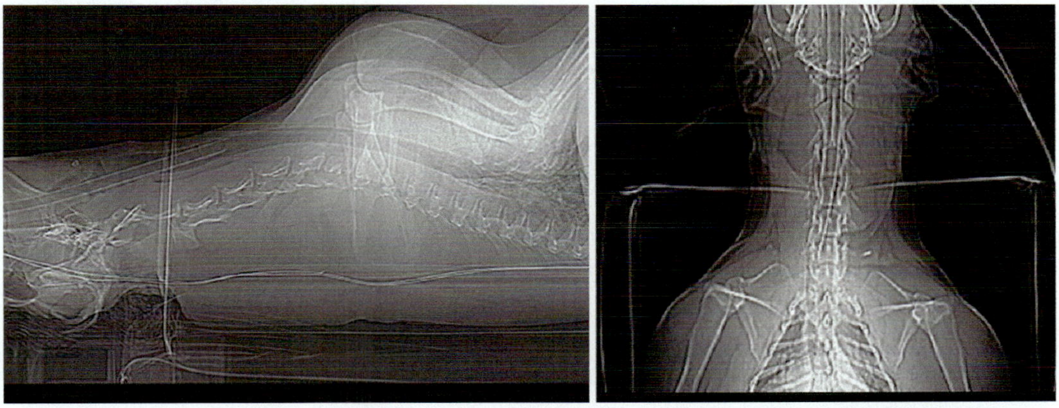

Figure A3 Lateral and dorsal scout views of cervical CT positioning. The head is slightly elevated with the forelimbs caudally extended.

Lumbosacral spine

- Same position as the ThL spine.
- The hind limbs should be tied caudally (Figure A6).
- Frog position (causing artefacts) should be avoided.

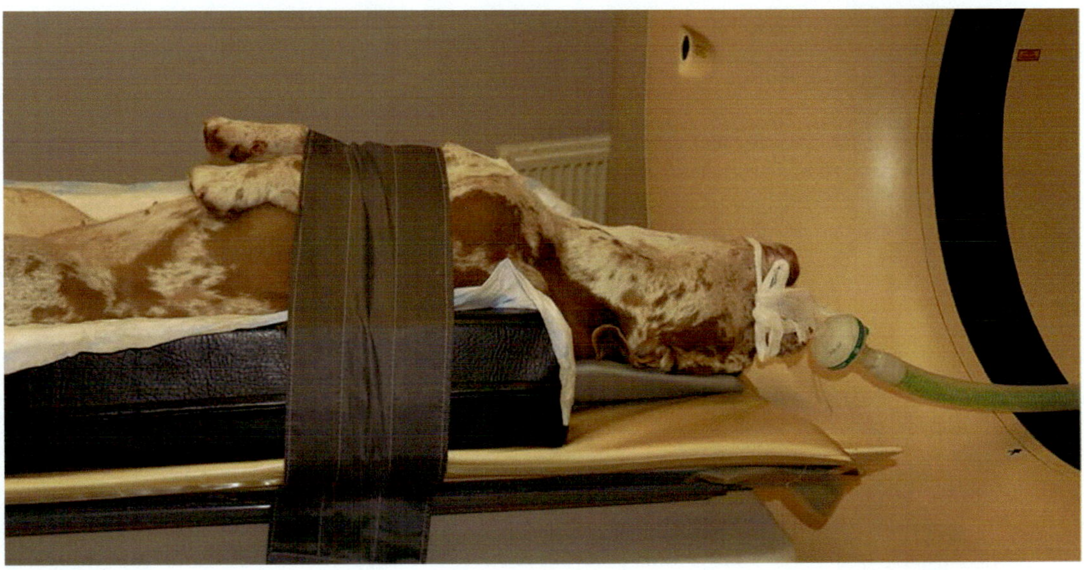

Figure A4 The patient is positioned stably and in a neutral, non-oblique alignment with the front limbs caudally for examination of the cervical spine.

PARAMETERS

KVP	MA	THICKNESS (MM)	INTERVAL (MM)	ALGORITHM	POSITIONING	CONTRAST
120	140	1.25	0.6	Bone	Dorsal recumbency with symmetrically extended forelimbs.	IV 2 mL/kg of 300–350 mg I/mL in bolus. CT myelography: 200 mg I/mL 0.2 mL/kg.
120	140	2	1	Soft tissue/ standard		

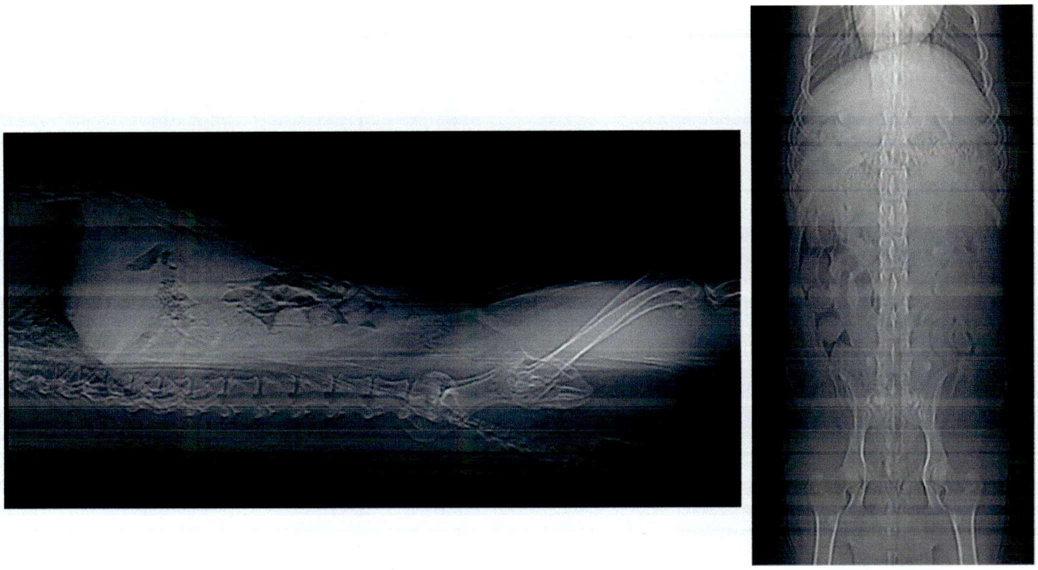

Figure A5 Lateral and dorsal scout views presenting the positioning of the thoracolumbar and LS CT. The forelimbs are extended cranially, and the hind limbs should be tied caudally. The frog position should be avoided.

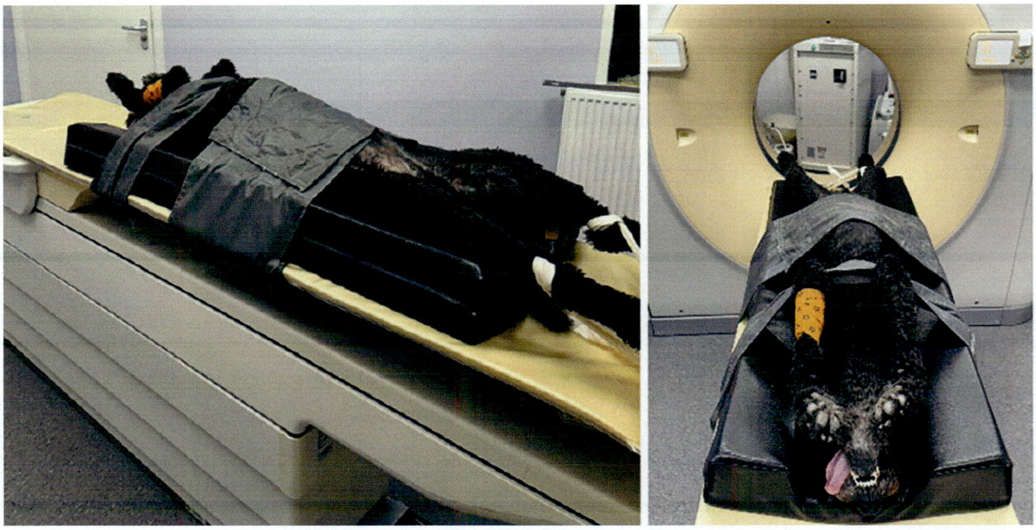

Figure A6 A patient positioned for vertebral column CT. Forelimbs are extended cranially, and hindlimbs are gently secured caudally. This position ensures a stable, symmetrical, and non-oblique alignment of the spine and minimizes movement.

PARAMETERS

KVP	MA	THICKNESS (MM)	INTERVAL (MM)	ALGORITHM	POSITIONING	CONTRAST
120 120	120-140 120–140	1–1.25 1–2	0.5–0.6 0.5–1	Bone Soft tissue/standard	Ventral or dorsal recumbency with the forelimbs symmetrically extended cranially.	IV 2 mL/kg of 300–350 mg I/mL in bolus in case of soft tissue pathology. IA 2–4 mL of 80 mg I/mL.

Slice thickness, always with 50% overlap, in cats and small dogs is about 1 mm and 2 to 3 mm for larger dogs. Thin slices (1–2 mm) in bone algorithm can be reformatted to thicker slices (2–3 mm) in soft tissue/standard algorithm for interpretation. For an overview of any spinal condition, 2 mm slice thickness is used for cats and small dogs and 3–5 mm for large dogs. In cases of discrete disc pathology and lumbosacral disease, 1-1.25 mm slices in bone and soft tissue algorithm are recommended. In cats and small dogs, 100 kVp and about 120 mA are used, while 120 kVp to 140 mA (based on the body volume) are required for larger dogs. IV contrast is useful in the majority of spinal pathologies. The study should extend through necessary vertebral regions based on pain and/or neurolocalization.

A3 CT PLEXUS BRACHIALIS/ LUMBOSACRALIS

The animal needs to be positioned straight and non-oblique in dorsal recumbency with symmetrically and cranially extended forelimbs (cfr. the thoracolumbar spine). Extended hindlimbs are recommended for scanning the lumbosacral plexus (cfr. the LS spine). This positioning allows comparison with the contralateral unaffected brachial/lumbosacral plexus. Positioning the front legs caudally bends the brachial plexus, which may cause partial volume artefacts and mimic recognition of early nerve thickening. The plexus brachialis region is scanned from C4 till Th2. The lumbosacral plexus is from L4 till S3.

The technical parameters should include a field of view (FOV) large enough to allow examination of the entire axillary or pelvic region and the vertebral canal. Using a large FOV is important, as specific clinical anatomic localization can be difficult. The slice thickness should be as thin as possible, 1.25 mm or thinner and slice increment of 50%. Settings should be 100–120 kVp and 100 to 140 mA or larger, depending on the animal's weight and size. Reconstruction parameters are soft tissue and bone window algorithms.

Contrast-enhanced CT images are obtained routinely in patients with clinically documented peripheral neuropathy. The contrast medium is administered as a bolus injection of 600 mg of iodine/kg body weight. CT myelography is best for showing intradural lesions.

A4 CT SHOULDER

The dog is positioned in ventral or dorsal recumbency with both front limbs maximally extended cranially (Figure A7). Shoulders, elbows, and carpal joints need to be positioned in a symmetrical position. If necessary, tie the feet with a string (Figure A8). A wedge, V-shaped, position aid is helpful in these examinations. Those are placed just caudally to the extended elbow joints. Both shoulder joints are included in the FOV. The joints are scanned from the acromion till the proximal third of the humerus. In case of necessity to examine the scapula, the extent of the scanning region needs to be enlarged.

PARAMETERS

KVP	MA	THICKNESS (MM)	INTERVAL (MM)	ALGORITHM	POSITIONING	CONTRAST
100–140 100–140	140 140	0.6–1 1–2	0.3–0.5 0.5–1	Bone Soft tissue/ standard	Left lateral or ventral recumbency with the head retracted caudally and the forelimbs symmetrically extended cranially.	IV 2 mL/kg of 300–350 mg I/mL in bolus. IA 2–4 mL of 80 mg I/mL

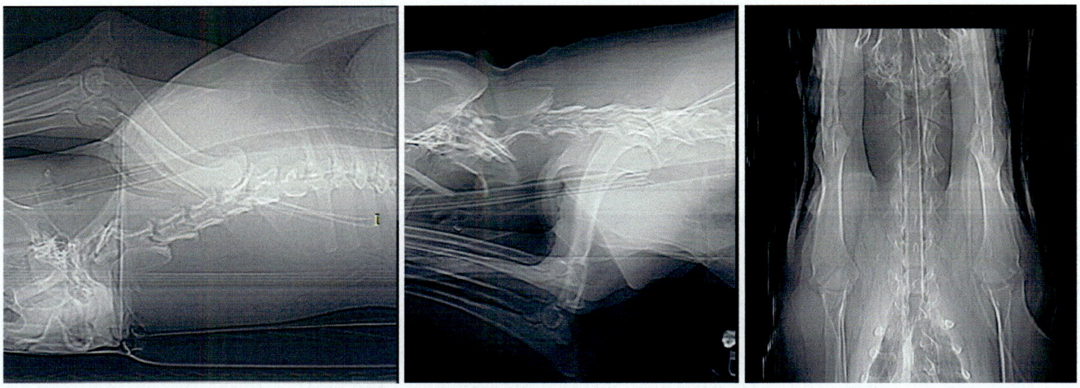

Figure A7 Lateral and dorsal scout views showing the position for shoulder CT. Both front legs are fully extended and parallel to each other to maintain symmetrical positioning.

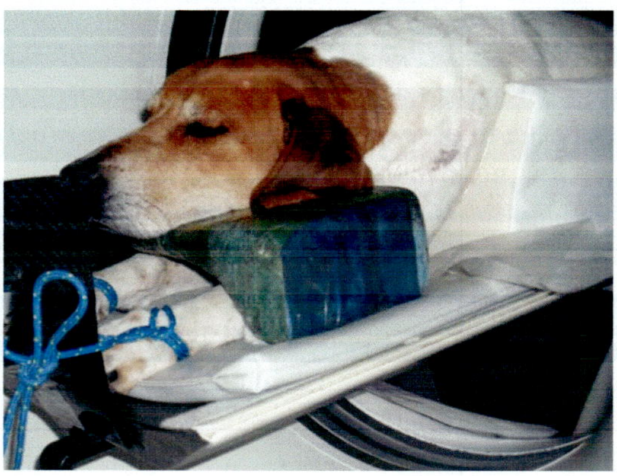

Figure A8 Positioning of shoulder CT. The animal is positioned in ventral recumbency with the front legs maximally extended and tied. A wedge, V-shaped, position aid is placed just caudally of the extended elbows.

PARAMETERS

KVP	MA	THICKNESS (MM)	INTERVAL (MM)	ALGORITHM	POSITIONING	CONTRAST
100–140 100–140	80–140 80–140	<1 1–2	0.5 0.5–1	Bone Soft tissue/ standard	Dorsal or ventral recumbency with the forelimbs symmetrically extended cranially. Scanning both joints.	IV 2 mL/kg of 300–350 mg I/mL in bolus.

Bone and soft tissue algorithms with slice thicknesses of 1–2 mm and with 50% overlap should be performed. The scan parameters are 120 kVp and 120–140 mA. IV contrast should be administered at 2 mL/kg of 300–350 mg I/mL in bolus. Sagittal and dorsal reconstructions of the standard algorithms are required. Arthro-CT can be performed at 2–4 mL of 80mg I/mL.

A5 CT ELBOW

The dogs are positioned in lateral or ventral recumbency on the computed tomographic scanning table with the elbow joints parallel and extended cranially (Figure A9). An angle of less than 120° in the elbow joint (flexion) should be avoided. It is essential to take into account the prevention of beam hardening artefacts due to the patient's head. This requires adjusting the patient's head laterally or caudally to prevent interference with the elbows, regardless of the patient's orientation. A recommended position is pulling the head of the dog caudally to the lateral side, in order to scan only both elbow joints in the gantry (Figure A10). Wedge sponges are placed between both elbow joints in order to avoid rotation. This position allows a perfect symmetry and comparison of both elbows at the same level. Scans are performed from the top of the olecranon to 3 cm distal to the cubital joint.

Bone and soft tissue algorithms with slice thicknesses of 0.6–1.5 mm with 50% overlap should be performed. The scan parameters are 120 kVp and 140 mA. IV contrast should be administered at 2 mL/kg of 300–350 mg I/mL in bolus. Sagittal and dorsal reconstructions of the standard algorithms are required. Eventually, an arthro-CT can be performed at 2–4 mL of 80 mg I/mL.

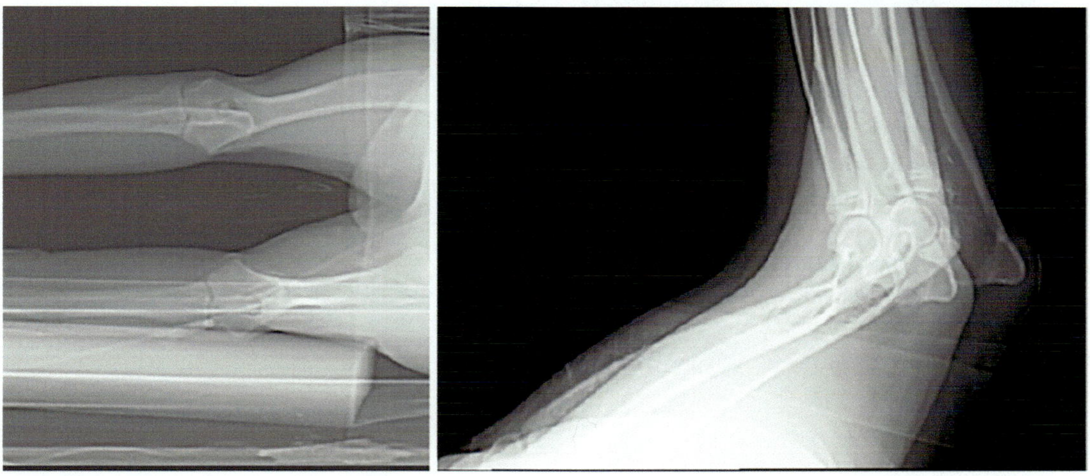

Figure A9 **Lateral and dorsal scout views showing the position for elbow CT with the elbow joints parallel and extended cranially.**

PARAMETERS

KVP	MA	THICKNESS (MM)	INTERVAL (MM)	ALGORITHM	POSITIONING	CONTRAST
100–140	120–160	0.6–1.25	0.3–0.5	Bone	Dorsal recumbency. Hindlimbs symmetrically extended caudally.	IV 2 mL/kg of 300–350 mg I/mL in bolus.
100–140	120–160	2	1	Soft tissue/ standard		

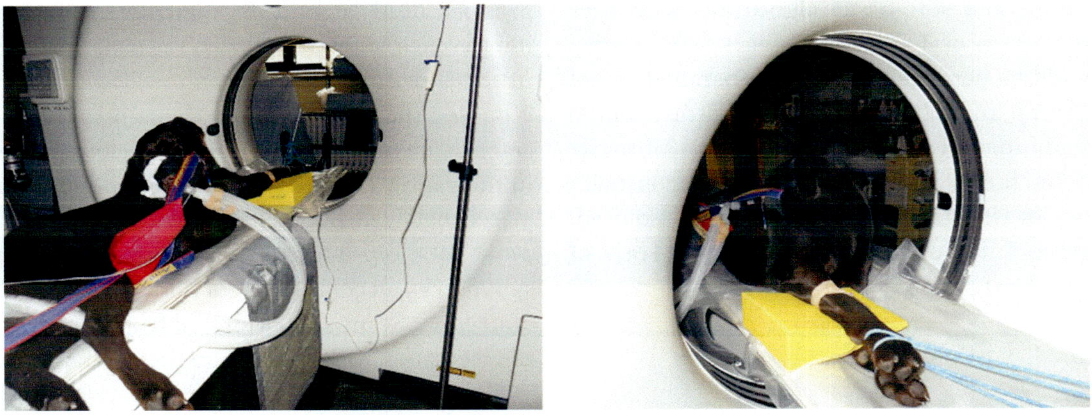

Figure A10 The animal is positioned in lateral recumbency for elbow CT with the front legs extended and the head withdrawn out of the gantry. A sponge wedge is placed between both elbows to avoid rotation.

Figure A11 Positioning of carpal CT. The animal is positioned in ventral recumbency. Forelimbs are symmetrically positioned with both joints at the same level.

A6 CT CARPUS AND DIGITS

Positioning in ventral or dorsal recumbency allows investigation of both limbs within a single scan. Care should be taken to position the antebrachium along the scan-axis with the limbs extended and the joints placed at the same level relative to the scan-plane (Figure A11). Additional functional studies ("stress-CT") help clarify the reasons for instability that could be demonstrated during the clinical examination. The carpal joint is scanned from the distal area of the radius and ulna till the toes.

The slice thickness should be as thin as possible, 1.25 mm or thinner with slice increment

PARAMETERS

KVP	MA	THICKNESS (MM)	INTERVAL (MM)	ALGORITHM	POSITIONING	CONTRAST
100–140 100–140	120–140 120–140	0.6–1 1–2	0.3–0.5 0.5–1	Bone Soft tissue/ standard	Ventral or dorsal recumbency. Hindlimbs symmetrically extended caudally.	IV 2 mL/kg of 300–350 mg I/mL in bolus in case of soft tissue pathology. IA 2–4 mL of 80 mg I/mL

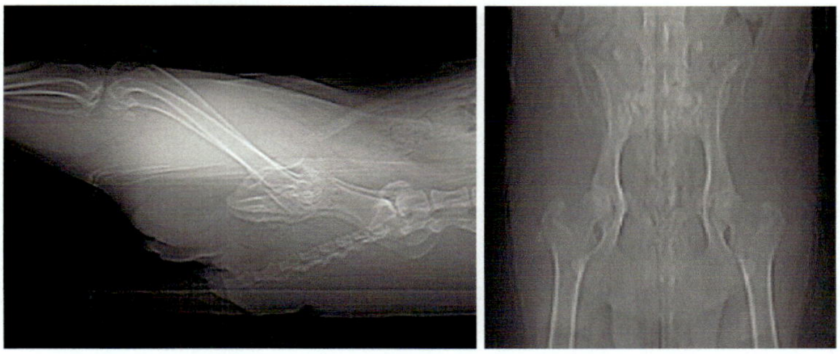

Figure A12 **Lateral and dorsal scout views of the positioning of the hip and pelvis CT. The animal is in dorsal recumbency in a special U-shaped positioning kit.**

of 50%. Settings should be 100–140 kVp and 80 to 140 mA, depending on the animal's weight and size. Reconstruction parameters are in soft tissue and bone window algorithms. Contrast studies (angiography, arthrography, fistulography, and lymphography) are methods to improve the differentiation of soft tissue structures.

A7 CT HIP AND PELVIS

The dog positioned in dorsal recumbency in a special U-shaped positioning kit is the easiest method to get a stable and reproducible position of a small animal's body (Figure A12). The hind legs are placed on a pillow in extension and carefully fixed parallel to each other similar to HD positioning, but less extended (Figure A13). A hip CT scan includes the hip joint and femurs,

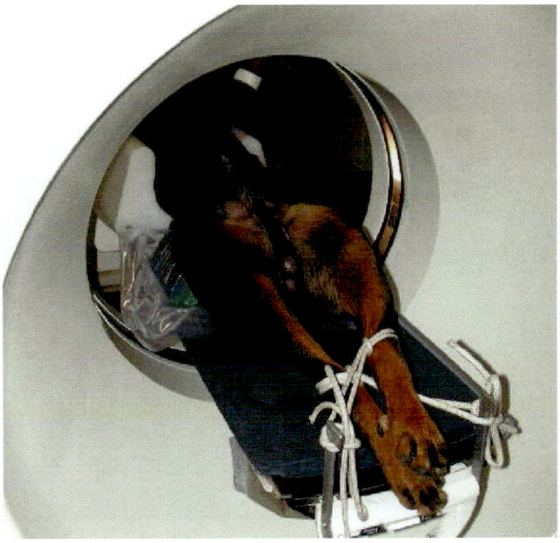

Figure A13 **Positioning of hip CT. The animal is positioned in dorsal recumbency. Hindlimbs are symmetrically extended caudally and carefully fixed parallel to each other similar to HD positioning, but less extended.**

PARAMETERS

KVP	MA	THICKNESS (MM)	INTERVAL (MM)	ALGORITHM	POSITIONING	CONTRAST
100–140 100–140	120–140 120–140	0.6–1 1–2	0.3–0.5 0.5–1	Bone Soft tissue/ standard	Ventral recumbency. Hindlimbs symmetrically extended caudally.	IV 2 mL/kg of 300–350 mg I/mL in bolus in case of soft tissue pathology. IA 2–4 mL of 80 mg I/mL

extending from the iliac crest through the mid femur. It may also include the surrounding soft tissues and, if needed, the entire pelvis.

Technical parameters are very much dependent on the type of scanner. Multidetector scanners usually scan all animals with 1- to 2-mm slice thickness, no matter the body size, to obtain thorough image resolution and also reformatted images, such as MPR or 3D-reconstruction.

A8 CT STIFLE

The patient is always "feet first" orientated to the CT gantry.

Some prefer to position the animal in dorsal recumbency. However, positioning the animal in sternal recumbency with the hind limbs extended caudally and parallel to each other and the tarsal joints rested on a pillow is often more stable (Figure A14). Both stifles must be positioned close to each other to minimize beam-hardening artefacts (Figure A15). The joint is scanned from the distal third of the femur to the proximal fifth of the length of the tibia.

The slice thickness should be as thin as possible, 1.25 mm or thinner and the slice increment of 50%. Settings should be 100–120 kV and 100 to 140mA, depending on the animal's weight and size. Reconstruction parameters are the soft tissue and bone window algorithm.

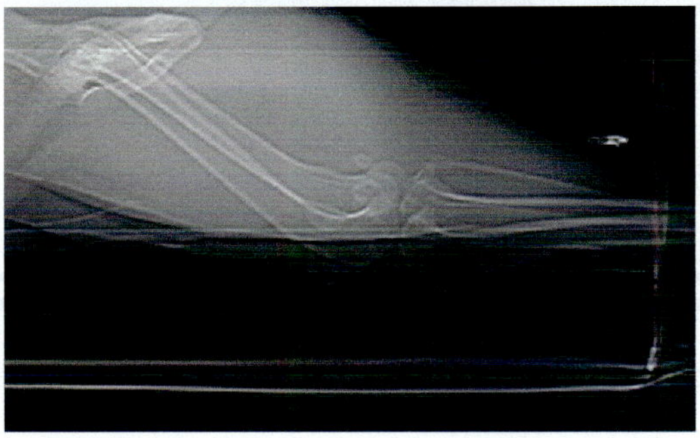

Figure A14 Lateral scout view of the positioning of the stifle CT. The animal is in sternal recumbency with the stifles parallel and extended.

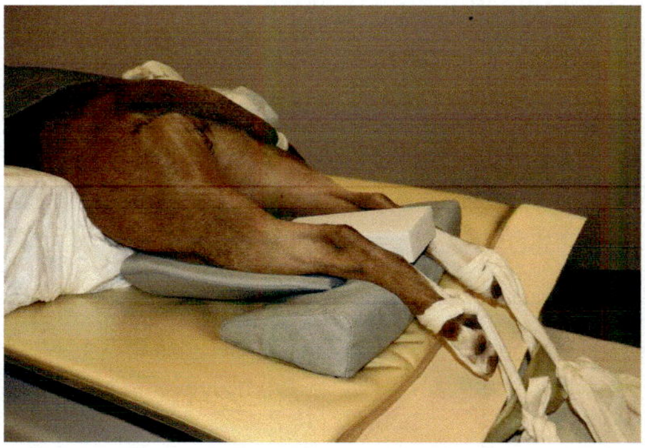

Figure A15 Positioning of the stifle CT. The animal is positioned in sternal recumbency with the hindlimbs symmetrically extended caudally. The body and stifles are on a positioning aid to maintain stability.

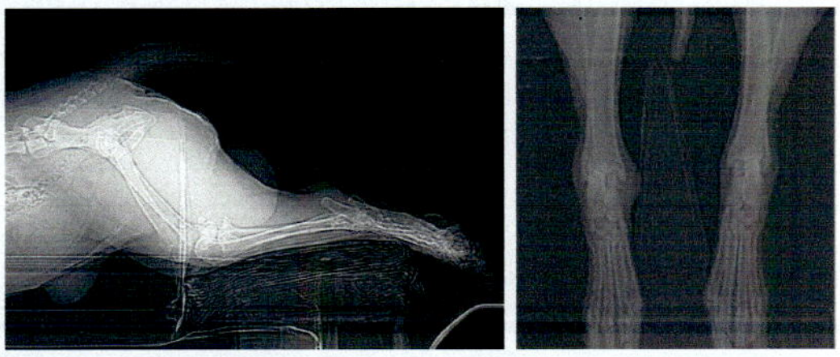

Figure A16 Lateral and dorsal scout views of the positioning of the tarsus CT. The animal is in sternal recumbency in a positioning kit. A sponge wedge can be placed between the joints to maintain stability.

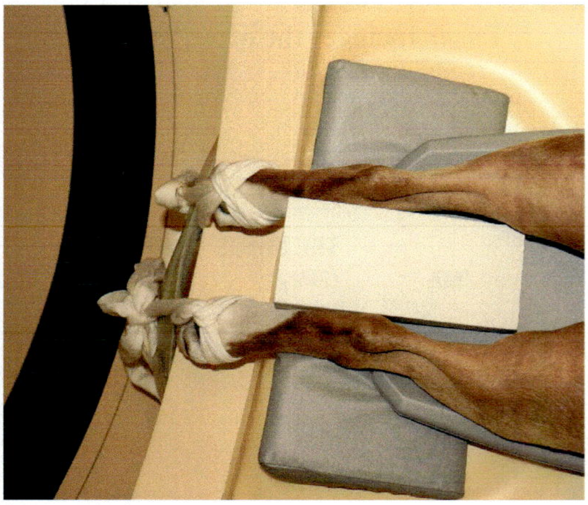

Figure A17 Positioning of tarsal CT. The animal is positioned in ventral recumbency with the hind legs on a position aid and the limbs extended.

For arthro-CT, the normal iodine concentration should be reduced to 80–100 mg I/mL to enhance visualization without contrast obscuring structures. The original contrast medium can be diluted with saline, and for an average knee joint, a volume of 4 ml is recommended. During administration of intra-articular contrast, care should be taken to avoid the fat pad. Gentle flexion and extension of the joint have been recommended to enhance contrast dispersion.

A9 CT TARSUS

The dog is positioned in ventral recumbency with the tarsal joints in extension. This position allows a perfect symmetrical scanning of both joints, allowing the evaluation of the CT images of both tarsi at the same level. A sponge wedge can be placed between the joints to maintain stability (Figure A16). Strings can be used to achieve extension of the limbs (Figure A17). The joint is scanned from 2 cm proximally of the calcaneus till 2 cm distally of the intertarsal bones.

A slice thickness of 0.6–1 mm in bone and of 1–2 mm in the soft tissue algorithm is recommended. It is preferred to scan with an overlap of 50%, resulting in detailed multiplanar reconstructions in additional planes, which are frequently utilized in tarsal joints. The parameters used for this joint are 100–120 kVp and 120 mA. The FOV needs to be adjusted to increase in-plane resolution with a maximum of 15 cm. IV contrast should be administered at 2 mL/kg of 300–350 mg I/mL in bolus. Sagittal and dorsal reconstructions are required. Arthro-CT can be performed at 2–4 mL of 80 mg I/mL. The authors thank Drs. Davina Böhm-Geerdink, Veterinary Clinic Den Heuvel, Best, NL, for her contribution in positioning.

APPENDIX B

HOUNSFIELD VALUES (HU)

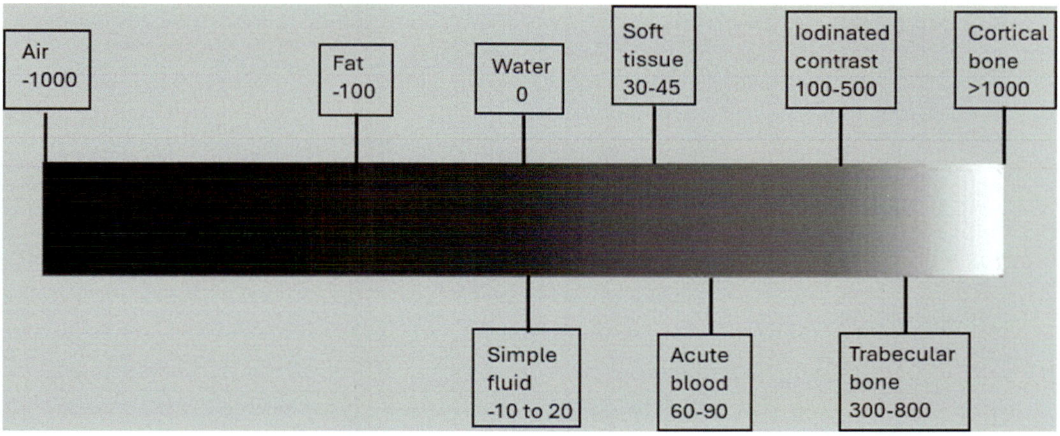

- Air: –1,000 HU
- Bone (cortical): >1,000 HU
- Bone (trabecular): 300–800 HU
- Brain (grey matter): 35–40 HU
- Brain (white matter): 20–30 HU
- Subcutaneous fat: –100 to –115 HU
- Metal: >3,000 HU
- Muscle: 45–50 HU
- Blood: 30–45 HU
- CSF: 15 HU
- Water: 0 HU

INDEX

Note: **bold** page numbers refer to tables and *italic* page numbers refer to figures.